Long-Range Transport of Airborne Pollutants

Edited by
H. C. MARTIN

Reprinted from
Water, Air, and Soil Pollution, Vol. 18 Nos. 1, 2, and 3

D. REIDEL PUBLISHING COMPANY

DORDRECHT : HOLLAND / BOSTON : U.S.A.

WATER, AIR, AND SOIL POLLUTION

EDITOR

ISBN-13: 978-94-009-7968-0   e-ISBN-13: 978-94-009-7966-6
DOI: 10.1007/978-94-009-7966-6

Order Ref. No. 90-277-9073-0

TABLE OF CONTENTS

Volume 18   Nos. 1, 2, and 3

TABLE OF CONTENTS

INTRODUCTION

Long-Range Transport of Airborne Pollutants and Acid Rain Conference

This issue of Water, Air, and Soil Pollution is devoted to the collection of papers presented at the Long-Range Transport of Airborne Pollutants and Acid Rain Conference held at Albany, N.Y., April 27–30, 1981. The issue includes most of the invited papers as well as a good number of the poster papers. The conference consisted of seven plenary sessions at which the invited papers were presented. After each session the participants discussed the session topic in the poster area where the subject was further explored and expanded. The seven technical sessions were:

(1) Networks.
(2) Models of Delivery.
(3) Interactions with Soils and Ground Water.
(4) Calibrated Water Sheds.
(5) Effects on Aquatic Biota.
(6) Effects on Terrestrial Biota.
(7) Health implications.

The closing session was devoted to the topic 'The Application of Scientific and Technical Data to the Development of Government Policy; Acid Rain – A Case Study'. The four papers given are not included here.

This conference was sponsored by the American Meteorological Society and the Canadian Meteorological and Oceanographic Society. Although the sponsors are both meteorological organizations, it was the intention of the program committee to bring together scientists from all disciplines involved in the LRTAP issue in order to focus on the interdisciplinary nature of this environmental problem. It was our intention to:

(1) Examine the interfaces between various disciplines such as soil-atmosphere interactions, atmosphere-vegetation interactions, etc.; and

(2) Examine available models of all components of the problem including, among others, atmospheric delivery, geochemistry and hydrology, and aquatic effects.

The papers contained herein should provide a valuable source of information on many of the broader problem areas being addressed. The high quality of these papers is a tribute to the authors. The important requirement of bridging disciplines in this unusual environmental problem, may not be entirely reflected here. Bearing in mind that the plenary sessions included a diversity of scientists who participated openly in the questions and discussions, I trust that that goal was achieved at Albany. Because it is critically important to follow the pathways taken by pollutants once they leave the emission point it would seem prudent to undertake similar gatherings from time to time in order to

Water, Air, and Soil Pollution **18** (1982) 5–6. 0049–6979/82/0181–0005$00.20.

provide an opportunity for researchers to reexamine the links between the 'residences' of pollutants along these pathways.

Federal LRTAP Liaison Office,   H. C. MARTIN
Atmospheric Environment Service,   *Program Chairman*
4905 Dufferin Street,
Downsview, Ontario, M3H 5T4,
Canada

ATMOSPHERIC MONITORING NETWORK OPERATIONS AND RESULTS IN CANADA

DOUGLAS M. WHELPDALE and LEONARD A. BARRIE

Atmospheric Environment Service, 4905 Dufferin Street, Downsview, Ontario, Canada, M3H 5T4

(Received 10 July, 1981; Revised 5 October, 1981)

Abstract. Atmospheric monitoring activities in Canada relevant to the long-range transport of atmospheric pollutants and the 'acid rain' problem are reviewed. Particular aspects examined are network objectives, station density and location, sampling protocol, and quality assurance. Results from a number of these networks are presented for the purpose of outlining the nature and extent of air and precipitation contamination by pollution released in eastern North America. Examples discussed include: the spatial distribution of acidic wet deposition, the temporal variation of acid-related substances in both air and precipitation, an episode of long-range transport, and the impact of acidic emissions on the Arctic atmosphere.

Acidic wet deposition is greatest in Canada east of the Manitoba-Ontario border. In 1978, it ranged from 18 to 46 mmol H^+ m^{-2} yr^{-1} in the southern half of eastern Canada, with maxima in southern Ontario (44 mmol H^+ m^{-2} yr^{-1}) and southwestern Quebec (46 mmol H^+ m^{-2} yr^{-1}). Western Canada receives less acidity in precipitation, but areas of some concern are the Pacific Coast (10 mmol H^+ m^{-2} yr^{-1}) and to a lesser extent northern Alberta and Saskatchewan (3 to 5 mmol H^+ m^{-2} yr^{-1}). Acidic emissions from mid-latitude sources which reach the Arctic in winter cause an increase in the acidity of snow from a pH of approximately 5.6 in the summer to values of 4.9 to 5.1 in January through March.

1. Introduction

The routine collection of reliable data on regional and global background air and precipitation quality has had a rather brief history in Canada. Prior to approximately 1970 precipitation composition data were collected to support scattered studies of nutrient budgets or smelter damage in areas of small geographical extent. Routine atmospheric measurements of gas and particle concentrations outside urban areas were virtually unheard of.

Concern about nutrient inputs led to the establishment of a network around the Great Lakes in 1969, and then, as interest in monitoring background air quality grew internationally, ten Canadian stations were established in the WMO Background Air Pollution Monitoring Network, beginning in 1973. In the following few years concern about acid rain increased rapidly in Canada, at first in connection with the impact of point sources, particularly in Ontario and Alberta, and then, as a larger scale phenomenon. This resulted in the establishment of networks and studies by several federal and provincial agencies. In the last few years, the number of measurement programs has increased greatly as has the number of chemical species being monitored.

Development of networks in the early part of the last decade was rapid and rather uncoordinated as far as sampling procedures, analysis methods, and quality assurance practices were concerned. However, with the establishment and use of central data banks and inter-laboratory sample exchange programs, coordination and data compatibility are improving rapidly.

Water, Air, and Soil Pollution **18** (1982) 7–23.   0049–6979/82/0181–0007$02.55.

Initially, most effort was concentrated on the measurement of the chemical composition of precipitation as a rough indicator of air quality. Now, however, interest is much broader. Air monitoring, an essential tool in the investigation of long-range transport, is being done at many locations, including several in the Arctic.

Network measurements are made across most of the country (Figure 1), with sampling density greatest in areas where problems are perceived to be greatest (e.g. in southeastern Canada and the Alberta Oil Sands region). The diversity of purpose remains large, ranging from background trends to calibrated watershed inputs. As geographical coverage has improved, it has become clear that more intensive sampling, both spatially and temporally, is required in many ecologically sensitive areas, both for model validation purposes, and to satisfy the needs of scientists in other disciplines.

During the last few years of intensive research on acid precipitation and long-range transport, we have learned the value and necessity of having high-quality, long-term data bases containing atmospheric concentration and deposition information. We have also learned how essential it is to use, analyze and interpret the data available.

The remainder of this paper contains a brief description of networks in Canada followed by a discussion of aspects of the long-range transport and acidic deposition phenomena in the country. The discussion is illustrated by current results from selected networks.

2. Description of Networks

Table I lists characteristics of major networks in Canada in which measurements relevant to acid precipitation and long-range transport are being made, and Figure 1 shows the locations of stations in these networks. For each network the table shows the sponsoring organization, the geographical area of interest, an abbreviated statement of purpose, the period of operation, the number of stations, measurements made, and sampling period. For further details, see NRCC (1981).

During the last decade, monitoring networks have been established for a variety of purposes, with the result that sampling procedures and results have not always been comparable. In broad terms, it is possible to identify three types of networks (not necessarily mutually exclusive):

(i) national networks, which operate primarily to survey and monitor large-scale distributions in both time and space – e.g., the Canadian Network for Sampling Precipitation, the WMO Background Air Pollution Monitoring Network, and the Canadian Network for Sampling Organics in Precipitation;

(ii) regional networks, which were established to undertake monitoring in particular regions (often political jurisdictions) of interest, and/or for specific problems – e.g., the Air and Precipitation Monitoring Network, the Canadian Arctic Air Pollution Program Network, the Great Lakes Precipitation Chemistry Network, and networks operated by provinces;

(iii) special interest networks, which were established to examine the impact of a specific source, industry ór industrial region – e.g., the Ontario Hydro Network, the

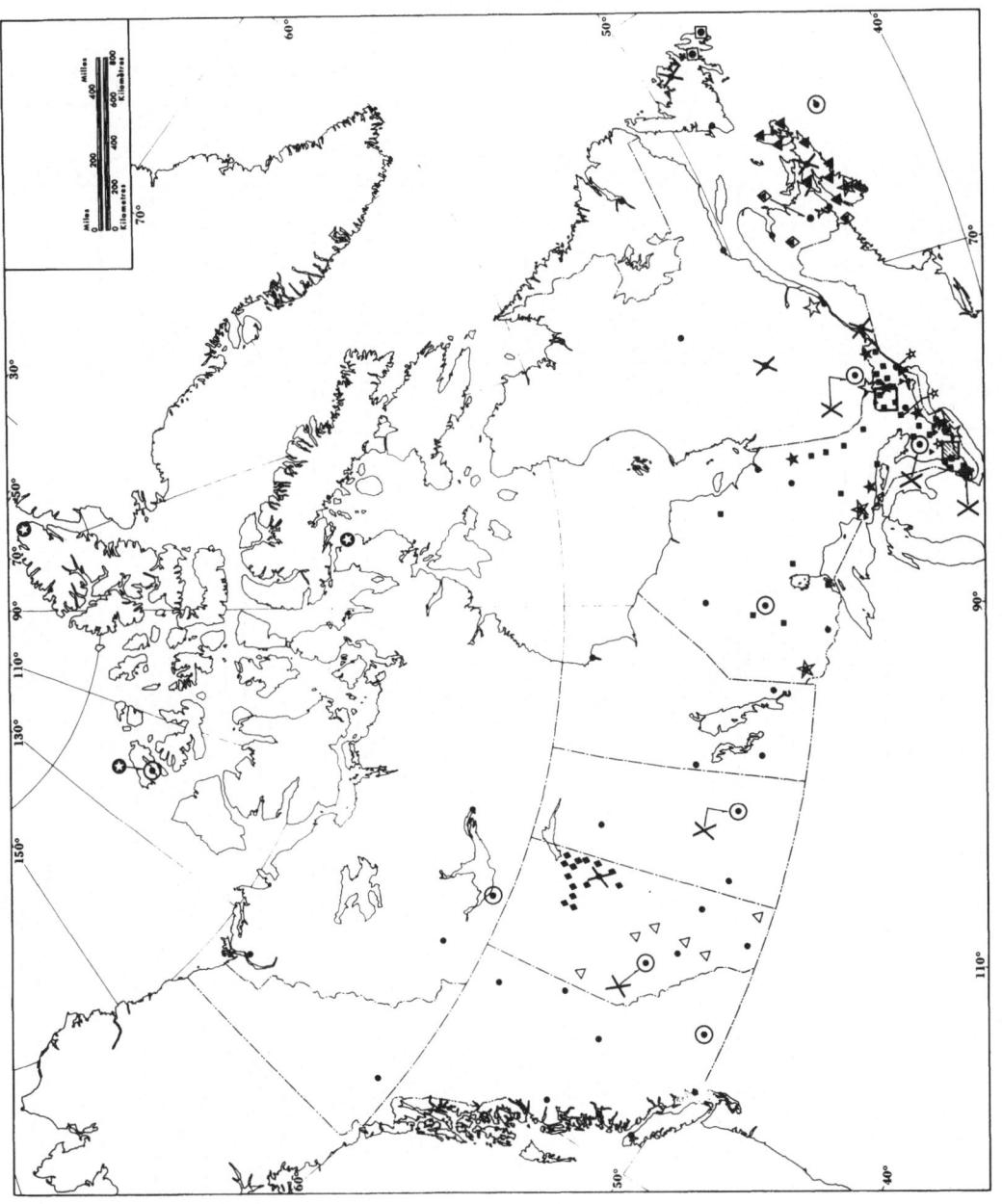

Fig. 1. Location of monitoring sites in Canada. (See Table I.)

TABLE I

Routine monitoring efforts in Canada relevant to the Long Range Transport of Air Pollutants

NETWORK	AGENCY	GEOGRAPHICAL AREA	PURPOSE	CURRENT NUMBER OF STATIONS	CODE ON MAP
Canadian Network for Sampling Precipitation (CANSAP)	Department of Environment Atmospheric Environment Service	Canada-wide	Large scale Fields and Trends of Concentration and Deposition	54	●
World Meteorological Organization Background Air Pollution Monitoring Network (WMO BAPMoN)	"	Canada-wide	Regional, Global Background Levels and Changes	9	◉
Air and Precipitation Monitoring Network (APN)	"	Eastern Canada	Regional Levels of Deposition on a Daily Basis Model Validation	5	☆
Canadian Arctic Air Pollution Program Network (CAAPP)	"	Canadian Arctic	Levels and Sources of Arctic Pollution (Haze)	3	✪
Canadian Network for Sampling Organics in Precipitation (CANSOC)	Department of Environment Environmental Management Service	Canada wide	Country wide Survey of Organics in Precipitation	12	✕
(A) Cumulative Acid Precipitation in Ontario Study (APOS)	Ontario Ministry of Environment Air Resources Branch	Ontario	Long Term Deposition of Acid-related Substances and Metals	32	■
(B) Event wet / dry	"	Ontario	Event Wet and Dry Deposition and Source Dependence	3	☼
Precipitation Quality Monitoring Program (PQMP)	Department of Environment Alberta	Alberta	Impact of Industrialization Historical Data Base	6	◁
Precipitation Chemistry in Nova Scotia	Department of Environment Nova Scotia	Nova Scotia	Levels and Sources of Acid-related Substances, Nutrients and Metals	5	◀
Quebec Provincial Precipitation Chemistry Network	Ministry of Environment Quebec	Quebec	Levels and Sources of Precipitation Constituents	46	▷
Great Lakes Precipitation Network	Department of Environment Environmental Management Service	Great Lakes Basin	Precipitation Composition and Input in Great Lakes Basin	7, 15	▶
Ontario Hydro Air and Precipitation Monitoring Network (OHAPM)	Ontario Hydro	Ontario	Spatial and Temporal Variations of Air and Precipitation Composition	6	★
Alberta Oil Sands Environ. Res. Program (AOSERP) Precip. Chemistry Network	Department of Environment Alberta	Fort McMurray	Assess Impact of Industrial Development in Area	16	◆
Nanticoke Environmental Management Program (NEMP)	Ontario Ministry of Environment Air Resources Branch	Nanticoke Area	Assess Impact of Industrial Development in Area	6	▨
Limnology Unit Precipitation Sampling Network (WRBPC)	Ontario Ministry of Environment Water Resources Branch	South-central Ontario	Determine Watershed Inputs	5	□
New Brunswick Precipitation Network	Environment New Brunswick Environmental Services Branch	New Brunswick	Chemistry of Precipitation in New Brunswick	3	◈
Newfoundland Provincial Wet Deposition Program	Department of Environment Newfoundland Air and Industry Division	Avalon Peninsula	To Measure Wet Deposition of Acidity	2	◘

W... wet deposition   D... dry deposition   B... bulk

Alberta Oil Sands Environmental Research Program Network, the Nanticoke Environmental Research Program Network and the Ontario Limnology Unit Network.

A fourth type of monitoring effort which is not discussed here is the local or urban network, attuned primarily to impact of air pollution on human health and property.

Most networks in Canada have been established to monitor the composition and deposition of constituents in precipitation. Only a few, four listed in Table I, measure particle-constituent concentrations, and three, gas concentrations. In most networks, precipitation is now sampled with collectors which are exposed only during precipitation periods (wet-only collection), although, in a few bulk sampling is still done for comparative purposes.

One of the main objectives of many networks is to determine the input rates of acidity to areas with sensitive receptors. Consequently, precipitation is usually analyzed for major ions that affect the acidity. Routine measurements of metals and organics are made less frequently, in part because collection, handling and analysis procedures for these substances are more complex than those for major ions.

One of the main stumbling blocks to comparable data in the various networks has been the different sample collection periods used. Most networks use monthly periods for reasons of economy, but some sample weekly and some daily or event. It is clear that a short collection period is required where the data are being used to understand atmospheric processes, to identify pollutant sources or to investigate effects caused by rapid fluctuations in precipitation deposition. The choice of monthly, weekly, daily or event sampling is contingent upon two factors: the objectives for which the network was established and the stability of the substances in precipitation under the conditions of collection and handling peculiar to the network. In view of the concentrated efforts now being made to test and intercompare the various networks, both inside and outside the country, more quantitative information on the compatibility of data sets can be expected in the near future. One of the most useful practices in this regard is the use of common sites by two or more networks. In this way, it has been possible to begin comparing CANSAP, APN, PQMP and APOS data. At a few sites, extensive testing programs are underway – e.g., at ELA, Dorset, Kejimkujik.

An intercomparison between the Canadian and American national networks, CANSAP and NADP (the U.S. National Atmospheric Deposition Program), has been organized to test the operation of both systems in their entirety to determine if they will provide a compatible data set for North America. Six sites are being used, three in Canada – Lethbridge, Alberta; Mount Forest, Ontario and Kejimkujik National Park, Nova Scotia – and three in the U.S.A. – Glacier National Park, Montana, University of Michigan Biological Station, Michigan, and Caribou, Maine. At Kejimkujik, an NADP weekly collector will be operated concurrently with the CANSAP monthly and the APN daily collectors presently there.

In addition to network intercomparisons, there are several quality control programs underway that are designed to test the performance of the sample analysis laboratories. Sample exchanges are being conducted between several Canadian networks and NADP; and between the CANSAP/WMO/APN laboratory and both the U.S. Environmental

Protection Agency Laboratory, and laboratories in the ECE European Monitoring and Evaluation Program (EMEP). Precipitation collector evaluations have also been carried out, most recently by the MAP3S program at Pennsylvania State University, and within the APOS program.

Results from each of these quality assurance activities deserve detailed reporting, and in fact some are described elsewhere in these proceedings. Quality assurance is a necessary part of any routine network operation and despite the additional costs, the benefits are not to be underestimated.

Each network produces reports of its data, which are available from the respective agencies. Many networks now submit their data to the national water quality data archive, NAQUADAT, and a few submit theirs as well to the EPA data base which contains data from, *inter alia*, MAP3S, NADP, WMO, and CANSAP. In Canada, there is a need for a coordinated quality control and data archiving system on a national scale supported both federally and provincially and involving all major networks.

3. Results

In this section, selected results from monitoring activities in Canada are presented in order to illustrate the insight afforded by network data into the nature and extent of the impact of man-made air pollutant emissions on the quality of air and precipitation in Canada. We purposely avoid consideration of urban and heavily industrialized areas, and concentrate instead on regional and continental background levels. On the regional scale in eastern North America, man-made emissions of S oxides are much greater than natural ones (Henry and Hidy, 1980; Galloway and Whelpdale, 1980); thus, the phenomena discussed below are predominantly of anthropogenic origin. These include spatial characteristics of acidic deposition, the temporal variability of air and precipitation composition, an episode of long-range transport, and some effects of acid-related emissions on the global scale, using the Arctic atmosphere as an example.

3.1. SPATIAL VARIATION OF ACIDIC WET DEPOSITION

Although pH or H^+ *concentration* is commonly used as the main indicator of the severity of the acid rain problem in North America, in fact, the *deposition* of H^+ is a more accurate measure of the impact on aquatic and terrestrial ecosystems. The deposition of chemical constituents in precipitation depends upon both the amount of precipitation and the concentration of the substance in precipitation. Figure 2 shows the annual wet deposition pattern of H^+ (as derived from measured pH), based on the 1978 CANSAP data. The influence of man's activity in southeastern Canada combined with the relatively high precipitation there, results in elevated deposition rates compared with the rest of the country. Marked regional differences in industrial and agricultural activity and in geography are reflected in the deposition pattern: the west coast and northern regions, generally far removed from sources (with the exception of the extreme southwestern portion of the country), show relatively low deposition levels as a result, respectively, of low concentrations and low precipitation amounts. Wet deposition in the central prairie

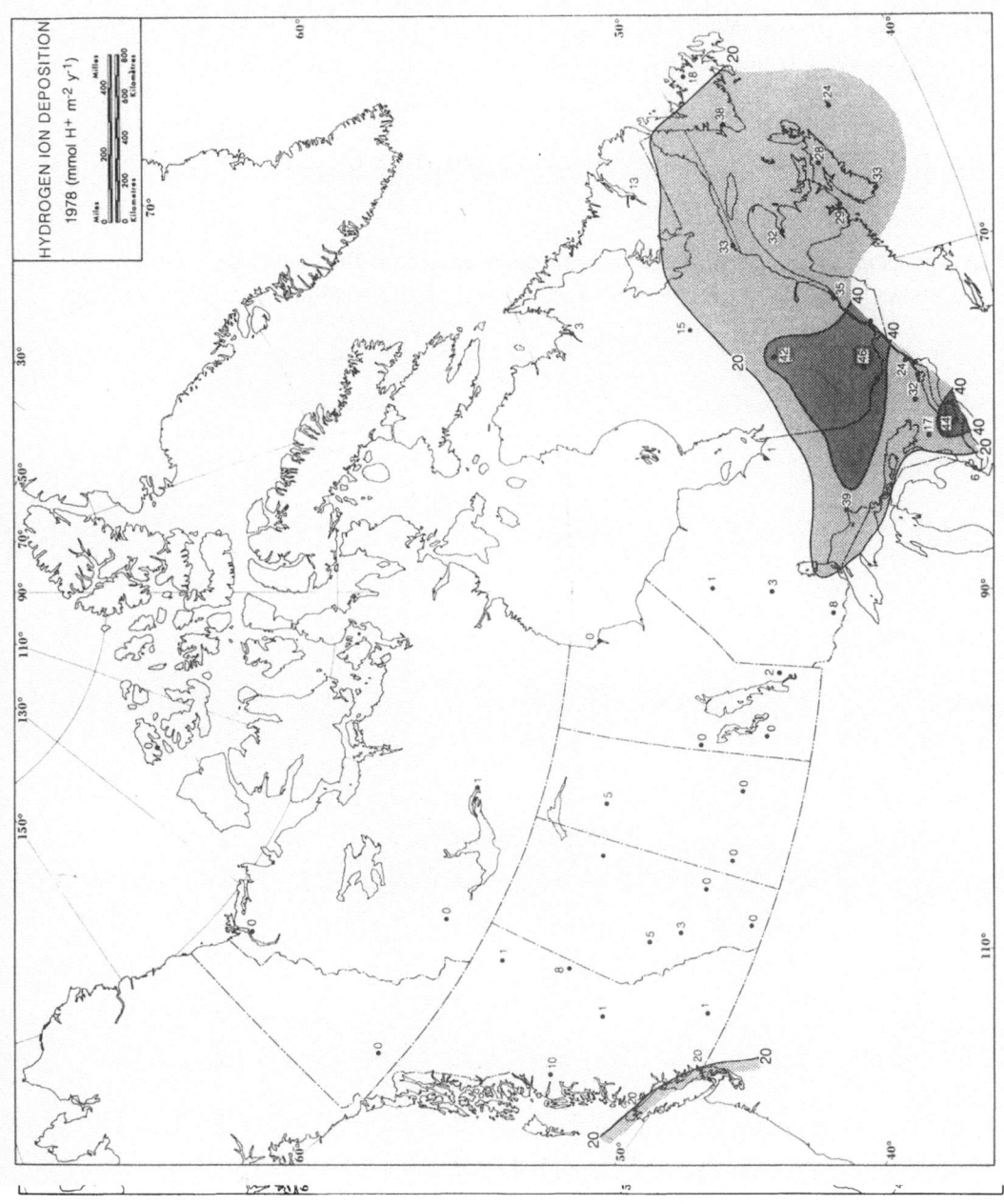

Fig. 2.   The spatial distribution of wet deposition of H⁺ in Canada during 1978 based on CANSAP data.

region is very low, showing the influence of calcareous prairie soils and dust on the composition of precipitation.

By way of comparison, the values of H^+ deposition ranging up to approximately 50 mmol H^+ m^{-2} yr^{-1} in the eastern part of the country are comparable to those measured in the northeastern United States (e.g. see NRCC, 1981). Deposition in remote continental areas is of the order of 1 mmol H^+ m^{-2} yr^{-1}, depending on the precipitation.

3.2. TEMPORAL VARIATION OF AIR AND PRECIPITATION QUALITY

3.2.1. *Seasonal variations*

In regions under the influence of anthropogenic emissions the concentrations of many trace constituents in the atmosphere and in precipitation exhibit a seasonal variation

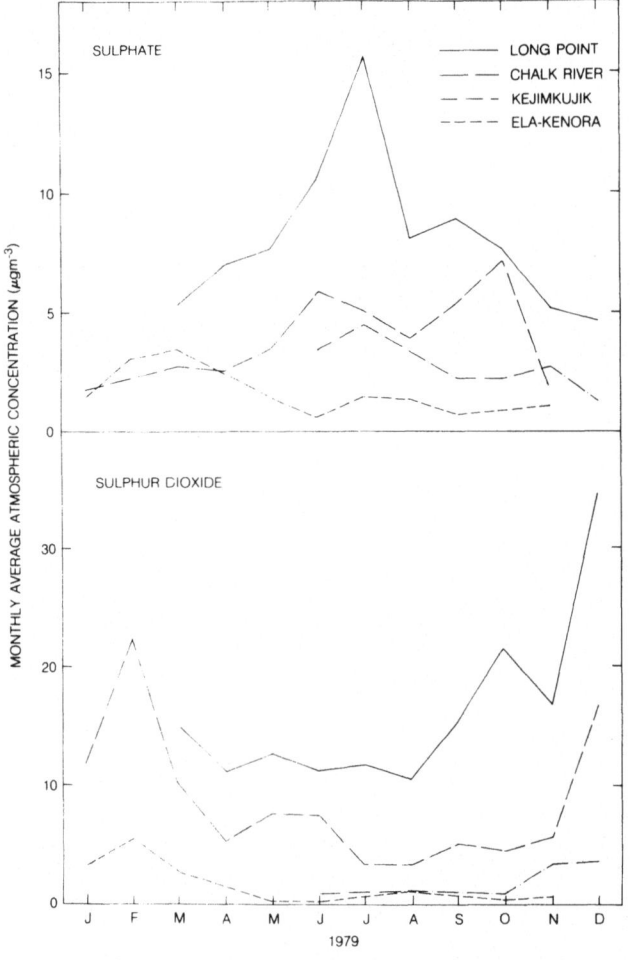

Fig. 3.   Temporal variations of the monthly average concentration of atmospheric sulphate and SO_2 at APN sites during 1979. (See Figure 7 for station locations.)

(Figures 3 and 8). A summer maximum in precipitation sulphate concentrations is observed in eastern Canada (Barrie, 1981) and in the northeastern United States (MAP3S/RAINE, 1982) at locations close to and downwind of the large S emission sources which are concentrated in the eastern half of the continent. In contrast, H^+ and nitrate ion concentrations in precipitation do not consistently show marked annual cycles.

Atmospheric SO_2 and particulate sulphate levels measured at APN sites (shown in Figure 7) in eastern Canada also vary seasonally (Figure 3). Sulphur dioxide has a winter maximum, and sulphate, with one exception, has a summer maximum. At ELA-Kenora, which is upwind of the continental plume of eastern North America, the winter maximum in both S species is a result primarily of meteorological factors. Maximum transport westward from eastern North America in winter coincides with a late winter peak of background S concentrations in arctic air masses (Barrie *et al.*, 1981) that prevail at this mid-continental location during winter.

3.2.2. *Daily variations*

Concentrations of acid-related substances in air and precipitation are highly variable from day-to-day or from storm-to-storm. It follows that dry deposition, which is proportional to air concentration, and wet deposition, which is proportional to precipitation concentration, are also highly variable. For example, consecutive rains may contribute H^+ loadings that differ by an order of magnitude. Figure 4 is a time series

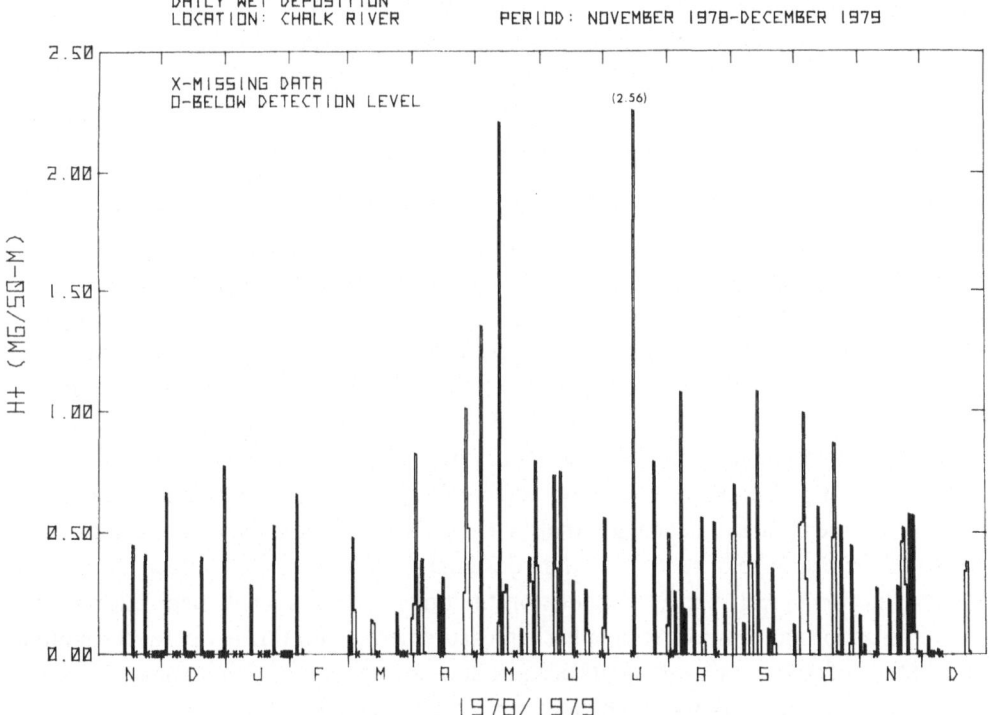

Fig. 4.   A time series of H^+ wet deposition at the Chalk River APN station.

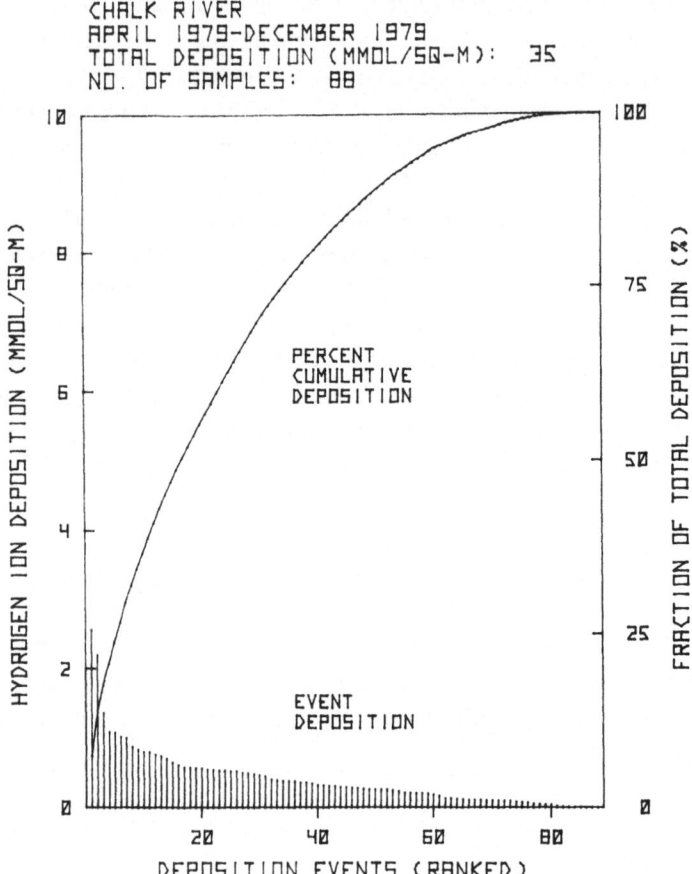

Fig. 5.   Hydrogen ion deposition ranked by event for Chalk River APN station.

of daily wet deposition of H⁺ at the Chalk River APN station for the period November 1978 to December 1979. A small number of events contribute a substantial amount of the total loading. This is demonstrated more clearly in Figure 5 where events are ranked in order of their H⁺ deposition. Cumulative deposition for the period is shown by the solid curve. Fifty per cent of the total wet deposition results from only 19% of the events.

As a further example, Figure 6 shows the day-to-day variability of the daily mean concentration of atmospheric sulphate and SO_2 at ELA-Kenora. This is also a fair representation of the temporal behavior of dry deposition which is usually calculated as the product of pollutant concentration and a relatively constant deposition velocity.

The episodic nature of deposition makes it important to know the temporal characteristics of the response of receptors when designing acidic deposition effects studies. If a particular biological species responds on the time-scale of a single deposition event, then it is important to make atmospheric measurements with this temporal resolution. In addition, of course, measurements of short duration lend themselves to meteorological interpretation, and thus to insight into responsible source regions and atmospheric processes.

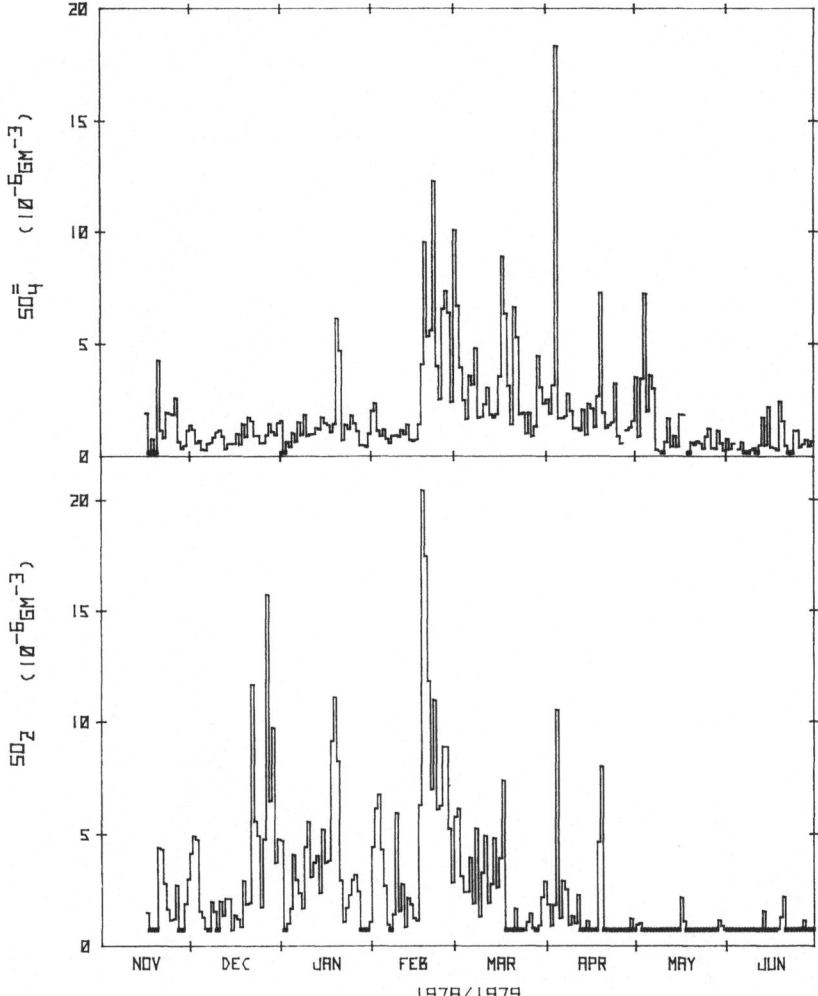

Fig. 6.   Daily sulphate and SO_2 concentrations in air at the Experimental Lake Area APN station (Barrie
et al., 1980).

3.3. A LONG-RANGE TRANSPORT EPISODE

Information about the sources of pollutants measured in event networks can be acquired
by analyzing available meteorological data with the assistance of models appropriate to
the larger scales. Daily particulate sulphate and gaseous SO_2 concentrations for the
period November 1978 to June 1979 at the Experimental Lakes Area (ELA) site have
been subjected to a detailed meteorological analysis (Barrie *et al.*, 1980). Large fluctuations in concentration are due to the presence, alternately, of clean background air
and of air polluted by large SO_2 sources in the lower Great Lakes area and the region
further south, as the data for February 1979 illustrate (Figure 6). Sulphur concentrations
were close to background for the first half of the month. On February 17, ambient
concentrations increased and for about two weeks remained high with an SO_2 maximum

of 20 µg m^{-3}. The beginning of this period was associated with the eastward movement of an extensive high pressure system which covered most of eastern North America.

Three-dimensional, four-day back-trajectories of air parcels (1000 mb) which arrived at ELA at 0 GMT on February 16, 17, and 18 are shown in Figure 7. On February 16, when low concentrations were measured, the trajectory was clearly from northern Canada. During the succeeding days, as the center of the high pressure system moved progressively eastward, the computed trajectories shifted southward into the United States and the particulate sulfate and SO$_2$ concentrations increased from minimum values on February 16 (0.76 µg m^{-3} SO$_4^=$ and 1.1 µg m^{-3} SO$_2$) to maximum values on February 18 and 19 (4–9 µg m^{-3} SO$_4^=$ and 18–20 µg m^{-3} SO$_2$).

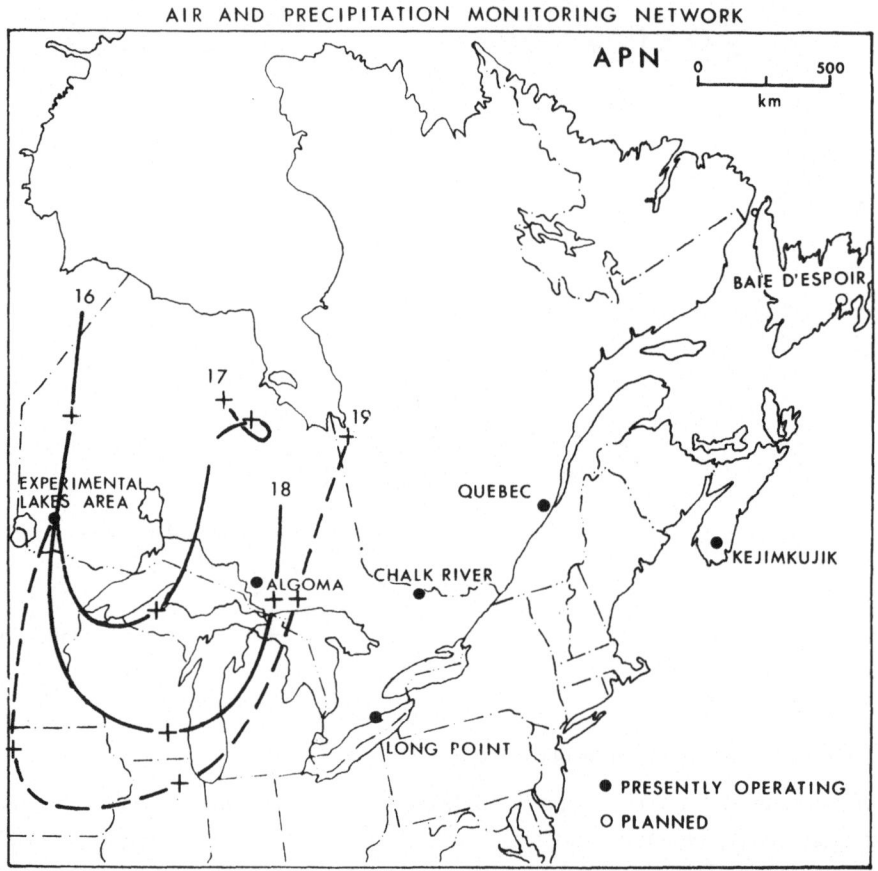

Fig. 7.   Locations of APN sites and 1000 mb back trajectories from ELA-Kenora for the February 16–19, 1979 episode of elevated oxides of S (Figure 6) from Barrie *et al.* (1980).

Pollution episodes are occasionally observed at ELA and Chalk River within a few days of one another even though the two sites are 2000 km apart. For instance, the February 18 episode at ELA discussed above was observed at Chalk River two days later. Trajectories show that as the center of the same high pressure system continued to move eastward, the southerly regional flow on the western side of the anticyclone brought polluted air, high in S concentrations, to southern Ontario from SO_2 source regions south of the lower Great Lakes. Similar analyses in the past (e.g. Whelpdale, 1978) have shown that the occurrence of large-scale episodes of elevated pollutant levels is usually associated with southerly flows behind large high pressure areas to the south and east. Following such episodes it is common for a frontal disturbance to pass through the lower Great Lakes – St. Lawrence Valley region and cause precipitation which cleans the air but results in high pollutant deposition rates.

3.4. POLLUTION IN THE ARCTIC ATMOSPHERE

The Arctic atmosphere is not exempt from the influence of man. Acidic substances released by fossil fuel combustion and smelting of metals at mid-latitudes in North America and Eurasia reach polar regions (Rahn and McCaffrey, 1979). Data from an international Arctic aerosol monitoring network involving Canada, the United States, Denmark, Norway and Iceland have revealed that the concentration of anthropogenic particulate matter containing sulphates, soot and hydrocarbons reaches a peak in late winter and early spring, and coincides with a maximum in the concentration of light-scattering aerosols and a minimum in visibility (see, e.g., Barrie et al., 1981).

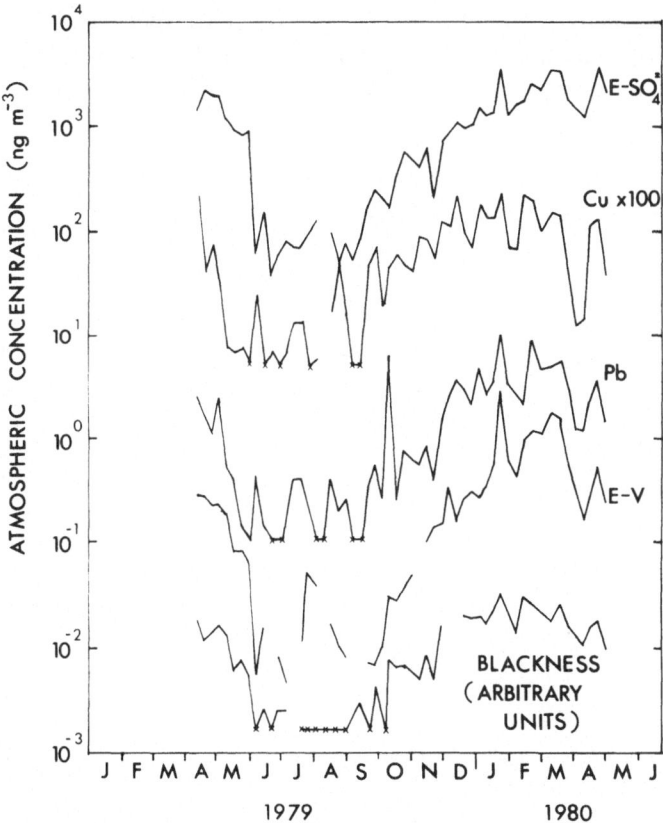

Fig. 8. Weekly average concentration of excess sulphate, Cu, Pb, non-soil V and filter blackness (soot) in the atmosphere at Mould Bay.

Arctic haze results from direct transport of air from mid-latitudinal source regions coupled with inefficient pollutant removal in winter. In summer, transport to the Arctic is less direct and scavenging en route is more efficient. Consequently Arctic aerosol levels are 20 to 40 times higher in winter than in summer. Arctic haze appears in the winter consistently from year-to-year and is widespread throughout the polar regions. Figure 8 shows the seasonal variation of some anthropogenic particulate constituents at Mould Bay (Barrie *et al.*, 1981). Peak weekly mean sulphate concentrations of 1 to 3.5 $\mu g\ m^{-3}$ are not much lower than those observed at mid-continent at ELA (Figure 6).

Arctic haze aerosols are acidic. An estimate of the acidity of precipitation and its seasonal variation in the North American Arctic has been made by Barrie *et al.* (1981) (Figure 9) on the basis of observed concentrations of aerosol acidity and an aerosol washout ratio (volumetric) of 2×10^5. Precipitation pH is expected to reach a minimum (maximum acidity) of about 5 to 5.1 in late winter. Until recently there has been little evidence to check this estimate. Recent analysis of samples from the Agassiz ice cap on Ellesmere Island provides such proof (Koerner and Fisher, 1981). Figure 10 shows the pH of ice cap snow for the period 1956 to 1978. The spring minima in pH in Figure 10, which are labelled with the year of occurrence, appear to confirm the seasonal variation of precipitation pH predicted from aerosol acidity (Figure 9). Cooler weather since 1964 has resulted in virtually no modification of the vertical distribution of H^+ concentration (pH) in the ice cap while prior to 1964, melting associated with somewhat milder weather has disturbed the actual deposition record. Minimum pH's near 5 are not uncommon.

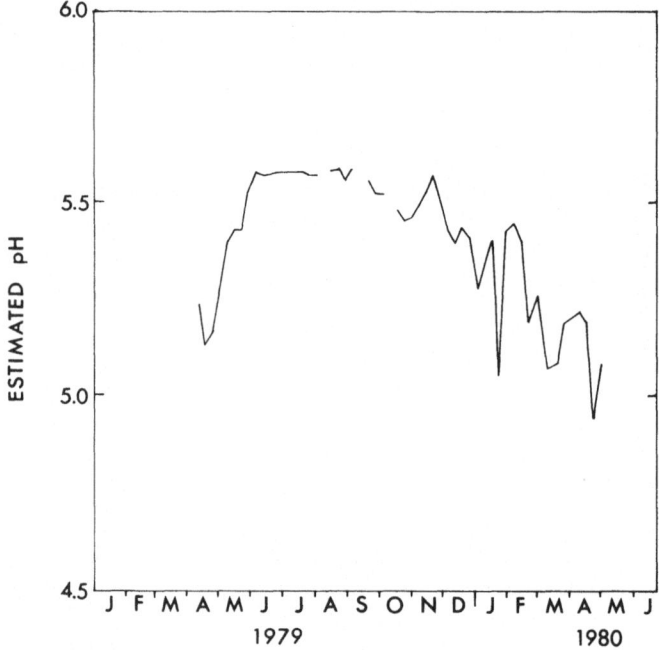

Fig. 9.   Estimated seasonal variation of precipitation pH in the North American Arctic (Barrie *et al.*, 1981) based on measured aerosol acidity and a washout ratio of 2×10^5.

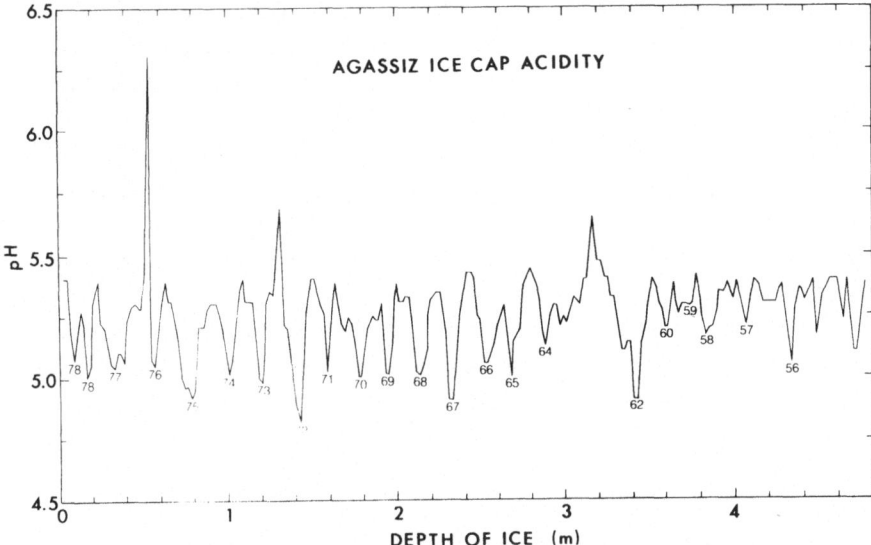

Fig. 10. Seasonal fluctuation of ice-pack pH on Ellesmere Island between 1956 and 1978 (Koerner and Fisher, 1981). Spring pH minima are marked with the year.

Attempts to draw conclusions from these data about trends in acidity should be made with caution.

4. Conclusions

Since the early 1970's considerable progress has been made in atmospheric monitoring in Canada. Although precipitation networks are well established and cover most of the country, there is still a scarcity of routine measurements of atmospheric gases and particles at regionally-representative, rural locations. The latter measurements are essential for determining dry deposition, an acknowledged important portion of the total deposition (Galloway and Whelpdale, 1980).

The status of Canadian precipitation monitoring activities is summarized as follows:
- Geographical coverage is adequate in most of Canada. In British Columbia, where topography is extremely varied, a denser network is desirable;
- The density of precipitation sampling stations is greater in those regions where acidic deposition is considered an environmental threat, namely, eastern Canada, areas downwind of pollutant emissions in Alberta, and on the west coast;
- Most networks collect monthly samples;
- There is a lack of precipitation sampling on an event or daily basis on a national scale. Such monitoring is now underway at nine sites in eastern Canada but not in western Canada. Event sampling is important for purposes of source identification, atmospheric processes research and assessment of effects of acidic deposition;
- Measurement of trace metals and organics in precipitation is inadequate at present. Based upon data which have been provided to date from the networks described, a

consistent picture emerges of the extent and severity of the large-scale air pollution problem affecting Canada. The southeastern portion of the country is most severely affected by acidic deposition, but the west coast and the Arctic are also affected. In the most severely impacted areas, the deposition of H^+, $SO_4^=$, and NO_3^- is as great as in southern Scandinavia and the northeastern U.S.A. Unfortunately, insufficient data are available with which to establish trends. Although pollution sources do exist in Canada, a significant part of this country's pollution originates from sources in the U.S.A. A smaller portion, particularly in the case of Arctic pollution, originates from outside the continent.

The data from monitoring networks in Canada fulfill several important needs – ranging from assessing the impacts of growing industrial regions to providing information on global background pollution levels. One of the most important of these uses is in support of research investigations. For example, network data have been used to a limited extent for:

– Deriving wet and dry deposition estimates;
– Investigating scavenging processes and improving model parameterizations of wet removal;
– Serving as a data base for model validation; and
– In conjunction with model results, investigating environmental processes and impacts.

Progress in research will continue to depend on the availability of a reliable data base from monitoring. Advances in network design, operation and data analysis depend to a very significant extent on the feedback resulting from the use, intercomparison and interpretation of network data. This is an area which requires more effort in the future.

Several actions which would improve our ability to quantify atmospheric inputs to the biosphere and to identify temporal trends in these inputs are recommended:

– First and foremost is the establishment of a central atmospheric data processing and quality control facility to ensure standardization of network operations and the compatibility of data sets. Such a facility should have federal and provincial support.
– Despite the great importance of dry deposition, no reliable routine measurements are being made. Development and testing of appropriate techniques must be encouraged at a few centers of expertise in the country. At the same time, an expansion of air monitoring at rural sites using appropriate techniques sensitive enough to determine 24 h average concentrations is needed.
– Although quality assurance and intercomparison activities continue to increase, an intensive effort is required to ensure that a compatible Canadian data base will be accurate enough to indicate temporal trends in precipitation quality.

The number of precipitation chemistry networks in operation is reaching a plateau. Improvements of existing networks are being made as experience dictates. In the next few years there is reason to expect that more emphasis will be placed on optimization of network configuration and procedures to produce required information. In many cases, this may well result in a reduction in number or relocation of stations (and networks). On the other hand, it is expected that the need for information on gaseous and particulate constituents will grow, and that much more effort will have to be placed on the establishment of viable networks for these purposes.

Acknowledgments

The authors are grateful for the support of the Canadian Atmospheric Environment Service and to Messrs J. Kovalick and B. Martin for technical assistance. One of the authors (DMW) acknowledges the support of the Norges Teknisk-Naturvitenskapelige Forskingsråd and the Norsk Institutt for Luftforskning during the preparation of this manuscript.

References

Barrie, L. A.: 1981, 'Environment Canada's Long-Range Transport of Atmospheric Pollutants Program: Atmospheric Studies, Proc. Conf.', *The Effects of Acid Precipitation on Ecological Systems in the Region of the Great Lakes*, March 31–April 3, 1981, Michigan State University (in press).

Barrie, L. A., Hoff, R. M. and Daggupaty, S. M.: 1981, *Atmospheric Environment* **15**, 1407.

Barrie, L. A., Wiebe, H. A., Anlauf, K., and Fellin, P.: 1980, 'The Canadian Air and Precipitation Monitoring Network. Atmospheric Pollution 1980', Proceedings of the 14th International Colloquium, Paris, France, May 5–8, 1980, M. M. Benarie (ed.). Studies in Environmental Science, Volume 8, Elsevier, Amsterdam, 355–360.

Galloway, J. N. and Whelpdale, D. M.: 1980, *Atmospheric Environment* **14**, 409.

Henry, R. C. and Hidy, G. M.: 1980, *Atmospheric Environment* **14**, 1095.

Koerner, R. M. and Fisher, D.: 1981, 'Acid Snow in the Canadian High Arctic', *Nature* (in press).

MAP3S/RAINE: 1982, 'The MAP3S/RAINE Precipitation Chemistry Network. Statistical review overview for the Period 1976–1980', by the MAP3S/RAINE Research Community, Coordinator, J. M. Hales, *Atmospheric Environment* (in press).

NRCC 1981: 'Acidification in the Canadian Aquatic Environment: Scientific Criteria for an Assessment of the Effects of Acidic Deposition on Aquatic Ecosystems'. Publication NRCC No. 18574, Publications, NRCC/CNRC, Ottawa, Canada.

Rahn, K. A. and McCaffrey, R. J.: 1979, 'Long-Range Transport of Pollution Aerosol to the Arctic; A Problem without Borders', *Proc. WMO Sympos., Rep. 538.* 25–36, Sofia, Oct. 1–5.

Whelpdale, D. M.: 1978, *Atmos. Environ.* **12**, 661.

THE INFLUENCE OF SURFACE STRUCTURE ON PREDICTED PARTICLE DRY DEPOSITION TO NATURAL GRASS CANOPIES

CLIFF I. DAVIDSON, JANEL M. MILLER, and MARK A. PLESKOW

Departments of Civil Engineering and Engineering and Public Policy, Carnegie-Mellon University, Pittsburgh, PA 15213, U.S.A.

(Received 10 July, 1981; Revised 2 October, 1981)

Abstract. Equations describing particle transport to surfaces by diffusion, interception, impaction, and sedimentation have been used to predict dry deposition onto five wild grass canopies. Detailed measurements of plant height and spacing, width of stems, leaves, and inflorescences, and wind data collected within and above each canopy have been applied as model input data. The resulting curves of deposition velocity versus particle diameter have then been used with size distribution data from the literature for Pb and sulfate to predict overall dry deposition. Results of these calculations suggest a wide range of dry deposition velocities of 0.05 to 1 cm s^{-1} for these species, the variation resulting from differences in surface structure and size distribution characteristics. At least 40% of the mass deposition of lead and sulfate results from the largest 10% of the airborne material.

1. Introduction

Dry deposition of airborne particles to natural surfaces is poorly understood, in spite of numerous theoretical, wind tunnel, and field studies. Factors responsible for our lack of understanding include difficulties in adequately parameterizing all of the relevant variables, problems associated with direct dry deposition measurement, and uncertainties in relating wind tunnel data to ambient atmospheric processes. Review papers by McMahon and Denison (1979) and Sehmel (1980) suggest that particle dry deposition data in the literature vary over several orders of magnitude; problems with measurement and theoretical prediction are at least partially responsible.

Existing dry deposition models are generally based upon wind tunnel data for monodisperse particles, where vegetation, soil, water, or artificial material is placed inside the tunnel as a deposition surface. The resulting models often express deposition as a function of friction velocity and roughness height (momentum sink), but rarely include details of the surface structure. In contrast, Wells and Chamberlain (1967) have shown that particle deposition onto a filter paper surface covered with fine hairs may exceed deposition onto a smooth brass surface by orders of magnitude in spite of comparable friction velocity and roughness height. Wind tunnel data for pine and oak shoots show that the former are much more efficient particle collectors, being composed of fine needles 0.2 cm in diameter rather than broad leaves (Belot and Gauthier, 1975). Calculations of impaction efficiencies for *Setaria viridis* (foxtail) have suggested that the presence or absence of awns of 60 μm diameter at the top of the plants can significantly affect the overall deposition to the canopy (Davidson and Chu, 1981).

The proposed dry deposition model incorporates new field data into a theoretical framework similar to that used by Davidson and Friedlander (1978) and by Slinn (1981):

Water, Air, and Soil Pollution **18** (1982) 25–43.   0049–6979/82/0181–0025$02.85.

(1) Cup and hotwire anemometry has been used to obtain wind profile data for use in estimating particle eddy diffusivities and filtration efficiencies.

(2) Measurements of the distribution of plant heights and of stem and leaf sizes have been used with the wind data to predict deposition by diffusion, interception, impaction, and sedimentation within the canopies.

Several theoretical curves of particle deposition velocity versus particle diameter have been thus calculated. The curves have then been used with airborne size distribution data for sulfate and Pb to calculate overall deposition velocities for each of these species.

2. Experimental

2.1. WIND PROFILES

Three cup anemometers (R. M. Young model 12–102) and a hot-wire anemometer (Datametrics 100VT) were used to obtain multipoint wind profiles above and below the vegetation height in five separate wild grass canopies in western Pennsylvania. The predominant grass in each canopy was *Tridens flavus* (tall redtop), *Dactylis glomerata* (orchard grass), *Phleum pratense* (timothy), *Andropogen virginicus* (broom sedge), and *Agrostis hyemalis* (hair grass). Each canopy also included a limited amount of other wild vegetation, for example *Andropogen virginicus, Dactylis glomerata,* and *Aster pilosus* (aster) in the second, third, and fourth canopies, respectively. All experiments were conducted during moderate winds and generally cloudy days where U.S. Weather Bureau soundings showed approximately adiabatic conditions. Several runs on two or three separate days were conducted for each canopy.

The cup anemometers were positioned at fixed heights above the vegetation, while the hot-wire probe was lowered incrementally within the canopy. Several times during these runs, hot-wire measurements were obtained at the same height as a cup anemometer to check for satisfactory agreement. Similarly, the cup anemometers were all placed at the same height before and after each run, and were frequently interchanged to minimize systematic error. All four anemometers were calibrated in a wind tunnel at 15 air velocities between 0.5 and 7 m s^{-1}.

Chart recorder outputs (25 cm width per pen) were integrated over periods of several seconds each using a manual planimeter. Although this method is less accurate than counters or electronic integration, the expanded chart width and large number of successive planimeterings are believed to have provided reliable data. This is supported by the resulting wind profiles, shown in Figure 1 for the five canopies.

The above canopy data were used to estimate friction velocity u_*, roughness height z_0 and zero plane displacement d, in the equation:

$$u(z) = \frac{u_*}{k} \ln \left(\frac{z - d}{z_0} \right) \tag{1}$$

where k is von Karman's constant (assumed equal to 0.4 in this analysis) and z is the height above ground. The hot-wire data were used to determine the value of the parameter n in the relation:

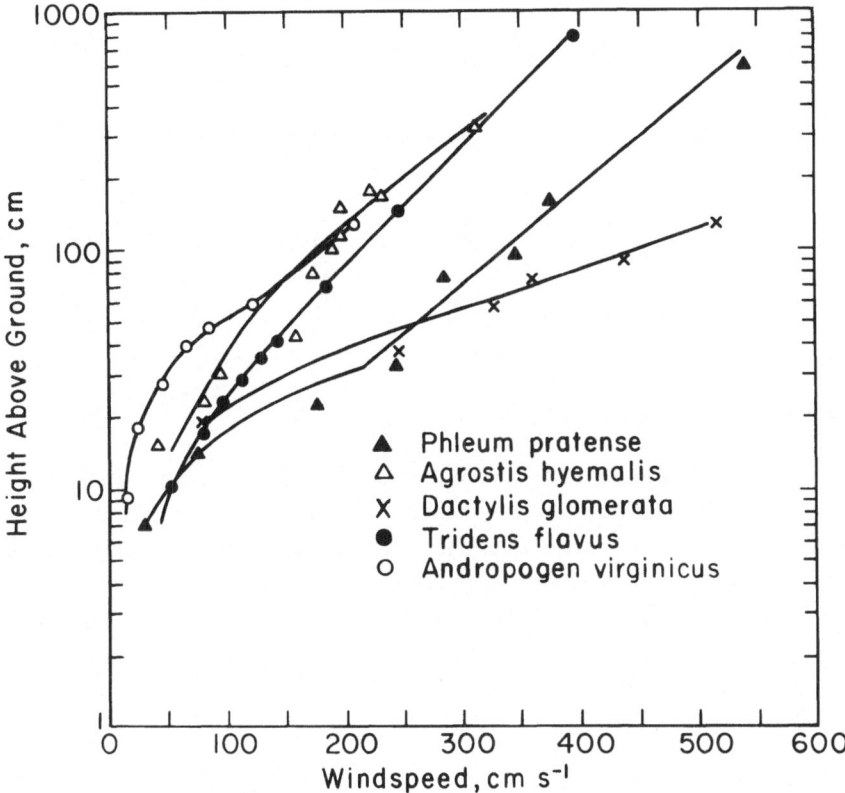

Fig. 1. Wind profiles for five natural grass canopies in western Pennsylvania, measured during October and November, 1980. The portion of each curve above the canopy height h (see Table II) corresponds to Equation (1), while that section below h is defined by Equation (2). Each profile is an average of 3 to 14 individual measurements, obtained on days of similar windspeed.

$$u(z) = u(h) \exp\left[-n\left(1 - \frac{z}{h} \right) \right] \qquad (2)$$

where h is the canopy height (in this context, the maximum height of those plants measured). Values of the various parameters associated with wind profiles from the five canopies are presented in Table I, calculated as a best fit to the measured data.

Values of 0 to 15 cm for z_0 and 0 to 30 cm for d have been previously reported for natural grass canopies in typical ambient winds (Calder, 1949; Sutton, 1953; Pasquill, 1962). Values of n in the range 2 to 4 have been measured in crop canopies (Tan and Ling, 1961; Lemon, 1965; Denmead, 1976), although Shinn and Cionco (1973) report values of 0.44 to 4.4 for a variety of vegetation types. The relatively small values of n in Table I indicate that airflow within the grass is significantly coupled to the above-canopy flow. In general, the wind data from this study are in agreement with other published wind profiles.

TABLE I

Wind profile data corresponding to the measurements of Figure 1

	Number of profiles measured	$\bar{u}$ (200 cm) cm s^{-1}	u_* cm s^{-1}	d cm	z_0 cm	n
Tridens flavus	6	270	34.5	0	8.1	1.25
Dactylis glomerata	14	580	63.8	14	5.0	1.75
Phleum pratense	8	420	42.1	0	4.0	2.5
Andropogen viriginicus	7	240	32.3	14	9.0	2.0
Agrostis hyemalis	3	260	40.4	13	14	1.0

2.2. CANOPY CHARACTERIZATION

Details of the vegetation geometry were measured for each canopy, including plant spacing, plant height, and width of leaves, stems, inflorescences, etc. In this manner, the total number of obstacles of each size as a function of height was determined. Frequency distributions of plant height and vegetative element width are plotted in Figure 2 for each of the five canopies, with the corresponding parameter values given in Table II.

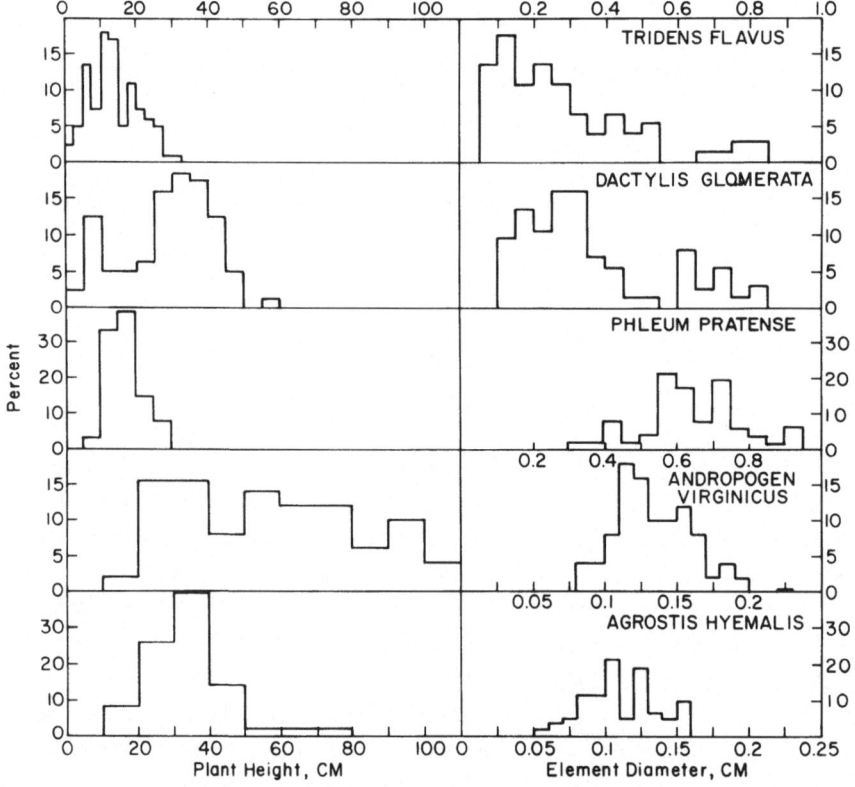

Fig. 2. Distributions of plant height and element diameter (or width) for each of the five canopies. The percent of the total plants measured having heights or diameters within each indicated range (2.5, 5, or 10 cm intervals for height; 0.01 or 0.05 cm intervals for diameter) are given for several size ranges.

TABLE II

Canopy characterization data for the distributions of Figure 2. Very thin, sparse vegetation in the *Andropogen* canopy required use of $\bar{h}$ rather than h in Equation (2) to obtain a reasonable fit to the wind profile data

	Number of plant elements examined	Maximum height h, cm	Average height $\bar{h}$, cm	Average element diameter, cm	Number of plant elements per cm² ground area N	Leaf area index cm²/cm² ground
Tridens flavus	78	32	14	0.28	0.44	1.7
Dactylis glomerata	78	57	29	0.35	0.17	1.7
Phleum pratense	60	30	17	0.64	0.96	10
Andropogen virginicus	50	100	57	0.13	0.087	0.6
Agrostis hyemalis	66	79	33	0.11	1.5	5.5

Previous characterization of wild grass and crop canopies incorporating total leaf area as a function of height (e.g., Waggoner and Reifsnyder, 1968; den Hartog and Shaw, 1975; Legg and Monteith, 1975; Ripley and Redmann, 1976) is not as useful for deposition model input data. As discussed earlier, particle collection efficiency is a strong function of the shape and size of individual elements.

3. Model Development

The goal of the proposed model is to derive estimates of particle dry deposition velocity v_g as a function of particle diameter d_p for several selected vegetative canopies under typical ambient atmospheric conditions. The defining equation is:

$$v_g = -\frac{J}{C} \tag{3}$$

where J is the flux in ng cm^{-2} s^{-1} (which has a negative value when net transport is downward) and C is the airborne mass concentration in ng cm^{-3}. Transport from the atmosphere into the canopy is assumed to result from sedimentation and eddy diffusion with the expression for flux given by:

$$J = -(D + \varepsilon_p)\frac{dC}{dz} - v_s C. \tag{4}$$

Here, D is the Brownian diffusivity in cm² s^{-1}, ε_p is the particle eddy diffusivity in cm² s^{-1}, and v_s is the sedimentation velocity in cm s^{-1}. In practice, $D \ll \varepsilon_p$ for typical ambient conditions. It is further assumed that the particle eddy diffusivity is analogous to that of momentum (Thom, 1975):

$$\varepsilon(z) = ku_*(z - d) \tag{5}$$

which applies only above the canopy. For $z < h$, the momentum eddy diffusivity can be considered an exponential function (Uchijima and Wright, 1964; Brown and Covey, 1966; Inoue, 1965), hence the analogy between particle and momentum transport yields:

$$\varepsilon_p(z) = ku_*(h - d) \exp\left[-n\left(1 - \frac{z}{h} \right) \right].$$

(6)

Another possible form for particle eddy diffusivity within the canopy, after Thom (1971), assumes ε_p = constant = $ku_*(h - d)$. Both expressions yield comparable results for v_g using the proposed theoretical framework, however.

The proper choice of boundary condition for Equation (4) also deserves comment. One possibility is to adopt a condition implying a non-perfect sink for particles at the surface. This leads to difficulties when attempting to quantify particle rebound with presently available data, however (Slinn, 1981). We therefore propose the simple boundary condition:

$$C(z = z_s) = 0$$

(7)

where z_s is the effective sink height, as an approximation to the true deposition. The error in using Equation (7) is likely to be small for submicron particles, although rebound of large particles can be significant (Chamberlain and Chadwick, 1972). The equation is also inappropriate for cases where sedimentation controls overall deposition onto the canopy; for these cases ($z_s < 0$), the concentration has been assumed invariant with height. As will be discussed later, canopy filtration efficiencies in this study are sufficiently small that a nearly constant concentration in the upper two-thirds of the canopy is a realistic approximation.

Particle collection by 'filtration' on the vegetative elements is assumed to occur by Brownian and eddy diffusion, interception, and inertial impaction. Electrostatic, thermophoretic, and other forces have been neglected in this analysis. For simplicity, the elements described in Figure 2 and Table II are taken as cylinders, whose diameters and lengths are represented by the measured frequency distributions. Transport equations for particle deposition on cylinders in transverse airflow can thus be taken from the literature for each regime of interest. A more accurate analysis would include details of the shapes and orientations of the plant elements, such as that performed by Chamberlain (1974) for bean leaves.

Aerosol collection by a cylindrical fiber of diameter d_f and height h can best be described in terms of an efficiency η:

$$\eta = \frac{\text{particles s}^{-1} \text{ depositing on cylinder}}{\text{particles s}^{-1} \text{ passing through projected cylinder area } d_f h}.$$

(8)

The total flux on a canopy containing N cylinders per unit area of ground is thus:

$$J = -Nd_f \int^h \eta(z)u(z)C(z)\, dz .$$

(9)

Diffusional transport can be expressed as:

$$\eta_{\text{diffusion}} = \frac{D\pi \, \text{Sh}(z)}{d_f u(z)} \tag{10}$$

where

$$\text{Sh} = \text{Sherwood number} = 0.683 \, \text{Re}^{0.466} \, \text{Sc}^{1/3} \tag{11}$$

$$\text{Re} = \text{Reynolds number} = ud_f / v \tag{12a}$$

$$\text{Sc} = \text{Schmidt number} = v/D \tag{12b}$$

$$v = \text{air viscosity in cm}^2 \text{ s}^{-1}.$$

The expression for Sherwood number applies for $40 < \text{Re} < 4000$ which represents the range of values for the present study.

The interception efficiency for potential flow is given by:

$$\eta_{\text{interception}} = \frac{2d_p}{d_f} \tag{13}$$

while the efficiency of impaction is expressed as:

$$\eta_{\text{impaction}} = \frac{\text{St}^3}{\text{St}^3 + 0.753 \, \text{St}^2 + 2.796 \, \text{St} - 0.202} \tag{14}$$

where

$$\text{St} = \text{Stokes number} = \frac{d_p^2 u(z) \, (\rho_p - \rho_a)K}{9\mu d_f} \tag{15}$$

$$\rho_p = \text{particle density in g cm}^{-3}$$

$$\rho_a = \text{air density in g cm}^{-3}$$

$$K = \text{slip correction factor}$$

$$\mu = \text{air kinematic viscosity in g cm}^{-1} \text{ s}^{-1}, \text{ equal to } v\rho_a.$$

The equations for diffusion and impaction have been taken from Davidson and Friedlander (1978), while the interception efficiency is given by Fuchs (1964). An overall efficiency for use in Equation (9) can be written as:

$$\eta = 1 - [(1 - \eta_{\text{diffusion}})(1 - \eta_{\text{interception}})(1 - \eta_{\text{impaction}})] . \tag{16}$$

Equations (4) and (9) can be solved for the flux J and the concentration sink z_s (thus completely defining the function $C(z)$). Equation (3) is used to determine v_g at any desired height.

The deposition velocity versus particle diameter curves of Figure 3 for *Tridens flavus* have been obtained by solving these equations numerically with a programmable

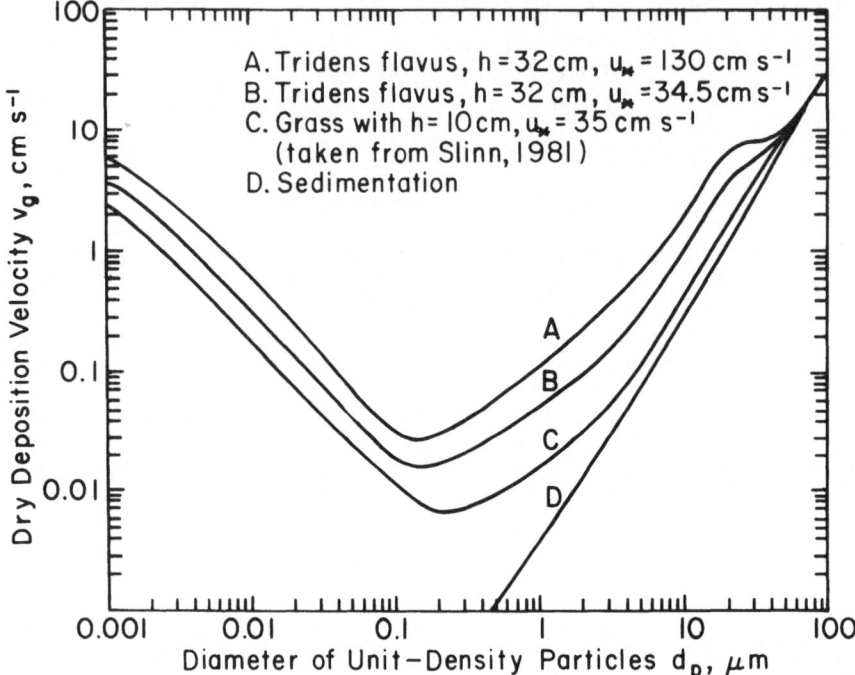

Fig. 3.   Dry deposition velocity at $z = 2$ m versus particle diameter for several conditions. Curves A and B use the model developed in this paper for the *Tridens flavus* canopy, corresponding to $h = 32$ cm and the appropriate distributions shown in Figure 2. Curve A assumes $u(2m) = 1000$ cm s^{-1} ($u_* = 130$ cm s^{-1}) while curve B uses the measured profile in Figure 1 with $u(2m) = 270$ cm s^{-1} ($u_* = 34.5$ cm s^{-1}). The deposition model of Slinn (1981) is represented by curve C for $h = 10$ cm and $u_* = 35$ cm s^{-1}. Sedimentation is depicted by curve D.

calculator. Distributions of plant characteristics shown in Figure 2 have been incorporated by integrating the equations over each range of plant heights and diameters. Curve B uses the measured wind profile for *Tridens*, where $u(2m) = 270$ cm s^{-1}, while hypothetical curve A for $u(2m) = 1000$ cm s^{-1} shows the sensitivity of v_g to ambient windspeed. Curve C is taken from Slinn (1981) for v_g at $z = 5$ m, for a grassy surface with $h = 10$ cm, $z_0 = 1.73$ cm, and $u_* = 35$ cm s^{-1}. Since $C(z)$ is nearly constant with height between 2 and 5 m, it is meaningful to compare Slinn's graph with curves from the present study. There are also similarities between the shape of these curves and the wind tunnel-derived plots of Chamberlain (1967), Clough (1975), and Sehmel and Hodgson (1976).

Deposition velocity plots for all five canopies are presented in Figure 4. Note that the curves for *Phleum* and *Agrostis* lie considerably above the other three, reflecting the relatively dense vegetation. Although *Phleum* has the greater leaf area index, larger element diameters result in poor impaction collection efficiencies compared with *Agrostis*. The smaller v_g values for *Tridens* are primarily due to the short canopy height, while the *Dactylis* and *Andropogen* canopies are tall but sparse. Differences in measured windspeeds among the five canopies are probably less important in influencing deposition than is the vegetation structure.

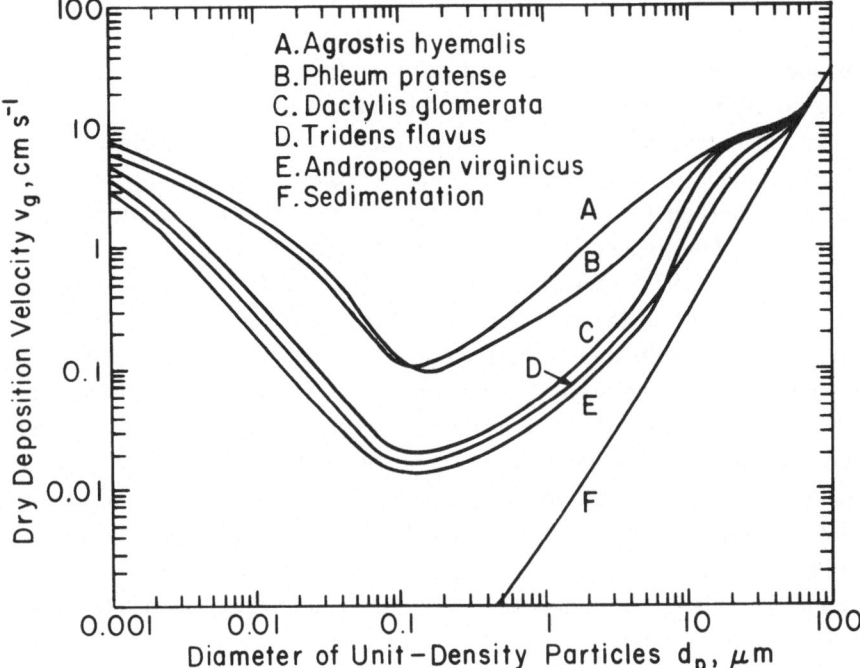

Fig. 4. Dry deposition velocity at $z = 2$ m versus particle diameter for the five canopies studied, using the wind profiles of Figure 1 and the canopy characterization data of Figure 2. Note that curve D in this Figure is identical to curve B in Figure 3.

The minimum *Phleum* and *Agrostis* v_g of 0.1 cm s^{-1} for 0.1 μm diameter particles is a factor of ten smaller than predictions of Sehmel and Hodgson (1976) for $z_0 = 10$ cm and $u_* = 30$ cm s^{-1}. However, the latter are only rough approximations for large z_0, being derived from wind tunnel measurements over surfaces with $z_0 \leq 0.1$ cm (Slinn *et al.*, 1978). It is of interest that interception as well as diffusion significantly affects 0.1 to 1 μm particle deposition on these canopies, based on Equations (10) to (16).

Equation (4) demands that for downward particle flux, the concentration must decrease as we move from the atmosphere toward the ground. However, the rate of decrease may be sufficiently small that $C(z)$ can be considered constant for use in Equation (9). This situation applies to the curves of Figures 3 and 4: the efficiency of the five canopies is so poor (especially for d_p near 0.1 μm) that the calculated function $C(z)$ is relatively constant for $0.3\,h < z < h$. In fact, curves very close to those in Figures 3 and 4 for the entire range of d_p can be computed by integrating Equation (9) over $0.3\,h < z < h$, with $C(z) = C(h)$ and $u(z)$ from Table I. A measured concentration just above the canopy can be used instead of $C(h)$ with little loss of accuracy. Values of $\eta(z)$ for diffusion, interception, and impaction are obtained from Equations (10) to (16); the resulting 'filtration flux' can be merely added to sedimentation as suggested by Slinn (1981). This simplification in the basic model only applies where aerosol filtration is inefficient, a condition apparently satisfied by certain natural grass canopies in moderate winds.

Before predicting overall deposition velocities based on ambient size distributions, it is worthwhile to highlight some of the weaknesses of the present study. These inadequacies include problems with the wind profiles, difficulties in characterizing particle transport by eddies within the canopy, and bounceoff.

The parameter values in Table I are strong functions of the measured wind data, the latter being subject to some uncertainty. Windspeeds above the canopy were probably influenced by short fetches in several instances: the ratio upwind fetch/top anemometer height is 60 for *Tridens* and *Phleum*, 130 for *Dactylis*, 80 for *Andropogen*, and 50 for *Agrostis*. Pasquill (1972) suggests a minimum ratio of 100 to insure representative profiles. The effect of obstacles at shorter distances upwind (i.e., a few 4 m trees in the *Tridens* canopy 150 m upwind of the anemometers) may also have affected the data, as discussed by Businger (1975) and Maki and Allen (1978). In-canopy measurements were probably influenced by the nonuniform nature of the vegetation. In addition, the calculated parameter n is sensitive to the value of h used in Equation (2) as well as the original wind data. A greater number of data points would have been desirable for some of the profiles. Finally, it should be noted that the wind measurements in the present study were obtained merely to provide example profiles for use in the model. Thus the uncertainties discussed above may be less important than weaknesses in the description of particle transport within the canopy.

The use of Equations (4) and (5) to represent particle eddy diffusivity has received considerable attention in the literature. The basic analogy between momentum and mass requires identical source/sink distributions of both properties, a condition obviously not fulfilled in many canopies. One problem is that bluff-body forces, which may augment momentum transport to the surface, have no analog in mass transport (Thom, 1972). In the case of natural grasses, however, Garratt and Hicks (1973) suggest that the bluff-body effect is minimal. Equating eddy diffusivities for mass and momentum also implies that particles follow the air eddies entirely, a weak assumption especially for supermicron aerosol with appreciable inertia (Friedlander, 1977). Legg and Monteith (1975) point out that changes in dC/dz over heights smaller than the characteristic mixing length make the use of Equations (3) to (5) within the canopy highly tenuous. They also argue that correlations between the instantaneous vertical velocity and particle concentration will be affected by variations in the particle source/sink distribution within the canopy, and that the divergence of the horizontal particle flux may not be negligible. In addition to spatial variations, Perrier (1975) discusses problems introduced by averaging over rapid changes in windspeed to obtain time-integrated values. Finally, the analogy between momentum and mass eddy diffusivities breaks down during nonadiabatic conditions, where a similar analogy between diffusivities of heat and mass may be more accurate (Denmead, 1976).

An adequate understanding of particle bounceoff has not yet been achieved. Slinn (1981) suggests as a rough approximation a reduction factor, to be multiplied by the deposition efficiency, which is an exponential function of Stokes number. It is also admitted in that study that a single function for the reduction factor is probably not sufficient for the wide variety of plant and particle characteristics involved. Legg and

Price (1980) have concluded that impaction may govern particle deposition on wet vegetation in light winds, but that bounceoff may reduce impaction on dry canopies resulting in increased importance of sedimentation. Poor characterization of bounceoff is a major weakness of this and other dry deposition models, necessitating additional research in this area.

In addition to these problems, other weaknesses include uncertainties in the overall shape of the concentration profile within the canopy (particularly near the ground), the inapplicability to the ambient atmosphere of using empirical Equations (10) to (15) derived in controlled laboratory experiments, insufficiently detailed canopy micro-structure used as model input data, (e.g., fine hairs which may increase impaction and interception efficiencies) and treatment of particle transport merely as a two-step eddy diffusion/filtration process. Finally, it must be emphasized that the wind profiles and canopy characterizations of Figures 1 and 2 are specific to the dates and locations sampled, and cannot be readily generalized.

4. Application of the Model to Ambient Aerosols

Several size distributions of Pb (Figure 5) and sulfate (Figure 6) have been taken from the literature for use with the deposition model. The geometric mean aerodynamic particle diameter on each impactor stage has been used with curves A and E of Figure 4 to calculate overall dry deposition velocities on these two vegetative canopies. Use of aerodynamic diameters is acceptable because the bulk of the deposition is due to sedimentation and impaction. Because the true distribution of particle sizes on the top impactor stage is unknown, it has been assumed that all of the mass in the uppermost range of each distribution is found in particles with the minimum d_p in that range. This will tend to underestimate v_g.

Two additional factors require that the calculated dry deposition rates be considered as underestimates. First, many of the distributions of Figures 5 and 6 employed vertically oriented impactors and less-than-optimal air inlets; this may have resulted in sampling losses for large particles (Wedding et al., 1977). In addition, some of the impactors used dry rather than adhesive-coated substrates. Dzubay et al. (1976) and Lawson (1980) have shown that particle bounceoff occurs under such conditions, resulting in a distribution biased toward smaller particles.

The resulting overall dry deposition velocities for each distribution are shown in Tables III and IV. Also shown for each distribution is the fraction of total mass deposition resulting from 10% of the airborne mass concentration found in particles with the greatest aerodynamic diameters.

Several conclusions can be drawn from these tables. Of primary importance is the order of magnitude difference in dry deposition velocities calculated for the two canopies: this is primarily due to the greater surface area of *Agrostis* compared with *Andropogen*. Also note the variations in deposition velocity among distributions, resulting primarily from differences in airborne mass concentration on the uppermost impactor stages. This is probably related to differences in sampling techniques as well as the true shapes of

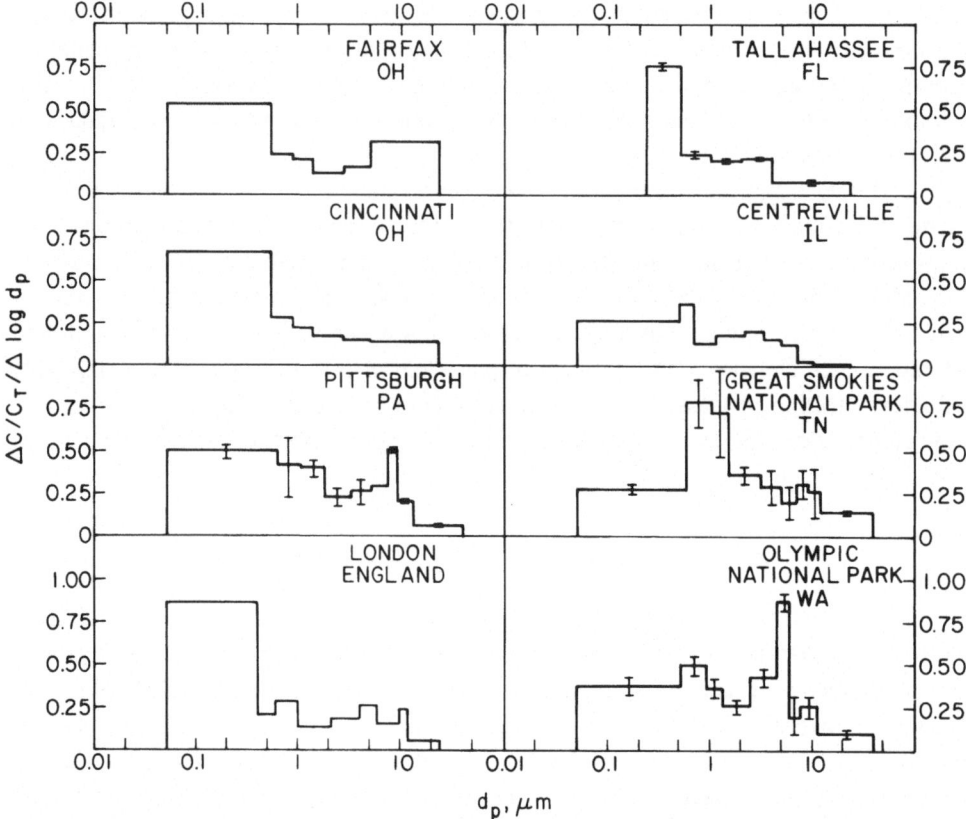

Fig. 5. Airborne size distributions of Pb obtained with cascade impactors, using data taken from the literature. The ordinate represents the normalized mass distribution function, where ΔC is the airborne concentration of Pb in each size range in $\mu g \, m^{-3}$, C_T is the total airborne concentration in all size ranges, and $\Delta \log d_p$ represents the size interval for each impactor stage. d_p refers to the aerodynamic diameter (equivalent unit density spheres). The area under each curve between any two values of d_p is thus proportional to the fraction of mass in that size range. Minimum and maximum diameters are assumed to be 0.05 and 25 μm, respectively, except where otherwise indicated in the original reference. Details of these distributions are given in Table III.

the size distributions. For example, the larger mass median diameter and deposition velocities for Pb in Pittsburgh compared with Fairfax and Cincinnati may be partially due to the use of adhesive-coated substrates and a wind-oriented horizontal air inlet in the former study. The latter sampling employed dry substrates and a vertical inlet. A difference in the d_p cutoff of the top impactor stage (13 μm for Pittsburgh, 5 μm for Fairfax and Cincinnati) is also partially responsible.

Values in the final column of Tables III and IV show the importance of accurate characterization of large particles when calculating dry deposition. Overall, roughly 40% of the Pb and sulfate mass deposition is due to only 10% of the airborne particle mass. This conclusion is probably conservative, because of the underestimation of large particle concentrations.

A considerable number of v_g measurements for Pb and sulfate are reported in the

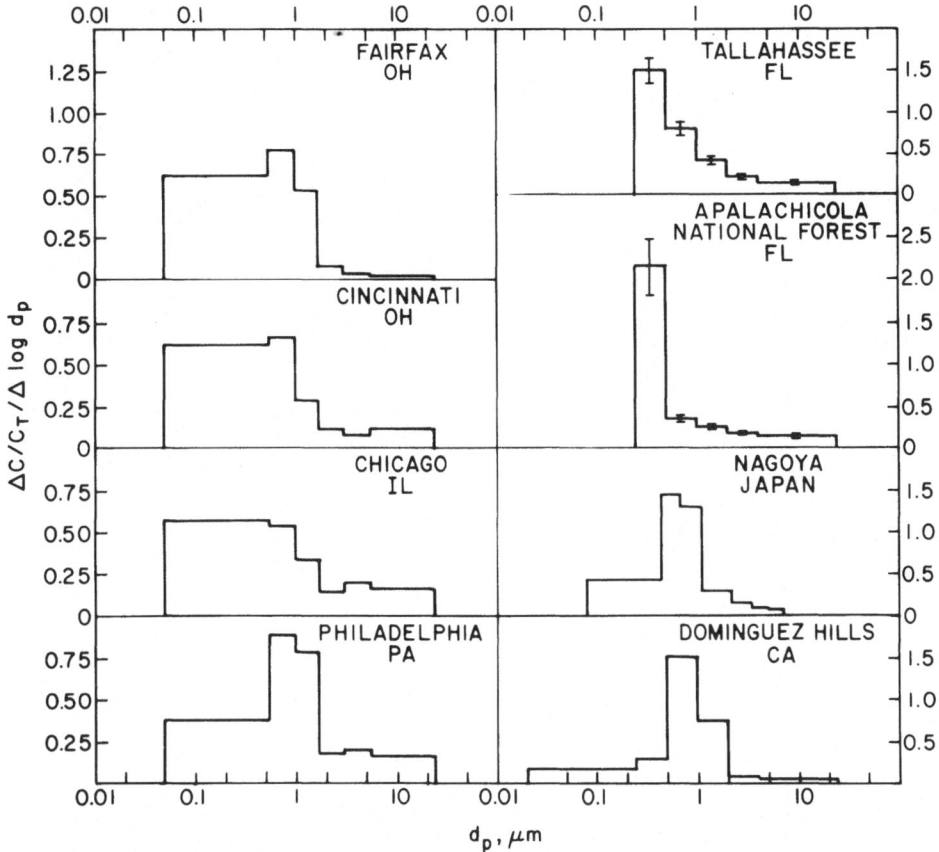

Fig. 6. Airborne size distributions of sulfate obtained with cascade impactors using data taken from the literature. The same assumptions as in Figure 5 apply here, with details of the distributions given in Table IV.

literature. Deposition velocities of Pb of 0.13 to 0.38 cm s^{-1} have been reported for various surrogate surfaces (Servant, 1976; Davidson, 1977; Crecelius *et al.*, 1978; Elias and Davidson, 1980), 0.16 to 2.0 cm s^{-1} for a snow surface (Dovland and Eliassen, 1976), 0.36 cm s^{-1} for a deciduous forest canopy (Lindberg *et al.*, 1979), and 1.6 to 4.4 cm s^{-1} for grass beside a heavily travelled roadway (Little and Wiffen, 1977). Sulfate deposition velocities have been determined from concentration gradient measurements over natural vegetation as 0.1 cm s^{-1} (Garland, 1978), 0.1 to 0.27 cm s^{-1} (Droppo, 1980), and 1.2 to 1.7 cm s^{-1} (Everett *et al.*, 1979). Wesely *et al.* (1977) have used eddy correlation to estimate v_g for sulfate of 0.1 to 1 cm s^{-1}. Based partially on experimental data (Wesely *et al.*, 1977; Wesely and Hicks, 1979; Everett *et al.*, 1979), Sheih *et al.* (1979) have estimated a sulfate dry deposition velocity of 0.8 to 0.9 cm s^{-1} for adiabatic conditions in Eastern U.S. on a regional scale. Sievering *et al.* (1979) estimate Pb and sulfate dry deposition velocities in the range 0.1 to 0.2 cm s^{-1} over Lake Michigan by comparing airborne concentrations in Chicago with midlake values. These investigators also report $v_g = 0.38$ cm s^{-1} for 0.15 to 0.30 μm diameter particles (> 90% containing

TABLE III

Size distribution data and calculated dry deposition velocities for Pb, corresponding to Figures 4 and 5

Sampling location and date	Total airborne concentration C_T μg m^{-3} Pb	Approximate mass median aerodynamic diameter μm	Andropogen virginicus v_g cm s^{-1}	Agrostis hyemalis v_g cm s^{-1}	Andropogen virginicus Fraction of deposition by largest 10% of airborne mass	Agrostis hyemalis Fraction of deposition by largest 10% of airborne mass	Reference
Fairfax, OH Feb. 1967	0.69	0.4	0.080	1.0	31	33	Lee et al. (1968)
Cincinnati, OH Sept. 1966	2.8	0.2	0.053	0.66	47	53	Lee et al. (1968)
Pittsburgh, PA July–Sept. 1979	0.59	0.6	0.21	1.1	69	46	Davidson et al. (1981)
Highway in London, England April, July, Oct.	5.6	0.2	0.097	0.64	76	64	Little and Wiffen (1977)
Tallahasse, FL June–July 1973	0.24	0.6	0.062	0.75	31	33	Johansson et al. (1976)
Centreville, IL Summer 1972	0.62	0.5	0.068	0.61	58	53	Peden (1977)
Great Smokies National Park, TN Oct. 1979	0.014	1.0	0.27	1.4	62	41	Davidson (1981)
Olympic National Park, WA July–Aug. 1980	0.0024	0.9	0.21	1.4	63	40	Goold and Davidson (1981)

TABLE IV

Size distribution data and calculated dry deposition velocities for sulfate, corresponding to Figures 4 and 6

Sampling location and date	Total airborne concentration C_T μg m^{-3} SO$_4$	Approximate mass median aerodynamic diameter μm	Andropogen virginicus v_g cm s^{-1}	Agrostis hyemalis v_g cm s^{-1}	Andropogen virginicus Fraction of deposition by largest 10% of airborne mass	Agrostis hyemalis Fraction of deposition by largest 10% of airborne mass	Reference
Fairfax, OH Aug. 1965	7.2	0.4	0.030	0.33	39	46	Wagman et al. (1967)
Cincinnati, OH Aug.–Sept. 1965	10	0.4	0.044	0.49	58	63	Wagman et al. (1967)
Chicago, IL Sept. 1965	8.7	0.4	0.062	0.70	49	50	Wagman et al. (1967)
Philadelphia, PA Oct. 1965	12	0.7	0.066	0.78	46	45	Wagman et al. (1967)
Tallahassee, FL June–July 1973	1.2	0.6	0.053	0.62	36	40	Johansson et al. (1976)
Apalachicola National Forest, FL May–July 1973	0.80	0.4	0.048	0.54	40	46	Kadowaki (1976)
Nagoya, Japan July 1974	19.6	0.6	0.036	0.37	35	35	Roberts and Friedlander (1976)
Dominguez Hills, CA Oct. 1973	45	0.7	0.039	0.45	29	33	(1976)

sulfate) by measuring profiles over 1 m height stands of rye and wheat canopies with
1 cm $< z_0 <$ 20 cm (Sievering, 1981a). These results, despite being obtained during light
and moderate windspeeds, appear to be affected by particle resuspension (Sievering,
1981b), so that the actual mean deposition velocity may have been greater than
0.38 cm s^{-1}.

The wide ranges of dry deposition velocity reported for Pb and sulfate reflect not only
actual variations in v_g but also the great uncertainty associated with these estimates. It
is important to note that none of the above data represents direct dry deposition
measurement on a vegetative canopy, as desired. Rather, deposition on vegetation is
inferred either from micrometeorological information or by analogy with surrogate
surface data. The questionable validity of existing dry deposition field data, and the
sensitivity of deposition to differing particle, atmospheric and surface conditions, make
it difficult to assess the accuracy of the present model. Proper model validation would
require measurement of monodisperse particle deposition directly on vegetative elements
with simultaneous measurement of those micrometeorological and surface parameters
controlling deposition.

Finally, it is appropriate to comment on the relative importance of aerodynamic and
surface resistances to dry deposition. The former refers to particle transport by wind
eddies from the reference height down into the canopy, while the latter determines
transport from the region immediately above the surface onto the vegetative elements.
For cases where sedimentation is negligible, the aerodynamic resistance (for a 2 m
reference height) is defined as:

$$r_a = \frac{(u2m)}{u_*}$$ (17)

while the surface resistance is expressed as:

$$r_s = \frac{1}{v_g} - r_a.$$ (18)

Note that $(v_g)^{-1}$ can be considered as the total resistance to particle transport.

The value of r_a for the *Tridens* canopy of Table I is 0.23 s cm^{-1}. Corresponding values
of r_s are 50, 20, and 0.8 s cm^{-1} for d_p = 0.1, 1, and 10 μm, respectively. *Agrostis* (curve A
of Figure 4) provides the greatest ratios of r_a/r_s, but even here r_s dominates for most
particle sizes. Other investigators have reported concentration profile or eddy correlation
data suggesting much smaller surface resistances and greater overall deposition velocities
for submicron particles. However, our model supports the contention of Slinn (1981)
that aerodynamic resistance is rarely if ever rate-limiting for 0.1 to 1 μm diameter
particles.

5. Conclusions

Differences in the physical structure of five species of wild grass play a major role in
influencing particle dry deposition from the atmosphere, according to the theoretical

model developed here. Canopies of *Phleum pratense* and *Agrostis hyemalis* have roughly an order of magnitude greater deposition velocities than *Tridens flavus*, *Dactylis glomerata*, and *Andropogen virginicus* due to greater surface areas available for deposition. Vegetation geometry is also important: impaction on *Agrostis* is greater than impaction on *Phleum* in spite of a smaller leaf area index. For the canopies and wind conditions studied, surface rather than aerodynamic resistance to particle transport is rate-limiting.

Application of the model to ambient Pb and sulfate data in the literature suggests that deposition of relatively few large particles may have major effects on overall deposition of these species. The bulk of the airborne mass concentration, generally submicron, is characterized by much smaller deposition velocities.

Because of uncertainties in existing dry deposition data in the literature, as well as uncertainties in the model developed here, it is difficult to predict deposition to better than an order of magnitude. Direct measurement of monodisperse particle deposition on canopies of interest, combined with assessment of surface and atmospheric parameters, is needed to improve our understanding.

Acknowledgments

The authors acknowledge the Pittsburgh Department of Parks and Recreation, J. Downie, and E. M. Krokosky for use of the canopies. Laboratory assistance of L. G. Cartwright and manuscript preparation of N. A. Ward are greatly appreciated. Identification of the grasses was accomplished by D. E. Boufford, W. E. Buker, E. W. Wood, and F. H. Utech of the Botany Section, Carnegie Institute of Pittsburgh. This work was supported in part by EPA grant V-0502-NTEX, Benedum Foundation grant 1-31327, and MPC grant 79-5.

References

Belot, Y. and Gauthier, D.: 1975, *Heat and Mass Transfer in the Biosphere*, Scripta Book Co., Washington, D.C.
Brown, K. W. and Covey, W.: 1966, *Agric. Meteor.* **3**, 73.
Businger, J. A.: 1975, *Heat and Mass Transfer in the Biosphere*, Scripta Book Co., Washington, D.C.
Calder, K. L.: 1949, *J. Mech. and Appl. Math.* **2**, 153.
Chamberlain, A. C.: 1967, *Proc. Roy. Soc. A.* **296**, 45.
Chamberlain, A. C.: 1974, *Boundary Layer Meteor.* **6**, 477.
Chamberlain, A. C. and Chadwick, R. C.: 1972, *Ann. Appl. Biol.* **71**, 141.
Clough, W. S.: 1975, *Atmos. Environ.* **9**, 1113.
Crecelius, E. A., Robertson, D. E., Abel, K. H., Cochran, D. A., and Weimer, W. C.: 1978, 'Atmospheric Deposition of [7]Be and Other Elements on the Washington Coast', Battelle Pacific Northwest Laboratory Report PNL-2500, Part 2, Richland, Washington.
Davidson, C. I.: 1977, *Powder Tech.* **18**, 117.
Davidson, C. I.: 1981, unpublished data.
Davidson, C. I. and Friedlander, S. K.: 1978, *J. Geophys. Res.* **83**, 2343.
Davidson, C. I. and Chu, L.: 1981, *Environ Sci. Tech.* **15**, 198.
Davidson, C. I., Goold, W. D., Nasta, M. A., and Reilly, M. T.: 1981, 'Airborne Size Distributions of Trace Elements in an Industrial Section of Pittsburgh', 74th Annual Meeting, Air Poll. Control Assoc., Philadelphia, Pa., June 21–26.

den Hartog, G. and Shaw, R. H.: 1975, *Heat and Mass Transfer in the Biosphere*, Scripta Book Co., Washington, D.C.

Denmead, O. T.: 1976, *Vegetation and the Atmosphere* 2, Academic Press, New York.

Dovland, H. and Eliassen, A.: 1976, *Atmos. Environ.* 10, 783.

Droppo, J. G.: 1980, 'Micrometeorological Profiles of Sulfur-Containing Particles', 73rd Annual Meeting, Air Poll. Control Assoc., Montreal, Quebec, June 22–27.

Dzubay, T. G., Hines, L. E., and Stevens, R. K.: 1976, *Atmos. Environ.* 10, 229.

Elias, R. W. and Davidson, C. I.: 1980, *Atmos. Environ.* 14, 1427.

Everett, R. G., Hicks, B. B., Berg, W. W., and Winchester, J. W.: 1979, *Atmos. Environ.* 13, 931.

Friedlander, S. K.: 1977, *Smoke, Dust and Haze*, Wiley, New York.

Fuchs, N. A.: 1964, *The Mechanics of Aerosols*, Pergamon Press, Oxford.

Garland, J. A.: 1978, *Atmos. Environ.* 12, 349.

Garratt, J. R. and Hicks, B. B.: 1973, *Quart. J. Roy. Met. Soc.* 99, 680.

Goold, W. D. and Davidson, C. I.: 1981, unpublished data.

Inoue, E.: 1965, *J. Agric. Meteor. (Tokyo)* 29, 137.

Johansson, T. B., Van Grieken, R. E., and Winchester, J. W.: 1976, *J. Geophys. Res.* 81, 1039.

Kadowaki, S.: 1976, *Atmos. Environ.* 10, 39.

Lawson, D. R.: 1980, *Atmos. Environ.* 14, 195.

Legg, B. and Monteith, J.: 1975, *Heat and Mass Transfer in the Biosphere*, Scripta Book Co., Washington, D.C.

Legg, B. and Price, R. I.: 1980, *Atmos. Environ.* 14, 305.

Lee, R. E., Jr., Patterson, R. K., and Wagman, J.: 1968, *Environ. Sci. Tech.* 2, 288.

Lemon, E. R.: 1965, *Plant Physiology* 4A, Academic Press, New York.

Lindberg, S. E., Harriss, R. C., Turner, R. R., Shriner, D. S., and Huff, D. D.: 1979, 'Mechanisms and Rates of Atmospheric Deposition of Selected Trace Elements and Sulfate to a Deciduous Forest Watershed', Oak Ridge National Laboratory Report ORNL/TM-6674, Oak Ridge, Tenn.

Little, P. and Wiffen, R. D.: 1977, *Atmos. Environ.* 11, 437.

Maki, T. and Allen, L. H., Jr.: 1978, *Proc. Soil and Crop Sci. Soc. of Florida* 37, 81.

McMahon, T. A. and Denison, P. J.: 1979, *Atmos. Environ.* 13, 571.

Pasquill, F.: 1962, *Atmospheric Diffusion*, Van Nostrand, New York.

Pasquill, F.: 1972, *Quart. J. Roy. Meteor. Soc.* 98, 469.

Peden, M. E.: 1977, 'Flameless Atomic Absorption Determinations of Cadmium, Lead and Manganese in Particle Size Fractionated Aerosols', National Bureau of Standards Spec. Publ. 464, 367.

Perrier, A.: 1975, *Heat and Mass Transfer in the Biosphere*, Scripta Book Co., Washington, D.C.

Ripley, E. A. and Redmann, R. E.: 1976, *Vegetation and the Atmosphere* 2, Academic Press, New York.

Roberts, P. T. and Friedlander, S. K.: 1976, *Atmos. Environ.* 10, 403.

Sehmel, G. A.: 1980, *Atmos. Environ.* 14, 983.

Sehmel, G. A. and Hodgson, W. H.: 1976, 'Atmosphere-Surface Exchange of Particulate and Gaseous Pollutants', ERDA Symposium Series 38.

Servant, J.: 1976, 'Atmosphere-Surface Exchange of Particulate and Gaseous Pollutants', ERDA Symposium Series 38.

Sheih, C. M., Wesley, M. L., and Hicks, B. B.: 1979, *Atmos. Environ.* 13, 1361.

Shinn, J. H. and Cionco, R. M.: 1973, 'A Note on Observations of Turbulence and Mean Flow in Vegetative Canopies', Eleventh Nat. Conf. on Agriculture and Forest Meteor., Durham, N. C., see den Hartog and Shaw (1975).

Sievering, H., Dave, M., McCoy, P., and Sutton, N.: 1979, *Atmos. Environ.* 13, 1717.

Sievering, H.: 1981a, 'Profile Measurements of Particle Dry Deposition Velocity at an Air/Land Interface', *Atmos. Environ.* 15 (in press).

Sievering, H.: 1981b, 'Impact of Particulate Matter Chemical and Physical Characterization on Reported Particle Deposition Velocities', *Atmos. Environ.* 15 (in press).

Slinn, W. G. N.: 1981, 'Predictions for Particle Deposition to Vegetative Canopies', *Atmos. Environ.* 15 (in press).

Slinn, W. G. N., Hasse, L., Hicks, B. B., Hogan, A. W., Lal, D., Liss, P. S., Munnich, K. O., Sehmel, G. A., and Vittori, O.: 1978, *Atmos. Environ.* 12, 2055.

Sutton, O. G.: 1953, *Micrometeorology*, McGraw-Hill, New York.

Tan, H. S. and Ling, S. C.: 1961, *Heat and Mass Transfer in the Biosphere*, Scripta Book Co., Washington, D.C.

Thom, A. S.: 1971, *Quart. J. Roy. Meteor. Soc.* **97**, 414.

Thom, A. S.: 1972, *Quart. J. Roy. Meteor. Soc.* **98**, 124.

Thom, A. S.: 1975, *Vegetation and the Atmosphere* **1**, Academic Press, New York.

Uchijima, Z. and Wright, J. L.: 1964, *Bull. Nat. Inst. of Agric. Sci. (Japan)*, Ser. A, **11**, 19.

Waggoner, P. E. and Reifsnyder, W. E.: 1968, *J. Appl. Meteor.* **7**, 400.

Wagman, J., Lee, R. E., Jr., and Axt, C. J.: 1967, *Atmos. Environ.* **1**, 479.

Wedding, J. B., McFarland, A. R., and Cermak, J. E.: 1977, *Environ. Sci. Tech.* **11**, 387.

Wells, A. C. and Chamberlain, A. C.: 1967, *Brit. J. Appl. Phys.* **18**, 1973.

Wesely, M. L., Hicks, B. B., Dannevik, W. P., Frisella, S., and Husar, R. B.: 1977, *Atmos. Environ.* **11**, 561.

Wesely, M. L. and Hicks, B. B.: 1979, 'Dry Deposition and Emission of Small Particles at the Surface of the Earth', Fourth Symposium on Turbulence, Diffusion, and Air Pollution, Reno, Nevada, January 15–18.

ESTIMATES FOR THE LONG-RANGE TRANSPORT OF AIR POLLUTION

W. G. N. SLINN

Atmospheric Sciences Department, Pacific Northwest Laboratory*, Richland, WA 99352 U.S.A.

(Received 10 July, 1981; Revised 5 November, 1981)

Abstract. Different atmospheric, source, and surface conditions can result in substantially different ranges of atmospheric transport of air pollutants; for example, even for anthropogenic S and N, the ranges can vary from about 10^1 to 10^5 km. In this report, the emphasis is on indicating some of the reasons for the great variability of these ranges. Thus, some of the complexities of dry deposition, atmospheric chemistry, and precipitation scavenging are described, and it is demonstrated how synoptic-scale meteorologic conditions can control both dry and wet deposition. On the other hand, it is suggested that the mean, tropospheric-residence time of particulate S and N from fossil-fuel combustion in temperate latitudes is probably about a week, but the amount of this material remaining airborne after a week can be large, since the amount is expected to have a log-normal distribution over an ensemble of realizations. Applications of the results to the U.S./Canadian acid-rain issue, to episodic-deposition events, and to global-scale atmospheric pollution are indicated.

1. Introduction

The purpose, here, is to demonstrate and comment upon some estimates for the long-range transport of air pollution, especially of anthropogenic S and N. In one of the simplest, crudest, and most-common estimates, account is taken only of dry removal from a well-mixed layer of fixed height h (see Figure 1). For this case, it is easy to obtain $\bar{u}h\, \partial\chi/\partial x = -v_d\chi$, and therefore

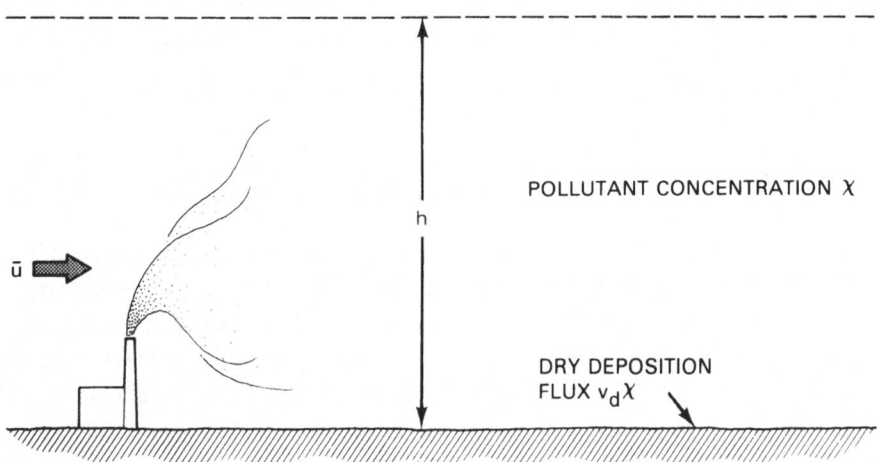

POLLUTANT CONCENTRATION χ

h

$\bar{u}$

DRY DEPOSITION FLUX $v_d\chi$

Fig. 1.   Schematic for the simple estimate of the long-range transport of air pollution with only dry deposition from a constant-height mixed layer.

* Operated for the U.S. Department of Energy under Contract DE-AC06-76RLO-1830.

$$\chi/\chi_0 = \exp\left\{-\frac{v_d x}{\bar{u}h}\right\} \equiv \exp\left\{-\frac{x}{\lambda_d}\right\} \equiv \exp\left\{-\frac{t}{\tau_d}\right\}, \tag{1}$$

where the e-fold, dry-removal length scale, λ_d, and time scale, τ_d, are given by

$$\lambda_d = \bar{u}h/v_d; \qquad \tau_d = \lambda_d/\bar{u} = h/v_d. \tag{2}$$

As examples, if for SO_2 and NO_2 $v_d \simeq C_D\bar{u}$ (which is applicable in some cases), then with use of the drag coefficients, C_D, given in Table I, there results: $\lambda_d \simeq 10^2\,\text{m}/10^{-2} = 10\,\text{km}$ for reactive gas deposition to grassland at night (with good mixing by mechanical turbulence to 100 m), and $\lambda_d \simeq 1\,\text{km}/10^{-3} = 10^3\,\text{km}$ for reactive gas deposition to lakes or oceans during the winter (when the water is warm relative to the air). The corresponding τ_d values, if $\bar{u} \sim 3\,\text{m s}^{-1}$, are $\tau_d \sim 1\,\text{h}$ and ~ 4 days.

TABLE I

Drag coefficient, $C_{D,10} = u_*^2/\bar{u}^2$ (10 m), for canopies of height h_c and aerodynamic roughness z_0. Tabulation from Plate (1971)

Surface cover	z_0 (cm)	h_c (cm)	$C_{D,10\,m}$
Snow	0.49	3	3.9×10^{-3}
Grassy surface	1.73	10	6.2×10^{-3}
Flat country	2.14	10	6.9×10^{-3}
Low grass	3.20	20	8.3×10^{-3}
High grass	3.94	30	9.3×10^{-3}
Wheat	4.5	130	10.7×10^{-3}

However, as will be seen, there are many complicating features of the long-range transport of air pollution, and these make the commonly-used estimate (Equations (1) and (2)) almost worthless. Consequently, in the rest of this report, improved estimates will be sought

2. Dry Deposition

Some of the complicating features of the long-range transport of air pollution can be seen by considering some details of dry deposition. First, consideration will be given to dry deposition from the atmospheric layer next to the earth's surface (the 'surface layer'), assumed to be well mixed by mechanical (in contrast to thermal) turbulence. After some aspects of gas and particle dry deposition from the surface layer are considered, then transfer from deeper in the atmosphere to the surface layer will be examined. If desired, and if gravitational settling is ignorable, then the overall transfer velocity, k_0, can be written in terms of separate transfer velocities past individuual layers, in a manner such as

$$k_0^{-1} = \sum_{i=1}^{n} k_i^{-1}; \tag{3}$$

i.e., the overall resistance, k_0^{-1}, is like the sum of transfer resistances in series (e.g., Slinn et al., 1978). However, this formalism will not be emphasized herein.

For gas deposition from the surface layer, it is only if the surface is a perfect sink for the gas that its v_d will be essentially as large as the transfer velocity for momentum:

$$v_{dm} = \frac{\text{Flux}}{\text{Conc.}} = \frac{\tau}{\rho \bar{u}} = \frac{\rho u_*^2}{\rho \bar{u}} = \frac{u_*^2 \bar{u}}{\bar{u}^2} = C_D \bar{u} , \tag{4}$$

where the same reference height is used throughout. (Actually, even if the surface is a perfect sink for the gas, v_d may not reach v_{dm} because some of the momentum may be transferred by pressure (or 'form') drag, which has no counterpart in the case of pollutant gas deposition.) For reactive gases such as SO_2, NO_2, H_2SO_4, HNO_3, etc., the perfect-sink case, with $v_d \simeq \frac{1}{2} C_D \bar{u}$ can occur for surfaces such as large (alkaline) lakes, oceans, moist soils, and rain-cleansed vegetation and other surfaces. For less reactive gases, such as CO, NO, CO_2, N_2O, etc., then even the surfaces listed in the previous sentence need not be perfect sinks for these gases, and their transfer is controlled by their behavior in the surface (e.g., Bennett et al., 1973; Slinn, 1977; O'Dell et al., 1977; Slinn et al., 1978).

But of more relevance to the acid-fallout issue of this conference is that unless there is mixing in the surface media (as for large lakes and the oceans), the surfaces can become saturated even with reactive gas molecules, and v_d will fall significantly below $C_D \bar{u}$. In important preliminary studies, Judeikis and Wren (1978) found that soils and building materials become saturated with SO_2 and NO_2 molecules at a concentration of about 0.1 g per projected m^2, if these surfaces were dry, and at ~ 1 g m^{-2} for moist surfaces. Then, further deposition must await other factors, such as rain cleansing of the surface, deposition of alkaline particles to oxidize the gases, or deposition of NH_3 to neutralize the oxidation products. A similar 'saturation effect' (for SO_2 depositing on grass) was seen in field studies by Fowler (1978), and possibly also for SO_2 depositing on water in a small reservoir by Garland (1978).

The practical significance of this concept of surface saturation can be seen from some simple calculations. If $v_d = \frac{1}{2} C_D \bar{u} \simeq \frac{1}{2} (5 \times 10^{-3}) (4 \text{ m s}^{-1}) = 1 \text{ cm s}^{-1}$ for SO_2 depositing on 'fresh' grass, and if during an episode, $\chi (SO_2) \simeq 100 \mu g \text{ m}^{-3}$, then a surface concentration of 0.1 g m $^{-2}$ would be achieved in about 10^5 s $\simeq 1$ day! Then, further dry deposition of SO_2 may be limited by the rate at which, say, NH_3 is deposited to neutralize the H_2SO_4 on the surface. Then, the additional flux of SO_2 will be (approximately) equal to the flux of NH_3; i.e., $v_d (SO_2) \simeq C_D \bar{u} [\chi (NH_3)/\chi (SO_2)]$, which might be only a tenth or less of $C_D \bar{u}$. In summary, it can be expected that dry deposition from the surface layer, even of reactive gases such as SO_2, NO_2, H_2SO_4, HNO_3, etc., can be suppresed under certain circumstances.

Dry deposition of particles from the surface layer can be even more complicated than the case of gases. Figures 2 and 3 suggest that v_d for submicron particles can be substantially smaller than $C_D \bar{u}$. Also, there is the obvious and extremely important point ignored by most authors that it is inappropriate to use a single, constant v_d for a

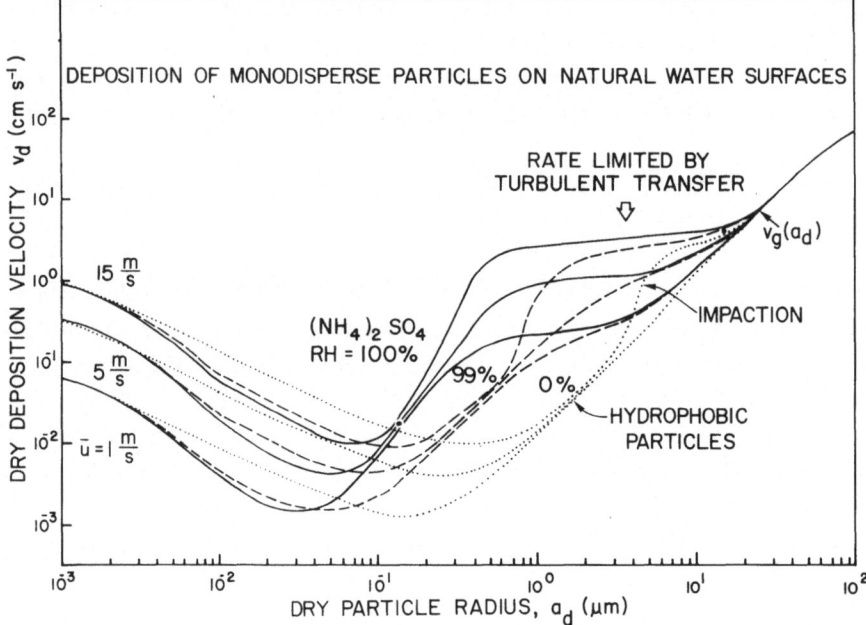

Fig. 2. Theoretical predictions for v_d for monodisperse particles depositing on natural waters, accounting for Brownian diffusion (which dominates for $a \lesssim 0.1\ \mu m$), gravitational settling (which dominates for $a \gtrsim 10\ \mu m$), impaction, and particle growth by water-vapor condensation. For hygroscopic particles, this growth when the particles are very close to the surface can be sufficiently great that the dry deposition of $\sim 1\ \mu m$ particles can be rate limited (at $v_d = C_D \bar{u}$) by transfer via turbulence through the constant flux layer. Figure reproduced from Slinn and Slinn (1980).

polydisperse aerosol such as $(NH_4)_2SO_4$ or NH_4NO_3. So long as v_d varies with particle size then the particle size distribution will evolve leaving the more-slowly-depositing particles airborne. As an example, if $v_d(a) = ca$ (see Figure 3 for $0.1 \lesssim a \lesssim 1\ \mu m$), then the mass-average v_d, $\langle v_d \rangle$, is v_d for the mass–average-size particle, $v_d(\langle a \rangle)$; i.e.

$$\langle v_d \rangle = \int_0^\infty cam(a)\,\mathrm{d}a = c\,\langle a \rangle = v_d(\langle a \rangle). \qquad (5)$$

But note that the mass-average radius changes with time: in Chicago, $\langle a \rangle$ for $SO_4^=$ may be $\sim 5\ \mu m$ (Everett *et al.*, 1979); in the Arctic it is $\sim 0.1\ \mu m$ (Shaw, 1981).

Furthermore, it obviously follows that dry deposition of a polydisperse aerosol does *not* lead to *exponentially* decreasing airborne particle mass. For example, if $v_d = ca$, if for each particle-size class the number distribution decreases according to

$$N(a, x) = N(a, 0) \exp\left\{ -\frac{v_d(a)x}{\bar{u}h} \right\}, \qquad (6)$$

and if (as a crude approximation to simplify the math) the initial mass distribution were uniform with particle radius (so that, for example, the number distribution of spherical

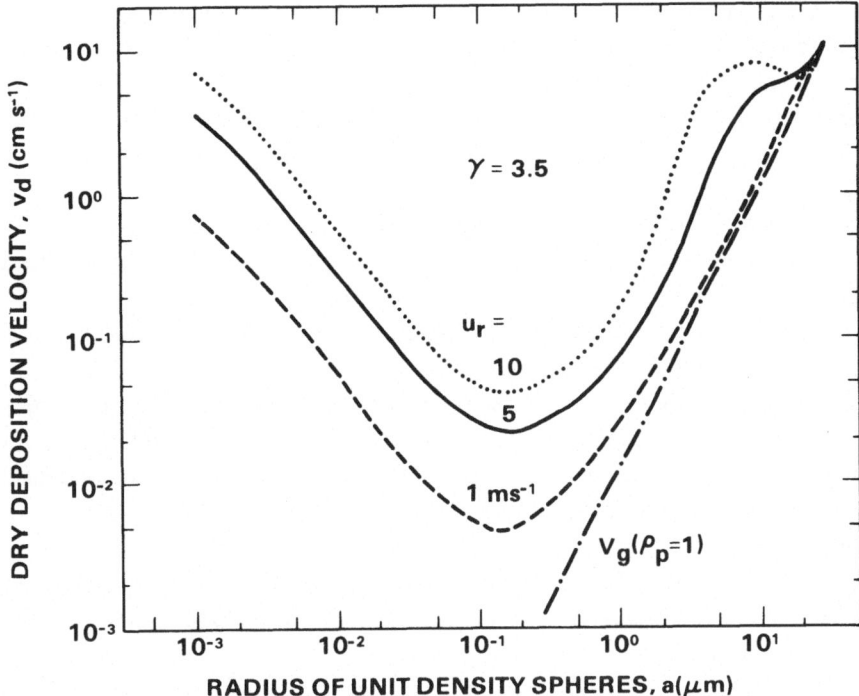

Fig. 3.   An illustration of predictions for v_d for monodisperse particles depositing on vegetation, for different reference wind-speeds, u_r, measured at the same height above the canopy as where χ is measured. v_g is the gravitational settling speed for unit-density particles; γ is one characteristic of the canopy-it is used to describe the wind speed in the canopy: $\bar{u}/u(h) = \exp\{-\gamma(1 - z/h_c)\}$, where h_c is the canopy height. For details, see Slinn (1981b), from which this figures has been reproduced.

particles in the radius range $0.1 \lesssim a \lesssim 1$ μm would be $N_T a_0^2/a^3$), then at x the remaining airborne mass would be

$$\chi(x) = \int_0^\infty \tfrac{4}{3}\pi a^3 \rho_P N(a, x)\, \mathrm{d}a$$

$$= \int_0^\infty \tfrac{4}{3}\pi a^3 \rho_P N(a, 0) \exp\left\{-\frac{cax}{\bar{u}h}\right\} \mathrm{d}a = \chi(0)\left[\frac{\bar{u}h}{v_d(a_0)x}\right]; \qquad (7)$$

i.e., the mass-average χ would decrease only as x^{-1}, which is very much slower than exponentially. Thus, if one returns even to the simple picture of Figure 1, with deposition from a well-mixed layer of fixed height h, then for the case of a polydisperse aerosol the characteristic length scale, l_d (*not* an *e*-fold length scale, but a 'linear-fold' length scale) can be $l_d \sim h\bar{u}/v_d(a_0) \sim (1\text{ km}) (5\text{ m s}^{-1})/(5 \times 10^{-2}\text{ cm s}^{-1}) \sim 10^4$ km!

3. Some Chemistry

Because submicron $SO_4^=$ and NO_3^- particles probably dry deposit more slowly than their precursor gases SO_2 and NO_2 (except if the surfaces become saturated with these gases, or if particle growth by water-vapor condensation is substantial, cf. Figure 2), then to estimate the long-range transport of anthropogenic S and N, it is relevant to examine the rate at which gaseous S and N convert to particles. In total, this conversion is extremely complicated; from a brief review of the literature, the following points are noted.

First, for S:

(1) Heterogeneous Oxidation of SO_2 (e.g., on particles)

(i) Oil-fired power plants may release more S as $SO_4^=$ than coal-fired plants because of catalytic oxidation by V_2O_5 (and possibly by Ni).

(ii) The entire acid-fallout issue may have been exacerbated by requirements to reduce (alkaline) particle emissions (e.g., from power plants).

(iii) Nevertheless, some particles pass through particle-control devices, and sub-micron soot particles, especially, can lead to substantial (1 to 10%) oxidation of SO_2 in the plume, the percentage depending on humidity (and therefore on water condensation in soot pores), and on the rate at which the plume diffuses.

(2) Homogeneous (Gas-Phase) Oxidation of SO_2

(i) For a plume, with OH scavenged by plume CO, NO, NO_2, SO_2, etc., the OH oxidation of SO_2 (and NO_2) tends to proceed at the plume's 'edges', and therefore is initially diffusion limited.

(ii) As the plume spreads, OH oxidation proceeds at a rate of ~ 0.1 to 1% h^{-1}, strongly dependent upon insolation (since the OH concentration is).

(iii) In photochemical smog, the presence of more OH and other oxidants may increase the SO_2 oxidation rate up to a few percent per hour.

(3) Aqueous-Phase Oxidation of SO_2

(i) This oxidation may be dominated in many cases by H_2O_2, whose concentration is extremely variable and difficult to measure, but it can also be promoted by O_3 and catalysts such as Mn.

(ii) Laboratory data for SO_2 aqueous-phase oxidation rates show great variability, and there is essentially no field data.

(iii) It is expected that SO_2 oxidation in cloud water can proceed as rapidly as 50 to 100% per h^{-1}, but the amount of SO_2 thereby oxidized depends on many factors: the amount of strong oxidant (e.g., H_2O_2) present; the time available for weaker oxidants (e.g., O_3, O_2) to react; the concentration of catalysts; and the amount of SO_2 that dissolves in the (liquid) cloud water, (which depends on the pH, and which in turn depends on how much SO_2 has already been oxidized, and on the presence of alkaline buffers such as Ca^{++}, Na^+, NH_4^+, etc.).

As complicated as the oxidation of SO_2 is, the case of NO_x seems even more complicated. While reviewing the literature, I have gained impressions:

(a) Of NH_3 and HNO_3 evaporating from NH_4NO_3 particles during summer days, and of NH_4^+ being replaced by Ca^{++}, Mg^{++}, etc. when NH_4NO_3, $(NH_4)_2SO_4$ etc. particles come in contact with coarse alkaline particles;

(b) Of NO_3^- being driven off particles by $SO_4^=$, but not when relative humidities are near 100%;

(c) Of NO_2 losing out to SO_2 in competition for aqueous-phase oxidizers, and perhaps even the aqueous-phase oxidation of SO_2 by HNO_3; and

(d) Of coarse NO_3^- particles and HNO_3 wet deposited locally, but of NH_3NO_3 and NO_x being transported even greater distances than similar S compounds.

If even just some of these impressions are correct, then I gain the overall impression that it may be useless to try to describe 'typical' behavior of SO_2 and NO_x oxidation and particle generation: the variance is quite likely larger than the mean, and it may therefore be necessary to consider different meteorologic conditions on a case-by-case basis. A particular case will be outlined later in this report. The range of possibilities seems to extend from ~ 0 to $\sim 10\%$ of the released S being $SO_4^=$ (and the larger particulate $SO_4^=$ depositing quite rapidly); from ~ 0 to 20% of the SO_2, and possibly of NO_2, being oxidized in the plume (the 'dirtier' the plume and the greater the humidity, and the more stable the atmosphere, then the greater the oxidation); from ~ 0 to 1% per h^{-1} oxidation by OH; from ~ 0 to 5% h^{-1} oxidation in photochemical smog; and from ~ 10 to 100% oxidation in cloud water. Thus, returning to the question posed at the start of this section, the fraction of the S or N in gaseous vs particulate form is very difficult to specify (it's $1/2 \pm 1/2!$); it depends on a host of aerochemetric variables!

4. Some Meteorology

But even if the problems of describing dry deposition from the surface layer of particulate vs gaseous S and N are resolved, there remain problems with meteorologic aspects of the simple model depicted in Figure 1. These can be classified as problems of transfer to the surface layer, but this classification and the associated mathematical development in terms of resistances will not be pursued here. Instead, a few examples wil be given to illustrate some points.

One example concerns the role of tall stacks in promoting long-range transport of air pollutants. Tall stacks can reduce pollution local to its source by injecting the pollution aloft, into a layer essentially decoupled from the surface (Figure 4). This decoupling can occur during daytime hours, and in Eastern North America it occurs (on an annual basis) during about 50% of the nights. Simultaneously with this decoupling, and because of it, the winds can reach quite high values aloft; this is the nocturnal, low-level jet (e.g., Smith et al., 1978). Thus, if there are 8 to 12 h of transport at 5 to 15 m s^{-1}, then the pollution can travel ~ 150 to 650 km before it even has a chance to deposit to a significant extent – although sudden bursts of turbulence during the night can break through the stable layer, fumigating the surface (e.g., Smith and Hunt, 1978). Otherwise, fumigation can occur the next day, if the sun's heating of the earth's surface destroys the surface-based inversion, and then the simple model of Figure 1 begins to become appropriate. How much S and N are then deposited depends (as has been described) on the surface characteristics, the depth of the mixed layer, and on the fraction of gaseous vs particulate S and N. The point, however, is that these frequently occurring meteorologic and source-characteristic variables can change the deposition from near zero to near $C_D\bar{u}$.

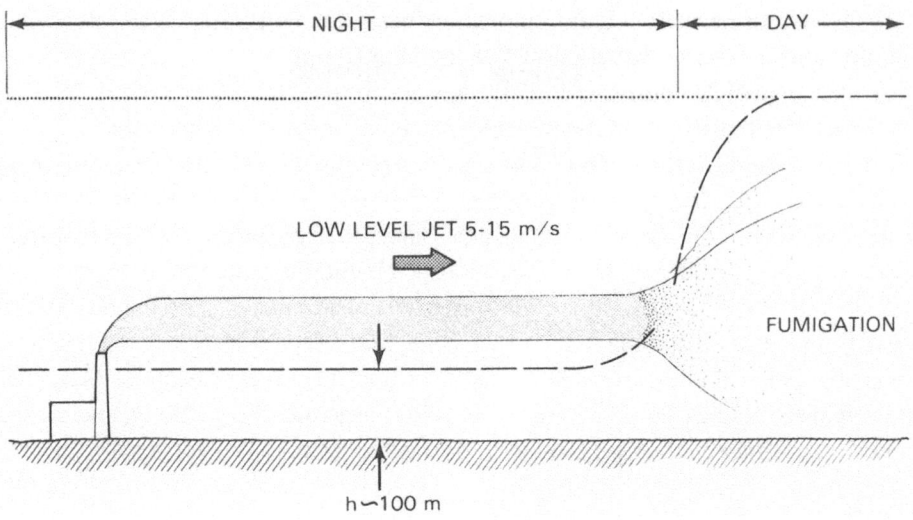

Fig. 4.   Schematic of the role of tall stacks in promoting the long-range transport of pollutants at night.

But *via* tall stacks is not the only way for pollution to become isolated from the surface. The same phenomenon occurs essentially every time the mid-afternoon, deep, mixed layer collapses, leaving $\sim 90\%$ of the pollution aloft, above a nocturnal, surface-based inversion. Also, this isolation occurs, both day and night, when warm air flows over relatively-cold air next to the surface; e.g., when warm air from the Ohio Valley, say, travels north over the Great Lakes during the summer (thereby resulting in U.S. air pollution of Canada), or Canadian (and U.S.) pollution travels across the North Atlantic during the summer to pollute Europe, and – to complete the circle of finger pointing! – when European pollution travels over Arctic snow in the winter, to pollute Northern Canada and the U.S. (in Alaska). A well-known example of this phenomenon is when warm, dust-laden air from North Africa travels in the Easterlies over the (relatively cold) South Atlantic to the Caribbean (Prospero and Nees, 1977). In such cases, the simple model of Figure 1 is totally irrelevant: dry deposition is essentially negligible (except for gravitational settling) for distances $\gtrsim 10^4$ km.

The final point I wish to make about dry deposition is actually about what I call 'dry ascension'; i.e., not about the transfer velocity down, but about the transfer velocity up, v_a. This ascension can arise from a number of different phenomena, including:

(i) Simultaneously with the pollution release, especially from fossil-fueled power plants, there can be significant heat releases, which add to the pollution's ascension;

(ii) The heat release from an entire urban area (or larger area), and the sun's heat absorbed by the pollution can also add to this ascension;

(iii) But, even without (i) and (ii), this ascension will proceed via vertical mixing. As sketched in Figure 5, this ascension to a height $\sim [K_z t]^{1/2}$ proceeds with a characteristic 'ascension velocity' $v_a \simeq [K_z/t]^{1/2}$, where K_z is the turbulent diffusivity and t is time. Thus, if $K_z = 10^5$ cm^2 s^{-1} and $t = 10^3, 10^5$ s, then $v_a \simeq 10, 1$ cm s^{-1}, which typically are

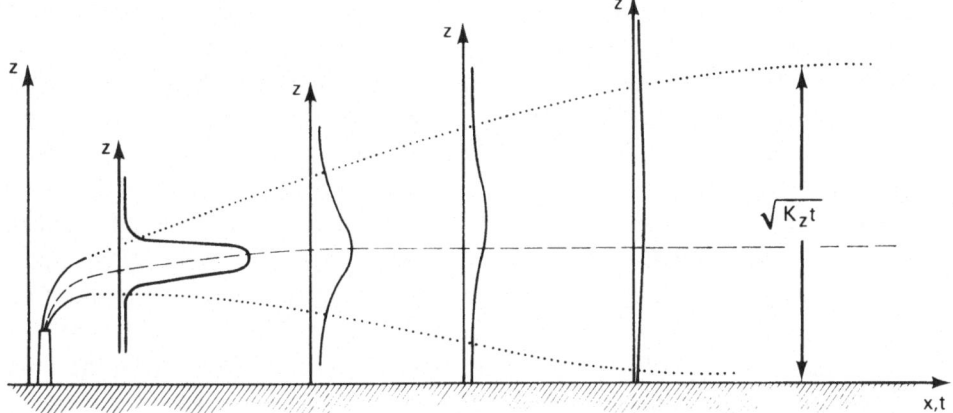

Fig. 5.   Schematic of the 'ascension' of air pollution, with characteristic ascension velocity
$$v_a \simeq [K_z/t]^{1/2}.$$

larger than a pollutant's dry deposition velocity. Thereby, the pollution can mix deep into the troposphere, as occurs for North American SO_2 before it reaches the middle of the Atlantic (e.g., Georgii, 1978). Consequently, if the only removal mechanism were dry deposition, it is not at all inconceivable that the characteristic removal time scale (e.g., for $\sim 0.1 \, \mu m \, SO_4^=$ particles) could be

$$\tau_d \simeq \frac{h}{v_g} \simeq \frac{3 \, km}{3 \times 10^{-2} \, cm \, s^{-1}} \simeq 10^7 \, s \simeq 100 \, days! \tag{8}$$

5. Precipitation Scavenging

Of course, what can dramatically change the picture sketched in the previous paragraph, and the estimated residence time for submicron $SO_4^=$ particles of ~ 100 days, is the occurrence of precipitation. It is my opinion that the dominant route for removal of anthropogenic $SO_4^=$ and $NO_3^=$ from the atmosphere is via precipitation scavenging, and not via dry deposition. This opinion is consistent with my skepticism of the inferences from some field studies that the dry deposition of $\sim 0.1 \, \mu m$ particles depositing on vegetation is greater than $0.1 \, cm \, s^{-1}$; such inferences are not consistent with results from other field studies (e.g., Garland, 1978). In contrast, precipitation's use of submicron particles as condensation nuclei, plus the expected rapid oxidation of SO_2 and NO_2 in cloud water [and Junge (1964) estimated that, in temperate latitudes, a given air mass will typically be involved in about 10, condensation-evaporation cycles (i.e., clouds) before it is involved in precipitation], plus precipitation's scavenging of gases, sum to give an indication of the importance of the atmospheric water cycle in purifying the atmosphere, and, simultaneously, in depositing pollution on the earth's surface.

An overview of the efficiency of precipitation scavenging can be obtained from estimating the average, tropospheric-residence (or turnover) times of some atmospheric trace constituents. These estimates can be obtained in the same manner as one can estimate, for example, the average lifetime of U.S. citizens: if the present population of U.S. were about 210×10^6 people, and if the birth rate (= death rate at a steady state with no immigration or emigration) were 3×10^6 people yr^{-1}, then the average lifetime $\bar{\tau} \simeq 210 \times 10^6$ people $\div 3 \times 10^6$ people $yr^{-1} \simeq 70$ yr. Similarly, Junge (1963) estimated that above the latitude belt 30 to $50°$ N there is about 1.8 g H_2O cm^{-2}. If typical precipitation in these latitudes is about 90 cm $yr^{-1} = 90$ g cm^{-2} yr^{-1}, and if dry deposition of H_2O and exchange between latitude belts are ignored (which are rather drastic assumptions), then $\bar{\tau}(H_2O) \simeq 1.8$ g $cm^{-2} \div 90$ g cm^{-2} $yr^{-1} \simeq 1$ week. For tropospheric S, if the global burden is between 1.4 and 2.0 Tg (Mesaros, 1978) and if the anthropogenic and natural sources sum to about 100 Tg yr^{-1} (Granat et al., 1976), then $\bar{\tau}(S) \simeq 1.7$ Tg $\div 100$ Tg $yr^{-1} \simeq 1$ week. For tropospheric, anthropogenic N, the situation is less clear. If the lightning-produced NO_x is ignored, because it is produced exactly where precipitation scavenging is most active, and if the mass mixing ratio for N (e.g., as NO_x and NO_3^-) in the northern hemisphere's troposphere is about 0.3 ppb, then by acknowledging only the anthropogenic source of about 30 Tg yr^{-1} of N in the northern hemisphere, there results $\tau(\text{Anthro. N}) \simeq (0.3 \times 10^{-9} \times 2 \times 10^{21}$ g$) \div 30$ Tg $yr^1 \simeq 1$ week. The reoccurrence of these residence times of about a week, (for H_2O, S, and N) is (to my mind) very significant: it suggests that the atmospheric water cycle/precipitation scavenging dominates cycling of these reactive gases and soluble particles.

However, just as for the simple estimate when only dry removal from a well-defined mixed layer was considered, many features enter to complicate the simple analysis of precipitation scavenging sketched in the previous paragraph. Some of these complications will be mentioned here; elsewhere, I have given additional details (Slinn, 1977, 1981a). Three introductory items that might be worth repeating to a broader audience are these: (i) it is recommended that attempts *not* be made to deduce scavenging rates from differences in airborne concentrations, before, during, and after precipitation – too many other phenomena change simultaneously; (ii) it is hoped that it would be recognized that there is no significant difference between below- and in-cloud scavenging – therefore it is hoped that the jargon 'rainout' and 'washout' will be abandoned; (iii) what appears to be many modelers' delight – precipitation falling through an uninvolved mixed layer of pollution – rarely occurs: precipitation is formed by the ascent of low-level, moist (polluted) air, and associated with the precipitation a downdraft must (by necessity of Newton's third law) be present. In summary, please never report on 'A Comparison Between Measured and Modeled Washout of Pollution from the Mixed Layer!'.

In the short space available here, only four points will be briefly mentioned. One concerns a simple derivation of the scavenging ratio for particles; scavenging ratios, s are the ratio of the concentration, κ_0, of the pollution in surface-level precipitation, to its concentration, χ_0, in surface-level air: $s = \kappa_0/\chi_0$. I prefer that the dimensions of mass per unit volume be used for both κ and χ, rather than mass mixing ratios, for then s is dimensionless (and not, for example, with units of g per g H_2O $\div$ g per g air), and for particles s then has a value near 10^6, which reflects the physically real concentrating ability of the distillation process known as precipitation scavenging. As will be illustrated later, there is a serious limitation with the scavenging-ratio concept because, for particles,

χ_0 should be measured in the updraft, whereas for gases, the appropriate χ_0 is the concentration in the downdraft. The other points to be briefly considered here concern rain scavenging of gases, snow scavenging, and the scavenging efficiency of entire storms.

To estimate particle scavenging, one can use Chamberlain's scavenging rates (e.g., Engelmann, 1968), which are valid both within and beneath clouds, or approximations to these rates (Slinn, 1977), and then integrate over all heights from which scavenging occurs. As an alternative, since most particulate mass will be incorporated into the cloud water (Junge, 1963), an estimate for the concentration of particulate mass in precipitation can be obtained by estimating the collection of cloud water by hydrometeors, and then by taking the ratio of the flux of a specific chemical or compound, $\kappa_0 p_0$ (where p_0 is the surface-level precipitation rate), to the flux of water from the storm:

$$\frac{\kappa_0 p_0}{\rho_w p_0} = \frac{\int_0^\infty dz \int_0^\infty da_c\, m(a_c)\, \{\, c\bar{E}(a_c, R_m)p(z)/R_m \,\}}{\int_0^\infty dz \int_0^\infty da_w\, m(a_w)\, \{\, c\bar{E}(a_w, R_m)p(z)/R_m \,\}} \tag{9}$$

where a_c is the radius of droplets containing the chemical of interest, a_w is the radius of the cloud droplets, $m(a_c)$ and $m(a_w)$ describe how the mass is distributed with droplet size, and the expressions in parentheses, $\{\ \}$, are approximations for the rain scavenging rates (Slinn, 1977). Expressions similar to Equation (9) can be written for cold or mixed-phase clouds. Since $\int_0^\infty da_c\, m(a_c) = \chi = \delta\chi_0$, where χ is the concentration of the pollution in cloud air, χ_0 is the pollution's concentration in the updraft, and δ is a dilution factor that depends on the storm's dynamics, and since $\int_0^\infty da_w m(a_w) = L$ is the liquid water content of the cloud, then Equation (9) gives

$$s = \left(\frac{\kappa}{\chi}\right)_0 \simeq \frac{\delta\rho_w}{L} \frac{\langle \bar{E}(\bar{a}_c)\rangle}{\langle \bar{E}(\bar{a}_w)\rangle} \tag{10}$$

where $\langle E\rangle$ is the mass-average collection efficiency. For rain scavenging of all droplets, inertial collection dominates, and an approximation for E is

$$E = \exp\{-1/\mathrm{St}\} \tag{11}$$

where, for rain, the Stokes number, St, is almost independent of raindrop size (radius, R):

$$\mathrm{St} = \frac{\tau_r v_t}{R} \simeq \frac{\tau_r (8000\,R)}{R}\ \mathrm{s}^{-1} \simeq 0.1\,(a/1\,\mu\mathrm{m})^2, \tag{12}$$

in which Stokes' relation has been used for the particle relaxation time, τ_r. To determine the mass-average radii of the pollution-bearing droplet, $\bar{a}_c$, and of the cloud droplets, $\bar{a}_w$, the particles entering the cloud can be assumed to grow via water-vapor condensation according to a simple formula such as

$$a = a_0 + 10 \; \mu m \; (gt)^{1/2} \tag{13}$$

where a_0 is the initial particle radius ($\gtrsim 0.01 \; \mu m$ for H_2SO_4 and HNO_3, and $\gtrsim 0.1 \; \mu m$ for most particles), and the growth rate, g, depends on the storm type: typically, $g \sim (10^3 \; s)^{-1}$ for convective storms, and $(10^4 \; s)^{-1}$ or slower for stable storms. In summary, Equation (10) gives the scavenging ratio simply as the ratio of the capture efficiencies for the polluted cloud water vs for the rest of the cloud water; Equation (10) differs from Junge's expression for s (Junge, 1963) only in that his 'capture efficiency' is here replaced by $\delta \overline{E}(\overline{a}_c)/\overline{E}(\overline{a}_w)$. Plots of Equation (10) will lead to favorable comparisons with data, as shown elsewhere (Slinn, 1977, 1981a).

For the long-range transport problem, a major simplification can be used to estimate the rain scavenging of most gases: it can be assumed that the concentration of the gas dissolved in the rain, κ_0, is the value in equilibrium with the local air concentration, χ_0 (Hales, 1978). Then $s = (\kappa/\chi)_0 = \alpha$ where α is the Ostwald solubility coefficient (or one of the many forms of the Henry's law constant). The wet flux of dissolved gas is then $W = \alpha p_0 \chi_0$. However, there are (at least two) major complications. One is to use the appropriate χ_0: for a convective storm, for example, the appropriate χ_0 is its value in the downdraft (usually mid-tropospheric air), not in the polluted updraft. A second complication is that many gases ionize, and α gives only the dissolved component of the gas. For SO_2, for example, and for $3 \lesssim pH \lesssim 6$, when most of the ionized SO_2 is HSO_3^-, then the scavenging ratio for the *total* S(IV) in the rain, is given by (Barrie, 1981):

$$\log_{10} s_{S(IV)} = (pH - 7.21) + (1.09 \times 10^4 \; K)/T. \tag{14}$$

For NO_x, I am not aware of a corresponding expression for s, but I have seen that even the solubility of NO_2 in water is not known at all well (e.g., see Schwartz and White, 1980). An additional problem in the scavenging of reactive gases [as can be seen in Equation (14)] is that the pH of the rain is needed, and/or a detailed specification of the chemical constituents of the rain. Clearly, much more research is needed here. For organic gases, my major problem is that I have been unable to find adequate tables of solubility coefficients! The best I have been able to find are those in Gerrard (1976).

For snow scavenging, there are similarities with rain scavenging, and differences. For snow scavenging of particles, an approximation for the scavenging rate similar to the one for rain scavenging used in Equation (9) can be developed (Slinn, 1977, 1981a):

$$\psi_s(a; r, t) = \gamma \overline{E}(a, \lambda) p(r, t)/D_m \tag{15}$$

where γ is a dimensionless constant of order unity, p is the precipitation rate (rainwater equivalent), and values for the characteristic volume-to-area length scale, D_m, are given in Table II. As with rain scavenging, too, there are major uncertainties about the collection efficiency, $\overline{E}$ (Slinn, 1977), although new data will soon be available (to be published by Magono and by Hallett and Vittori). Snow scavenging ratios for particles are becoming available (Rahn and McCaffrey, 1979; Davidson et al., 1981), and the values are quite similar to the values for rain scavenging ratios. Snow scavenging of gases, however, has not been adequately explored; on the one hand, it is true that few gases

TABLE II

D_m for use in the snow scavenging rate given by Equation (15);
from Slinn (1981a)

Crystal type	D_m (cm)
Graupel particles	1.4×10^{-2}
Rimed plates and stellar dendrites	2.7×10^{-3}
Powder snow and spatial dendrites	1.0×10^{-3}
Plane dendrites	3.8×10^{-4}
Needles	1.9×10^{-3}

will be incorporated into the ice lattice, but on the other hand, snowflakes are frequently formed by capture and freezing of (gas-laden) cloud droplets, gas molecules can physically attach to the (large) surfaces of snowflakes, and snow can partially melt in the atmosphere. Generally, though, since even the (simple-solution) rain scavenging of gases is typically small compared to the gases' dry deposition (e.g., Slinn et al., 1978), I expect that snow scavenging of gases does not influence the long-range transport of pollutants such as SO_2 and NO_2 so much as the oxidation and then scavenging of the reaction products.

The final point to be addressed in this brief survey of precipitation scavenging concerns the very difficult problem of specifying the overall scavenging efficiency of a storm, ε, by which I mean the fraction of the ingested pollution that is deposited. The root cause of the difficulty is that there is comparable difficulty in specifying the efficiency with which storms remove their water. Compounding-difficulties include: a variety of definitions for the 'precipitation efficiency', evaporation of water in the downdraft, and of course the great variability in precipitation efficiencies for different storms. From a recent survey (Slinn and Hales, 1982), I came to the conclusion that the best estimate for the cloud-water removal efficiency may again be the $\frac{1}{2} \pm \frac{1}{2}$ (!), though I later yielded to the estimate $\frac{1}{2} \pm \frac{1}{4}$.

To appreciate some of the difficulties involved in determining ε, suppose that a storm processes an STP volume of air V_{in}, and that the (absolute) humidity is ρ_v. Then the mass of water vapor entering the storm is $\rho_v V_{in}$. Also, let the mass of precipitation leaving the storm be M_{out} (obviously related to the precipitation rate, duration, and aerial extent). Then the efficiency with which the storm removes processed water can be defined as $\varepsilon_{wv} = M_{out}/\rho_v V_{in}$. Meanwhile, the mass of pollution entering the storm is $\chi_0 V_{in}$, and the mass deposited is $\kappa_0 M_{out}/\rho_w$, where ρ_w is the density of water. Therefore, the scavenging efficiency of the storm can be written as

$$\varepsilon = (\kappa_0 M_{out})/(\rho_w \chi_0 V_{in}) = (\rho_v/\rho_w)(\kappa_0/\chi_0)\varepsilon_{wv}, \qquad (15)$$

and if the result in Equation (10) is used for the scavenging ratio for particles,

$$\varepsilon = \frac{\rho_v}{\rho_w} \frac{\delta\rho_w}{L} \frac{\langle \overline{E}(\bar{a}_c) \rangle}{\langle \overline{E}(\bar{a}_w) \rangle} \varepsilon_{wv} \simeq \frac{\langle \overline{E}(\bar{a}_c) \rangle}{\langle \overline{E}(\bar{a}_w) \rangle} \varepsilon_{wv}. \qquad (16)$$

Equation (16) is useful in that it suggests that a storm's efficiency for removing a specific pollutant can be more or less than the storm's efficiency for removing its water vapor (depending on the relative efficiency with which polluted vs unpolluted cloud droplets are scavenged), but Equation (16) gives no information about ε_{wv}. For long-term average estimates, it may be best to force ε_{wv} to have whatever value is necessary, given the frequency of storms to yield a mean, tropospheric-residence time for water vapor of ~ 1 week (in temperate latitudes). Otherwise, to estimate ε_{wv} for a specific storm, it appears necessary to rely on data for the amount of moisture processed by the storm, and for the amount of rainfall from it.

6. Residence Times

The concept of pollutant residence times in atmosphere reservoirs is probably with us to stay, but there are very significant limitations in this concept. Rigorously (Eriksson, 1971; Rodhe and Grandell, 1972; Rodhe, 1978; Baker et al., 1979; Slinn, 1980), the reservoirs should be well mixed, and this rarely occurs for reactive pollutants in the troposphere (e.g., see the estimate in the previous section for the tropospheric residence time of NO_x). Frequently, then, the residence or turnover time does not represent an e-fold relaxation time since the pollutant's average removal is not proportional to its average concentration. There is also the difficulty, described earlier, that the residence time for aerosol particles should *not* be treated as an integral property of the aerosol: different-size particles typically reside in the atmosphere for different durations. Nevertheless, as stated, the concept is probably with us to stay, if for no other reason than that it reflects the first Fourier coefficient in a (time- or spatial-) analysis of a pollutant's concentration (Slinn et al., 1978). Alternatively, as will be pursued here, one can seek a probability density function (pdf) to describe a pollutant's airborne concentration, and from this pdf obtain a mean concentration and its variation in time. From this time variation, an estimate for a 'mean residence time' may be possible.

As an illustration of the method, consider particulate $SO_4^=$ and NO_3^-, one or two days downwind of a source region. The desirability of starting the analysis downwind of the source region is that there are just too many possible details that can occur during the first day or so. Some of these possibilities were sketched in earlier sections; as a summary statement, the analysis here starts with an atmospheric burden q_0, which (say for $SO_4^=$) may be somewhere between 10 and 90% of the released S, and typically might be 50% of the released S (and similarly for N). This S is then available for significant, long-range transport, and it is assumed here (consistent with the earlier estimate that τ_d for $SO_4^=$ can be as large as 100 days), that the dominant removal path for submicron particles is via precipitation scavenging.

To estimate the amount of this pollution, say q, remaining airborne after time t, it is assumed that when the pollution encounters precipitating storm i, then a random fraction $\tilde{\varepsilon}_i(\frac{1}{2} \pm \frac{1}{2}!)$ of the pollution is scavenged. Consequently, after n-storm encounters, (and ignoring dry deposition) the remaining airborne pollution is the random variable

$$\tilde{q}_n(t) \, q_0 = \prod_{i=1}^{n} (1 - \tilde{\varepsilon}_i).$$

(17)

From the central-limit theorem applied to $\ln (\tilde{q}/q_0)$ (which, incidentally, is a negative number), then the pdf for $\tilde{q}$ will, for large n, asymptotically tend to a log-normal distribution regardless of the pdf for $\tilde{\varepsilon}$. Thus (e.g., see Aitchison and Brown, 1976) the pdf for $\tilde{q}$ will be approximately

$$f(q/q_0) = [(q/q_0)\sigma_T\sqrt{2\pi}]^{-1} \exp\left\{-[\ln (q_0/q) + \mu_T]^2/2\sigma_T^2\right\},\tag{18}$$

where I have used $\ln (q_0/q)$ in the exponent of Equation (18), and where $\mu_T = +\bar{v}t\mu_\varepsilon$ and $\sigma_T^2 = \bar{v}t\sigma_\varepsilon^2$ (with $n = \bar{v}t$, $\bar{v} =$ the average, Lagrangian-frequency of encounters with storms, and with μ_ε the mean of $\ln (1 - \varepsilon_i)$ and σ_ε^2 its variance).

The result of Equation (18) suggests that at a time t downwind of a source region (for t greater than a day or so), $\tilde{q}/q_0$ should have a log-normal distribution over an ensemble of events. The mean value of $\tilde{q}/q_0$ over this ensemble, all at time t, would then be (Aitchison and Brown, 1976):

$$\alpha = \exp\left\{\mu_T + \tfrac{1}{2}\sigma_T^2\right\} = \exp\left\{\bar{v}tE\left[\ln (1 - \tilde{\varepsilon}_i)\right] + \frac{\bar{v}t}{2}\sigma_\varepsilon^2\right\} \approx$$

$$\approx [1 - \bar{\varepsilon}_g]^{\bar{v}t} \simeq \exp\left\{-\bar{v}\bar{\varepsilon}t\right\}.\tag{19}$$

For example, if $\bar{\varepsilon} = 1/2$ and $\bar{v} \simeq (100\text{ h})^{-1}$, then the e-fold residence time, $\bar{\tau} = [\bar{\varepsilon}\bar{v}]^{-1}$ is about 200 h $\simeq$ 1 week (an answer which guided the estimate for $\bar{\varepsilon}$ and $\bar{v}$!). Similarly, the variance $\beta^2 = \alpha^2 [\exp (\sigma_T^2) - 1]$, and the coefficient of deviation, f (the standard deviation divided by the mean), is

$$f = [\exp (\sigma_T^2 - 1)]^{1/2} \simeq \sigma_T \simeq \sqrt{\bar{v}t}\,\sigma_{\varepsilon_i} \simeq (t/\bar{\tau})^{1/2},\tag{20}$$

which yields a $\bar{\tau}^{-1/2}$ dependence for f, which is consistent with other studies (Gibbs and Slinn, 1973; Baker et al., 1979). However, the main result here, and which will be used in the next section, is that it is typical that there will be a substantial variability in the amount of pollution remaining airborne; i.e., it is a 'long-tail' log-normal distribution!

7. Some Applications

The long-range transport of S and N, and the U.S./Canadian acid-rain issue are of particular interest at this conference. If SO_2 or NO_x pollution become involved in a slow, large-scale ascent of stable, warm, moist air over a colder air mass (Figure 6), then the

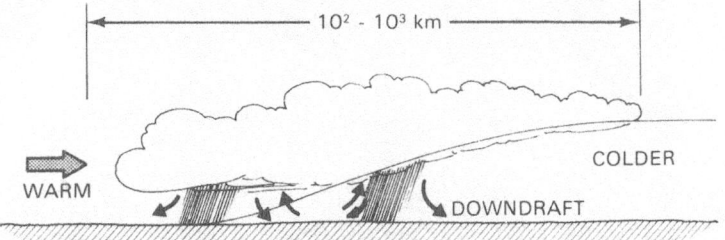

Fig. 6. Schematic of large-scale, 'stable', warm-frontal storm, although instabilities may be associated with the rainbands.

pollution may travel 10^2 to 10^3 km (e.g., within cloud water), and then a substantial fraction of the pollution can be precipitated as $SO_4^=$ and NO_3^- with the cloud water. In eastern North America, though, only about 10% of the summer-semester storms, and about 30% of the winter storms, involve this slow, stable ascent (Huff, 1971). More common are unstable storms, and even for stable storms it may be that most precipitation results from unstable 'rainbands' within the storm (see Figure 6). Figure 7 shows a sketch of an isolated, unstable convective storm; Figure 8 shows a sketch of the destabilization of air, and the resulting airflows and precipitation caused, for example, by the frequently occurring interaction of Canadian polar air and U.S./Gulf Coast tropical air. Such storms would typically precipitate $\frac{1}{2} \pm \frac{1}{4}$ of their ingested S and N particulate pollution, and oxidize and precipitate a substantial portion of their ingested SO_2 and NO_x.

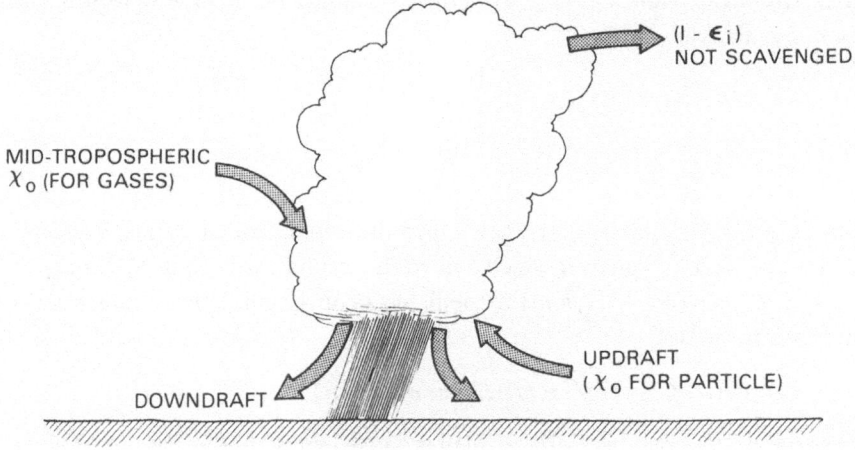

Fig. 7.   Schematic of an isolated convective storm, indicating the appropriate χ's for scavenging ratios, and emphasizing that not all ingested pollutants are scavenged.

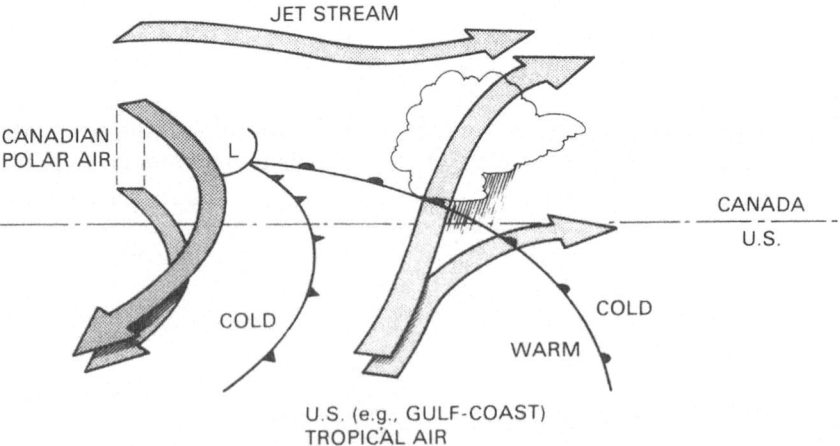

Fig. 8.   Schematic of a typical, unstable frontal storm near the Canadian/U.S. border, caused by the interactions of polar and tropical air masses, and that can lead to substantial acid precipitation in Canada, especially for a slow-moving, low-pressure area (after Palmén and Newton, 1969).

Even these simple sketches of different storm types can contribute to an understanding of the acid-rain issue. For example, there are the observations of maxima of $SO_4^=$ deposition in the Northeast U.S. during the summer, and in Scandinavia during the winter and spring. These results are quite to be expected from consideration of differences in space-scales of the storms: for the smaller-scale ($\sim 10^2$ km), isolated, typically-summertime convective storms, local sources dominate the deposition, and therefore during the summer, more $SO_4^=$ would be expected to deposit on the Northeast U.S., and less on Scandinavia, because of the relative amounts and contributions from local vs distant pollution sources. In contrast, for the larger-scale ($\sim 10^3$ km), typically-winter-semester storms, distant sources would contribute more, leading to a minimum for deposition in the Northeast U.S. (e.g., sources in the Southeast U.S.), and a maximum in Scandinavia (sources in South Europe). More generally (and more succintly) in a source region, maximum wet deposition will arise from small-scale storms; other regions will receive more pollution from distant source regions during larger-scale storms. Of course, there are other factors involved (e.g., more SO_2 oxidation because of more OH, more liquid cloud water, and more NH_3 during the summer), but it is essential to appreciate, also, the differences in the space-scales of different storm types.

Of particular importance in the acid-rain issue are wet deposition 'episodes': $\sim 50\%$ of the acid wet deposition can result from ~ 10 to 20% of the storms. Smith and Hunt (1978) identify two main types of synoptic situations that can lead to episodic deposition: (i) when slow-moving active frontal storms (e.g., Figure 8) process a substantial quantity of polluted air, and lead to heavy and prolonged rainfall, and (ii) when anticyclonic,

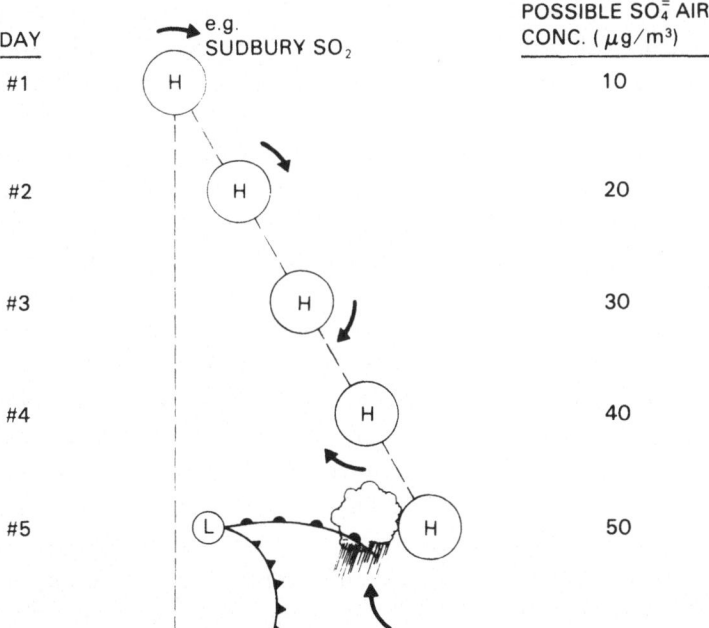

Fig. 9.   Schematic of a second, synoptic-situation that can lead to episodic, acid-deposition, with polluted air from a persistent anticyclone drawn into an active frontal system.

high-pressure areas, with light winds, persist for several days over major source areas, subsequently drawing polluted air into active frontal zones.

In Figure 9 is sketched a possible scenario leading to episodic deposition at the periphery of a persistent anticyclone. Vukovich *et al.* (1977) suggest that for warm-season anticyclones in the eastern U.S., typical residence times of air masses in the anticyclone, for air entering the NE quadrant (e.g., from southeastern Canada) can be 3 to 5 days, and Vukovich (1979) describes an episode case with airborne $SO_4^=$ concentrations increasing at a rate of about $10 \, \mu g \, m^{-3} \, day^{-1}$ for 5 days. Complicating features of anticyclones include: (i) little or no ascension (associated with large-scale subsidence); (ii) little dry deposition because of low wind speeds, poor vertical mixing, and dry surfaces; (iii) typically, a lot of insolation, and therefore OH, O_3, etc. production and therefore enhanced gas-phase oxidation of SO_2; (iv) low ventilation, and (v) little or no rain. As a result, when the polluted air mass moves to a new region, and/or is lifted to form precipitation, substantial (episodic) deposition occurs. For example, if there were nothing to buffer $50 \, \mu g \, m^{-3}$ of H_2SO_4, and if this enters $\sim 0.3 \, g \, m^{-3}$ of cloud water, then (ignoring other factors) rains with pH ~ 2.5 could result.

With regard to the U.S./Canadian acid-rain issue, three items are reasonably clear: (i) the whole problem is exceedingly complex, (ii) U.S. SO_2 and NO_x emissions result in substantial acid rain in Southeast Canada; and (iii) it is going to be difficult to solve the problem. In so far as wet deposition is episodic in character, and so long as the U.S./Canada border is frequently near the border between polar and tropical air masses, it is inevitable that Southeast Canada will receive substantial rain from air flowing over the U.S. The pollution in this air may contain, for example, Sudbury SO_2 released many days earlier and after it has travelled to near the U.S. Gulf Coast and back, but in the main, the pollution will be from the eastern U.S. Thereby, as a crude estimate of how much of Southeast Canada's acid rain is from the U.S., I expect it is at least 50% of the total. Technological 'fixes' include deducing U.S. SO_2 and NO_x emissions (at a *net* cost that may not be too large if *all* pollution costs are evaluated); possibly seeding storms that would lead to episodic deposition, using alkaline particles or NH_3 (but, for example, to counteract $50 \, \mu g \, m^{-3}$ of $SO_4^=$ entering a 1×500 km storm at $8 \, m \, s^{-1}$ for 10^5 s would require $\sim 10^4$ tonne of a buffer!); releasing more alkaline material with all SO_2/NO_x releases (Slinn, 1976); moving U.S. fossil-fuel plants to the U.S. eastern seaboard; piping U.S. emissions direct to the deep ocean; and/or building more nuclear plants. However, as we have seen throughout the 'environmental movement', technological 'fixes' can cause new problems, and it may be necessary to address the root problem that there are too many people devouring too many resources. I have opinions about political and social actions that might be taken to solve this root problem, but will describe these opinions elsewhere. Nevertheless, I would like to mention here the self-explanatory title I have given to the underlying philosophy, 'rational survivalism', and a possibility for its widespread acceptance: Spinoza's or a similar synthesis of rationalism and mysticism.

At a larger space-scale than for the U.S./Canadian or Southern/Northern European acid-rain issues, more important than the fractions, ε_i, of ingested pollution scavenged 'locally' by storms, are the fractions $(1 - \varepsilon_i)$ not scavenged. Intense convective and unstable frontal storms can penetrate deep into the stratosphere, as is familiar to anyone who has skimmed through stratus clouds, or seen Cb clouds towering above them, while flying on commercial airlines at $\sim 35\,000$ feet. The $\sim 10\%$ of the pollution ingested but not scavenged by these intense storms is then available for global-scale transport, e.g., in the stratosphere. Figure 10 sketches a possibility. A possible example of this global-scale transport is the findings by Barrie *et al.* (1981) of U.S. ragweed pollen in the Arctic,

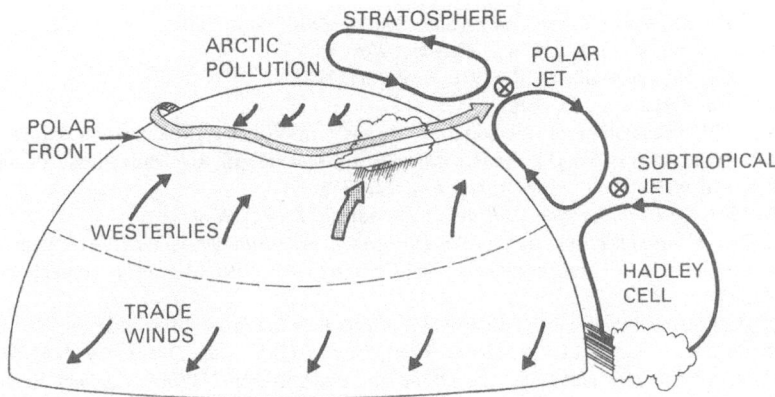

Fig. 10. Schematic of global wind patterns (after Palmén and Newton, 1969), suggesting pathways for global-scale pollution of the atmosphere.

during the spring, before/after the summertime ragweed pollen season in the U.S. Quite conceivably, this pollen was imbedded into the stratosphere by summertime convective storms during the previous summer, then travelled north in the stratosphere, and descended to the Arctic during the next spring. This time- and space-scale of the long-range transport problem should also be of concern to all of us.

Acknowledgments

My training was funded to a large degree by the citizens of Canada; this research has been funded by citizens of the U.S. via the U.S. Department of Energy and Environmental Protection Agency, under U.S. DOE Contract DE-AC06-76RLO-1830. Special thanks are given to Walter H. Gage, Robert W. Beadle, David S. Ballantine, and Charles E. Elderkin, and to J. M. Hales for inviting this presentation, arranging for partial funding, and improving an earlier draft of the paper.

References

Aitchison, J. and Brown, J. A. C.: : 1976, *The Lognormal Distribution*, Cambridge University Press, New York.
Baker, M. B., Harrison, H., Vinelli, J., and Erickson, K. R.: 1979, *Tellus* **31**, 39.
Barrie, L. A.: 1981, *Atmos. Env.* **15**, 31.
Barrie, L. A., Hoff, R., and Daggupaty, S.: 1981, *Atmos. Env.* **15**, 1407.
Bennett, J. H., Hill, A. C., and Gates, D. M.: 1973, *J. Air Poll. Control Assoc.* **23**, 957.

Davidson, C. I., Chu, L., Grimm, T. C., Nasta, M. A., and Qamoos, M. P.: 1981, *Atmos. Env.* **15**, 1429.

Engelmann, R. J.: 1968, 'The Calculation of Precipitation Scavenging', in D. H. Slade (ed.), *Meteorology and Atomic Energy*-1968, U.S. AEC, Division of Tech. Info., Oak Ridge, TN and available as TID-24190 from NTIS, Springfield, VA, pp. 208–221.

Eriksson, E.: 1971, *Ann. Rev. Ecology and Systematics* **2**, 67.

Everett, R. G., Hicks, B. B., Berg, W. W., and Winchester, J. W.: 1979, *Atmos. Env.* **13**, 931.

Fowler, D.: 1978, *Atmos. Env.* **12**, 369.

Garland, J. A.: 1978, *Atmos. Env.* **12**, 349.

Georgii, H. W.: 1978, *Atmos. Env.* **12**, 681.

Gerrard, W.: 1976, *Solubility of Gases and Liquids*, Plenum Press, New York.

Gibbs, A. G. and Slinn, W. G. N.: 1973, *J. Geophys. Res.* **78**, 574.

Granat, L., Rodhe, H., and Hallberg, R.: 1976, *Ecol. Bull.* **22**, 89.

Hales, J. M.: 1978, *Atmos. Env.* **12**, 389.

Huff, F. A.: 1971, 'Evaluation of Precipitation Records in Weather Modification Experiments', in H. E. Landsberg and J. Van Mieghem (eds.), *Advances in Geophysics* **15**, Academic Press, NY, pp. 60–136.

Judeikis, H. S. and Wren, A. G.: 1978, *Atmos. Env.* **12**, 2315.

Junge, C. E.: 1963, *Air Chemistry and Radioactivity*, Academic Press, NY.

Junge, C. E.: 1964, *The Modification of Aerosol Size Distributions in the Atmosphere*, Final Tech. Report, Meteor. Geophys. Inst., Johannes Guttenberg Universität, Contract DA-91-591-EVC 2979, available from NTIS, Springfield, VA.

Liu, S. C.: 1978, 'Possible Nonurban Environmental Effects Due to Carbon Monoxide and Nitrogen Oxide Emissions', in J. S. Levine and D. R. Schryer (eds.), *Man's Impact on the Troposphere*, NASA Reference Pub. 1022, NASA Scientific and Tech. Info. Office, and available from NTIS, Springfield, VA, pp. 65–80.

Mészáros, E.: 1978, *Atmos. Env.* **12**, 699.

O'Dell, R. A., Taheri, M., and Kabel, R. L.: *J. Air Poll. Control Assoc.* **27**, 1104.

Palmén, E. and Newton, C. W.: 1969, *Atmospheric Circulations Systems*, Academic Press, NY.

Plate, E. J.: 1971, *Aerodynamic Characteristics of Atmospheric Boundary Layers*, U.S. DOE, Div. Tech. Info., Oak Ridge, TN and available as TID-25465 from NTIS, Springfield, VA.

Prospero, J. M. and Nees, R. T.: 1977, *Science* **196**, 1196.

Rahn, K. A. and McCaffrey, R. J.: 1979, *Nature* **280**, 479.

Rodhe, H.: 1978, *Atmos. Env.* **12**, 671.

Rodhe, H. and Grandell, J.: 1972, *Tellus* **24**, 442.

Schwartz, S. E. and White, W. H.: 1980, *Equilibrium Solubility of the Nitrogen Oxides and Oxyacids in Aqueous Solution*, BNL 27102, Brookhaven National Laboratory, Upton, NY.

Shaw, G. E.: 1981, *Atmos. Env.* **15**, 1483.

Slinn, S. A. and Slinn, W. G. N.: 1981, 'Modeling of Atmospheric Particulate Deposition to Natural Waters', in S. J. Eisenreich (ed.), *Atmospheric Input of Pollutants to Natural Waters*, Ann Arbor Science Publishers, Inc., Ann Arbor, MI, Chapter II.

Slinn, W. G. N.: 1976, *Bull. Am. Met. Soc.* **57**, 1166.

Slinn, W. G. N.: 1977, *Water, Air, and Soil Poll.* **7**, 513.

Slinn, W. G. N.: 1980, *AIChE Symposium Series* 196 **76**, 185.

Slinn, W. G. N.: 1981a, 'Precipitation Scavenging', in D. Randerson (ed.), *Atmospheric Sciences and Power Production*, U.S. DOE Tech. Info. Center, Oak Ridge, TN, in press and to be available from NTIS, Springfield, VA, Chapter 11.

Slinn, W. G. N.: 1981b, *Atmos. Env.* **16** (in press).

Slinn, W. G. N., Hasse, L., Hicks, B. B., Hogan, A. W., Lal, D., Liss, P. S., Munich, K. O., Sehmel, G. A., and Vittori, O.: 1978, *Atmos. Env.* **12**, 2055.

Slinn, W. G. N. and Hales, J. M.: 1982, 'Wet Removal of Atmospheric Particles', in A. P. Altshuller (ed.), *Atmospheric Fine Particles*, U.S. EPA Special Publication, available from NTIS, Springfield, VA (in press).

Smith, F. B. and Hunt, R. D.: 1978, *Atmos. Env.* **12**, 461.

Smith, T. B., Blumental, D. L., Anderson, J. A., and Vanderpol, A. H.: 1978, *Atmos. Env.* **12**, 605.

Vukovich, F. M.: 1979, *Atmos. Env.* **13**, 255.

Vukovich, F. M., Bach, W. D., Crissman, B. W., and King, W. J.: 1977, *Atmos. Env.* **11**, 967.

SOURCE REGIONS OF SUMMERTIME OZONE AND HAZE EPISODES IN THE EASTERN UNITED STATES

GEORGE T. WOLFF, NELSON A. KELLY, and MARTIN A. FERMAN

Environmental Science Department, General Motors Research Laboratories, Warren, MI 48090, U.S.A.

(Received May 1, 1981; Revised August 13, 1981)

Abstract. During the summer of 1979, the mobile Atmospheric Research Laboratory (ARL) was sited near the Louisiana Gulf Coast to monitor the concentrations of air contaminants in air masses moving northward from the Gulf. Using the ARL data in conjunction with O_3 and visibility measurements across the entire eastern United States as well as synoptic meteorological data and satellite photographs, major source areas were identified. The haze which was observed on the Louisiana Gulf Coast appears to be primarily due to aged sulfate aerosols. The evidence presented strongly suggests that the sulfates were largely due to SO_2 emissions in the northeastern and midwestern United States. The haze initially formed over and downwind of these source areas and was transported to the Gulf Coast area. In the last two episodes, the haze was subsequently transported back to the Midwest source region. Elevated levels of O_3 were also associated with the haze. Again, the Northeast and Midwest appear to be the most significant source areas for the regional O_3 episodes. Occasionally, O_3 formed from emissions in the east Texas area also appear to affect a large portion of the Gulf Coast. In addition, each episode appeared to have a significant stratospheric air component based on ^7Be measurements. All three episodes followed the passage of upper levels troughs which produce stratospheric intrusions due to tropopause folding.

1. Introduction

During the warmer half of the year, the climate in the eastern United States is characterized by the succession of continental polar and maritime tropical air masses moving through the region (Brunnschweiler, 1952). A typical sequence begins as a high pressure system containing relatively clean continental polar air moves out of Canada into the upper Midwest. As it crosses the United States-Canadian border, typical background concentrations of O_3 range from 30 to 50 ppb (Wight *et al.*, 1978; Kelly *et al.*, 1981) and sulfates from 1 to 2 $\mu g\ m^{-3}$ (Kelly *et al.*, 1981). As these air masses move southeastward and pass over the large pollutant source regions of the Midwest, O_3 and sulfate-producing haze accumulate primarily on the backside of the high pressure system (Wolff *et al.*, 1977b, 1979a, 1980; Hidy *et al.*, 1978; Vuckovich *et al.*, 1977). By the time the center of the high pressure system reaches the Appalachian Mountains, maritime tropical (mT) air from the Gulf of Mexico is frequently advected into the clockwise circulation of the high. This mT air, which arrives in the Midwest on southerly winds, generally already contains high concentrations of sulfate and O_3, typically in excess of 15 $\mu g\ m^{-3}$ (Hidy *et al.*, 1978) and 80 ppb, respectively (Wolff *et al.*, 1977b, 1979a). Since the studies cited above analyzed data only from the midwestern and northeastern United States, the sources of the high O_3 and sulfate in mT air could not be determined.

When the geographic boundaries of the O_3 study area were expanded to encompass the entire eastern United States, Wolff and Lioy (1980) observed that during 3 episodes in July 1977 the area of high O_3 in mT air extended from the Midwest to New Orleans and Houston on the Gulf Coast.

Similarly, we studied the distribution of haze over the same geographic area during August and September 1979, and our preliminary findings have been reported (Wolff *et al.*, 1981a). These findings included four important results. First, the hazy areas also extended from the Gulf Coast to the Midwest. Second, the area of haze did not form from emissions near the Gulf Coast. The haze was formed in the Midwest and Northeast in continental polar (CP) air which was transported to the Gulf Coast area and modified by the maritime tropical environment so that its temperature and relative humidity became similar to those observed in mT air. From the Gulf Coast, the haze was then advected back to the Midwest. Third, sulfates appeared to be the predominant visibility-reducing species in the haze and, fourth, high O_3 was generally associated with the hazy air masses.

Since these preliminary results were reported, additional analyses, which support the initial findings, have been conducted and they will be reported here. In particular, this paper contains a quantitative evaluation of the contributions of the various particulate and gaseous species to the visibility-reducing properties of the haze. Further documentation that emissions from the Midwest and Northeast source areas are the principal causes of the haze in the 'mT' air masses is also presented. Finally, new results on the sources of the O_3 associated with the haze will also be discussed.

2. Experimental

For the analyses presented here, we compiled four types of data sets for the period August to September 1979. The first included ambient measurements of the important gaseous and particulate species made by our Atmospheric Research Laboratory (ARL) which was situated near the Gulf Coast from August 5 to September 9, 1979. It was located in the rural Louisiana Bayou Country, about 10 km south of Abbeville and 15 km north of the Gulf of Mexico. New Orleans was 150 km to the east. The pertinent measurements made are listed in Table I. Procedures used by the ARL were originally described by Groblicki *et al.* (1975) and later modification to the continuous instrumentation was described by Kelly *et al.* (1981) and Ferman *et al.* (1981a). The sampling and analytical procedures used for particulates are described by Countess *et al.* (1980) except for the GMR sequential filter sampler (Wolff *et al.*, 1979b) which was modified to include a rain hat and a Sierra constant volume sampler.

The second and third data sets were daily 1300 h EST visibility and 1 h maximum O_3 values from sites in the eastern United States. Visibility measurements were obtained from some 250 National Weather Service Stations while the O_3 data was obtained from 200 state and local monitoring sites. Only visibility observations made in the absence of precipitation or fog were included. The fourth data set was the meteorological data transmitted on the NAFAX National Weather Service Facsimile System. These data were supplemented with additional satellite photographs and air parcel trajectories (Heffter and Taylor, 1975).

TABLE I

Selected measurements made in Louisiana[a]

Parameter	Instrument
Meteorological	
Wind speed	Climet 011-213 cup anemometer
Wind direction	Climet 012-6C wind vane
Temperature	Climet 015-74 thermometer
Dew point	Climet 015-12(7) LiCl dew cell
Ultraviolet radiation	Eppley radiometer
Gaseous	
O_3	Dasibi 1003AH UV photometer
NO_x, NO, NO_2	Monitor Labs 8440/E chemiluniscent analyzer
CH_4[b]	Beckman 6800 flame ionization GC
CO[b]	Above + hydrogenation
Total hydrocarbons (THC)	Beckman 400 FID
SO_2	Meloy SA 285 flame photometric analyzer
Individual hydrocarbons[c]	Perkin-Elmer Sigma I FID GC
Fluorocarbon-11 (F-11)[d]	Perkin-Elmer Sigma I EC GC

Parameter	Sampler	Filter Media	Analytical Technique
Particulate			
Total suspended particulates[e]	High volume sampler	Quartz	Gravimetric
Total suspended particulates[f]	GMR SFS[g]	Ghia Teflon	Gravimetric
Fine particulate mass (d $\leqslant$ 3.5 μm)[e]	Sierre dichotomous	Ghia Teflon	Gravimetric
Coarse particulate mass (3.5 < d < 15 μm)[e]	Sierre dichotomous	Ghia Teflon	Gravimetric
Sulfate (SO_4^{2-}), nitrate (NO_3^-)[f]	GMR SFS[g]	Ghia Teflon	Ion chromatography
Ammonium (NH_4^+)[f]	GMR SFS[g]	Ghia Teflon	Spectroscopy
Elemental (Cae) and organic carbon (Cao)[f]	GMR SFS[g]	Micro-Quartz	GMR combustion technique
^{7}Be	High volume sampler	Quartz	γ-ray spectroscopy
Light scattering coefficient[h] (bsp)	MRI 1550		Integrating Nephelometer
Light absorption coefficient[f] (bap)	GMR SFS[g]	Nuclepore membrane	Integrating plate

[a] All measurements continuous unless otherwise noted.
[b] Instantaneous sample collected every 10 min.
[c] 5-min sample collected every 90 min.
[d] Instantaneous sample collected every 15 min.
[e] 24-h samples.
[f] 4-h samples.
[g] GMR sequential filter sampler.
[h] Heated inlet.

3. Results and Discussion

During the sampling program, wide variations in some pollutant species were experienced at the Louisiana sampling site. This is demonstrated for O_3, $SO_4^=$ and b_{sp}^* in Figures 1a-c. While precursor gases such as NO_x and SO_2 showed no multiday periods

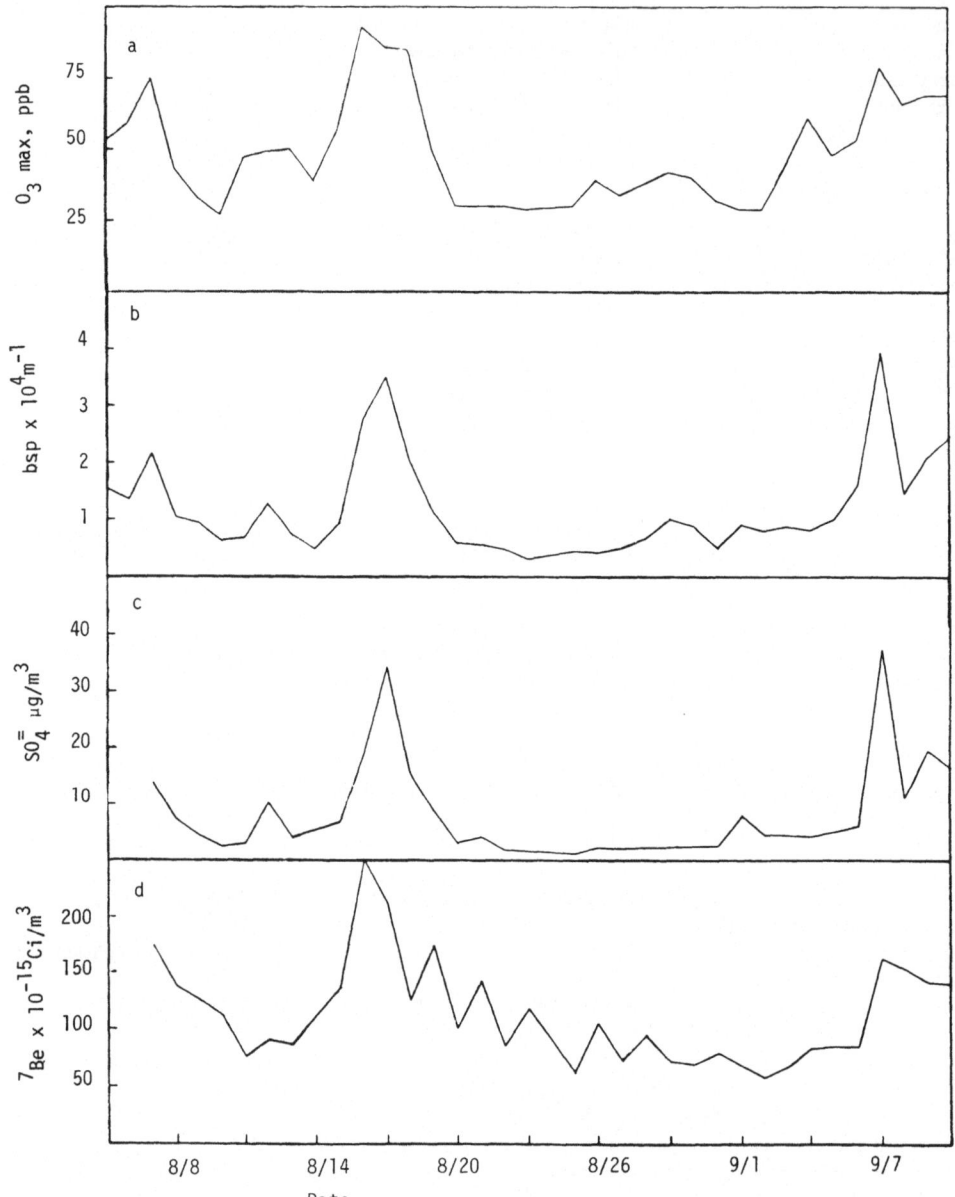

Fig. 1.   Daily variation of O_3 max, bsp, $SO_4^=$ and ^7Be in Louisiana during August and September, 1979.

* It should be noted that all bsp data prior to August 10 were estimated using an extrapolated calibration. Their accuracy, however, is sufficient to demonstrate the temporal trends.

of elevated concentrations, three episodes or periods of high pollution were evident from the species plotted in Figure 1. They occurred on August 5–8 (episode I), August 16–19 (episode II) and September 6–10 (episode III). The three episodes were characterized by simultaneous elevated concentrations of all three species. In addition, the levels of ^{7}Be, a tracer of stratospheric air, followed a similar temporal pattern (Figure 1d). To determine the sources of the elevated levels of O_3 and bsp, we examined the spatial patterns of O_3 and visibility as well as the synoptic-scale circulation across the entire eastern half of the United States during each of the episodes and for several days prior to each episode. Before the individual episodes are discussed, however, the nature of the haze needs to be established.

3.1. THE NATURE OF THE HAZE

When a black object is viewed against a bright background of uniform brightness, the visual range or 'visibility' is inversely proportional to the light extinction coefficient, bext (Middleton, 1963). The extinction coefficient is the sum of several terms:

$$bext = bsg + bsp + bap + bag . \tag{1}$$

The first term, bsg, is Rayleigh scattering, or scattering due to gases. It is a constant at a given temperature and pressure and is equal to approximately $2.1 \times 10^{-5} \, \mathrm{m}^{-1}$ in Louisiana. The scattering due to particles is denoted by bsp. However, since we used a nephelometer with a heated inlet, we measured only the fraction of bsp that was due to the dried particles and that fraction is represented by bsdp. Heating the inlet stream 10 to 15°C above ambient reduced the relative humidity to less than 40%. The implications of this will be discussed below. Light absorption by particles, bap, is essentially due to fine elemental carbon (Rosen et al., 1978). Finally, bag, is absorption due to NO_2 and can be calculated at 550 nm from bag = 3.3 $[NO_2]$ (Hodkinson, 1966), with NO_2 in units of ppm. The mean values for these parameters are given in Table II.

TABLE II

Components of the extinction coefficient

Parameter	Mean value $\times 10^4 \, \mathrm{m}^{-1}$		
	Episodes II and III	Non-episodes	All periods
bsp	2.23	0.55	1.03
bap	0.19	0.13	0.15
bag	0.01	0.01	0.01
bsg	0.21	0.21	0.21
Total	2.64	0.90	1.40

It is evident that the extinction is dominated by particulate light scattering. As mentioned above, the fraction of bsp due to the associated water at ambient conditions has not been included. Previous studies have shown that the water component, bsw, is a function of the concentration of the hygroscopic particulate concentration and the

TABLE III

Summary of concurrent aerosol chemical measurements during episodes II and III

Specie	Specie concentration ($\mu g\ m^{-3}$)		% of fine particulate mass
	Total	Fine	
Mass[a]	65.9	45.8	–
$SO_4^=$	21.6	21.0	45.9
NH_4^+	6.9	6.9[b]	15.0
Organic	17.3	12.1[b]	26.4
Elemental Carbon	2.5	2.0[b]	4.4
NO_3^-	1.1	0.2	0.4
			Total 92.1
$NH_{4_i}^+/SO_{4_i}^=$ [c]		0.87	

[a] Based on filters weighed at 40% RH.
[b] Estimated from previous data. See text for explanation.
[c] Equivalent ratio.

relative humidity (Groblicki *et al.*, 1981; Ferman *et al.*, 1981b). Table III indicates that $SO_4^=$ and the associated NH_4^+ are the only significant hygroscopic species. This was also the case in the Blue Ridge Mountains where a mean RH of 70% resulted in 47% of the bsp being due to water associated with the sulfate (Ferman *et al.*, 1981b). Since the mean RH during the study in Louisiana was 88%, it is probable that the total bsp was more than double the value of bsdp reported in Table II.

The dry fraction, bsdp, is a function of the fine particulate mass (Waggoner and Weiss, 1980; Groblicki *et al.*, 1981). This relationship for the Louisiana data is shown in Figure 2. The correlation coefficient is 0.94 and the slope corresponds to a specific scattering efficiency of 5.93 $m^2\ g^{-1}$. This is considerably higher than previously reported values of 3.13 to 3.45 (Waggoner and Weiss, 1980; Groblicki *et al.*, 1981) but is in better agreement with the values of 7.3 $m^2\ g^{-1}$ found in the Blue Ridge Mountains (Ferman *et al.*, 1981b). The reason for this higher efficiency is suspected to be due to the dominance of sulfate in the fine particulate mass and the Blue Ridge. A statistical technique used in previous studies has provided evidence that sulfates are more efficient light scatterers than the other principal species of the fine particulate mass (Groblicki *et al.*, 1981; Ferman *et al.*, 1981b). This technique will be applied to the Louisiana data below.

This technique utilizes multiple regression analysis to relate the dependent variable, bsdp, to independent variables which are the principal components of the fine particulate mass. The data in Table III illustrate that during episodes II and III sulfate, ammonium and carbon accounted for about 92% of the fine particulate mass (FPM). Episode I was not included because of incomplete data. The fine $SO_4^=$ and NO_3^- values were obtained directly from the dichotomous sampler filters while, based on our previous experience in Denver (Countess *et al.*, 1980), we assumed that all the NH_4^+ was in the FPM. The fractions of carbonaceous species in the fine fraction were assumed to be 0.7 for organics

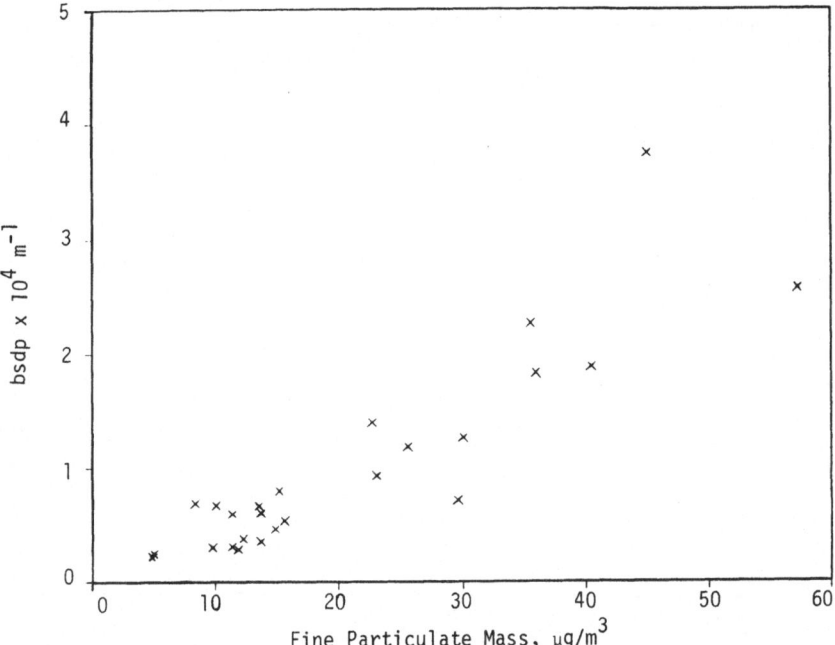

Fig. 2.   Relationship between bsdp and fine particulate mass (< 2.5 µm in diameter) in Louisiana during August and September, 1979)

and 0.8 for elemental carbon. These fractions were average values obtained from other studies in which we measured the coarse/fine split (Countess *et al.*, 1980; Wolff *et al.*, 1981b; Ferman *et al.*, 1981b). It should also be noted that the organic values in Table III include a factor of 1.2 to account for the H and O atoms associated with Cao (Countess *et al.*, 1980).

If we consider only sulfates (including the associated ammonium) and C as the two principal FPM constituents, the form of the multiple regression equation is:

$$bsdp \times 10^4 = a\,S + b\,C \tag{2}$$

where S is $SO_{4f}^=$ plus NH_{4f}^+ and C is the sum of the estimated concentrations of fine Cao and Cae. Using data from 42 concurrent measurements, the coefficients were calculated to be: a = 0.65 ± 0.004 and b = 0.035 ± 0.008, with an R^2 value of 0.87 (see Figure 3). Within the standard error, the coefficients are similar to those calculated for the Blue Ridge Mountains (a = 0.078 ± 0.006; b = 0.060 ± 0.030 (Ferman *et al.*, 1981b)) and for Denver (a = 0.066 ± 0.005; b = 0.044 ± 0.010 (Groblicki *et al.*, 1981)).

The fraction of bsdp due to sulfates, fs, and C, fc can be estimated from:

$$fs = \frac{0.065\,S}{bsdp \times 10^4}. \tag{3}$$

$$fc = \frac{0.035\,C}{bsdp \times 10^4}. \tag{4}$$

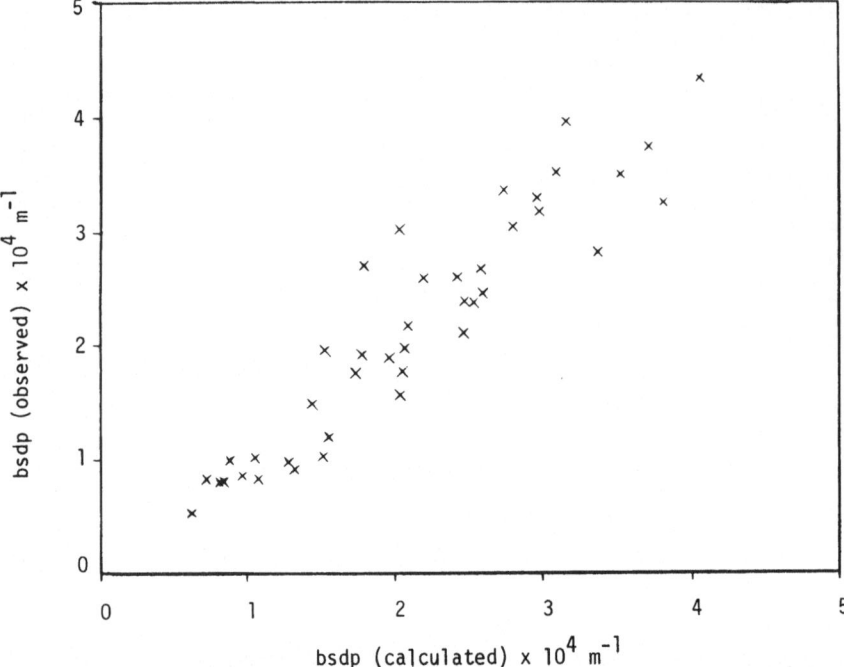

Fig. 3.   A comparison between the observed and predicted dry particulate scattering coefficients.

These equations ignore the contribution from the 'other' 8% of the FPM which was not sulfate or C. The effect of this should be negligible, however, since the 'other' in the Blue Ridge Mountains also accounted for 8% of the FPM but only 5% of the bsdp (Ferman *et al.*, 1981b). Using these equations then, the sulfates account for 79% of the bsdp. Since it is reasonable to assume that bsw is at least equal to bsdp, sulfates and the associated water would account for at least 82% of the total extinction while the C would account for only 14%. This is consistent with our results from the Blue Ridge Mountains which showed that sulfates accounted for 76 to 88% of the total extinction (Ferman *et al.*, 1981b).

There is also chemical evidence which suggests that the haze is due to an aged rather than a freshly emitted aerosol. The ratio of NH_4^+ to $SO_4^=$ indicates that the sulfate was 85% neutralized by NH_3 which is suggestive of an aged aerosol (Weiss *et al.*, 1977). In addition, the aerosol in the Blue Ridge Mountains, which are closer to the industrialized source areas, was only 50% neutralized (Ferman *et al.*, 1981b).

3.2. Episode 1

Although this episode did not affect the Louisiana site until August 7, the area of high O_3 associated with this episode began to develop over Kansas and Western Missouri on August 1 (see Figure 4 which depicts monitoring sites reporting $O_3 \geqslant 80$ ppb). Peak O_3 concentrations in this area were 111 ppb in Wichita, Kansas. This area moved to the east-northeast and reached Illinois by August 2 and western Ohio by August 3. Maximum O_3 concentrations within this area on August 2 were similar to August 1, but by

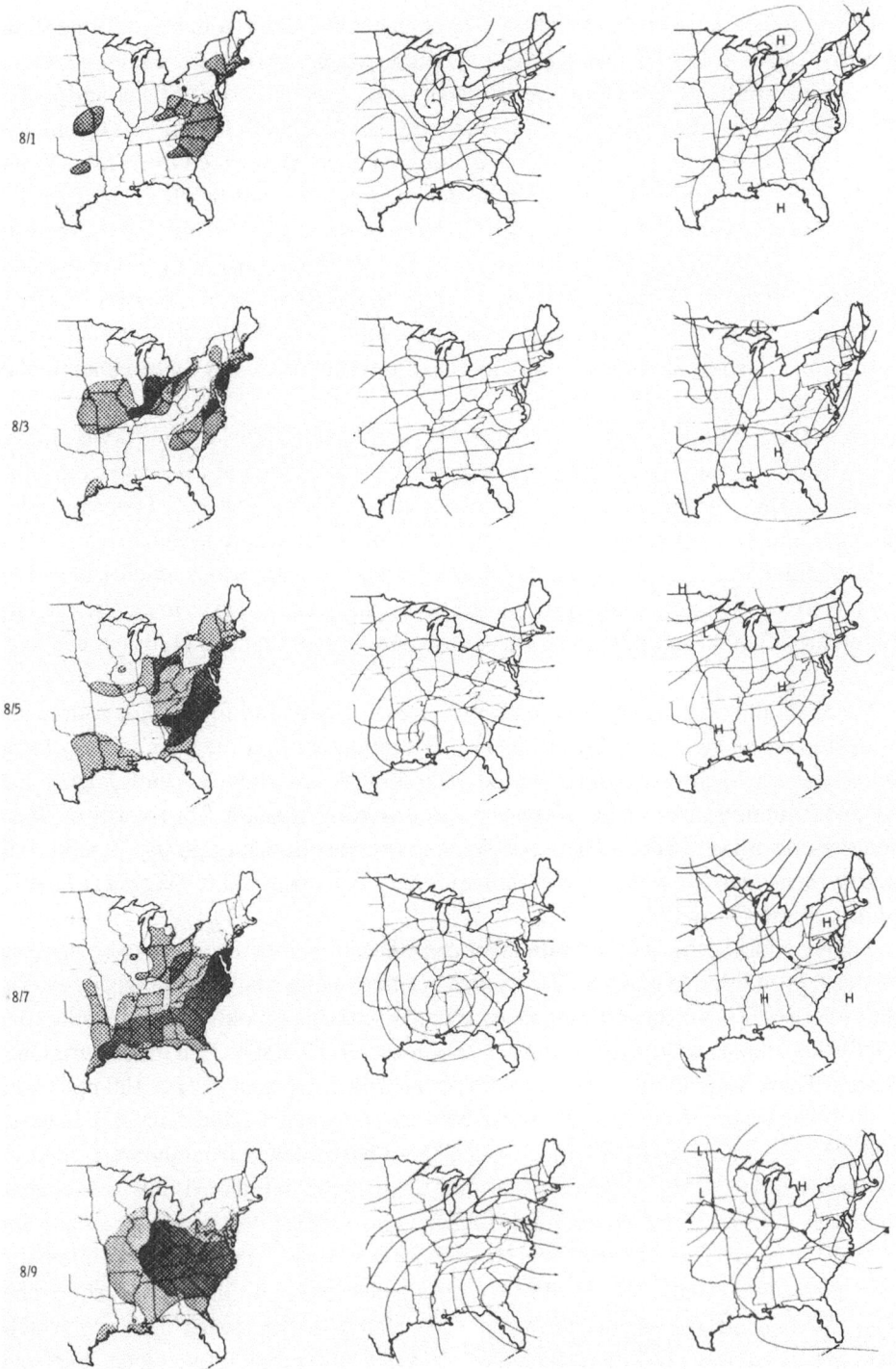

Fig. 4.   Column 1 – Spatial pattern of haze (≤ 10 km denoted by [   ]) and O$_3$ (≥ 80 ppb denoted by [   ]);
Column 2 – 850-mbar wind fields; column 3 – surface weather maps for Episode I.

August 3 maximum concentrations across Indiana reached 131 ppb (we use Indiana data with some caution because it was always higher than data from neighboring states). Also by August 3, low visibility ($\leqslant$ 10 km) began to develop in the area of high O_3 that had reached Indiana and Ohio. At the same time, a second area of haze and high O_3 developed in a northeast-southwest line from Boston to South Carolina. This area was advected clockwise around the high-pressure system which formed over West Virginia on August 3. By August 4 the two areas merged and covered much of the East from Indiana to the Atlantic Coast. Supported by a closed upper air high-pressure system which extended through the 500-mbar level, the West Virginia surface high remained nearly stationary until August 9. Consequently, the movements of the areas of high O_3 and haze were governed by the clockwise circulation around the high-pressure system. As demonstrated by Wolff *et al.* (1981a), these movements closely followed streamlines constructed at the 850-mbar level.

On subsequent days (August 5 to 9), the O_3 and haze traveled in a clockwise manner from the Northeast to the Southeast by August 5 and arrived at our Louisiana site on August 7. This is reflected in the temporal traces shown in Figure 1. From Louisiana, the haze and O_3 moved north-eastward on the backside of the system.

In addition to the areas of high O_3 discussed above, there was a smaller area which persisted over portions of the western Gulf States and eastern Texas from August 3 to 6. From the 850-mbar streamlines on these days, it appeared that this O_3 originated in east Texas.

Based on this episode, two major source areas of haze and four major source areas of ozone can be identified. From Figure 4, it is apparent that the haze initially formed in the eastern Midwest (east of central Illinois) between August 2 and August 3. On August 3, another area of haze appeared in a narrow corridor from north of Boston south-westward into North Carolina. With a northeast flow on that day, it is probable that the dense emission areas between Boston and Washington, DC (Wolff *et al.*, 1977a) produced the O_3 and haze.

For regional O_3, the eastern Midwest and northeast corridor also seem to be significant sources. In addition, the eastern Texas area appears to affect large portions of Texas and the Gulf Coast. According to regional emission inventories (Wolff and Lioy, 1980; Clark, 1980), these three areas have the highest emissions of HCs and NO_x in the eastern United States. These source areas, however, cannot explain the high O_3 ($>$ 100 ppb) which appeared in eastern Kansas and western Missouri on August 1 and August 2. In eastern Kansas, the air flow from the surface to the 850-mbar level was from the west-northwest on August 1. Local photochemistry certainly contributed but there is also evidence that there was a stratospheric component. Johnson and Viezee (1981) and Viezee and Singh (1980) observed that in virtually every upper air trough, regardless of intensity, a tropopause fold event occurs and brings stratospheric O_3 into the middle troposphere. The O_3 concentration in the intrusion was related to the trough intensity which is determined by the maximum wind speed at the 500-mbar level. Once it is in the mid-troposphere, the O_3 can be brought to the surface by convective mixing (Johnson and Viezee, 1981; Wolff *et al.*, 1979c).

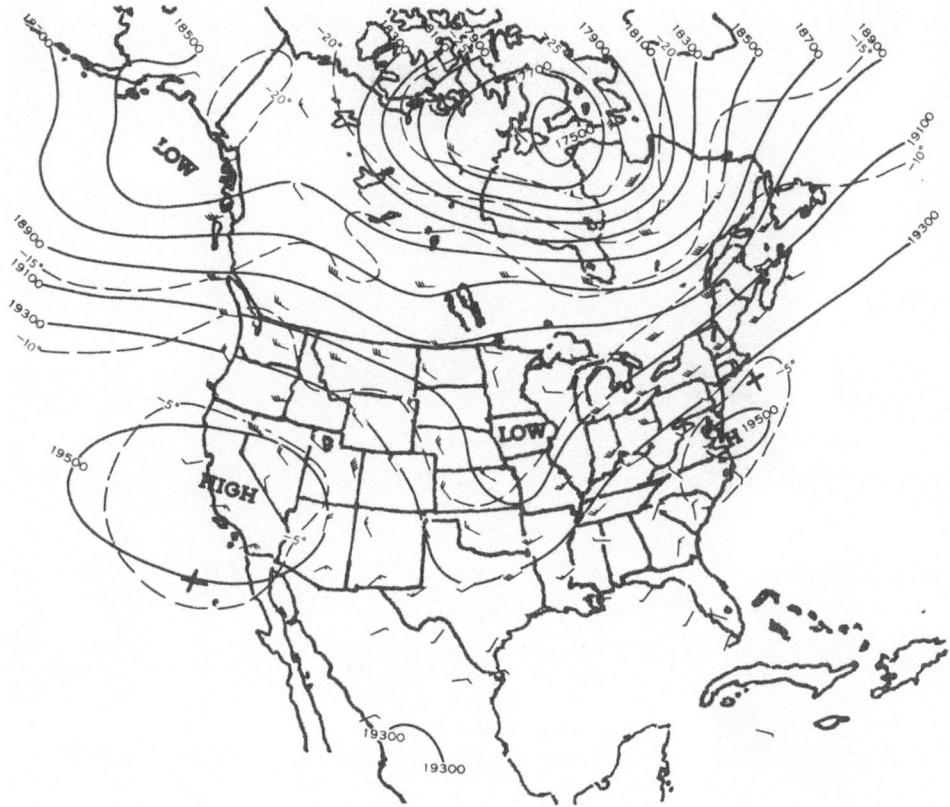

Fig. 5. 500-mbar map for 8/1/79 showing trough over the middle of the United States.

On August 1, a rather intense (for summer) 500-mbar trough passed over Kansas and western Missouri (Figure 5) and reached the western Michigan-Indiana area by August 2. Following the passage of the trough, daytime temperatures soared to the mid-80's and low 90's across Kansas and Missouri. Such temperatures have the potential for producing the convective activity needed to transport the O_3 from the mid-troposphere to the surface. The presence of isolated thunderstorm activity in the area on August 2 indicates that vertical exchange did in fact occur.

Based on the 850-mbar streamlines, the air over Kansas and west Missouri reached our Louisiana site on August 7 and resulted in the relatively high ^{7}Be value of 174×10^{-15} Ci m^{-3}. This is almost two times higher than the expected mean ^{7}Be value for 30° N latitude in the summer (Viezee and Singh, 1980). Consequently, all of the above evidence suggests that the air mass which arrived in Louisiana on August 7 had a significant stratospheric air component. Without a comprehensive 3 dimensional trajectory-photochemical model, however, it is not possible to calculate the stratospheric O_3 contribution to the high O_3 levels.

We should also point out that the contribution of tropospheric photochemistry to the long-term concentration of O_3 in the free troposphere is still uncertain. Recent work by

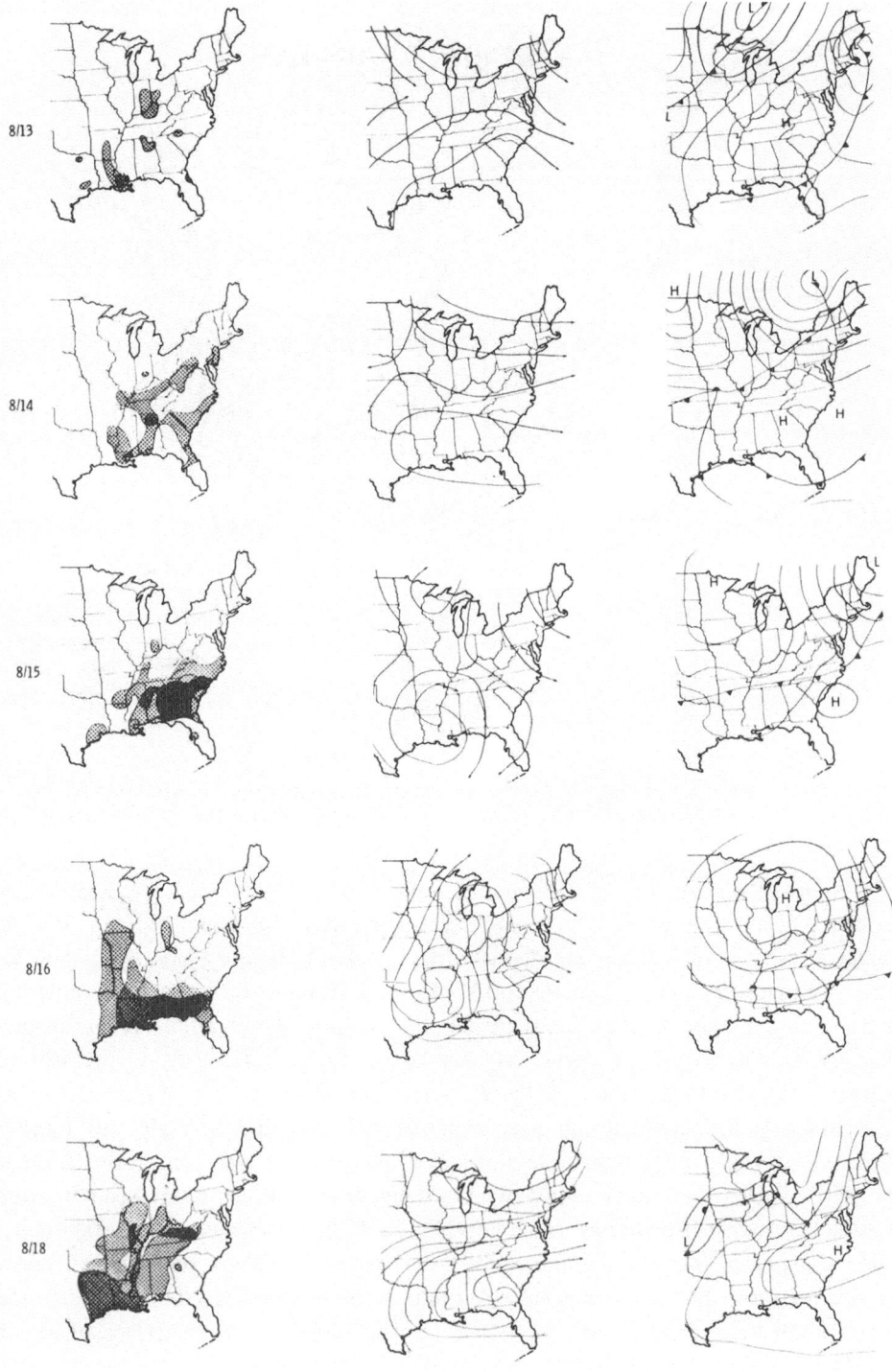

Fig. 6. Same as Figure 4 except for Episode II.

Liu *et al.* (1980), who use the most recent NO_x profile, suggests that much of the tropospheric O_3 is formed photochemically via the methane-CO-NO_x system in the upper troposphere with NO_x that is injected from the stratosphere. From our vantage point, O_3 formed in the upper troposphere would have been indistinguishable from O_3 formed directly in the stratosphere via photodissociation of O_2, since the mechanisms responsible for transporting both to the surface would be similar and both would be accompanied by [7]Be excursions. Consequently, the discussions in this paper on stratospheric O_3 also apply to O_3 formed in the upper troposphere.

3.3. EPISODE II

The synoptic meteorology associated with this episode was quite similar to that in Episode I. For this reason and because the meteorology and haze have been described previously (Wolff *et al.*, 1981a), this episode will only be discussed briefly.

High O_3 developed over the Indiana-Ohio-Kentucky area on August 13 (see Figure 6). By August 14, an expanded area of high O_3 moved to the Southeastern States preceding an area of haze which had formed over the Ohio Valley and downwind areas. The combination of an upper air high-pressure system over the Southeast States and weak

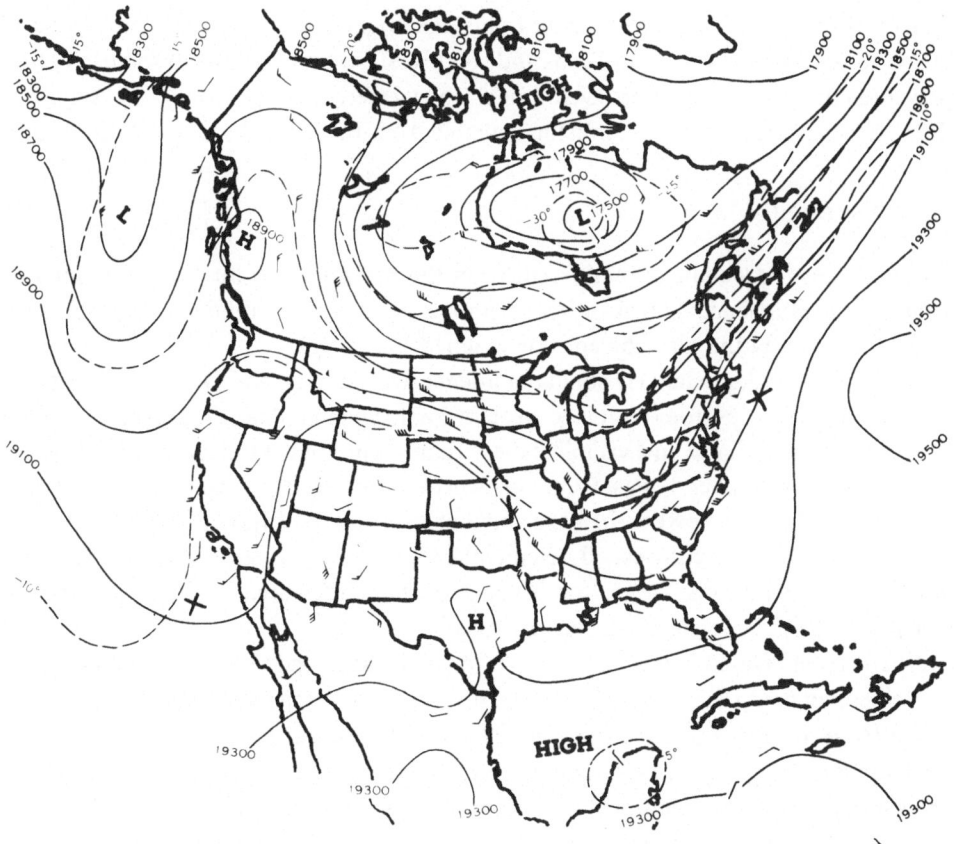

Fig. 7.   500-mbar map for 8/12/79 showing trough over the eastern United States.

surface systems caused the high O_3 and haze areas to move in a clockwise manner and arrive at the Louisiana site on August 16. Maximum O_3 in this episode at the Louisiana site was 90 ppb and occurred on August 16 (see Figure 1). Twenty-four hour average sulfates and bsp peaked the following day at 34 μg m^{-3} and 3.5 × 10^{-4} m^{-1}, respectively. By August 18 and 19, the areas of high O_3 and haze were returning back into the Midwest.

As shown in Figure 1, the ^{7}Be also peaked on August 16 and remained relatively high on August 17. The peak ^{7}Be value of 250 × 10^{-15} Ci m^{-3} is more than 2.5 times greater than the average for 30° N latitude during the summer (Viezee and Singh, 1980) and suggests that there was a stratospheric contribution to the observed O_3. An examination of the 500-mbar charts shows that a fairly intense trough moved through the eastern United States from August 11 to 14 (see Figure 7). Unlike Episode I, where the intrusion appeared to reach the ground in a relatively low emission density area, the high O_3 in Episode II initially developed over the Ohio-Indiana-Kentucky area. Consequently, it appears that the intrusion reached the ground somewhere between the Midwest (August 13) and Louisiana (August 16) but the precise location cannot be determined.

The influence of the east Texas source area can also be seen in Louisiana from August 13 to 15. With persistent 850-mbar westerly winds from August 11 to 14, it appears that the haze and high O_3 over Louisiana originated in east Texas on August 11. Heavy thunderstorm activity in Texas and southwestern Louisiana from August 12 to 14 prevented additional transport even though the westerly flow persisted. By August 15, the east Texas plume became indistinguishable from the larger haze and high O_3 area.

3.4. Episode III

This episode was significantly different from the previous two due to the dominating influence of Hurricane David on the synoptic air flow. In addition, O_3 levels were generally lower by the time the episode reached Louisiana.

On September 3, a relatively small but intense area of haze and O_3 was situated over the Ohio Valley states from Illinois to western Pennsylvania. Peak O_3 exceeded 200 ppb in southeastern Indiana and visibilities less than 8 km covered much of the area. The air flow on September 3 was from the west-southwest but this changed 180° by September 4 (see Figure 8) as the haze and high O_3 from the Ohio Valley and Northeast moved towards the southwest. By September 5, the leading edge of the haze was in northern Louisiana while the O_3 (for this episode the shaded areas of O_3 are greater than or equal to 60 ppb) reached central Arkansas.

As Hurricane David moved up the East Coast, the haze and O_3 moved southeastward on September 6. On September 7, the flow diverged toward the southeast and southwest and arrived at our Louisiana site. The 24 h sulfate concentration was 37 μg m^{-3} on September 7. The strong northerly flow, associated with the advancing Canadian high-pressure system on September 8, pushed most of the haze blob off the Gulf Coast and considerable improvement in the visibility was experienced (see Figure 1). The presence of the haze over the Gulf was also confirmed by satellite photographs. By September 10,

Fig. 8.   Episode III – Same as Figure 4 except for $O_3 \geq 60$ ppb.

the return flow around the Canadian high began advecting the haze northward through Louisiana and back towards the Midwest.

This episode again illustrates that midwestern and northeastern United States emissions contributed to the haze which covered most of the eastern United States. A more localized effect from the east Texas area was evident on September 5 over the Gulf Coast extending to the Florida Panhandle. On September 7, the ^{7}Be also increased, suggesting that this episode too, had a stratospheric contribution to the ground level O_3. Furthermore, an intense 500-mbar trough over the East Coast was associated with Hurricane David on September 4 to 9.

4. Summary and Conclusions

The haze which was observed on the Louisiana Coast appears to be primarily due to aged sulfate aerosols. The evidence presented strongly suggests that the sulfates were largely due to SO_2 emissions in the Northeastern and Midwestern United States. The haze initially formed over and downwind of these source areas and was transported to the Gulf Coast area. In the last two episodes, the haze was subsequently transported back to the Midwest source region.

Elevated levels of O_3 were also associated with the haze. Again, the Northeast and Midwest appear to be the most significant source areas for the regional O_3 episodes. Occasionally, O_3 formed from emissions in the east Texas area also appears to affect a large portion of the Gulf Coast.

In addition, each episode appeared to have a significant stratospheric air component because of the high levels of ^{7}Be. All three episodes followed the passage of upper-level troughs which produce stratospheric intrusions due to tropopause folding.

In conclusion, it appears that in order to model regional scale haze and sulfate episodes in the eastern United States, emissions for an area at least as large as the ones which appear in Figure 4 must be included. In addition, the model must be 3-dimensional and include vertical transport from the stratosphere.

Acknowledgments

The authors are grateful for the technical support provided by G. Morris, D. Pierson, M. Ruthkowsky, W. Scruggs, D. Stroup, and J. Zemla. Ammonium and ^{7}Be analysis were conducted by Robert Kohn and Robert Hill, respectively, from the Analytical Chemistry Department. Dr Walter Lyons and his staff at Mesomet provided some of the synoptic meteorological data while Dale Conventry of EPA kindly provided backward trajectories for our site.

References

Brunnschweiler, D. H.: 1952, *Vierteljahrsschrft der Naturf. Gesellschaft in Zurich* **97**, 42.
Clark, T. L.: 1980, *Atmos. Environ.* **14**, 961.
Countess, R. J., Wolff, G. T., and Cadle, S. H.: 1980, *J. Air Pollut. Control Assoc.* **30**, 1194.

Ferman, M. A., Wolff, G. T., and Kelley, N. A.: 1981a, *J. Environ. Sci. and Health* **A16**, 315.

Ferman, M. A., Wolff, G. T., and Kelley, N. A.: 1981b, *J. Air Pollut. Control Assoc.* **31**, 1074.

Groblicki, P. J., Eisinger, R. S., and Ferman, M. A.: 1975, 'Design of a Mobile Atmospheric Research Laboratory', General Motors Res. Labs. Publication GMR-1814, Warren, MI.

Groblicki, P. J., Countess, R. J., and Wolff, G. T.: 1981, 'Visibility Reducing Species in the Denver "Brown Cloud". Pt. I. Relationships between Extinction and Chemical Composition', *Atmos. Environ.* (in press); also available as GMR-3417.

Hidy, G. M., Mueller, P. K., and Tong, E. Y.: 1978, *Atmos. Environ.* **12**, 735.

Heffter, J. L. and Taylor, A. D.: 1975, 'A Regional-Continental-Scale', Tranport, Diffusion, and Deposition Model; NOAA Tech. Memo., ERL ARL-50, U.S. Dept. of Commerce, Silver Springs, MD.

Hodkinson, R. J.: 1966, 'Calculations of Color and Visibility in Urban Atmospheres Polluted by Gaseous NO_2', *Air and Water Pollut. Int. J.* **10**, 137.

Johnson, W. B. and Viezee, W.: 1981, 'Stratospheric Ozone in the Lower Troposphere: Part I – Presentation and Interpretation of Aircraft Measurements', *Atmos. Environ.* (in press).

Kelly, N. A., Wolff, G. T., and Ferman, M. A.: 1981, 'Background Pollutant Measurements in Air Masses Affecting the Eastern U.S. Part I. Air Masses Arriving from the Northwest', *Atmos. Environ.* (in press).

Liu, C. L., Kley, D., McFarland, M., Mahlman, J. D., and Levy II, H.: 1980, *J. Geophys. Res.* **85**, 7546.

Middleton, W. E. K.: 1963, *Vision Through the Atmosphere*, Univ. of Toronto Press.

Rosen, H., Hansen, A. D. A., Gundel, L., and Novakov, T.: 1978, *Applied Optics* **17**, 3859.

Viezee, W. and Singh, H. B.: 1980, *Geophys. Res. Let.* **7**, 805.

Vuckovich, F. M., Bach, W. D., Crussman, B. W., and King, W. J.: 1977, *Atmos. Environ.* **11**, 967.

Waggoner, A. P. and Weiss, R. E.: 1980, *Atmos. Environ.* **14**, 623.

Weiss, R. E., Waggoner, A. P., Charlson, R. J., and Ahliquist, W. C.: 1977, *Science* **195**, 979.

Wight, G. D., Wolff, G. T., Lioy, P. J., Meyers, R. E., and Cederwall, R. T.: 1978, in Morris, A. L. and Barras, R. C. (eds.), *Air Quality Meteorology and Atmospheric Ozone*, American Society of Testing and Materials, Philadelphia, PA, p. 445–457.

Wolff, G. T., Lioy, P. J., Meyers, R. E., Cederwall, R. J., Wight, G. D., Pasceri, R. E., and Taylor, R. S.: 1977a, *Environ. Sci. Technol* **11**, 506.

Wolff, G. T., Lioy, P. J., Wight, G. D., Meyers, R. E., and Cederwall, R. T.: 1977b, *Atmos. Environ.* **11**, 797.

Wolff, G. T., Lioy, P. J., Leaderer, B. P., Bernstein, D. M., and Kleinman, M. T.: 1979a, *Ann. NY Acad. Sci.* **322**, 57.

Wolff, G. T., Monson, P. R., and Ferman, M. A.: 1979b, *Environ. Sci. Tech.* **13**, 1271.

Wolff, G. T., Ferman, M. A., and Monson, P. R.: 1979c, *Geophys. Res. Letters* **6**, 637.

Wolff, G. T. and Lioy, P. J.: 1980, *Environ. Sci. Tech.* **14**, 1257.

Wolff, G. T., Lioy, P. J., and Wight, G. D.: 1980, *J. Environ. Sci. Health* **A15**, 183.

Wolff, G. T., Kelly, N. A., and Ferman, M. A.: 1981a, *Science* **211**, 703.

Wolff, G. T., Groblicki, P. J., Cadle, S. H., and Countess, R. J.: 1981b, in Wolff, G. T. and Klimisch, R. L. (eds.), *Particulate Carbon: Atmospheric Life Cycle*, Plenum Press, New York, NY, p. 297 (in press).

LAKE ONTARIO ATMOSPHERIC DEPOSITION
1969–1978

C. H. CHAN and K. W. KUNTZ

Water Quality Branch, Environment Canada, P.O. Box 5050, Burlington, Ontario, Canada, L7R 4A6

(Received May 20, 1981; Revised July 29, 1981)

Abstract. Nine years of atmospheric deposition data have been analyzed from six locations along the Canadian shore of Lake Ontario. Results indicate that atmospheric deposition is affected by large scale air masses which influence the entire northern shore, and local inputs which at times could mask the large scale air masses effects. The observed increase in acidity in atmospheric deposition in the early seventies may be due to an imbalance caused by a greater reduction in total suspended particles than in SO_2 emissions.

1. Introduction

Problems of eutrophication and chemical pollution in the lower Great Lakes have resulted in efforts to identify and control nutrient inputs into the Great Lakes. Investigations in Scandinavia, Northern Europe and the United States have demonstrated that atmospheric deposition is an important source of nutrients (Eriksson, 1966). The establishment of a network of precipitation stations along Lake Ontario in 1967 was intended to estimate the atmospheric contribution to the overall chemical loadings to Lake Ontario. Due to recent concerns over the adverse effects of acid precipitation, the long range transport of toxic organic contaminants, and the general deterioration of air quality, this network has taken on added significance. Bulk precipitation chemistry data have been recorded for over 10 yr at six stations on the Canadian shore of Lake Ontario. Initial results from the 1970 and 1971 data reported by Shiomi and Kuntz (1973) have confirmed that bulk precipitation is a significant source of N and P. This report summarizes 10 yr of chemical data and discusses the findings in relation to the temporal and spatial variation in the composition of atmospheric deposition in the Lake Ontario Basin.

1.1. Sampling location

The six sampling locations are shown in Figure 1. Two locations, Burlington and Toronto Island, are situated in large urban centers while the remaining four are located on the outskirts of small urban centres. The Burlington station is located near the Burlington Skyway Bridge, about 2 km away from the Stelco Steel Industrial Park. Traffic on the Bridge is heavy. The Toronto station is located at the Toronto Island Airport, not far away from the heavy traffic and the air pollution of downtown Toronto. The Woodbridge station is located on a metereological research station about 36 km north of Toronto's water front. Both the Trenton and Kingston stations are situated at the local airport with only light air traffic. The Ancaster station is in a rural area surrounded by farmland.

At the outset of this investigation, the overriding factor in siting criteria was the

Water, Air, and Soil Pollution **18** (1982) 83–99. 0049–6979/82/0181–083$02.55.

Copyright © 1982 by D. Reidel Publishing Co., Dordrecht, Holland, and Boston, U.S.A.

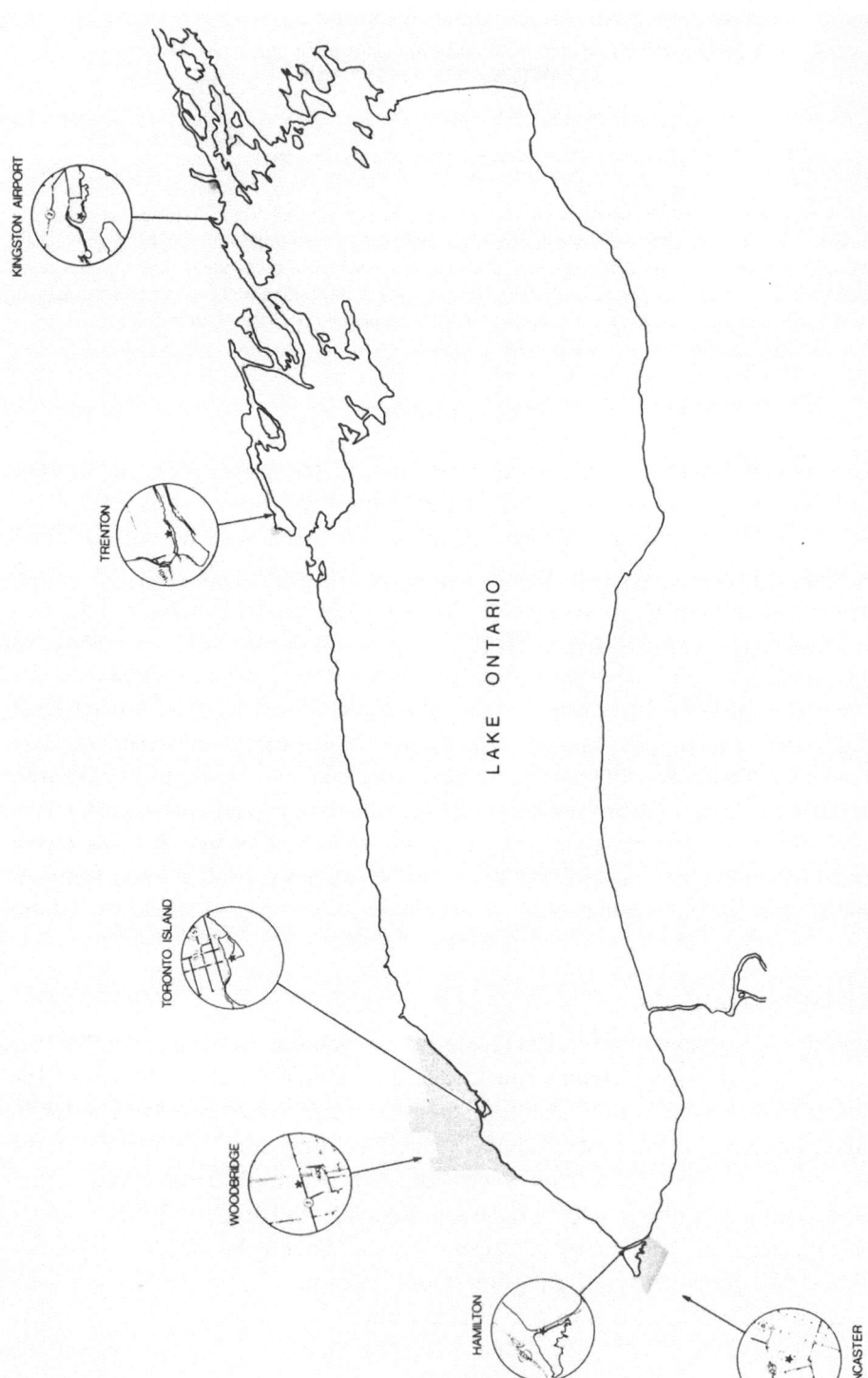

Fig. 1. Bulk precipitation sampler locations.

availability of power and accessability and proximity to a metereological station. Some of the locations do not meet the siting criteria of a chemical precipitation station as suggested in the WMO Manual (1971), and data from these stations were at times strongly influenced by local effects. It is intended that a number of these stations will be relocated.

1.2. SAMPLING METHODS

The sampling apparatus used in this study is the bulk precipitation sampler similar to that used by Gambell and Fisher (1966). Sketches of the sampler are shown in Figures 2 and 3. (Many of these samplers are now gradually being replaced by modified Finnish-type samplers which collect both wet and dry fallout.) The polyethylene storage bottle was rinsed with 50% nitric acid and then thoroughly rinsed with deionized water. A glass wool prefilter was placed at the base of colleccing funnel to filter out the coarser particulate matter. This practice was discontinued in mid 1974 because of possible ion exchange between the glass wool and the sample. The polyethylene storage bottle in each sampler was changed at the end of each month and the collected samples sent to the Burlington laboratory for chemical analysis. All the analysis were performed with no further filtration in the laboratory. Details of the analytical methodologies are described in Analytical Methods Manual (Environment Canada, 1974).

2. Results

The methodologies of collecting and analyzing the chemical composition of precipitation have been thoroughly discussed by Galloway and Likens (1978). An indepth evaluation of this type of sampler was discussed by Matheson (1975). Since this type of sampler is open at all times, it collects both wet and dry fallout, along with many other extraneous materials. The sample collected in the container is also subjected to evaporation and re-entrailment of gaseous and particulate matter into the atmosphere. These and other factors contribute to a large inherent sampling error in the measurements. To facilitate data interpretation it was often necessary to exercise some form of subjective data screening in the treatment of extreme values. In this process, each extreme value was examined in relation to the average values, the frequency of occurrence, sample history and evidence and possibility of contamination. Data points which were orders of magnitude above or below the average value or low in frequency, were included in the discussion below but not in the computation. If there were sufficient reasons to believe that the sample had been contaminated (i.e., high conductivity, low pH and high chloride due to improper rinsing of containers), the data point was deleted.

The mode of continuous open sampler in reality measures atmospheric deposition. The proper unit for deposition is mass per unit area per unit time. The amount of deposition is dependent on many factors, and an important one is the amount of precipitation. The annual average concentrations discussed in this report have not been volume weighted because the Great Lakes Basin does not have any marked seasonality in precipitation (Phillips and McCulloch, 1972). Annual precipitation over the study period shows little variation from year to year (Table I).

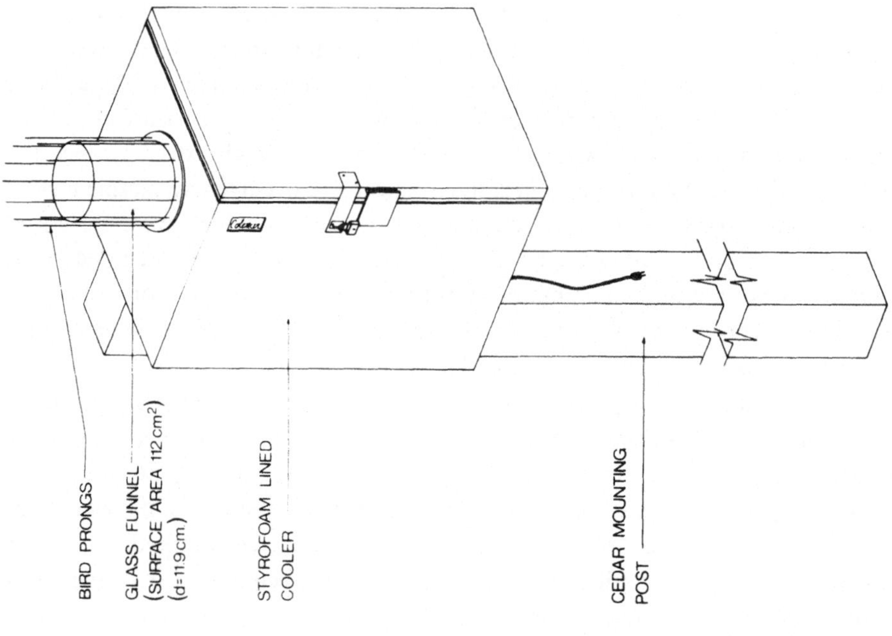

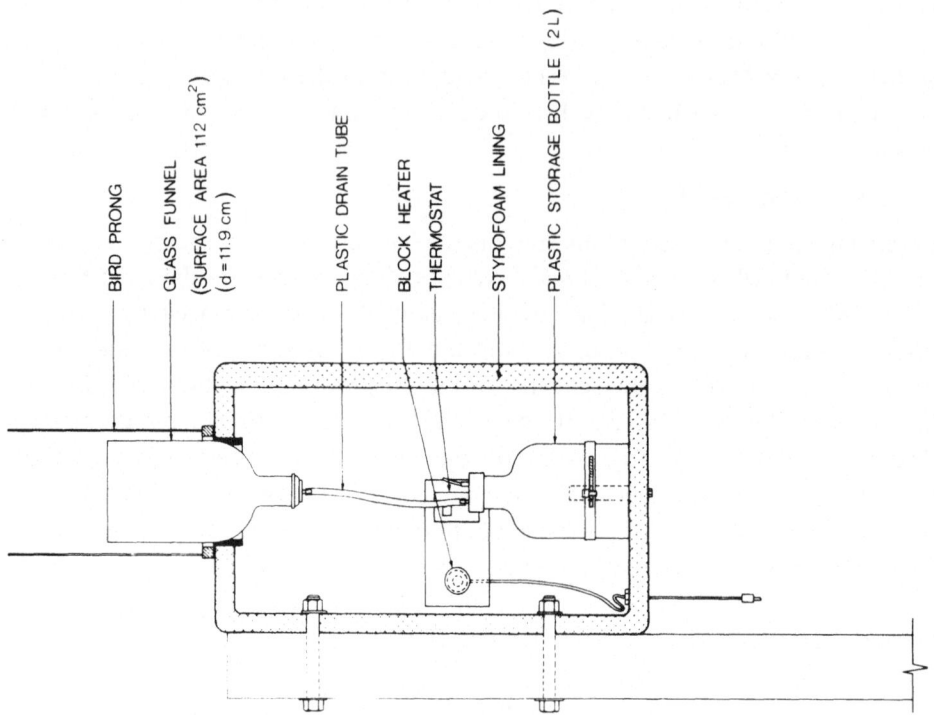

BIRD PRONG

GLASS FUNNEL
(SURFACE AREA 112 cm²)
(d = 11.9 cm)

PLASTIC DRAIN TUBE

BLOCK HEATER

THERMOSTAT

STYROFOAM LINING

PLASTIC STORAGE BOTTLE (2 L)

Fig. 3. Bulk precipitation sampler (interior).

BIRD PRONGS

GLASS FUNNEL
(SURFACE AREA 112 cm²)
(d = 11.9 cm)

STYROFOAM LINED
COOLER

CEDAR MOUNTING
POST

Fig. 2. Bulk precipitation sampler (exterior).

TABLE I

Annual average precipitation and catching efficiencies of bulk sampler 1972–1978

Station	1972	1973	1974	1975	1976	1977	1978
Average Standard Rain Guage (A) (cm)							
Ancaster	8.56	7.83	7.33	8.71	7.95	7.45	7.07
Toronto	8.57	7.70	6.75	6.70	6.67	9.17	6.11
W'Bridge	7.49	6.48	6.00	5.56	6.60	8.57	6.93
Trenton	8.79	8.20	7.14	6.68	7.20	—	—
Kingston	9.51	8.02	7.94	7.73	9.40	8.19	7.91
Annual Catch of Bulk Sampler (B) (cm)							
Ancaster	6.99	6.53	5.78	7.43	6.72	7.36	5.28
Toronto	7.65	6.23	5.57	6.37	—	8.03	4.88
W'Bridge	5.24	4.95	4.34	3.96	5.77	7.06	5.47
Trenton	6.45	6.68	5.28	5.20	4.37	—	—
Kingston	7.38	6.75	6.03	6.03	6.95	5.69	5.09
Ratio B/A of Catch							
Ancaster	0.73	0.83	0.79	0.85	0.84	0.98	0.75
Toronto	0.82	0.81	0.82	0.95	—	0.88	0.80
W'Bridge	0.70	0.76	0.72	0.71	0.86	0.82	0.78
Trenton	0.73	0.81	0.74	0.78	0.61	—	—
Kingston	0.80	0.84	0.76	0.78	0.74	0.69	0.64

2.1. PRECIPITATION

Table I compares the precipitation catch of each sampler with those collected by standard rain gauges. The data showed that in every case, the catch of the sampler was lower than the standard rain gauge. The catching efficiency of the sampler was between 75 to 90%. This was to be expected since some loss of precipitation would occur due to evaporation over the 1 mo sampling period.

2.2. Specific conductance

Specific conductance of rain water samples is very sensitive to the level of salt content, particularly at low concentrations. Typical specific conductance values in wet atmospheric precipitation are between 30 to 80 μSie cm^{-2}. Specific conductance exceeding 100 μSie cm^{-2} can usually be traced to either road salts or acid contamination. Stations located near urban centers usually have higher readings and display a prominent seasonal pattern as shown in Burlington (Figure 4) which has the highest average specific conductance.

2.3. pH

Annual averages (arithmetic) of pH and the ranges of atmospheric deposition are shown in Figure 6. Average annual pH lies between 4.5 to 7.0. Typical pH readings taken from wet precipitation in Lake Ontario Basin between 1977–1980, were between 4.5 to 5.0 (WQB, 1980). pH readings can vary by 3 pH units within any single year. This wide range

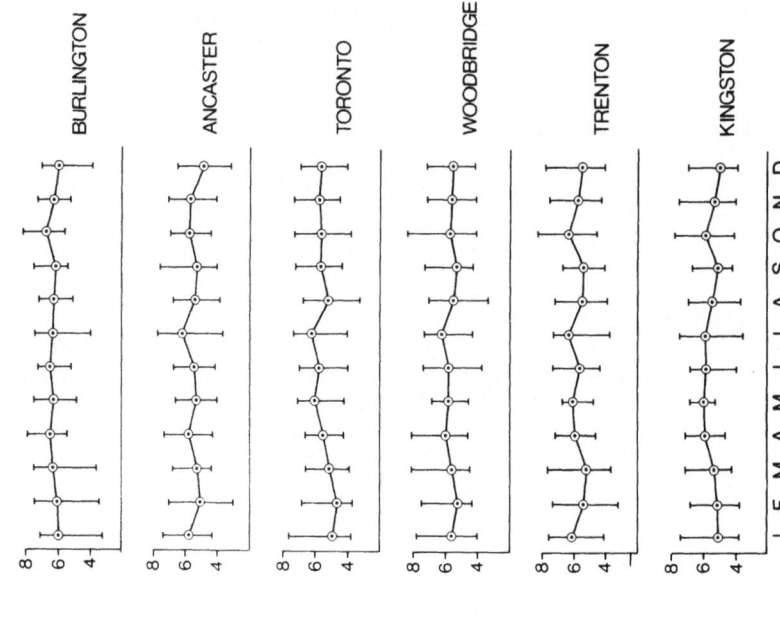

Fig. 5.   Seasonal variation of pH, 1969–1978 (⊙ mean, I range).

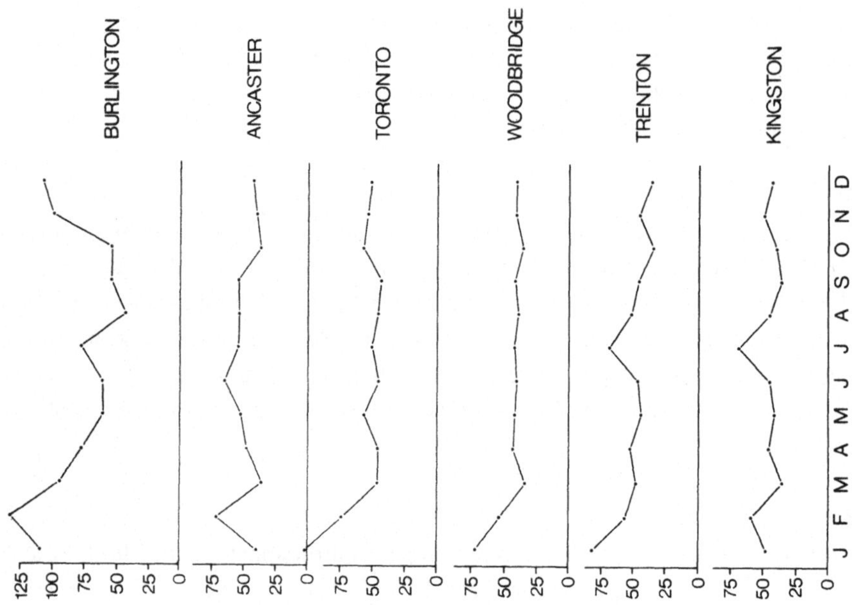

Fig. 4.   Seasonal variation of specific conductance (µSie cm⁻¹) for 1969–1978.

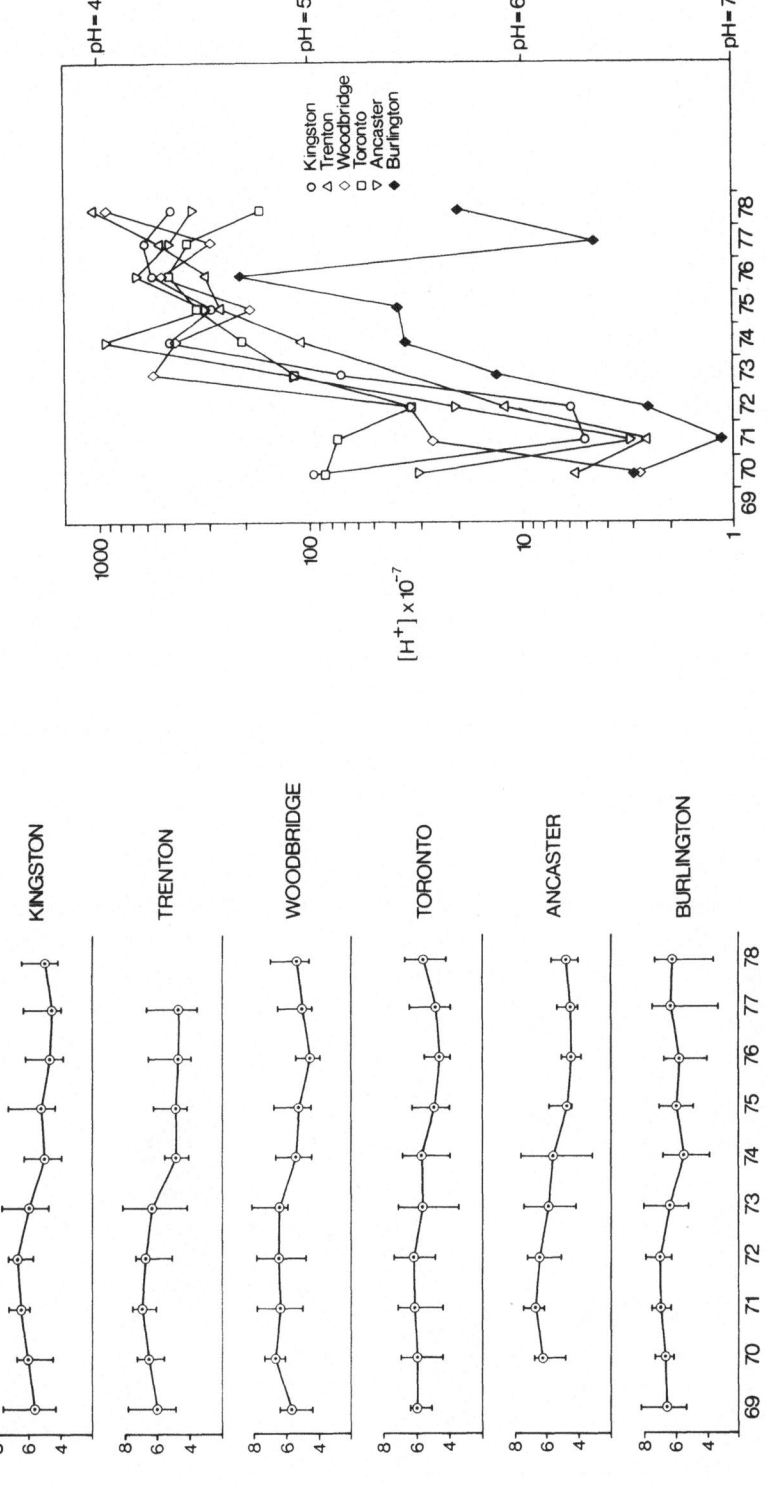

Fig. 7.   Annual average Hydrogen Ion concentration.

Fig. 6.   Annual average pH in Lake Ontario atmospheric deposition (⊙ mean, I range).

of pH is undoubtedly caused by the influence of wind blown dust and other alkaline particles which raises the pH in the sample. The amount of soil originated dust will be related to the amount of soil cover. A seasonal plot of pH (Figure 5) shows slightly higher pH in the summer months.

The time plots (Figure 6) of pH show decreases in pH during 73 to 75 at all stations. The changes in pH are more dramatic when these changes are plotted as H^+ concentrations (Figure 7). The biggest change occurred around 1974 which coincided with a procedural change of removing glass wool filters from the funnel. This raises the question whether the change in pH is an artifact or a real change. Laboratory experiments have shown that the pH of a rain sample was not affected by the same type of glass wool filter. Stensland (1979) reported a pH change from 6.05 in 1954 to 4.1 in 1977 in atmospheric precipitation collected in Central Illinois. He attributed the higher pH in 1954 to higher Ca^{++} and Mg^{++} levels in atmospheric aerosols which acted as neutralizing agents. He further suggested that the higher cations levels were due to dry weather in the midwest States at the time. A change of pH from 6.5 to 4.5 amounts (Figure 7) to 30.6×10^{-6} equivalence of H^+ which translates into 0.612 mg of Ca. Pre-1974 data showed Ca and Mg were higher by about 1.5 mg l^{-1}, which is more than sufficient to buffer the pH change. Therefore, it is possible that the observed change in pH is a consequence of changes in alkaline dust concentration. The removal of the glass wool filter should enhance the entry of dry deposition thereby raising the pH. However, this was not the case. Post-1974 pH actually declined after the glass wool had been removed. If there were any particulates being deposited in the sampler, the glass wool filter was probably not very effective in screening them out.

2.4. MAJOR CATIONS

Essentially, the same time trend is characterized by all the cations at all locations (except Burlington) with sharp decreases in the annual average concentrations between 1972 and 1973. Pre-1974 annual average Na concentration ranged from 1.0 to 4.0 mg l^{-1} while post 1974 average Na concentrations were consistently below 1.0 mg l^{-1} (Figure 8).

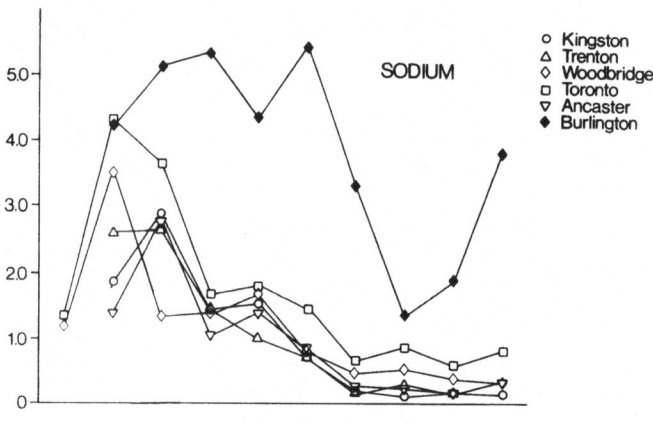

Fig. 8.   Annual average concentration of Na (mg l^{-1}).

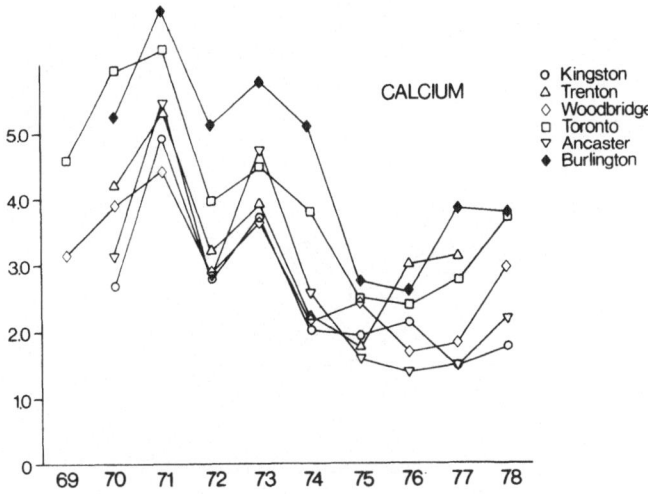

Fig. 9.   Annual average concentration of Ca (mg l⁻¹).

Annual average calcium concentration before 1974 ranged from 3.0 to 5.0 mg l⁻¹ but seldom exceeded 3.0 mg l⁻¹ after 1974 (Figure 9). Similar behavior is evident for K and Mg (Figures 10, 11). Rutherford (1967) in 1965 recorded the chemical composition of

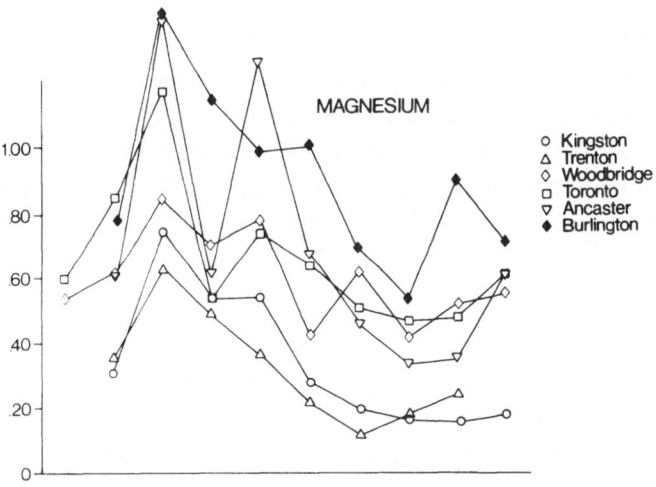

Fig. 10.   Annual average concentration of Mg (mg l⁻¹).

precipitation in south eastern Ontario near Kingston. Sodium contents varied from trace amounts to 6.3 mg l⁻¹ in rainfall, and from 0.6 to 14 mg l⁻¹ in snow, while Ca contents varied from trace amounts to 12.0 mg l⁻¹ in rainfall and 2.0 to 13.0 mg l⁻¹ in snow. Although the ranges of these two data sets were comparable, the post 1974 cations concentrations seem low by comparison since they were bulk samples. Sources of major ions in the atmosphere are wind blown dust, sea spray, volcanic eruption and emission from industrial activities. Because of the distance from the coastline, it is doubtful that

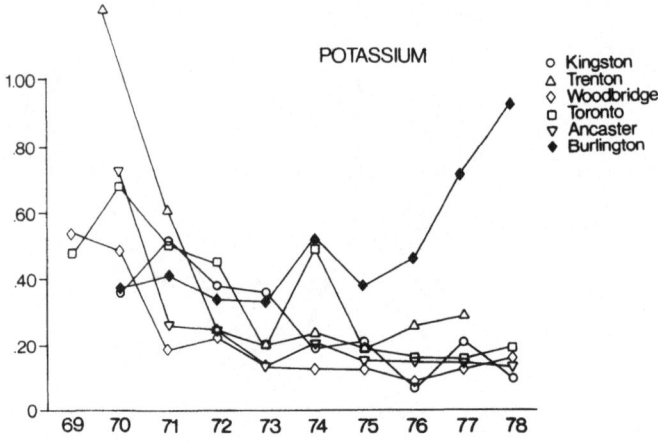

Fig. 11.   Annual average concentration of K (mg l^{-1}).

sea spray would be a major contributing factor in the Lake Ontario Basin. A search through meteorological records from regions North West and South West of Lake Ontario failed to uncover any metereological phenomena which would increase soil dust concentrations in the atmosphere. Thus, it seems most likely that the observed changes were humanly induced.

2.5. ANIONS

Annual average sulphate concentration varied between 5.0 to 8.0 mg l^{-1} over most of the study period with slightly higher levels in 1970 and 1971 (Figure 12). Unlike the major cations, the abrupt decrease in concentrations between 1972 and 1974 is not found for anions. Sulphate in the atmosphere is largely related to SO_2 emission from the burning of fossil fuels. This is reflected in the higher sulphate levels in Burlington and Toronto.

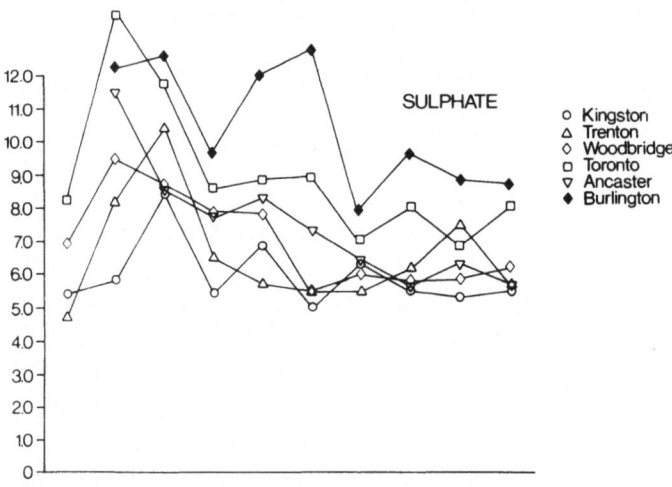

Fig. 12.   Annual average concentration of sulphate (mg l^{-1}).

The levels of sulphate in wet precipitation have been demonstrated to have a positive correlation with acidity (Cogbill and Likens, 1974). Linear correlation calculations of the data collected in this study showed that sulphate and pH are poorly correlated whereas sulphate and Ca are highly correlated with a correlation coefficient of about 0.8. These results indicate that most of the sulphate being deposited is mostly associated with alkaline dust particles.

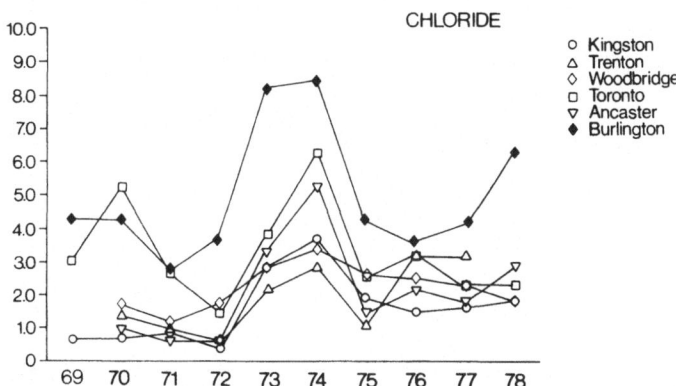

Fig. 13.   Annual average concentration of chloride (mg l^{-1}).

Time plots of chloride (Figure 13) showed little resemblance to any other cations. They show a definite increase in concentration from 1.0 mg l^{-1} in the early seventies to 2.0 mg l^{-1} in the late seventies with peak concentrations of 3.0 to 6.0 mg l^{-1} in 1974. This temporal behavior of chloride concentration is difficult to explain. Ocean sea spray is the major source of chloride in the atmosphere, and the Cl/Na ratio has been used extensively to interpret the origins of other cations and anions in the atmosphere (Junge, 1963). The Cl/Na ratio varied from 0.3 in the early seventies to 4.0 in the late seventies. The expected Cl/Na ratio for sea water is 1.8. Most of the chloride data is probably affected by the use of road salts in the winter months as illustrated by the strong seasonal pattern (Figure 14). The strong seasonal pattern in the chloride data suggests that there is contamination of the precipitation samples by the road salts used extensively in Lake Ontario during the winter months.

2.6. REACTIVE SILICATE

Annual average reactive silicate concentrations show the biggest decline during the study period (Figure 15). Average silicate concentrations were higher than 3.0 mg l^{-1} before 1974 and abruptly decreased to less than 0.2 mg l^{-1} after 1975, when the use of the glass wool prefilter was discontinued. The only other reported silicate concentration measurements were those by Rutherford, in south eastern Ontario, in 1965. The values ranged between 0.08 to 0.56 mg l^{-1}. At first glance it seemed that the high pre-1974 silicate data were due to silicate contamination from the pyrex glass wool. However, laboratory experiments have been unable to support this theory. Wind blown dust of soil origin, composed mainly of mineral silicate could be another contributing factor. The fact that this change occurred over the whole basin would suggest that the dust is of distant origin.

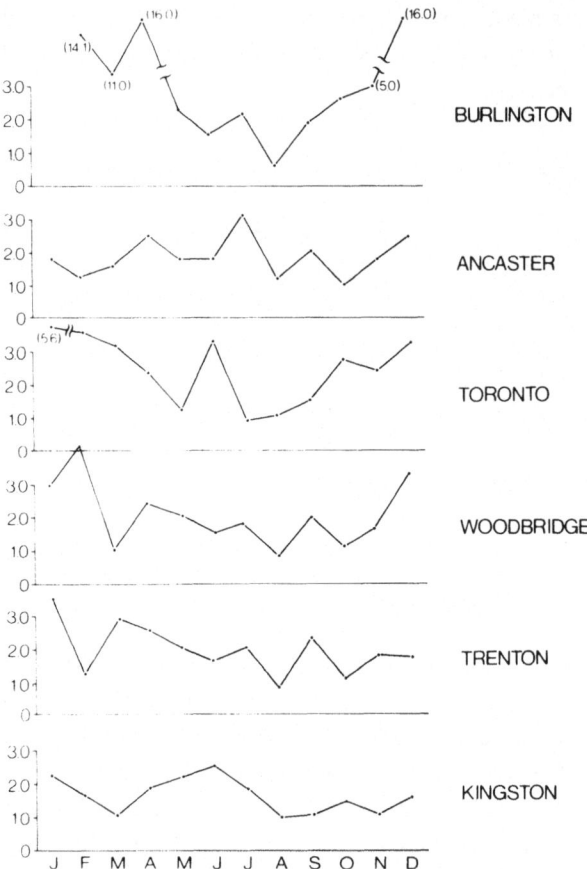

Fig. 14.   Seasonal variation of chloride, 1969–1978 (mg l^{-1}).

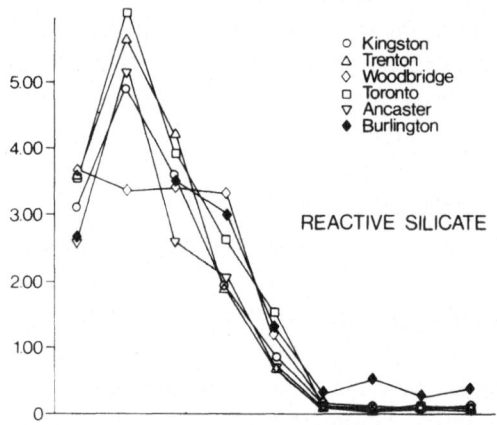

Fig. 15.   Annual average concentration of reactive silicate (mg l^{-1}).

2.7. PHOSPHORUS

Atmospheric P loading to the Great Lakes has been a subject of many investigations (Acres, 1975; Eisenrich *et al.*, 1977; Murphy, 1976). Murphy (1974) in his Lake Michigan study found that the sources of the P in the atmosphere are general in nature and he was not able to identify any single large source. Levels of P found by this study varied from $0.020 \, \text{mg} \, l^{-1}$ P to over $0.300 \, \text{mg} \, l^{-1}$ (Figure 16) with no identifiable temporal and spatial

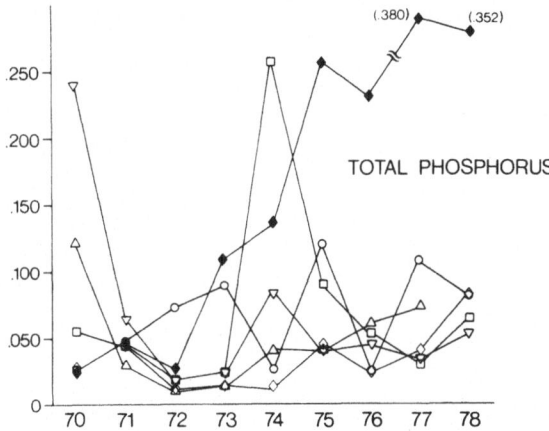

Fig. 16. Annual average concentration of total P ($\text{mg} \, l^{-1}$).

patterns. Bulk deposition samples are very susceptible to contamination from organic debris such as bird droppings, pollen, insects, etc., which are rich in P. This problem of high P contamination from organic sources is illustrated in the seasonal plots of P (Figure 17). In the April to October period, P readings are many times higher than the winter months. Phosphorus concentrations at Burlington approached the ppm level. This seasonal dependency will have a significant impact on estimating atmospheric P loading to the open lakes using data collected from bulk samplers. This type of contamination should decrease considerably some distance off the shore. How much of the P from the organic fraction can be considered as legitimate inputs? A more realistic estimate may be represented by those values recorded in the winter months when sample contamination is minimal.

2.8. NITROGEN

The N components analyzed in the samples are nitrate plus nitrite, and ammonium ions, both of which are major and active species in the N cycle. Unless the samples are kept at low temperatures and away from sunlight, transformation between various N species can occur. Geographically, Toronto and Burlington showed higher concentrations than the other stations. Both of the two N fractions show little variation with time and no apparent seasonal pattern (Figure 18). In view of the changes which occurred with the other chemical parameters, this static pattern is unexpected. It is possible then with the

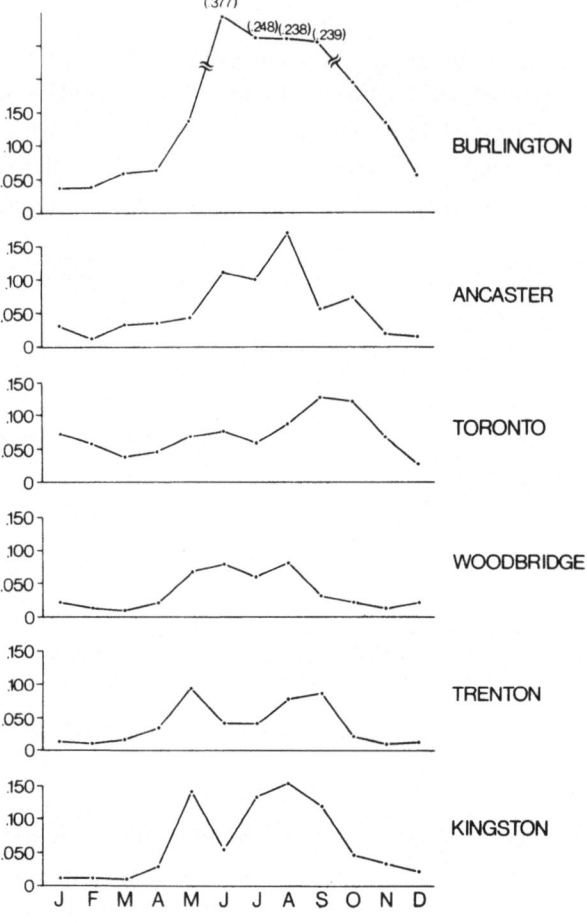

Fig. 17.   Seasonal variation of total P, 1969–1978 (mg l^{-1}).

sample left under field conditions for weeks, the N measured at the time of analysis was equilibrated values rather than the true value at the time of deposition.

3. Discussion

There are two features which are evident in these data. The first is that all the six locations have very similar variations over the study period, and the second is the abrupt change in the composition of atmospheric deposition in the early seventies.

The similarities in the temporal pattern at all the six locations suggest that the samples collected have similar origins and history and come from identical air masses. These air masses most likely exert their influences over the whole Lake Ontario Basin. The trajectories and origins of these air masses are important factors governing the transport and deposition of materials in the Great Lakes Basin. Most of the air masses reaching Lake Ontario come from the north west and south west of Ontario (Phillips and McCulloch, 1972). Materials being deposited within Lake Ontario most likely originated

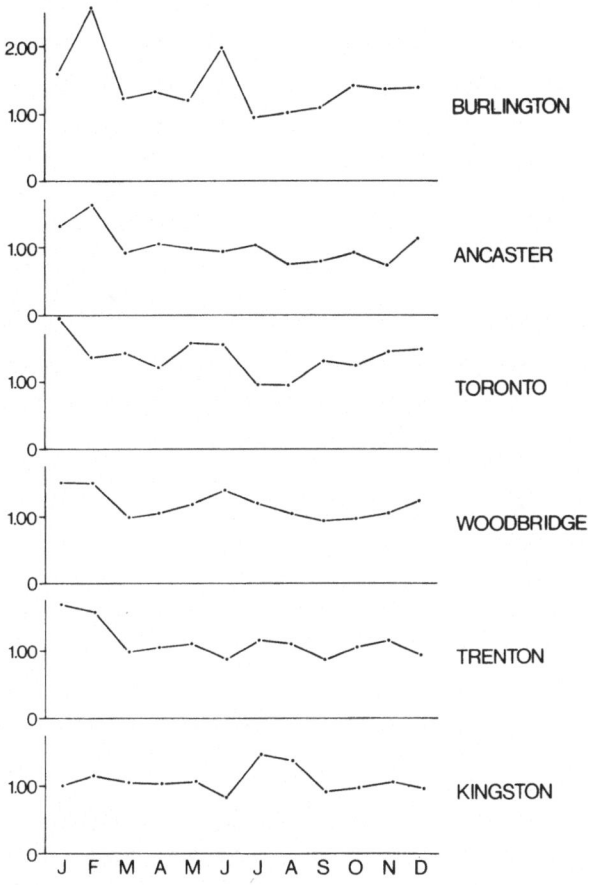

Fig. 18.   Seasonal variation of Nitrate + Nitrite, 1969–1978 (mg l^{-1}).

from outside the Basin. In addition to the long range transport of material, there are emissions from regional and local sources. The effects of these localized or regional inputs are superimposed on the effects of large scale air masses. These local inputs can, at times, mask the large scale air mass effect. The aerial extent of these local inputs is difficult to delineate. Unless some kind of weighting function is used to proportion the effects of urban and industrial inputs, atmospheric loading estimates could be biased high. Thus, depending on the objective of a particular study, siting of a station can be critical.

Determining the cause of the changes in the composition of atmospheric deposition in the early seventies is complicated by a procedural change in collection that was implemented at about the same time. However, it is doubtful that fiber glass wool would affect the chemistry of rainwater to such a large degree. This is confirmed by laboratory tests which show that the glass wool fiber had little effects on pH or major ions concentration although silicate concentration was shown to be higher. Furthermore, stoichiometric consideration show no inconsistencies in the chemical analyses. Thus, it appears that the observed changes were not due to sampling or analytical errors but were real. Air circulation and other meteorological conditions do change from year to year, and it is conceivable that the observed changes were legitimate short term variations. Climatic changes could bring about variations in air dust and aerosol concentrations in the atmosphere and thus alter the transport and deposition of material via the atmosphere. Although a search through meteorological records did not reveal any abnormal climatic condition which could bring about changes in dust concentrations in the air, a U.S. EPA report (1977) on air quality trends has shown that total suspended solids (TSP) in air have dropped significantly between 1970–1976 (Table II). The Great Lakes Basin showed the largest decrease from 1970 to 1972. Similar findings showing a 25% decrease in suspended particles from 1970 to 1977 were also reported in Canada (Environment Canada) (1978). Likens and Bormann (1974) suggested that the increase in rain acidity may have been associated with the augmented use of natural gas and with the installation of particle removal devices in smokestacks. Gorham (1955) and Overrein (1972) found that various basic particles in smoke can neutralize acids. Emission data (Table II) estimates show that TSP emission was reduced by 41.0% while SO_2 only

TABLE II

Summary of U.S. National Emission Estimate[a] (10^6 tonne yr^{-1})

	TSP	SO_x	NO_x	HC	CO
1970	22.6	29.1	20.4	29.7	99.8
1971	21.4	27.9	21.3	29.3	100.2
1972	20.3	28.8	22.2	29.7	102.0
1973	19.9	29.7	22.9	29.8	98.3
1974	17.5	28.2	22.6	28.6	91.5
1975	14.4	25.7	22.2	26.2	85.9
1976	13.4	26.9	23.0	27.9	87.2

[a] Taken from U.S. EPA report EPA-450/1-77-002

showed an 8% reduction. TSP have been removed more effectively than SO_2 from air emissions. A reduction of TSP would lower basic particles in the atmosphere and reduce the buffering capacity in the atmosphere. This creates an imbalance in the overall buffering capacity of atmospheric aerosols. Therefore, a similar reduction in SO_2 is necessary to maintain a balance between alkaline and acidic particles in the atmosphere.

Acknowledgments

The authors wish to thank Mr F. C. Elder and Mr M. T. Shiomi for their advice and review of this paper.

References

Acres Consulting Services: 1975, 'Atmospheric loading of the Upper Great Lakes'. Report to Canada Centre for Inland Waters.

Cogbill, C. V. and Likens, G. E.: 1974, *Water Resources Res.* **10**, 1133.

Eisenreich, S. J., Emmling, P. J. and Beeton, A. M.: 1977, *Internat. Assoc. Great Lakes Res.* **3**, 291.

Environment Canada: 1974, 'Analytical Methods Manual', Water Quality Branch, Inland Waters Directorate, Ottawa, Ontario.

Environment Canada: 1978, 'National Air Quality Trends 1970–1977', Air Pollution Control Directorate, Report No. EPS 5-AP-78-27.

Eriksson, E.: 1966, 'Air and Precitation as Source of Nutrients', in *Handbuch der Pflanzenernahrung und Dungung*, Band II: Boden und Dungemittel, 774–792, Springer-Verlag. New York 774–792.

Galloway, J. N. and Likens, G. E.: 1978, *Tellus* **30**, 71.

Gambell, A. W. and Fisher, D. W.: 1966, 'Chemical Composition of Rainfall Eastern North Carolina and Southeastern Virginia', Geological Survey Water-Supply Paper 1535-K, 41 pp Washington, D.C.

Gorham, E.: 1955, *Geochim. Cosmochim. Acta* **7**, 231.

Junge, C. E.: 1963, '*Air Chemistry and Radioactivity*', Academic Press, New York.

Likens, G. E. and Bormann, F. H.: 1974, *Science* **184**, 1176.

Matheson, D. H.: 1975, 'Measurement of Atmospheric Inputs to the Great Lakes', Report to Canada Centre for Inland Waters. Contract OSR3-0130.

Murphy, J. J.: 1974, 'Sources of Phosphorus Inputs from the Atmosphere and their Significance to Oligotrophic Lakes', University of Illinois, Research Report No. 92 UJLC-WRC-74-0092.

Murphy, J. J.: 1976, *Internat. Assoc. Great Lakes Res.* **2**, 127.

Overrein, L. N.: 1972, *Ambio.* **1**, 145.

Phillips, D. W. and McCulloch, J. A. W.: 1972, *Atmospheric Environment*, Climatological Studies 20.

Rutherford, G. K.: 1967, *Can. J. Earth Sci.* **4**, 1151.

Shiomi, M. T. and Kuntz, K.: 1973, 'Great Lakes Precipitation Chemistry, Part I, Lake Ontario Basin', Proc. 16th Conf. Great Lakes Res., Internat. Assoc. Great Lakes Res. 581.

Stensland, G. J.: 1979, 'Precipitation Chemistry Trends in Northeastern United States', presented at the 12th Annual Rochester International Conference on Environmental Toxicity, May 1979.

U.S. EPA: 1977, 'National Air Quality and Emissions Trends Report 1976', EPA-450/1-77-022, December 1977.

Water Quality Branch, Ontario Region: 1980, Unpublished data.

WMO: 1971, 'WMO Operations Manual, Sampling and Analysis Techniques for Chemical Constituents in

ESTIMATION OF WET AND DRY DEPOSITION OF POLLUTANT SULFUR IN EASTERN CANADA AS A FUNCTION OF MAJOR SOURCE REGIONS

J. D. SHANNON*

Radiological and Environmental Research Division, Argonne National Laboratory, Argonne, IL 60439 U.S.A.

and

E. C. VOLDNER

Atmospheric Dispersion Division, Atmospheric Environment Service, Downsview, Ontario

(Received 6 July, 1981; Revised 2 October, 1981)

Abstract. The contributions of major anthropogenic source regions to wet and dry deposition of total S in eastern Canada are estimated for a winter month and a summer month with the ASTRAP model. Results indicate that the U.S. and Canada contribute approximately equal amounts to total S deposition in Canada; Canadian sources contribute more than one half of dry deposition and less than one half of wet deposition.

The significance of the 'acid rain' issue is well established, and need not be elaborated upon here, except to note that the delivery of acidic pollutants to the surface by precipitation is but one component of the problem; dry deposition should also be considered. A comprehensive comparison and verification of several models of long-range transport and deposition of air pollution is underway by a joint U.S./Canadian work group, in preparation for a treaty on transboundary air pollution (U.S. Department of State and Canadian Deprtment of External Affairs, 1980). An important goal of the work group is determination of source/receptor relationships. The approach taken is to estimate, through simulation, the contributions of a number of emission source regions to pollutant S concentrations and deposition amounts at specific receptor points located in sensitive areas. Our approach here is to integrate S deposition over the eastern Canadian provinces. While pollutant S is not the sole source of acid deposition, it is probably the main source.

The Advanced Statistical Trajectory Regional Air Pollution (ASTRAP) model used here has been described elsewhere (Shannon, 1981). This application differs from past uses in that the meteorological data are from 1978 rather than 1974–1975, emission data are more complete, wet deposition is constrained to be a maximum of 80% of the airborne mass during any 6-hr period and is proportional to $(h/10)^{1/2}$ where h is the 6-hr precipitation in mm h^{-1}, and the primary sulfate emission factor approximates variations with fuel type and usage (Homolya and Cheney, 1978).

The emission inventory used in these simulations is the Multistate Atmospheric Power Production Pollution Study (MAP3S) inventory (Benkovitz, 1980), which is representative of the late 1970's. Point sources have seasonal emission factors and area sources

* Work supported by the U.S. Environmental Protection Agency,

TABLE I

Contribution to eastern Canada S deposition in January

Source	Emissions (kT S)	Receptor regions												
		E. Canada			Ontario			Quebec			Atlantic provinces		
		Wet	Dry (kT S)	Total	Wet	Dry (kT S)	Total	Wet	Dry (kT S)	Total	Wet	Dry (kT S)	Total
Ohio Valley	400	23.0	3.6	27.0	12.0	2.4	15.0	9.1	1.0	10.0	1.5	0.3	1.8
Upper Midwest	270	12.0	4.4	16.0	7.8	3.2	11.0	3.6	1.0	4.6	0.4	0.2	0.6
Mid-Atlantic	100	4.9	0.3	5.2	1.4	0.1	1.5	2.7	0.9	3.6	0.9	0.1	1.0
New England	26	1.7	0.4	2.1	0.1	0.0	0.1	1.3	0.2	1.5	0.2	0.3	0.5
South	230	0.2	1.8	2.0	1.3	0.1	1.4	0.5	0.1	0.6	0.1	0.0	0.1
Eastern U.S.A.	1000	42	10	51	23	6	29	17	3	20	3	1	4
Ontario	87	18.0	8.2	26.0	13.0	6.6	20.0	5.0	1.4	6.4	0.2	0.2	0.4
Quebec	52	13.0	9.1	22.0	5.1	2.6	7.7	7.7	5.9	14.0	0.3	0.6	0.9
Atlantic provinces	10	1.1	1.0	2.1	0.0	0.0	0.0	0.6	0.1	0.7	0.5	0.9	1.4
Eastern Canada	150	32	18	50	18	9	27	13	7	20	1	2	3
Eastern N.A.	1200	73	28	100	41	15	56	30	10	40	4	3	7

TABLE II

Contribution to eastern Canada S deposition in July

Source	Emissions (kT S)	Receptor regions											
		E. Canada			Ontario			Quebec			Atlantic provinces		
		Wet	Dry (kT S)	Total	Wet	Dry (kT S)	Total	Wet	Dry (kT S)	Total	Wet	Dry (kT S)	Total
Ohio Valley	340	19.0	5.0	24.0	7.3	3.4	11.0	8.8	1.3	10.0	2.4	0.3	2.7
Upper Midwest	260	29.0	8.9	38.0	17.0	7.0	24.0	12.0	1.8	14.0	0.6	0.1	0.7
Mid-Atlantic	93	3.4	1.1	4.5	0.3	0.2	0.5	1.4	0.6	2.0	1.7	0.4	2.1
New England	21	1.1	0.8	1.9	0.0	0.0	0.0	0.5	0.2	0.7	0.6	0.6	1.2
South	230	1.3	0.2	1.4	0.6	0.1	0.7	0.4	0.0	0.4	0.2	0.0	0.2
Eastern U.S.A.	940	54	16	70	25	11	36	23	4	27	6	1	7
Ontario	82	32.0	17.0	49.0	13.0	10.0	23.0	18.0	6.1	24.0	1.6	0.5	2.1
Quebec	44	13.0	12.0	25.0	1.6	1.2	2.8	11.0	10.0	21.0	0.7	0.7	1.4
Atlantic provinces	8	0.5	1.2	1.7	0.0	0.0	0.0	0.1	0.1	0.2	0.4	1.1	1.4
Eastern Canada	130	46	30	76	15	11	26	29	16	45	3	2	5
Eastern N.A.	1100	100	46	150	40	22	62	52	20	72	9	3	12

are assumed to have emission factors of 1.2 in January and 0.8 in July. The sources in eastern North America are combined into five U.S. regions and three Canadian regions, as shown in Tables I or II. The Ohio Valley region here consists of Ohio, Indiana, Kentucky, West Virginia, and western Pennsylvania; definitions of the other source regions listed are straightforward.

Tables I and II show the source/receptor relationships for S deposition as simulated by ASTRAP for January and July, 1978, months chosen because of the availability of convenient meteorological data sets. The simulations indicate that total S deposition in eastern Canada can be attributed about equally to the U.S. and to Canada; the relative share of Canadian sources in Canadian dry deposition is somewhat greater than in wet deposition. The relative Canadian contributions for the two months are about the same. The major U.S. source regions contributing to Canadian S deposition are the Ohio Valley and the upper Midwest, primarily because their positions are adjacent to and upwind (southwest) of southeastern Canada and their emission densities are high. The South is more distant, and the Mid-Atlantic and New England source regions are usually downwind of most of eastern Canada. (The latter regions do account for almost 30% of the S deposition in the Atlantic Provinces during July, however). The most influential Canadian source region is Ontario, which has the greatest emissions as well as a location to the west or upwind side.

It is extremely difficult to establish the confidence range of the simulation results. Net horizontal mass flux and dry deposition are not monitored. Wet deposition is sampled at a growing number of sites, but problems of collector efficiency and sample contamination make many of the data suspect. In addition, normal meteorological variations, particularly in precipitation and winds, add uncertainty to extrapolations from single months. It is the authors' opinion that the accuracy of ASTRAP simulations of areally integrated total S deposition over periods of a month or longer, under the assumption that the emission inventory is complete, is within a factor of two. Model comparisons in progress will serve to quantify better the accuracy of simulations.

References

Benkovitz, C.: 1980, 'MAP3S/RAINE Emission Inventory Progress Report', Brookhaven National Laboratory Report BNL 53178.

Homolya, J. B. and Cheney, J. L.: 1978, 'Workshop Proceedings on Primary Sulfate Emissions from the Combustion of Fossil Fuels', Vol. 2, EPA-600/9-78-020b, U.S. Environmental Protection Agency, Research Triangle Park, NC, 1978, pp. 3–13.

Shannon, J. D.: 1981, *Atmos. Environ.* **15**, 689.

United States Department of State and Canadian Department of External Affairs: 1980, 'Memorandum of Intent on Transboundary Air Pollution', (annex).

THE EFFECTS OF SULFATE AND NON-SULFATE PARTICLES
ON LIGHT SCATTERING AT THE MAUNA LOA OBSERVATORY

ALLEN C. DITTENHOEFER

Mauna Loa Observatory, P.O. Box 275, Hilo, HI 96720 U.S.A.

(Received 6 July, 1981; Revised 5 November, 1981)

Abstract. The results of a 1 yr sampling program designed to quantify the effects of sulfate and non-sulfate particles on light scattering measurements routinely made at the Mauna Loa Observatory in Hawaii are described. Aerosol sampling with a cascade impactor was conducted at an altitude of 3400 m above sea level near the summit of a large, gently sloping volcano during nocturnal downslope flow, when particles are believed representative of clean, mid-tropospheric background conditions. A microchemical spot test using a transmission electron microscope was applied to quantitatively identify individual sulfuric acid, ammonium sulfate, and non-sulfate particles. Simultaneous measurements of integrated light scattering with a four-wavelength nephelometer were also made.

The results of the study indicate that significant variations in aerosol chemistry and morphology occur at Mauna Loa Observatory over an annual period. Episodes of high particle scattering caused by large non-sulfate particles, presumably soil dust, occur mainly in the spring and are generally associated with strong northwesterly large scale flow. Evidence has been gathered suggesting that dust storms over the Asian continent largely account for the springtime turbidity peak in Hawaii. During other times of the year, the effects of sulfate and non-sulfate particles on light scattering are approximately equal.

1. Introduction

Continuous monitoring of atmospheric transmission, Aitken particle concentration, and particle light scattering has been an integral part of the NOAA Geophysical Monitoring for Climatic Change (GMCC) program for a number of years, with the purpose of establishing long term trends in global background turbidity for investigation of possible climatic effects. The Mauna Loa Observatory, located on the island of Hawaii on a gently sloping volcano at an altitude of 3400 m, is the oldest baseline station of a world-wide network. Previous investigations have shown that volcanically related secular trends exist in long term records of atmospheric transmission at Mauna Loa (Mendonca *et al.*, 1978). Distinct annual cycles in atmospheric transmission and aerosol light scattering have also been detected (Mendonca *et al.*, 1979; Hansen *et al.*, 1979) with no absolute explanation. Knowledge of the chemical and physical properties of particles in the optically active size range is crucial to our understanding of the causes of light scattering variations at baseline stations and their possible impacts on the global radiation budget.

Only a limited number of studies designed to investigate the nature of the aerosol at the Mauna Loa Observatory have been conducted in the past (Simpson, 1972; Bodhaine and Pueschel, 1972; Pueschel *et al.*, 1973; Bigg, 1977). These investigations have revealed distinct differences in the mass concentration and chemical composition between daytime and nighttime particle samples, due to a diurnally-varying local wind circulation pattern which is thermally induced by the solar heating and radiative cooling

* Current affiliation: Enviroplan, Inc., 59 Main Street, West Orange, N.J. 07052 U.S.A.

Water, Air, and Soil Pollution **18** (1982) 105–121.   0049–6979/82/0181–0105$02.55.
Copyright © 1982 by D. Reidel Publishing Co., Dordrecht, Holland, and Boston, U.S.A.

of Mauna Loa. Daytime upslope air was generally found to be humid and relatively rich in particles of anthropogenic, oceanic, vegetative, or volcanic origin, showing an abundance of sea salt, ammonium sulfate, organics, and crustal elements. During nocturnal subsidence, the air was found to be considerably cleaner and drier, containing a continentally-derived, aged aerosol representative of background tropospheric conditions. Downslope flow showed a predominance of mixed sulfuric acid aerosols having crystalline or amorphous solid centers (Bigg, 1977), implicating atmospheric sulfur (S) compounds of possible anthropogenic origin as important components of the global tropospheric background aerosol.

Recent evidence of the long range atmospheric transport of Asian desert aerosol to the tropical North Pacific has also been gathered (Duce *et al.*, 1980; Bodhaine *et al.*, 1981). These studies have shown that dense hazes due to high concentrations of soil dust over this region are observed most often in the spring, when meteorological conditions favor the frequent occurrence of dust storms in northern China and Mongolia.

Thus, both anthropogenic and natural sources are believed to contribute to the tropospheric background aerosol observed at Mauna Loa Observatory. In order to gain a better understanding of the chemical and physical properties of these aerosols, which will aid in elucidating the relative importance of man-made versus natural sources and possible climatic impacts, the following study was conducted.

2. Experimental Methods

Approximately 75 aerosol samples were collected during the nighttime downslope flow at the Mauna Loa Observatory over a 12-mo period. Sampling was conducted using a Casella cascade impactor situated on a raised platform approximately 5 m above the ground in a remote area of the observatory.

Particles within the size range 0.1 to 2.0 μm diam were collected for sizing, morphological identification, and sulfate analysis using a modification of the technique described in Mamane and de Pena (1978). Three copper electron microscope screens which had been coated with a formvar membrane and a carbon film were placed onto the fourth stage of the impactor directly below the jet nozzle. A sampling time of 1 h was generally found to be sufficient to obtain a representative sample of aerosol. After particle collection, a layer of approximately 300 Å thick of $BaCl_2$ was vacuum evaporated onto two of the screens. The fist screen (screen No. 1) was developed for 1 h over a NaCl-saturated solution (75% RH) prior to examination under a transmission electron microscope. At this relative humidity it is expected that all chemical forms of sulfate aerosol will react with the $BaCl_2$ to produce the characteristic reaction spots, or halos. The other $BaCl_2$-coated screen (screen No. 2) was kept in a sealed desiccator containing $CaSO_4$ desiccant and received no humidification. At such a low relative humidity only sulfuric acid will react with $BaCl_2$. The third screen (screen No. 3) did not receive treatment of any kind and was placed directly into the electron microscope and examined at low magnification and low beam intensity to minimize particle evaporation.

A series of electron micrographs were taken across the line of impacted particles, and the particles were counted, sized, and categorized by type. Particle concentrations within a specific size interval were computed from the number of particles counted on the photomicrographs, the total picture area, the dimensions of the jet nozzle, the sampling time (generally 1 h), the flow rate through the impactor ($17.5\ l\ min^{-1}$), and the particle impaction efficiency. The impaction efficiency curve for the fourth stage of the Casella impactor was constructed using the method described in Dittenhoefer and de Pena (1980).

The size distribution of non-sulfate particles observed on screen No. 1 was subtracted from the size distribution of all particles as measured from screen No. 3 to give the size distribution of sulfates. This approach precludes the use of halo-to-particle size ratios which, for the case of sulfuric acid, is a function of particle size (Mamane and de Pena, 1978). The relative number of sulfuric acid droplets compared to total sulfate particles was obtained by comparing the number of reaction spots on screens No. 1 and No. 2.

Computer-generated averages of particle light scattering, total Aitken nuclei concentration, wind speed and direction, temperature, and dew point corresponding to the exact impactor sampling time period were taken from the observatory ICDAS computer. A four-wavelength integrating nephelometer provided continuous light scattering measurements at wavelength bands centered at 450, 550, 700, and 850 nm (Bodhaine, 1978). In order to separate the aerosol scattering from the molecular scattering of air and instrument background, the Mauna Loa nephelometer operates in an AIR CHOP mode, i.e., the scattering coefficient measured when the instrument is filled with filtered air is subtracted from the scattering coefficient measured with ambient air. This allows the instrument to detect aerosol scatter as low as 1% of the molecular scattering of air. A. G.E. condensation nucleus counter, which is calibrated against a Pollak CN counter, provided continuous measurement of Aitken nuclei concentration.

3. Results and Discussion

In order to first characterize the annual variation and long range transport of atmospheric aerosols contributing to particle light scattering under clean, background conditions at the Mauna Loa Observatory, approximately 5 yr of nephelometer data covering the period April 1975 through January 1980 were analyzed. Average scattering coefficients for the hour beginning at 1200 Z (i.e., 0200 LST) were selected if all of the following conditions were satisfied: (1) Hourly average Aitken nuclei concentration, which is a sensitive index of local contamination at the observatory, showed no evidence of the effects of outgassing from the Mauna Loa volcano, (2) Mean hourly relative humidity was less than 20%, and (3) ten-day meteorological back-trajectories using wind data from the 3000 to 5000 m layer exceeded 3000 km in length.

With regard to the first requirement, Bodhaine et al. (1980) found high Aitken nuclei concentrations coincided with episodes of volcanic contamination identified by high SO_2 concentrations and nocturnal downslope flow which passed over the Mauna Loa caldera. A concentration of $840\ cm^{-3}$ was determined to be the best estimated nuclei concentration above which conditions at the observatory are considered to be contaminated.

A relative humidity of less than 20% at the observatory is considered representative of a stable air mass containing an aged aerosol having a history of long range transport and is generally associated with a well-defined trade wind inversion (Mendonca and Pueschel, 1973). In the event of a breakdown or complete absence of the inversion caused by a synoptic disturbance, relative humidity at the observatory generally ranges from 20 to 100% for an extended period of time, due to the upward entrainment of high humidity subinversion air. These mixed air masses most probably contain particles having relatively recent contact with the Pacific Ocean or the island and are therefore not representative of mid-tropospheric background conditions. Elimination from the data set of scattering coefficients corresponding to ambient relative humidities of 20% or greater also reduces relative humidity effects on light scattering due to particle hygroscopicity.

To investigate long range transport effects on the background aerosol at Mauna Loa, ten-day back-trajectories using the air trajectory model developed by Heffter *et al.* (1975) were analyzed. Back-trajectories ending at 1200 Z at Mauna Loa Observatory were calculated using transport winds averaged in the vertical layer between 3000 and 5000 m. Ten-day trajectories of 3000 km or more in length were chosen in this study as being most representative of mid-tropospheric air that had undergone long range transport. The accuracy of trajectories computed from higher velocity winds is also expectedly greater.

Accounting for instrument downtime, the availability of meteorological input data for trajectory calculation, and the data screening criteria outlined above, the final edited data set for the 5 yr period contained 188 sets of 1200 Z four-channel nephelometer measurements. Monthly averages of the scattering coefficient in units of m^{-1} for a wavelength of 550 nm are plotted in Figure 1. Also shown on the graph is the annual variation of the Ångström exponent α, which describes the wavelength dependence of aerosol scattering according to the relation.

$$b_{SP} \sim \lambda^{-\alpha}. \tag{1}$$

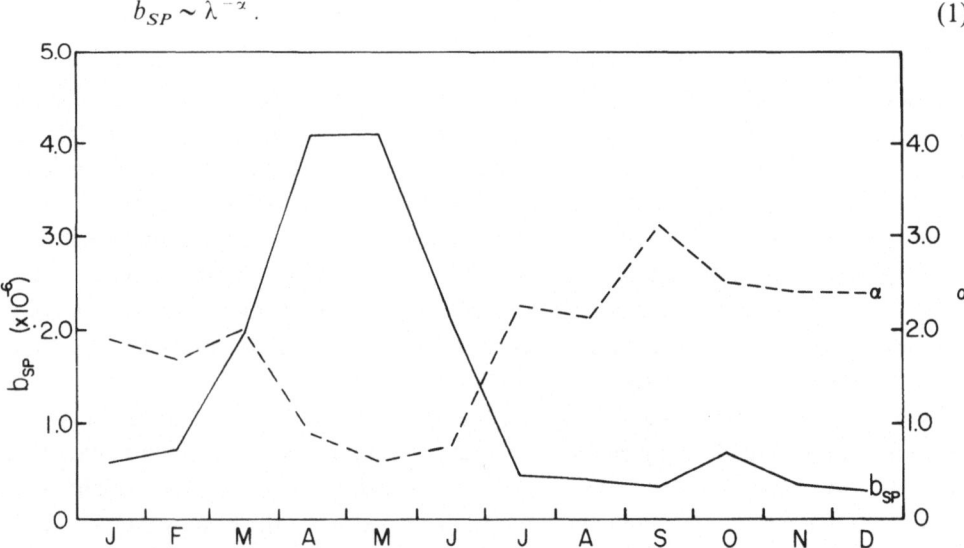

Fig. 1.   The annual variation of particle light scattering coefficient and Ångström exponent at the Mauna Loa Observatory.

The value of α was obtained by finding the negative slope in the linear regression between the logarithm of the scattering coefficient b_{SP} (dependent variable) and the logarithm of wavelength λ (independent variable), using the set of four measured scattering coefficients corresponding to each nephelometer wavelength channel. If we assume a power law aerosol size distribution of the form

$$\frac{dN}{d \log D} = CD^{-\beta} \tag{2}$$

where dN is the number of particles in the diameter increment d log D, C is a constant, and β is the slope of the size distribution, a simple relationship between the Ångström exponent and the slope of the size distribution exists (Butcher and Charlson, 1972; Shaw et al., 1973), i.e.,

$$\beta = \alpha + 2 . \tag{3}$$

Thus, measurement of the scattering coefficient at several different wavelengths provides an estimate of the slope of the aerosol size distribution.

Figure 1 shows a pronounced peak in aerosol light scattering during March through June, with maximum monthly average scattering coefficients almost one order of magnitude greater than those measured during other months of the year. A minimum in the Ångström exponent also occurs during this period, indicating the dominance of larger particles. The inverse relationship between aerosol scattering coefficient and the Ångström exponent, or the slope of the particle size distribution, that is apparent from Figure 1 suggests that the highest light scattering is associated with a comparatively flat particle size distribution favoring larger particles, while low light scattering occurs in combination with a relatively steep size distribution dominated by smaller particles.

Ten-day back-trajectories were divided into five categories, corresponding to the NE, SE, SW, and NW quadrants and a special case, NW/E, according to the typing scheme used by Miller (1981). The latter case was chosen because many air trajectories arriving at the observatory under the influence of easterly flow actually originated from the northwesterly sector. Average scattering coefficients and Ångström exponents, along with the number of cases N, for each trajectory type are shown in Figure 2. Flow originating from the northwest is clearly associated with highest light scattering coefficients and low Ångström exponents, indicating relatively high concentrations of large particles. Particle light scattering associated with air trajectories from the SW and SE quadrants is roughly half that for northwesterly flow and considerably less influenced by larger particles. Trajectories originating from the NE quadrant show average scattering coefficients of intermediate magnitude and Ångström exponents favoring larger particles.

An analysis similar to that shown in Figure 2 was conducted including ten-day back-trajectories with origins of less than 3000 km. Although this resulted in a greater relative frequency of trajectories from the NE and SE sectors, since easterly winds dominate the less vigorous summertime flow regime over Hawaii (Miller, 1981), the

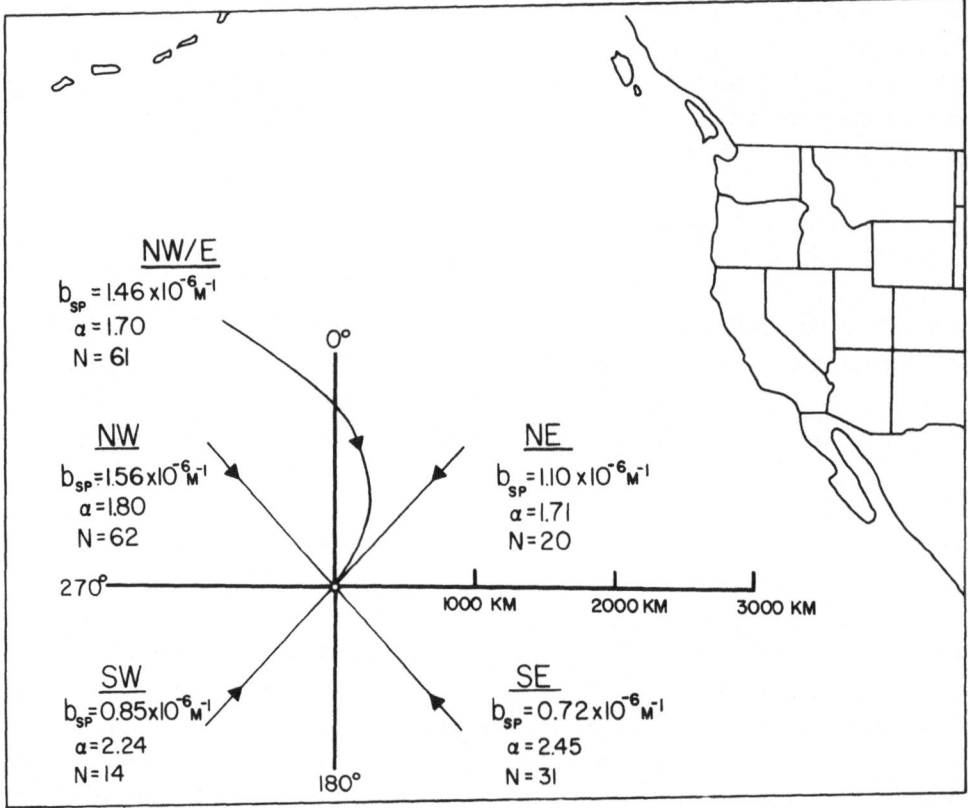

Fig. 2. Averages of particle light scattering coefficient and Ångström exponent by trajectory type.

variations in the magnitude and wavelength dependence of particle light scattering with trajectory type were not significantly different than those shown in Figure 2.

The sampling program during the months of April 1980 through April 1981 yielded only 16 impactor samples that were collected during the nighttime downslope flow when average relative humidity was less than 20% and no evidence of volcanic contamination was present. Aerosol samples were especially prone to local volcanic contamination during the summer and early fall, when light drainage winds dominated the circulation and the influence of synoptic-scale flow was small. Periods of relative humidity exceeding 20% and particle instrument downtime also limited the number of samples. Table I lists the date, average relative humidity, the mean nephelometer scattering coefficient for 550 nm, and the slope of the particle size distribution as computed from the Ångström exponent using Equations (1) and (3) for each sample. Also included in Table I is the ratio of the number of sulfuric acid droplets to the total number of sulfate particles.

Inspection of Table I shows that the particle scattering coefficient, the slope of the aerosol size distribution, and the frequency of sulfuric acid relative to total sulfate underwent wide variations from sample to sample. Examples of electron micrographs of particle samples that exhibited sharp contrasts in aerosol chemistry and morphology are presented in Figures 3 and 4.

TABLE I

Date, relative humidity, scattering coefficient, particle size distribution slope, and number ratio of sulfuric acid to total sulfate for each aerosol sample

Sample	Date	RH	b_{SP} (m^{-1})	β	H$_2$SO$_4$/SO$_4$
1	4–28–80	8	34.68×10^{-6}	2.09	1.00
2	6–6–80	14	2.760×10^{-6}	3.32	1.00
3	8–20–80	13	0.517×10^{-6}	5.19	1.00
4	10–29–80	18	0.123×10^{-6}	4.27	0.10
5	12–17–80	18	1.546×10^{-6}	3.55	0.54
6	12–28–80	19	0.161×10^{-6}	7.13	0.38
7	12–31–80	9	1.262×10^{-6}	3.85	0.45
8	1–4–81	7	0.272×10^{-6}	4.45	0.85
9	1–30–81	6	1.613×10^{-6}	5.52	0.16
10	1–31–81	10	0.472×10^{-6}	5.93	0.60
11	2–2–81	13	0.143×10^{-6}	6.35	0.31
12	3–3–81	7	1.607×10^{-6}	2.77	1.00
13	3–18–81	7	1.310×10^{-6}	3.47	0.19
14	3–25–81	9	1.050×10^{-6}	2.94	0.41
15	3–30–81	7	5.843×10^{-6}	2.84	0.11
16	4–18–81	12	1.317×10^{-6}	2.78	1.00

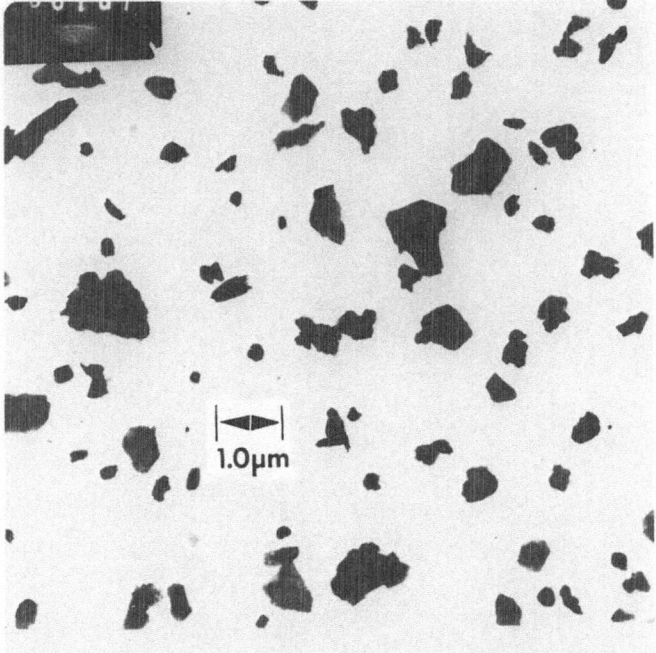

Fig. 3.   Untreated particles collected on 4–28–80 (sample No. 1).

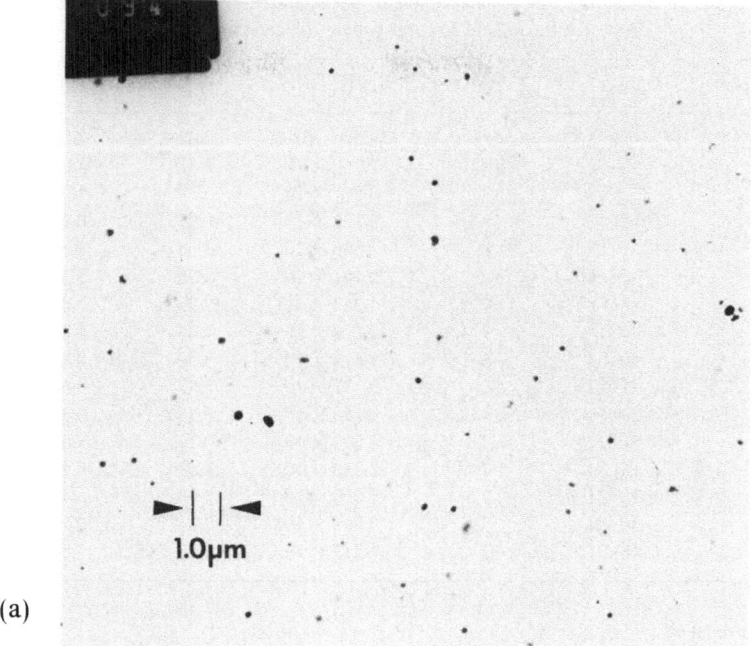

(a)

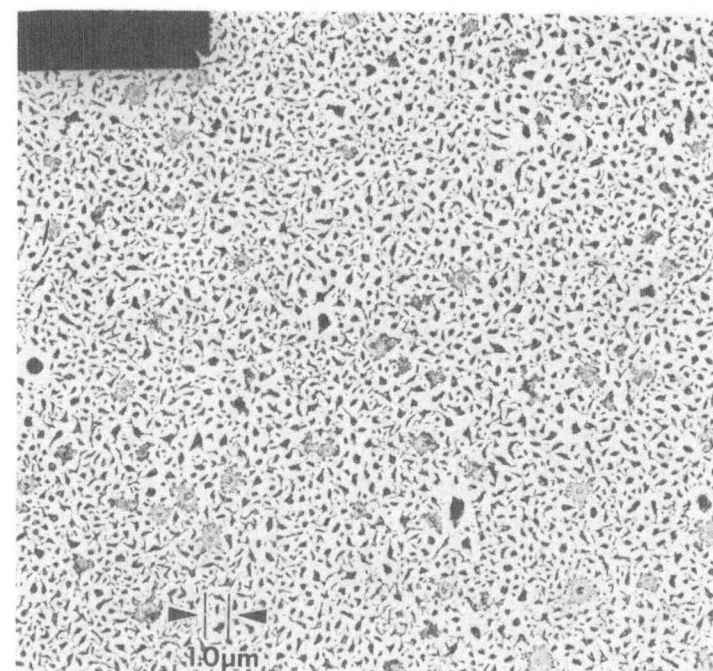

(b)

Fig. 4.   Untreated particles (a) and $BaCl_2$ – treated particles (b) collected on 8–20–80 (sample No. 3).

The aerosol collected on April 28, 1980 (sample No. 1) was predominantly composed of large, irregularly-shaped particles resembling soil dust, as shown on the untreated sample in Figure 3. Impactor sampling time on this day had to be reduced to only 10 min due to the abnormally high concentrations of large particles, as evidenced by the extremely high scattering coefficient and the small negative slope of the particle size distribution. When treated with $BaCl_2$, relatively few particles reacted, indicating that most of the particles did not contain sulfate.

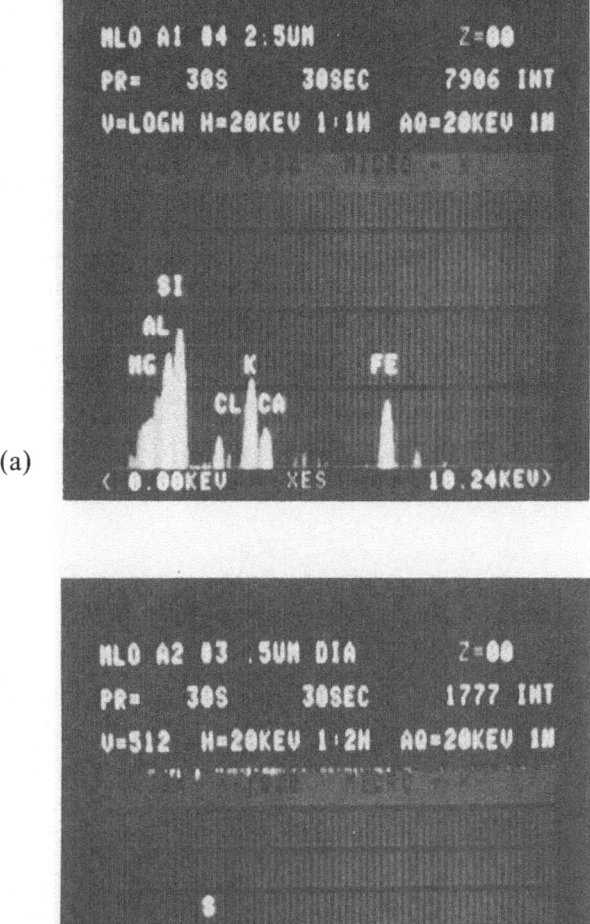

Fig. 5. X-ray spectra of particles collected on 4–28–80 (a) and 8–20–80 (b).

The particles of sample No. 3, collected for a period of 1 h, were much smaller and mainly spherical in shape, as illustrated in Figure 4a. Treatment with BaCl$_2$ at low relative humidity resulted in the formation of numerous reaction spots, indicating a predominance of sulfuric acid, as viewed in Figure 4b (the very numerous oblong-shaped grains that appear in Figure 4b are BaCl$_2$ crystals). The wet removal of these unneutralized sulfuric acid droplets may account for the relatively high acidity of rainwater collected at Mauna Loa Observatory (Hanson, 1977). A detailed analysis of the possible effects of background concentrations of sulfuric acid on the precipitation chemistry at Mauna Loa will be reported separately.

Samples No. 1 and No. 3 were analyzed by G. Shaw of the University of Alaska using a scanning electron microscope x-ray microprobe. Figure 5a shows the predominance of the crustal elements Mg, Al, Si, K, Ca, and Fe found in the aerosol collected during the dust episode of April 28, 1980. Under clean background conditions, S is the principal aerosol component (Figure 5b).

For most samples, the agreement between the particle size distributions observed on the impactor samples and that computed from the Ångström exponent was good (the overall correlation coefficient was significant at the 95% confidence level). The four-channel nephelometer data thus serve as a reliable means of determining the aerosol size distribution on a continuous, routine basis at the observatory.

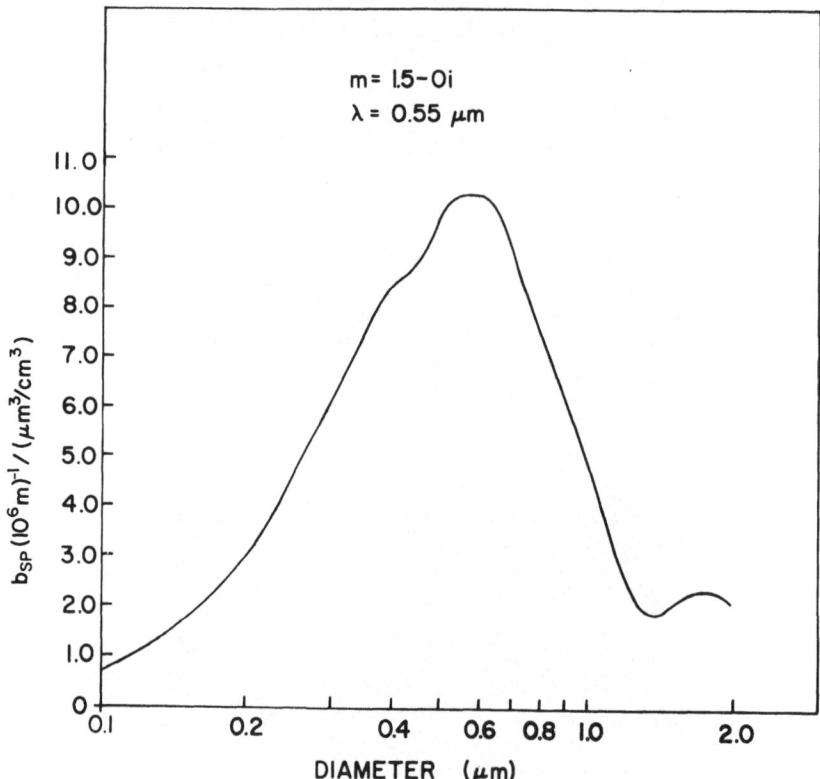

Fig. 6.   The theoretical relationship between light scattering per unit particle volume and particle diameter.

The theoretical relationship between particle light scattering per unit particle volume concentration and particle diameter for a wavelength of 0.55 μm (550 nm) and an aerosol refractive index of 1.0–0i is shown in Figure 6. This relationship was derived from a curve of Mie scattering efficiency as a function of the dimensionless optical particle size parameter as given by Friedlander (1977). Figure 6 clearly shows that the most efficient particle size for light scattering on a volume basis for this wavelength and refractive index occurs in the size range between 0.5 and 0.6 μm. Scattering per unit particle volume is considerably less for particles smaller than 0.1 μm and larger than 2.0 μm.

Using the curve in Figure 6, together with the particle volume size distributions obtained from the electron microscope analyses, it is possible to compute for each sample the total light scattering due to particles in the 0.1 to 2.0 μm size range, as well as the relative contributions of sulfates and non-sulfates. For roughly half of the samples collected, particles within the size range 0.1 to 2.0 μm accounted for practically all of the light scattering, and the agreement between the computed scattering coefficient and the nephelometer was excellent (in most cases well within a factor of 2). For the other samples, particles larger than 2.0 μm, most likely non-sulfate, accounted for a sizeable portion of the light scattering.

The scattering coefficient computed for all particles is plotted against particle diameter in Figure 7, where the 16 samples have been categorized according to slope of the size distribution as computed from the four-wavelength nephelometer data. It is seen that for the samples having a relatively steep particle size distribution, submicrometer particles account for most of the light scattering. When the slope of the size distribution is small, however, the coarse particle mode contributes significantly.

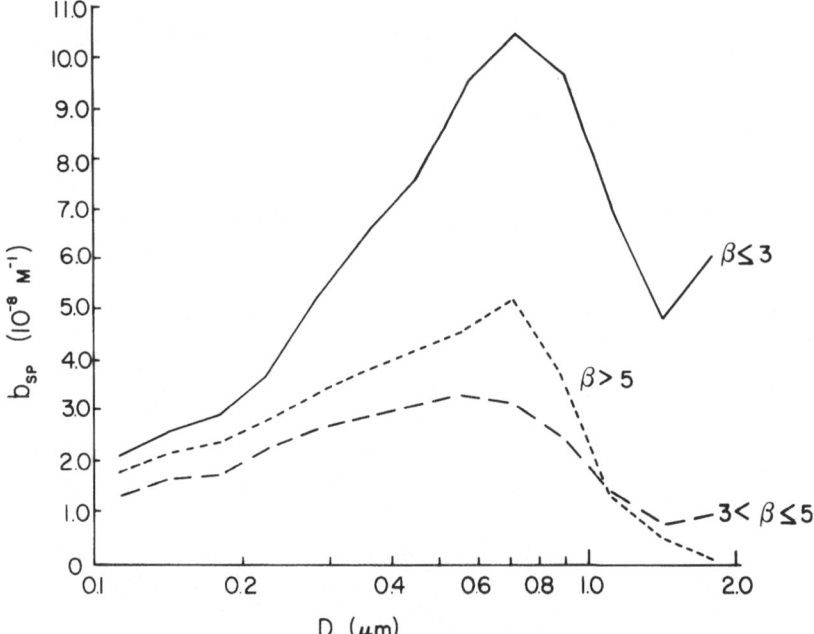

Fig. 7.   Computed particle scattering coefficient as a function of diameter, categorized by the slope of the particle size distribution.

The ratio of the scattering due to sulfate to the scattering due to all particles is shown in Figure 8 as a function of particle diameter. Sulfates clearly dominate the light scattering up to 0.4 μm for all categories. The influence of non-sulfates becomes important at larger particle sizes, especially when the slope of the size distribution is small.

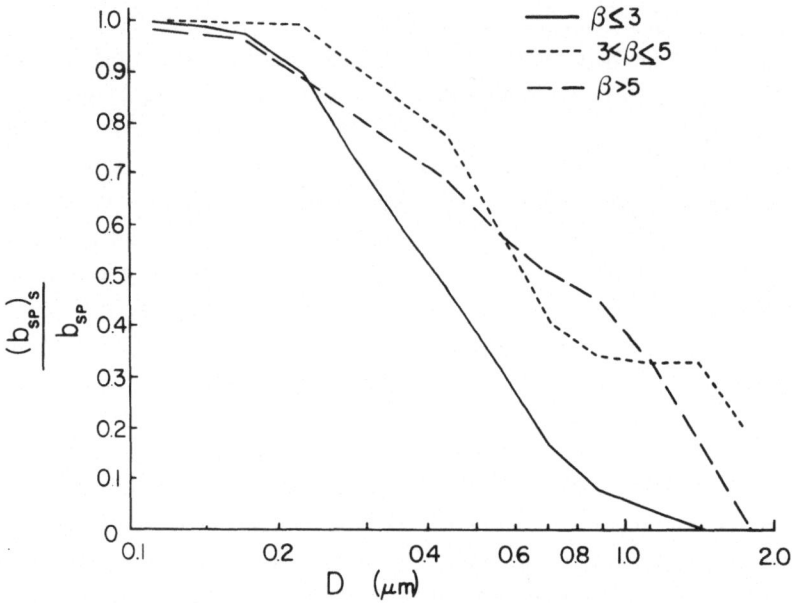

Fig. 8.   The ratio of light scattering due to sulfates to total scattering as a function of diameter, categorized by the slope of the particle size distribution.

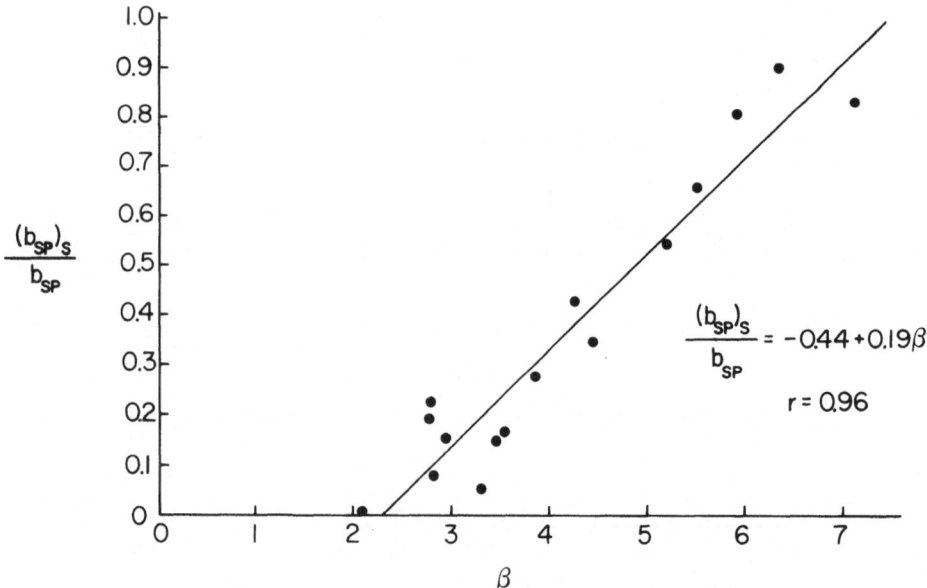

Fig. 9.   The ratio of light scattering due to sulfates to total scattering as a function of the slope of the particle size distribution.

The sulfate-to-total-particle scattering ratio and the slope of the particle size distribution are shown in Figure 9 to be linearly related and highly correlated. For those samples in which particles larger than 2.0 μm contributed to the total light scattering, it was assumed that these particles were non-sulfate, as suggested by the particle size distributions.

Use of the regression line drawn in Figure 9 represents a means by which to estimate the relative contribution of sulfate particles to observed light scattering at the Mauna Loa Observatory using only the four-channel nephelometer data. This was applied to each component of the 5 yr historical data set to indirectly determine the annual variation of scattering due to sulfate and non-sulfate particulate matter, displayed in Figure 10. It is plainly seen that the peak in light scattering at the Mauna Loa Observatory which occurs in the springtime is due to non-sulfate particles, with scattering due to sulfate undergoing comparatively little variation over the annual period. The contributions to total light scattering of sulfate and non-sulfates are comparable during the latter half of the year.

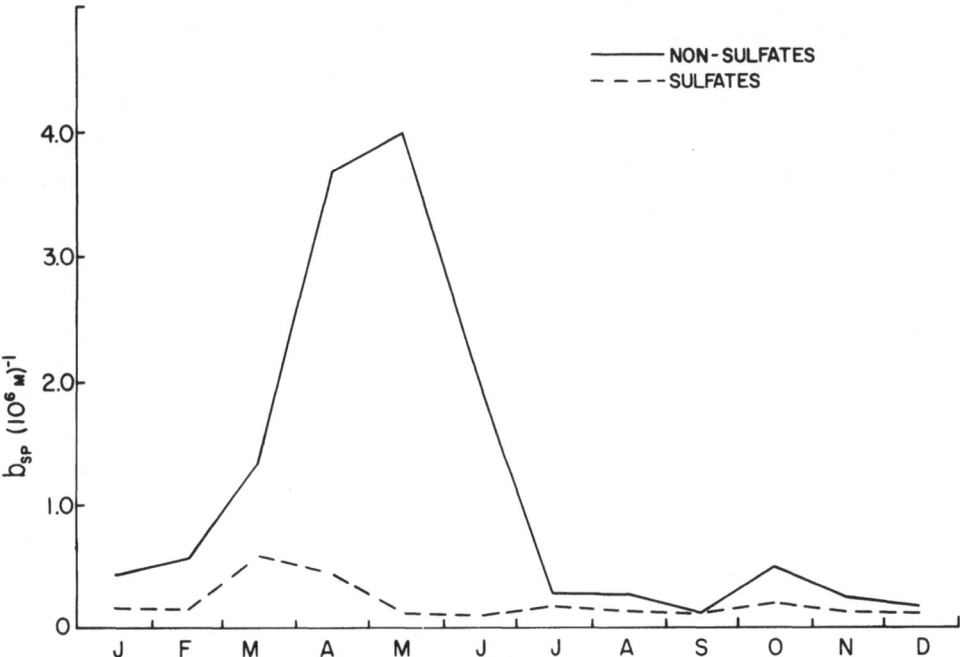

Fig. 10.   The annual variation of particle light scattering due to non-sulfate and sulfate particles.

When analyzed with respect to air mass trajectory, as shown in Table II, it is seen that the scattering due to sulfate is highest for meteorological transport from the NW quadrant. The influence of sulfate scattering associated with northeasterly flow is minimal, according to this analysis. It appears that the direct transport to Hawaii of anthropogenic sulfur emitted from North America is of minor importance. If the results of this analysis are real, they imply that these emissions are subject to long range westerly transport, where they mix with other emissions in the Northern Hemisphere and undergo chemical transformation and atmospheric removal before reaching Hawaii.

TABLE II

The light scattering coefficient due to sulfate and its ratio to
the total scattering according to trajectory type

Trajectory type	$(b_{SP})_s$, (m^{-1})	$(b_{SP})_s/b_{SP}$
NW	2.97×10^{-7}	0.16
NW/E	1.78×10^{-7}	0.12
SW	1.30×10^{-7}	0.15
NE	1.39×10^{-7}	0.13
SE	1.65×10^{-7}	0.23

The results thus far suggest that large, non-sulfate particles resembling soil dust
(Figure 3) dominate the aerosol light scattering during episodes which occur most
frequently in the spring. These dust episodes most likely occur during strong northwesterly large-scale flow, supporting the theory that dust storms over the Asian continent
are the principal source of these particles.

During one such episode in April–May 1979, which lasted over a week, a Florida State
University streaker aerosol sampler was in operation at the observatory. This sampler,
unlike an impactor, provides a continuous time record of aerosol concentration along
a 0.4 μm pore diameter Nuclepore filter strip (Woodward *et al.*, 1977). The streaker
sampler is highly efficient for the collection of particles smaller than several micrometers
in diameter, but relatively insensitive to coarser particles (Darzi and Winchester, 1981).
Concentrations of Si, K, Ca, Fe, Al, and S determined by PIXE (proton induced X-ray

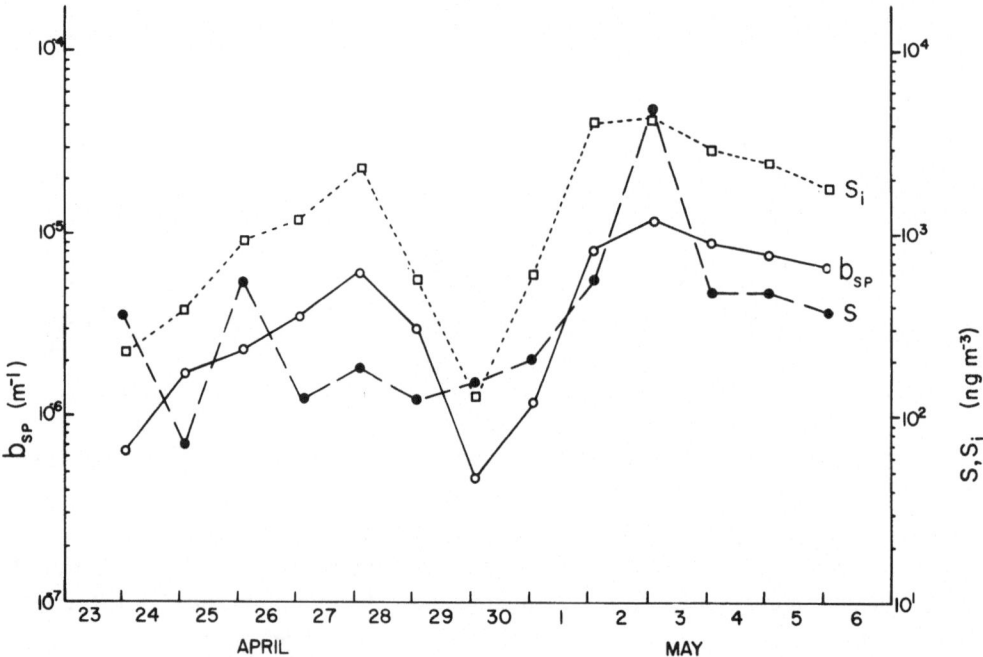

Fig. 11.   A time record of nighttime scattering coefficients and particulate S and Si elemental concentrations
during a dust episode at Mauna Loa Observatory, April 24–May 6, 1979.

emission) analysis were compared to simultaneous observations of particle light scattering, the Ångström exponent, and Aitken nucleus concentration routinely made at the observatory.

Figure 11 shows the time trace of average nighttime (2200–0600 LST) scattering coefficients (550 nm) and elemental concentrations of S and Si. Although the curves exhibit approximately the same time trends, best agreement exists between the scattering coefficient and Si, a primary constituent of soil dust.

The peak in S concentration which occurred on the early morning of May 3 was probably due to local volcanic contamination. Average Aitken nucleus concentrations during this time were greater than 1000 cm^{-3}, well above the concentrations which were observed on the other nights during the period, typically 100 to 300 cm^{-3}. High concentrations of sulfuric acid aerosol have been observed during conditions of volcanic contamination at Mauna Loa Observatory (Dittenhoefer, unpublished results).

A linear correlation matrix of all elemental concentrations and aerosol parameters is shown in Table III. All of the crustal elements, i.e. Si, K, Ca, Fe, and Al, are highly intercorrelated, and each is highly positively correlated with the scattering coefficient.

TABLE III

Linear correlation matrix (correlation coefficients significant at the 95% confidence level are in bold face)

	b_{SP}	α	Si	K	Ca	Fe	Al	S
b_{SP}	—							
α	**−0.66**	—						
Si	**0.96**	**−0.63**	—					
K	**0.90**	**−0.67**	**0.99**	—				
Ca	**0.95**	**−0.67**	**0.97**	**0.94**	—			
Fe	**0.96**	**−0.64**	**1.00**	**0.99**	**0.98**	—		
Al	**0.96**	**−0.62**	**1.00**	**0.98**	**0.96**	**1.00**	—	
S	**0.71**	−0.25	**0.65**	**0.62**	0.57	**0.65**	**0.70**	—
AN	−0.45	0.26	−0.40	−0.40	−0.49	−0.42	−0.39	−0.06

These elements are also significantly negatively correlated with the Ångström exponent, indicating that they are found primarily in the coarse particle mode. Sulfur, on the other hand, displays a lower, yet statistically significant, correlation with respect to the scattering coefficient and an insignificant correlation with the Ångström exponent, suggesting a more uniform distribution over the optically active particle size range. Crustal elemental ratios remained rather constant throughout the sampling period and closely approximated the average earth crust composition (Darzi and Winchester, 1981). The ratio K/Fe = 0.45 found for this study, for example, is very close to the earth crustal average of 0.43 (Mason, 1966), but much higher than the ratio for Mauna Loa basalt, i.e., K/Fe = 0.04 (Macdonald and Katsura, 1964).

These results indicate that the entrainment of Hawaiian basaltic rock by local winds apparently exerted little influence on the aerosol measurements during this dust episode. The long range transport of soil dusts from Asia represents a more likely source. The observed time-dependent associations of sulfur with particle light scattering and various crustal elements suggest a continental origin of aerosol sulfur at Mauna Loa during this period.

During this same period, Shaw (1980) monitored the temporal variation of optical thickness above Mauna Loa and found the maximum to be one order of magnitude higher than the average for that time of year. Inversion of the optical scattering and extinction measurements showed the particles to be mainly within the 1.0 to 10.0 μm diameter size range. Isobaric trajectory analysis at the 500 mb level traced the origin of these dust particles to the Gobi Desert of eastern Asia, where sand storms had occurred about one week before the dust was first observed over Hawaii.

4. Summary and Conclusions

This study has shown that wide variations in the chemistry and morphology of mid-tropospheric, natural background aerosol particles collected at the Mauna Loa Observatory can occur over an annual period. Episodes of high particle scattering coefficients caused by large non-sulfate particles, presumably soil dust, occur mainly in the spring and are generally associated with strong northwesterly large scale flow. Evidence has been gathered suggesting that dust storms over the Gobi desert in eastern Asia largely account for the springtime turbidity peak in Hawaii.

Sulfuric acid comprises a major portion of the particulate S in the clean nocturnal downslope air at the observatory, although significant neutralization by ammonia to form ammonium sulfate occurs on occasion. Sulfate particles primarily occupy the submicrometer size range and represent the dominant light-scattering aerosols for sizes less than 0.4 μm.

The fraction of the total light scattering coefficient accounted for by sulfate is directly related to and highly correlated with the slope of the overall particle size distribution. Measurements show that for a size distribution slope more negative than -5, sulfate particles contribute over half of the total light scattering.

Over an annual period, light scattering due to sulfates probably undergoes a relatively small variation compared to the scattering caused by non-sulfate particles. Non-sulfates clearly dominate light scattering during the late winter and spring, while the effects of sulfates and non-sulfates are more nearly equal during the latter half of the year. There is some indirect evidence for the long range atmospheric transport to Mauna Loa of sulfur derived from continental sources in the Northern Hemisphere.

During a dust episode, particle elemental ratios were found to remain relatively constant and closely resembled average earth crust composition, different significantly from ratios found in Hawaiian basaltic rock. Particulate concentrations of crustal elements such as Si, K, Ca, Fe, and Al were each highly correlated with aerosol light scattering during this event.

In conclusion, particles derived from natural sources clearly dominate global background light scattering observed at the Mauna Loa Observatory, Hawaii. The relative importance of atmospheric sulfates is significant only under the cleanest conditions.

Acknowledgments

This project was conducted while the author was a National Research Council Post-Doctoral Research Associate. The author gratefully acknowledges the helpful assistance of the entire staff at the Mauna Loa Observatory, including Dr K. Coulson, director, Mr T. Defoor, and Mrs J. Pereira. Use of the electron microscope at the University of Hawaii-Hilo was made possible through the generosity of Dr D. Hemmes. Special thanks is offered to Mr M. Darzi and Dr J. Winchester of Florida State University for their streaker sampler data. Drs B. Bodhaine and G. Herbert of GMCC kindly provided MLO aerosol and meteorology data for use in this study, while ARL trajectories were supplied by Dr J. Miller. Finally, the author thanks Dr G. Shaw of the University of Alaska for conducting the x-ray energy dispersive analyses.

References

Bigg, E. K.: 1977, *J. Appl. Meteor.* **16**, 262.

Bodhaine, B. A.: 1978, 'The Mauna Loa Four Wavelength Nephelometer: Instrument Details and Three Years of Observations', Air Resources Laboratories/GMCC, Boulder, Colorado, 39 pp.

Bodhaine, B. A. and Pueschel, R. F.: 1972, *J. Geophys. Res.* **77**, 5106.

Bodhaine, B. A., Harris, J. M., Herbert, G. A., and Komhyr, W. D.: 1980, *J. Geophys. Res.* **85**, 1600.

Bodhaine, B. A., Mendonca, B. G., Harris, J. M., and Miller, J. M.: 1981, *J. Geophys. Res.* **86**, 7395.

Butcher, S. S. and Charlson, R. J.: 1972, *An Introduction to Air Chemistry*, Academic Press, New York, 241 pp.

Darzi, M. and Winchester, J. W.: 1981, *J. Geophys. Res.* (submitted).

Dittenhoefer, A. C. and de Pena, R. G.: 1980, *J. Geophys. Res.* **85**, 4499.

Duce, R. A., Unni, C. K., Ray, B. J., Prospero, J. M., and Merrill, J. T.: 1980, *Science* **209**, 1522.

Friedlander, S. K.: 1977, *Smoke, Dust and Haze*, John Wiley and Sons, New York, 317 pp.

Hanson, K. J. (ed.): 1977, Geophysical Monitoring for Climatic Change No. 5, Summary Report 1976, NOAA/ERL, Boulder, Colorado, 110 pp.

Hansen, E. B., Caniparoli, D. G., and Charlson, R. J.: 1979, 'Aerosol Light-Scattering Characteristics at Mauna Loa Observatory', presented at the WMO Technical Conference on Regional and Global Observation of Atmospheric Pollution Relative to Climate, 20–24 August 1979, Boulder, Colorado.

Heffter, J. L., Taylor, A. D., and Ferber, G. J.: 1975, 'A Regional-Continental Scale Transport, Diffusion, and Deposition Model', NOAA Tech. Memo. ERL ARL-50, Silver Spring, MD, 29 pp.

Macdonald, G. A. and Katsura, T.: 1964, *J. Petrol.* **5**, 82.

Mamane, Y. and de Pena, R.: 1978, *Atmos. Environ.* **12**, 69.

Mason, B.: 1966, *Principles of Geochemistry*, 3rd ed., Wiley, New York.

Mendonca, B. G. and Pueschel, R. F.: 1973, *J. App. Meteor.* **12**, 156.

Mendonca, B. G., Hanson, K. J., and De Luisi, J. J.: 1978, *Science* **202**, 513.

Mendonca, B. G., De Luisi, J. J., Hanson, K. J., and Peterson, J. T.: 1979, 'Signatures in Atmospheric Transmission Variations at Mauna Loa Observatory, Hawaii', presented at the WMO Technical Conference on Regional and Global Observation of Atmospheric Pollution Relative to Climate, 20–24 August 1979, Boulder, Colorado.

Miller, J. M.: 1981, 'A Five-Year Climatology of Back Trajectories from the Mauna Loa Observatory, Hawaii', *Atmos. Environ.* **15**, 1553.

Pueschel, R. F., Bodhaine, B. A., and Mendonca, B. G.: 1973, *J. Appl. Meteor.* **12**, 308.

Shaw, G. E., Reagan, J. A., and Herman, B. M.: 1973, *J. Appl. Meteor.* **12**, 374.

Shaw, G. E.: 1980, *J. Appl. Meteor.* **19**, 1254.

Simpson, H. J.: 1972, *J. Geophys. Res.* **77**, 5266.

Woodward, A. P., Jensen, B., Leslie, A. C. D., Nelson, J. W., Winchester, J. W., Ferek, R. J., and Van Espen, P.: 1977, 'Aerosol Characterization by Impactors and Streaker Sampling and PIXE Analysis', Proceedings of Symposium, Recent Advances in Air Pollutant Analysis, American Institute of Chemical Engineers, New York.

IMPACT OF NEW YORK STATE EMISSION SOURCES ON CLASS 1 AREAS

GOPAL SISTLA, ALAN J. DOMARACKI, and SURY N. PUTTA

New York State, Department of Public Service, The Governor Nelson A. Rockefeller, Empire State Plaza, Albany, NY 12223 U.S.A.

(Received June 29, 1981; Revised November 16, 1981)

Abstract. The ARL-ATAD (Air Resources Laboratory-Atmospheric Transport and Dispersion) Model is used to calculate trajectories of air parcels leaving New York City, Albany, and Buffalo airshed regions and terminating near Federally-mandated Class 1 areas in the Northeastern United States, for which visibility is protected from degradation under the Clean Air Act. The purpose of this study is to provide an estimate of the frequency of occurrence of trajectory end points terminating over or near these environmentally sensitive regions from data for a one year period. Results indicate that these regions are not substantially effected by the air parcels either on an annual or seasonal basis.

1. Introduction

Under the Section 169A of the Clean Air Act, the United States Environmental Protection Agency (USEPA) has proposed regulations for visibility protection in Class 1 areas. The pollutants, which cause a degradation of visibility, are gases and particulate matter released into the atmosphere from human activities, as well as natural phenomenon. The impairment to visibility occurs due to absorption and scattering of light by the pollutants SO_x, NO_x, and particulate matter comprising of sulfates and nitrates, etc. Since industrial activities, such as electric power generation units burning coal or oil, are one of the major sources for these pollutants; their effect, if any, on the nearby Class 1 areas will be of interest to regulatory agencies. In this study, the ARL-ATAD (Air Resources Laboratory Atmospheric Transport and Dispersion) Model is used to provide estimates of the air parcel trajectories leaving the urban airsheds of New York City, Albany, and Buffalo and passing over or near the Class 1 areas of the Northeast. These estimates will be useful to assess the rate at which a Class 1 area is traversed by the plumes originating from the airshed, as well as the age of air parcel. The results indicate that less than 1% of the trajectories leaving New York City, Albany, and Buffalo airsheds cross over a Class 1 area of the Northeast. Also, no discernible seasonal trends were found.

2. Data Base

The Class I areas in the Northeast are mainly wilderness areas and are shown in Figure 1. The areas are in the states of New Jersey, Vermont, New Hampshire, and Maine, and in the Province of New Brunswick, Canada, and range in size from 2700 acres to 37 000 acres. The latitude and longitude coordinates listed in the Figure 1 are the approximate centers of the areas.

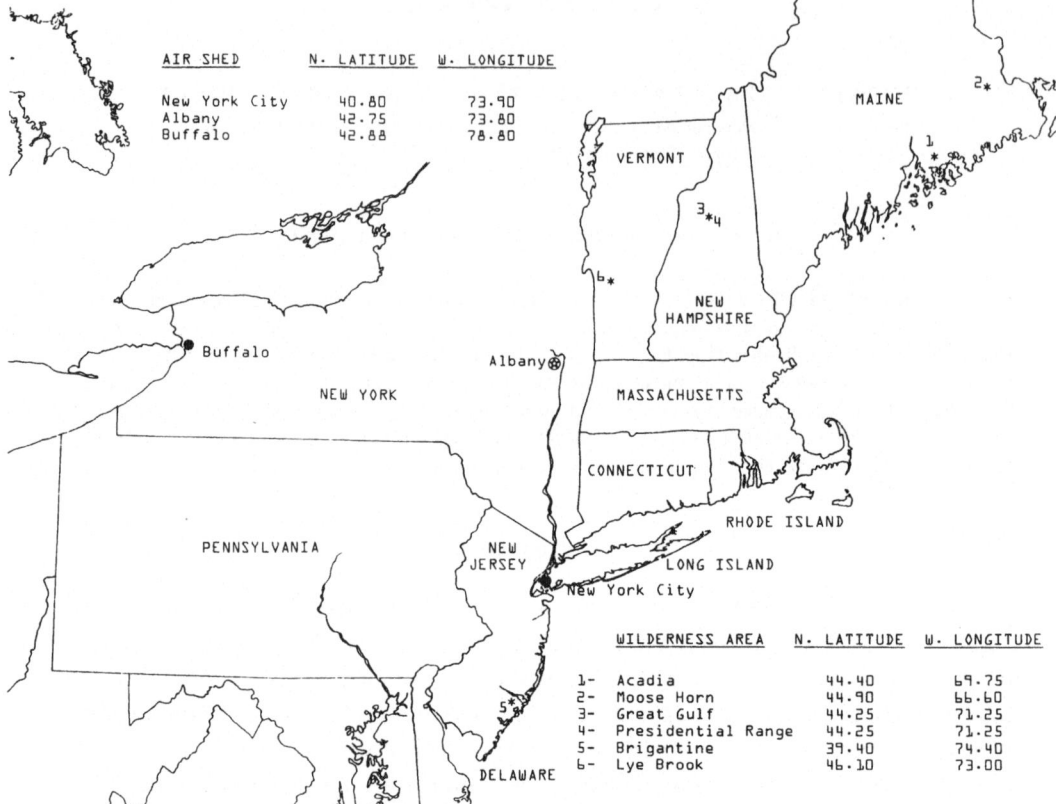

Fig. 1. Location of the New York State air shed points and Class 1 wilderness areas of the Northeast.

Each of the locations in the urban airshed, from which the air parcel trajectories are computed, is shown in Figure 1. Each site is the location of a National Weather Service Station and was selected so as to be a representative point of the airshed. The meteorological data base for this study was upper air data from rawinsonde and pibal stations for North America (from surface to 500 mb) for the calendar year 1976. These data were obtained on a tape called the NAMER-WINDTEMP from the National Weather Service, Asheville, NC.

The ARL-ATAD Model, described in detail elsewhere (Heffter, 1980), was used to compute forward trajectories of 24-h, duration for each day of the year. The model computes four trajectories per day from a selected origin at 00Z, 06Z, 12Z, and 18Z. Each trajectory is segmented into eight successive 3-h periods, thereby providing the air parcel location by latitude and longitude every three hours. These 3-h end points form the data base for the analysis reported in this study.

3. Analysis

One of the common procedures employed to account whether an air parcel trajectory has 'hit' an area is to establish if the 3-h trajectory end point is within a predefined boundary or zone. However, this method will yield only a lower limit of estimate, since it does not take into account those air parcel segments which 'pass over' the area, thus resulting in a 'no hit'. In the present study, both methods have been applied to estimate the total frequency of air parcel movement over a given area. Since each of the Class 1 areas are considerably smaller than a circle of $0.1°$ radius (the area of such a circle being $\approx 90\,000$ acres), it was selected as a representative zone of the Class 1 area. The total number of trajectory segments for each of the airsheds is listed in Table I for the year

TABLE I

Annual and seasonal counts for each of the airsheds

Season	Air shed	Total no. of counts in each 3-h segment period							
		0–3	3–6	6–9	9–12	12–15	15–18	18–21	21–24
Annual	NYC	1426	1418	1406	13337	1222	1087	981	862
	ALB	1426	1416	1415	1402	1348	1217	1110	982
	BUF	1425	1412	1410	1401	1396	1372	1340	1274
Winter	NYC	120	120	117	107	96	81	77	69
	ALB	120	120	120	119	110	97	91	76
	BUF	120	120	120	120	120	117	113	105
Spring	NYC	119	118	115	107	83	74	62	49
	ALB	119	118	118	117	110	93	73	62
	BUF	119	118	117	116	116	115	114	106
Summer	NYC	120	120	120	119	119	115	102	94
	ALB	120	120	120	120	120	120	118	111
	BUF	120	120	120	120	120	120	119	118
Autum	NYC	120	120	120	116	108	97	88	74
	ALB	120	120	120	120	118	104	98	89
	BUF	120	120	120	120	120	119	118	114

1976, as well as per season. In the latter case, the months of January, April, July, and October were used as representative of winter, spring, summer, and autumn seasons, respectively. Using the coordinates given in Figure 1 as the center of a circle of $0.1°$ radius, the 3-h segments from each airshed were examined if it 'passed' over the Class 1 area and the resulting frequency counts are determined, and listed in Tables II and III on a climatological and seasonal basis respectively.

4. Discussion

On a climatological basis (Table II), only the wilderness area in Vermont appears to have a high number of 'pass overs' from air parcels leaving the Albany airshed. However, on a percentage basis, the counts are less than 5% for the 0 to 3, 3 to 6-h segments and about 1% if all the segments are summed up. It should be noted that these counts will be an upper limit since the circular area considered is about seven times larger than the size of the wilderness area. The passover counts from the remaining airsheds on these Class 1

TABLE II

Number of segments 'passing over' the Class 1 areas for 1976

Class 1 area	Air shed	No. of counts in each 3-h segment period – annual							
		0–3	3–6	6–9	9–12	12–15	15–18	18–21	21–24
Acadia	NYC	0	0	1	4	6	5	4	3
	ALB	0	0	5	5	3	3	0	0
	BUF	0	0	0	0	1	4	1	0
Moose	NYC	0	0	0	1	1	3	0	0
Horn	ALB	0	0	0	1	3	1	0	4
	BUF	0	0	0	0	0	0	1	2
Great Gulf	NYC	0	0	2	2	2	6	6	1
and Pres.	ALB	0	5	13	7	3	5	2	0
Range	BUF	0	0	0	2	9	4	4	2
Brigantine	NYC	0	2	4	0	3	1	1	0
	ALB	0	0	0	3	5	1	0	0
	BUF	0	0	0	0	3	6	0	0
Lye Brook	NYC	0	3	7	9	3	1	0	1
	ALB	52	36	11	3	3	2	4	1
	BUF	0	0	2	6	11	5	7	0

areas are generally less than 1% for any 3-h segment period, indicating that their contributions to these areas, if any, would be minor.

The seasonal counts listed in Table III again indicate that the 0 to 3-h segment air parcel from the Albany airshed passes over the Lye Brook area with a frequency less than 5% for all the four seasons. The percentage counts for the remaining segments from this airshed on the area are generally less than 1%. A similar anlysis for the remaining airsheds shows that the percentage counts on a given Class 1 area are less than 1%.

5. Conclusions

From the above analysis, we can conclude that:

(1) Air parcels leaving the New York City, Albany, and Buffalo airshed regions have minimum impact on the Class 1 areas of the Northeast; and

(2) There is no evidence of a seasonal pattern as these air parcels pass over the Class 1 areas.

Acknowledgment

The authors wish to thank the New York State Energy Research Development Authority and the New York State Department of Environmental Conservation for supporting a portion of this study.

TABLE III

Number of segments 'passing over' the Class I areas—seasonal

Class I area	Air shed	No. of counts in each 3-h segment period																														
		0–3				3–6				6–9				9–12				12–15				15–18				18–21				21–24			
		w	s	s	f	w	s	s	f	w	s	s	f	w	s	s	f	w	s	s	f	w	s	s	f	w	s	s	f	w	s	s	f
Acadia	NYC	—	—	—	—	—	—	—	—	—	—	—	—	—	1	—	—	—	—	—	—	—	1	—	—	—	1	—	—	—	—	—	—
	ALB	—	—	—	—	—	—	—	—	—	—	—	—	—	1	—	1	—	1	—	—	—	1	—	1	—	1	—	—	—	1	—	—
	BUF	—	—	—	—	—	—	—	—	—	—	—	—	—	—	—	—	—	—	—	—	—	—	1	—	—	—	—	—	—	—	—	—
Moose Horn	NYC	—	—	—	—	—	—	—	—	—	—	—	—	—	—	—	—	—	—	—	—	—	—	—	—	—	—	—	—	—	—	—	—
	ALB	—	—	—	—	—	—	—	—	—	—	—	—	—	—	1	—	—	1	—	—	—	—	—	—	—	—	1	—	—	1	—	—
	BUF	—	—	—	—	—	—	—	—	—	—	—	—	—	—	—	—	—	—	—	—	—	—	—	—	—	1	—	—	—	1	—	—
Great Gulf and Pres. Range	NYC	—	—	—	—	—	—	—	—	1	—	—	—	—	—	—	—	—	—	—	—	—	1	—	—	—	3	—	—	—	—	—	—
	ALB	—	—	1	—	—	—	—	—	—	—	—	—	—	1	1	1	—	1	1	1	—	2	1	1	—	1	—	—	—	1	—	—
	BUF	—	—	—	—	—	—	—	—	—	—	1	—	—	1	—	1	—	1	1	1	—	—	—	—	—	1	—	—	—	1	—	—
Brigantine	NYC	—	—	—	—	—	1	—	—	—	2	—	—	—	1	1	—	2	3	—	—	—	1	—	—	—	1	—	—	—	—	—	—
	ALB	—	—	—	—	—	2	—	—	—	—	—	—	1	1	1	1	2	3	—	—	1	1	1	—	—	—	—	—	—	—	—	—
	BUF	—	—	—	—	—	—	—	—	—	—	—	1	—	—	—	1	—	3	1	1	1	1	1	—	—	1	1	—	—	—	—	—
Lye brook	NYC	—	—	—	—	—	—	1	1	—	1	1	—	1	1	—	1	—	2	—	—	—	1	—	—	—	1	—	—	1	—	—	—
	ALB	5	3	2	5	3	2	3	1	2	1	3	—	1	—	1	1	—	—	—	1	1	1	—	1	—	—	1	1	1	—	—	—
	BUF	—	—	—	—	—	—	—	—	—	—	—	1	1	—	—	—	—	1	1	1	—	1	1	—	—	—	—	—	—	—	—	—

Notice

This report reflects the opinions of the authors and does not necessarily represent the views, opinions, or policies of the New York State Department of Public Service, New York State Energy Research Development Authority, or the New York State Department of Environmental Conservation.

Reference

Heffter, J. L.: 1980, 'Air Resources Laboratories – Atmospheric Transport and Dispersion Model (ARL-ATAD)', NOAA Tech. Me. ERL-ARL-81, Air Resources Lab., Silver Spring, MD.

MODELING OF FIRST PLUME ENCOUNTERS WITH PRECIPITATION

DANIEL J. McNAUGHTON

*North American Weather Consultants, 1141 East 3900 South,
Salt Lake City, UT 84117, U.S.A.*

(Received 22 May, 1981; Revised 14 September, 1981)

Abstract. A regional air pollutant transport model was used to simulate the fate of S emissions in the northeast United States. Hourly calculations were analyzed to describe trajectory characteristics and mass balances as pollutant trajectories first encounter precipitation during transport away from emission sources. Model results include air concentrations of SO_2 and sulfate, dry deposition values, and sulfate concentrations scavenged by precipitation, along with trajectory statistics. Results of sensitivity tests are compared to base case simulations which consider all precipitation events as a means of suggesting priorities for future regional transport model development.

1. Introduction

Recent activity in regional air pollutant transport modeling in support of acid deposition studies has been steadily increasing, but additional studies will be required to understand the complexities of the problem. Research priorities are being reassessed (e.g. Hilst, 1981) and this reassessment can be aided by analyses using existing models. The objective of this paper is to provide some estimates of the importance of pollutant scavenging during a plume's first encounter with a precipitation system as it is transported away from a source. The study uses results of the Regional Air Pollutant Transport model (RAPT) (McNaughton, 1980) developed by the Pacific Northwest Laboratory as part of the U.S. Environmental Protection Agency/Department of Energy Multistate Atmospheric Power Production Pollution Study (MAP3S). The tests reported are part of a series on model sensitivity and verification. Tests of first precipitation encounters are important in understanding the needed degree of complexity in transport components. Specifically, if a majority of sulfate mass in a plume is removed during its first encounter with rain, then the need to simulate transport through complex storm and frontal structures could be reduced and models could be simplified.

The following sections describe briefly the RAPT model, the cases studied, results, and conclusions and recommendations. Mass balance summaries provided by the model for monthly simulations provide some insight into the disposition of emissions as represented by the model.

2. The Model

The RAPT model simulates the airborne transport, diffusion, transformation and removal of SO_2 and sulfate ($SO_4^=$) from point emission sources in the northeastern

Water, Air, and Soil Pollution **18** (1982) 129–137. 0049–6979/82/0181–0129$01.35.
Copyright © 1982 by D. Reidel Publishing Co., Dordrecht, Holland, and Boston, U.S.A.

United States. Transport is simulated by following trajectories in a 100- to 1000-m layer average windfield with reference to a fixed grid. Trajectories originate hourly at each point source. Hourly calculations of diffusion, transformation and wet and dry deposition are made using coupled mass conservation equations (McNaughton, 1980). The resultant mass distributions are accumulated over a month at each grid square to which the plumes contribute. For these tests, the model provides results in the form of monthly averages of SO_2 and $SO_4^=$ concentrations and monthly deposition totals.

Some significant model features as used in this study include:

– Horizontal dispersion is given by the spread of the trajectories with effects of eddy diffusion neglected.

– The rates of dry deposition of SO_2 and $SO_4^=$ are constant, with deposition velocities of 1.0 cm s^{-1} and 0.1 cm s^{-1}.

– SO_2 to $SO_4^=$ transformation is simulated as linear with a daytime rate of 2% h^{-1} and a night rate of 0.25% h^{-1}. The need for increased in-cloud conversion rates, identified by McNaughton and Scott (1980) is accounted for by specifying a 5% rate during precipitation events.

– Wet removal of SO_2 is described by: $w_1 = 0.005\ P(t)$ (μg) where $P(t)$ is an hourly precipitation rate (mm h^{-1}). Wet removal of $SO_4^=$ is determined from a formulation by Scott (1978) for Bergeron-type clouds as $w_2 = 0.232\ P(t)^{0.625}$ (μg).

– Mixing height is specified by a sinusoidally-varying cycle representing an average building of a stable nocturnal layer and a daytime mixed layer from the surface with a maximum mixed layer depth of 1500 m and a nocturnal layer depth of 200 m.

– Hourly wind data for calculations are provided by a linear interpolation from observations taken at 12-h intervals.

Model inputs include wind data, hourly precipitation data and average SO_2 emissions. The data are specified on a grid composed of 24 elements of the National Meteorological Center (NMC) Octagonal Northern Hemisphere Grid. Transport calculations are carried out on a grid of twice the NMC resolution ($\sim$ 170 km), whereas dispersion, transformation and removal calculations are carried out on a one-tenth NMC grid ($\sim$ 35 km). Data selected for the tests were rawinsonde and hourly precipitation data for October 1977 and a composite emissions inventory representing annual average SO_2 levels in 1973–1974.

3. Analysis Cases

Sensitivity testing was performed using four cases and additional supporting analyses. These cases are:

3.1. *Case 1 – Base Case*

Simulations were performed to provide results most closely matching observed SO_x concentration and deposition fields using the best available input parameters. The tests used data for October 1977 and verification using this case was reported by McNaughton (1980), McNaughton and Scott (1980), and McNaughton *et al.* (1981). The model parameters were selected to best simulate general cases rather than being adjusted in response to the specific month of study.

3.2. Case 2 – Total $SO_4^=$ Removal Case

Base case simulations were modified so that each time the trajectory of a pollutant parcel encountered precipitation, scavenged $SO_4^=$ was recorded and the remaining $SO_4^=$ was deleted. Remaining SO_2 was carried along the trajectory where it could later be converted to $SO_4^=$ or removed. Mass balances for this case describe total removal the first time newly formed $SO_4^=$ encounters precipitation.

3.3. Case 3 – Total SO_x Removal Case

Base case simulation trajectories were terminated on the first encounter with precipitation so that no SO_2 or $SO_4^=$ was carried forward. Actual scavenged SO_2 and $SO_4^=$ were accumulated in deposition arrays while the remaining parcel mass was accumulated for a mass balance summary. The case differs from Case 2 in that no SO_2 is transported after the first encounter thus additional transformation to $SO_4^=$ was prevented.

3.4. Case 4 – Pre-Encounter Case

Case 4 accumulates deposition values only up to the time of first precipitation. It therefore describes the dry deposition that occurs along a trajectory before any precipitation removal occurs.

Simulations were performed using two sets of emissions data. The first was a SO_2 emissions inventory consisting of the major SO_2 sources in the northest U.S. and southern Canada. The inventory consists of 63 sources which represent 60 to 65% of total emissions in the area. This data set was used to give broad spatial coverage in estimating the number of trajectories which would encounter precipitation over the U.S. portion of the grid. For economy, most simulations were made using ten SO_2 emission sources representing the largest utility emission sources in the area of calculation (Figure 1). The modeled sources are combinations from multi-stack plants or multiple plants located in the same vicinity. The total emissions for the 10 sources represented approximately 23% of the total SO_2 emissions for the area of calculations. The magnitude and distribution of these sources are sufficient to represent $SO_4^=$ air concentrations over the Northeast. One other set of simulations was performed using a gridded 1973–74 SO_2 and NO_2 emissions inventory to calculate rainfall pH.

Emissions in the model are transported using trajectories originating hourly from each source. Simulations were performed using 720 h of consecutive October 1977 data which resulted in 7200 trajectories for analysis in the 10 source test cases and 45 360 trajectories for the 63 source test.

4. Results

Simulations were designed to provide estimates in response to the following questions
 (1) How many trajectories leaving a source are subject to precipitation scavenging?

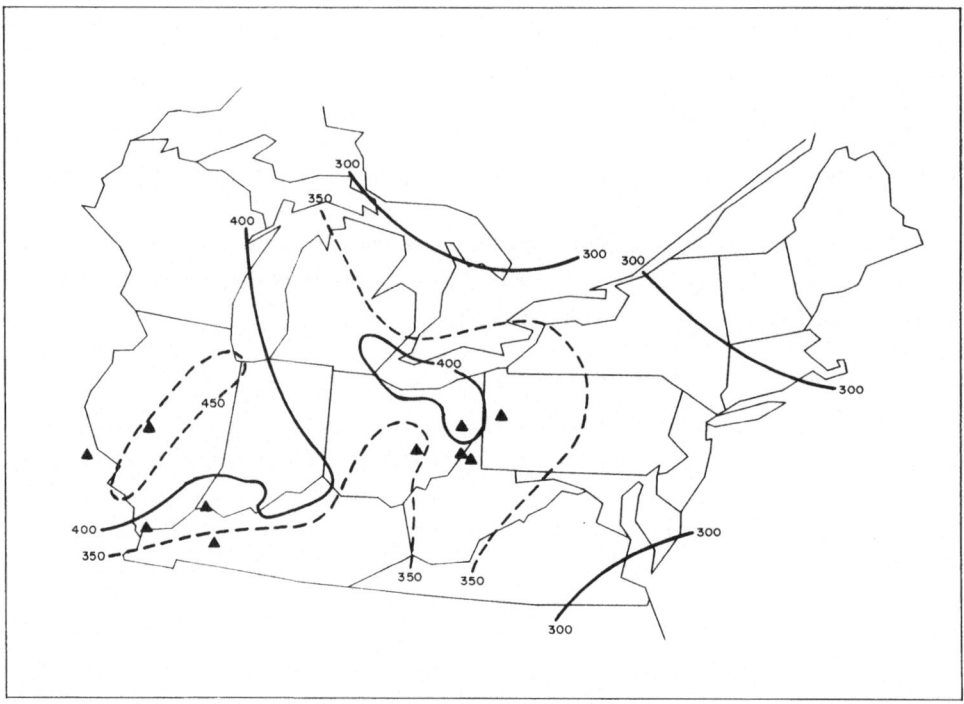

Fig. 1.   Number of hourly trajectories (out of 720) per month from each of 63 sources per month encountering
precipitation over U.S. land areas in the Northeast. The 10 largest sources are indicated by triangles.

(2) How much mass is involved in the following deposition processes:
– precipitation scavenging during the first hour of a plume encounter with a precipitation system
– dry deposition before any precipitation encounter
– dry deposition scavenging over the first hour of a $SO_4^=$ plume encounter with
precipitation after its formation from SO_2?

Results of the simulations are presented primarily using mass balance summaries from
the model (Table I) and sulfate deposition patterns (Figure 2). Mass balance results are
given in Table I by eight categories. SO_2 and $SO_4^=$ deposition represent the sum of wet
and dry deposition. SO_2 and $SO_4^=$ off grid, represent the mass which is transported off
the grid, or deleted in the different study cases. Wet deposition of $SO_4^=$, total wet
deposition and the sum of SO_2 and $SO_4^=$ left in the air at the end of the run are the
remaining categories. Approximately 6% of trajectories remain over the grid at the end
of a one month simulation. Table II presents the fractional decrease of concentration and
deposition maxima from the base case.

Prior to discussing the disposition of emissions from the ten primary sources used in
testing, it is informative to examine Figure 1, the frequency of trajectories encountering
precipitation. Contours in the figure represent the total number of trajectories from each
of 63 major source locations which encounter precipitation before being transported out
the area of calculation. For example, the 450 contour over central Illinois indicates that

TABLE I

Mass balance summary (percent of initial SO_2 emissions)

	Case			
	1 Base case	2 Total $SO_4^=$	3 Total SO_x	4 Pre-removal
Categories				
Total SO_2 deposition	41.8	41.8	30.3	29.5
Total $SO_4^=$ deposition	9.5	2.5	1.5	0.7
Total SO_2 off grid	28.3	28.3	50.3	52.8
Total SO_x off grid	16.1	23.2	13.7	12.8
SO_2 left on grid	3.0	3.0	2.8	2.9
$SO_4^=$ left on grid	1.4	1.3	1.3	1.3
Total	100	100	100	100
Subcategories				
Wet deposition of $SO_4^=$	8.3	1.6	0.8	0
Total wet deposition	9.2	2.5	1.0	0

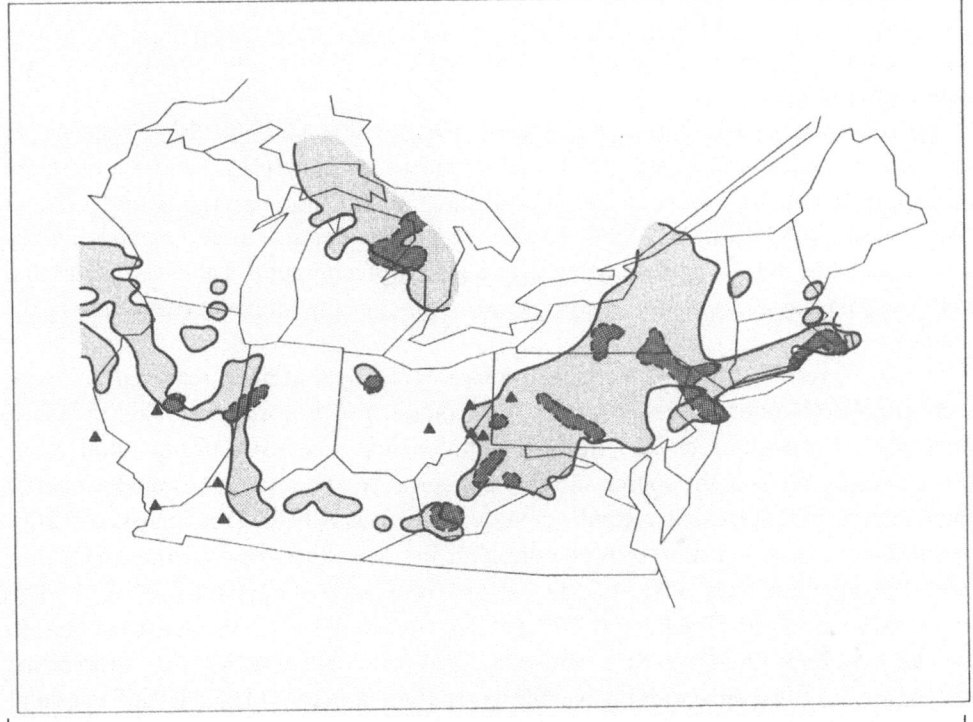

Fig. 2.   Wet deposition of $SO_4^=$ for October 1977 for 10 sources (triangles) (base case simulation results $< 40 \, g \, m^{-2}$ light shadowing; Case 3 deposition $< 20 \, g \, m^{-2}$, dark shading).

TABLE II

Fractional decrease in maximum values from base case (percent)

	Case		
	2 Total $SO_4^=$ removal	3 Total SO_x removal	4 Pre-encounter
SO_2 concentration	0.0	5.7	7.0
$SO_4^=$ concentration	15.6	20.9	21.6
SO_2 deposition	0.0	6.6	8.7
$SO_4^=$ deposition	64.4	65.5	98.0
Total wet deposition	65.1	66.3	100.0
$SO_4^=$ wet deposition	64.4	65.4	100.0

450 of 720 trajectories calculated from a source in the area during October 1977 would have encountered precipitation before being transported out of the U.S. land area shown. In the study month, 52% ($\pm$ 8%) of trajectories from the 63 major SO_2 emitting sources encounter precipitation over U.S. land areas on the grid. Trajectories from sources in the northern and eastern portions of the grid show a slightly less chance of hitting precipitation because of their nearness to the grid boundaries and the occurrence of prevailing westerly flows, but the numbers of these trajectories which encounter precipitation are still significant.

An examination of the base case in Table I indicates that 44.4% of SO_2 emissions are transported off the grid as either SO_2 (28.3%) or as sulfate (16.1%) and are therefore deposited elsewhere. Total deposition in the area of simulation accounts for 51.3% of SO_2 emissions (SO_2, 41.8%; $SO_4^=$, 9.5%). SO_2 deposition is predominately by dry processes while $SO_4^=$ is predominately scavenged by precipitation. Table I indicates that $SO_4^=$ scavenging accounts for $\sim$ 90% (8.3%/9.2%) of the total wet deposition in the model.

In Case 2 results, the $SO_4^=$ deposition represents only that mass removed during or before the first hour of the first precipitation encounter after its formation. This, therefore, includes dry deposition of $SO_4^=$ and precipitation scavenged $SO_4^=$ formed both along the trajectory between the source and first precipitation and also that formed from SO_2 between precipitation events. Results of these calculations (Table I) indicate that if $SO_4^=$ scavenging were assumed to be total during the first encounter newly formed $SO_4^=$ had with precipitation, then 1.6% of SO_2 emissions would be deposited as $SO_4^=$. This represents only a small fraction ($\sim$ 20%) of $SO_4^=$ deposited wet in the base case. Results also show a small change in $SO_4^=$ left on the grid which along with a 16% drop in the maximum $SO_4^=$ air concentration over the base case simulations (Table II) indicates that on the average $SO_4^=$ air concentrations are not very sensitive to the level of $SO_4^=$ scavenging in the model.

Case 3, results in Table I show the impact of the first scavenging of a plume after leaving a source. During the first hour of encounter, 0.8% of total SO_2 emissions are removed by precipitation as sulfate. This indicates that in the model 90% of wet $SO_4^=$ deposition as shown by the base case occurs within a storm system rather than on the fringes of the system.

Case 4, giving pre-encounter results, indicates by definition no precipitation scavenging but dry deposition of $SO_4^=$ before first encounter accounts for 0.7% of emissions. Average SO_2 dry deposition before first plume encounter is 29.5% of initial emissions. The disparity between SO_2 and $SO_4^=$ pre-encounter dry deposition is a result of their difference in being primary and secondary pollutants. Comparison of dry deposition for the base case and case 3 indicates that combined dry deposition of SO_2 and $SO_4^=$ along trajectories before precipitation events is very significant.

Figure 2 illustrates the impact of $SO_4^=$ wet deposition during the first precipitation encounter as compared to the base case monthly deposition pattern. The figure shows that first encounter deposition (Case 3) maxima are located at some distance from sources and in limited bands. These bands contribute to the maximum case pattern and in some cases, for example in the high deposition area over New Jersey, can contribute the majority of the $SO_4^=$ deposition.

Patterns of maximum sulfate wet deposition are loosely tied to the monthly precipitation patterns as is indicated by Figure 3 which gives monthly precipitation totals. Work

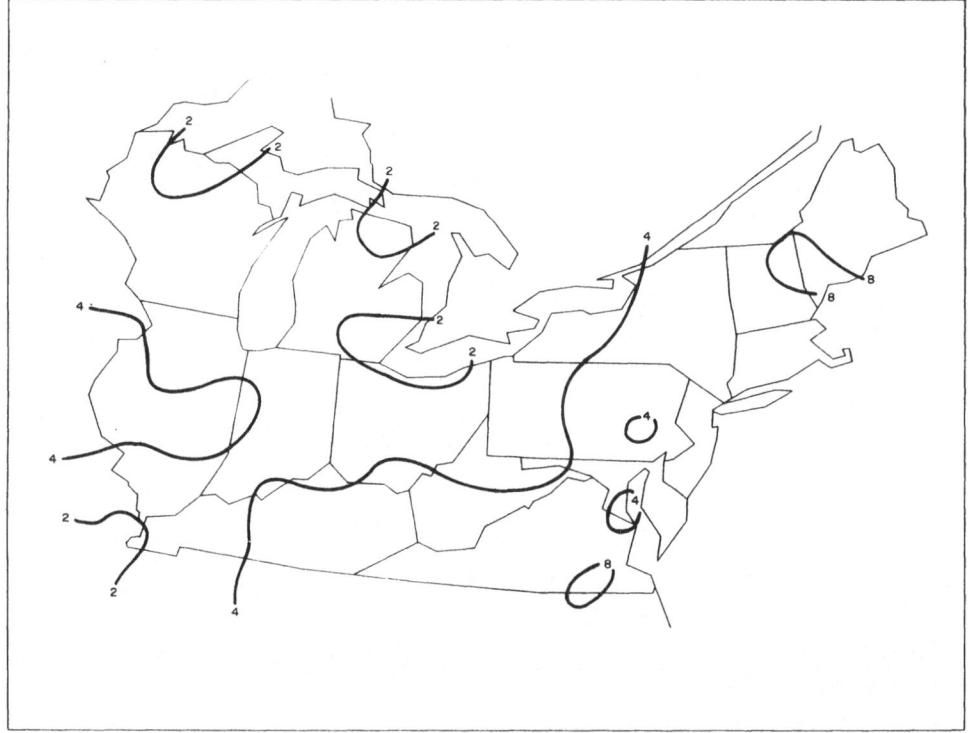

Fig. 3.   Precipitation totals for October 1977 (inches).

by Wendell *et al.* (1977) indicated the need for high resolution precipitation data to show variability of deposition demonstrated by the model in Figure 2. Figure 4 ties deposition and precipitation patterns in a figure showing mean rainfall acidity expressed as pH. This map from an analysis by McNaughton (1981) shows simulated pH patterns based on calculations from all U.S. sources in the Northeast using NO_2 and SO_2 emissions and an empirical relationship between deposition values and total rainfall and pH. pH calculations consider scavenged $SO_4^=$ and NO_3^- and precipitation volume to estimate concentrations in the rainwater. Figures 2 and 4 show common features of high $SO_4^=$ deposition and high acidity (low pH) predominantly over northern Michigan, New York, and Pennsylvania. This figure along with Case 3 patterns in Figure 2 indicate that first encounters do contribute to regions of low precipitation acidity.

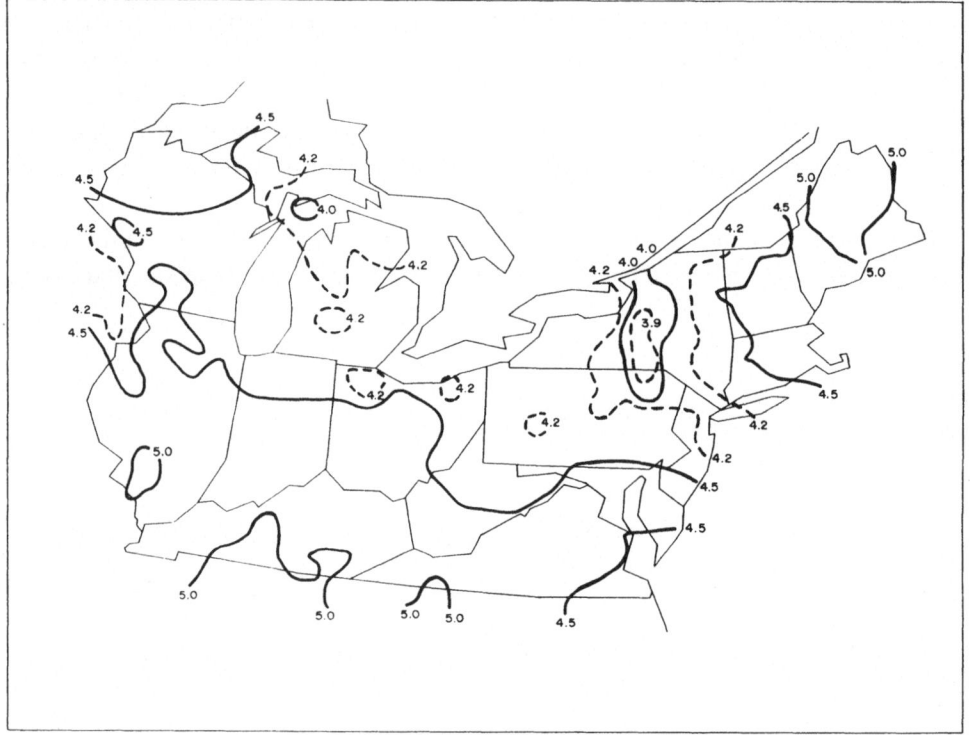

Fig. 4.   Predicted rainfall acidity for October 1977 (pH).

5. Summary

Several useful results were derived from the October 1977 study for gauging model sensitivity. These results represent an estimate of some components of the long range pollutant transport problem as represented by a model for a 1 mo period. Some observations from the tests are:

— 50 to 60% of emissions parcels released over the Northeast encounter precipitation over the northeast United States. In the study, approximately half of the trajectories calculated from 63 emission sources distributed over the Northeast passed through a precipitation area at least once over the U.S.

 – 1.0% of SO_2 emissions are deposited wet during a plume's first encounter with precipitation. This represents approximately 10% of the total wet deposition caused by scavenging in all precipitation areas.

 – Approximately 30% of SO_2 emissions are deposited dry as SO_2 before precipitation is encountered. Since total dry deposition of SO_2 normally accounts for 41.8% of emissions this indicates that 70% of SO_2 dry deposition occurs before a plume reaches precipitation.

 – Approximately 1.5% of SO_2 is deposited dry as $SO_4^=$ before first encounter with precipitation.

From the results it appears that airborne $SO_4^=$ concentrations may be satisfactorily simulated by a simple model such as RAPT without a sophisticated transport scheme for flow within storms. The sensitivity test case results showed that total airborne $SO_4^=$ mass was equivalent to that of the base case as well as showing a decrease in maximum $SO_4^=$ concentrations of only 15 to 20% (Table II). SO_2 deposition is predominantly described by dry processes ahead of storm systems again showing low sensitivity to wet scavenging and in-storm transport. To provide estimates of $SO_4^=$ deposition and precipitation acidity, in storm transport simulations should be considered since first encounter precipitation scavenging accounts for only 10% of total wet $SO_4^=$ deposition.

Acknowledgments

Model runs were supported by the U.S. Environmental Protection Agency under a related services agreement with the U.S. Department of Energy (Contract DE-AC06-76RLO-1830).

References

Hilst, G. R.: 1981, 'Regional Air Quality Studies: The Needs and Priorities', Electric Power Research Institute Report EA-1650-SR, Palo Alto, CA.
McNaughton, D. J.: 1980, *Atmos. Envir.* **14**, 55.
McNaughton, D. J.: 1981, *Atmospheric Environment* **15**, 1075.
McNaughton, D. J. and Scott, B. C.: 1980, *Journal of Air Pollution Control Association* **30** (30).
McNaughton, D. J., Berkowitz, C. M., and Williams, R. C.: 1981, *J. Appl. Met.* **20**, 61.
Scott, B. C.: 1978, *J. Appl. Met.* **17**, 1375.
Wendell, L. L., Powell, D. C., and McNaughton, D. J.: 1977, 'Joint Conf. on Applications of Air Pollution Meteorology', SLC, Ut.

A COMPUTED SULPHUR BUDGET FOR THE
EASTERN CANADIAN PROVINCES

M. P. OLSON, E. C. VOLDNER, and K. K. OIKAWA

Atmospheric Environment Service, 4905 Dufferin Street, Downsview, Canada M3H 5T4

(Received 22 June, 1981; Revised 10 September, 1981)

Abstract. Sulphur budgets for Ontario, Quebec and the Atlantic Provinces have been computed using the Long-Range Transport of Air Pollutants model (LRTAP) which has been developed within the Atmospheric Environment Service of Canada. Meteorological data from 1978 and a North American SO_2 emissions inventory for 1970–1974 form the basic model input.

The S budgets for the eastern Canadian regions were computed for large-scale emission scenarios. The budget shows the trans-boundary mass transport, S deposition and S concentrations within the regions for each scenario and shows the relative contribution to the deposition in each region.

For eastern Canada, the model shows an annual S transboundary input of about 2 Tg S, an emission of about 1.8 Tg S, a deposition of about 2.4 Tg S and an output of about 1.4 Tg S. For southwestern Ontario, the model shows an annual average SO_2 concentration of 25 to 30 µg m^{-3} (10 ppb), an annual sulphate concentration of about 8 µg m^{-3}, an annual wet deposition of S of about 15 kg S ha^{-1} and an annual sulphate concentration in precipitation of about 5 to 6 mg l^{-1}.

1. Introduction

Much concern has been expressed recently about the release and transport of S compounds into the air and the deposition of these compounds in the form of acid precipitation in eastern Canada. The Atmospheric Environment Service Long-Range Transport of Air Pollutants model has been applied to the problem of estimating the annual S budget of eastern Canada. The present Lagrangian model consists of a trajectory model component and a concentration/deposition model component. The trajectory model computes the atmospheric pathways followed by the pollutants and the concentration model computes the concentration and deposition of pollutants in the air parcels as they move across emission and precipitation fields toward specified receptor points.

The S mass budget is based on the concept that the mass inflow across the boundaries of a region plus the emissions within the region should approximately balance the total deposition within the region plus the mass flow out of the region. The model has been adapted to provide estimates of the individual budget terms. The results, which are discussed in this paper, should be viewed with some caution since the parameterization of the physical and chemical processes and the emissions inventory itself have not yet been fully evaluated and stabilized. In addition, the model is presently being evaluated and the model results may be revised. Figure 1 shows the modelling area.

Water, Air, and Soil Pollution **18** (1982) 139–155. 0049–6979/82/0181–0139$02.55.
Copyright © 1982 *by D. Reidel Publishing Co., Dordrecht, Holland, and Boston, U.S.A.*

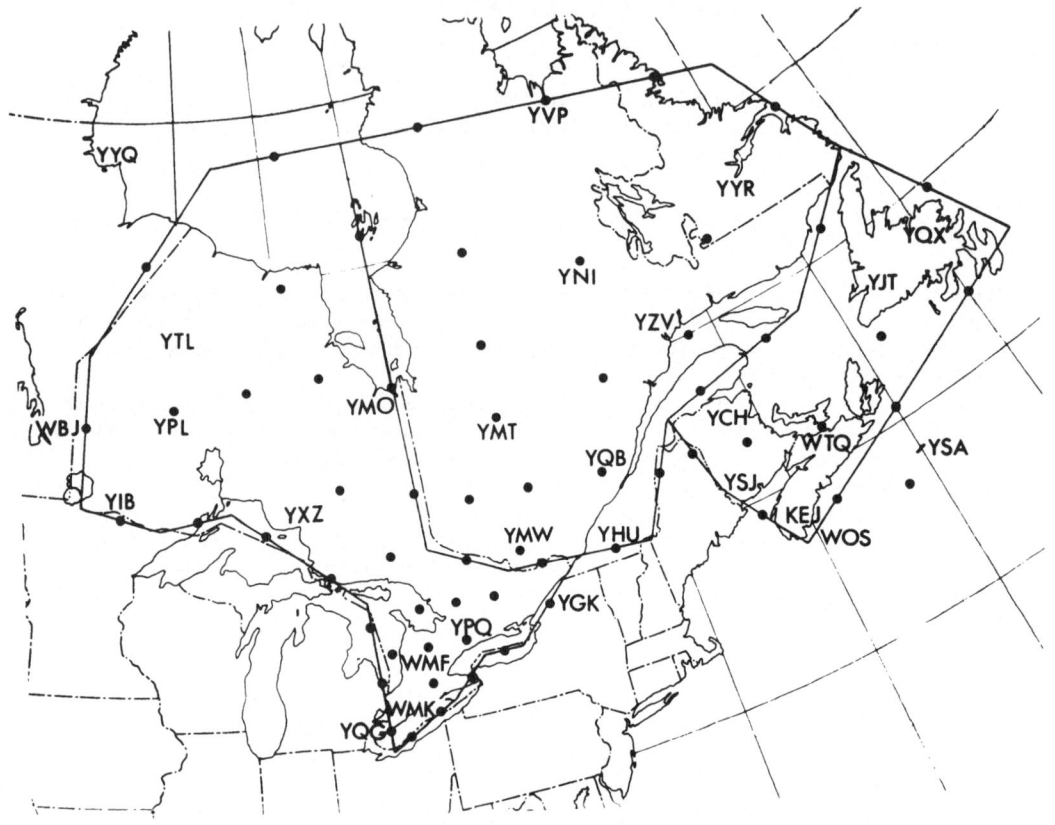

Fig. 1.   Eastern Canadian modelling area.

2. Trajectory Model

The trajectory model (Olson *et al.,* 1978) uses objectively analyzed wind fields and computed vertical motions at the following four pressure levels: 1000, 850, 700, and 500 mb. Input wind fields are available every 6 h and interpolation routines are used to obtain winds at intermediate positions in time and space. The wind data are objectively analyzed at the Canadian Meteorological Centre (CMC) every 6 h onto a 381 km grid superimposed on a polar stereographic projection true at 60° N. The analysis procedure is essentially a three dimensional scheme that incorporates hydrostatic and height-wind balance routines and produces gridded u and v wind components (Rutherford, 1977). The trajectory computations are performed on the Canadian Meteorological Centre grid and the analyzed winds are interpolated to a 127 km grid length. The trajectory model assumes a constant acceleration between segment end-points so that an iterative scheme is used to determine the end-point positions. The starting points can be located anywhere in the North American domain between 1000 and 500 mb. Trajectories can be computed either forward or backward for up to 5 days at a specified time-step interval and are normally computed backward for four days at 6 h intervals.

In order to simulate the ascent of parcels through low pressure areas and descent through high pressure areas, grid point fields of vertical motion are computed every 6 h. In our estimation, the Haltiner technique adequately represents atmospheric vertical motions and allows air parcels to rise and fall, thereby passing through changing wind regimes and permitting the important and substantial influence of vertical wind shear to affect the parcel trajectory.

No modifications are made to the analyzed wind components (u, v) at the upper levels (850, 700, 500 mb). However, due to the occasional lack of observed wind data at the 1000 mb level, the geostrophic assumption has a strong influence in the objective wind analysis and a frictional turning term is applied to the 1000 mb wind components to take partial account of the ageostrophic component or cross-isobar flow.

A numerical analysis of the trajectory model has been conducted and a report has been prepared (Walmsley *et al.*, 1981). An analytic non-divergent wind field was formulated to simulate a real atmospheric wind field and was used to intercompare the model trajectory positions with the analytic solutions to the trajectory equation. No serious deficiencies were found in the formulation of the model and the largest source of error seemed to be in the horizontal interpolation routines. The use of cubic interpolation gave smaller errors than the present linear interpolation but required a significant increase in computation time.

The modelling procedure may produce errors especially in shearing flows and light, variable flows. Many trajectory modelling techniques are being used but all are subject to inaccuracies in the wind fields and numerical techniques. However, the assumption is that over a month to a year time scale, the errors will tend to compensate and that the large scale features of the flow will dominate and govern the basic transport of the pollutants.

3. Concentration/Deposition Model

The model (Olson *et al.*, 1979) is based on the CMC grid with a grid length of 127 km at 60° N. The one-layer model parameterizes the physical and chemical processes within a box of unit area cross-section which extends vertically from the ground to the mixing height. The mixing layer is regarded as being capped by an inversion and pollutant removal is parameterized by wet and dry deposition and chemical transformation. Pollutant input to each box is provided from an annual, North American, SO_2 gridded emissions inventory on a 127 km grid (Voldner and Shah, 1980) and instantaneous mixing occurs throughout the box.

The boxes follow trajectories that have been previously computed and stored by the trajectory model. At each timestep (3 h), there is a pollutant input from the inventory, a chemical transformation and a surface deposition. The combination of these processes results in a new concentration value within the box. The new concentration value is carried over to the next point where the process is repeated. An improved numerical procedure using the trapezoidal rule, which assumes a linear parameter change during the timestep, has recently been satisfactorily tested. The final concentration in the box is assumed to be the final surface concentration at the receptor.

The simplified conservation equations describing the movement, emissions input, transformation and net removal of SO_2 and sulphate are given in parameterized form in terms of pollutant concentration by:

$$\frac{dC_1}{dt} = -\frac{(V_{d1} + \alpha_1 P)}{H} C_1 - k_t C_1 + f_1 \frac{Q}{H} \tag{1}$$

$$\frac{dC_2}{dt} = -\frac{(V_{d2} + \alpha_2 P)}{H} C_2 + \tfrac{3}{2} k_t C_1 + f_2 \frac{Q}{H}. \tag{2}$$

The coupled equations (1) and (2) are solved in Lagrangian form and represent the change of pollutant concentration in an air parcel following a trajectory due to deposition, transformation and source input.

C_1 and C_2 are the atmospheric concentrations of SO_2 and sulphate respectively. H represents the climatological mixing height field which has been gridded over North America on the 127 km polar stereographic grid on a monthly basis. A typical grid point has a monthly value which averages about 500 m in winter and about 1500 m in summer over eastern North America.

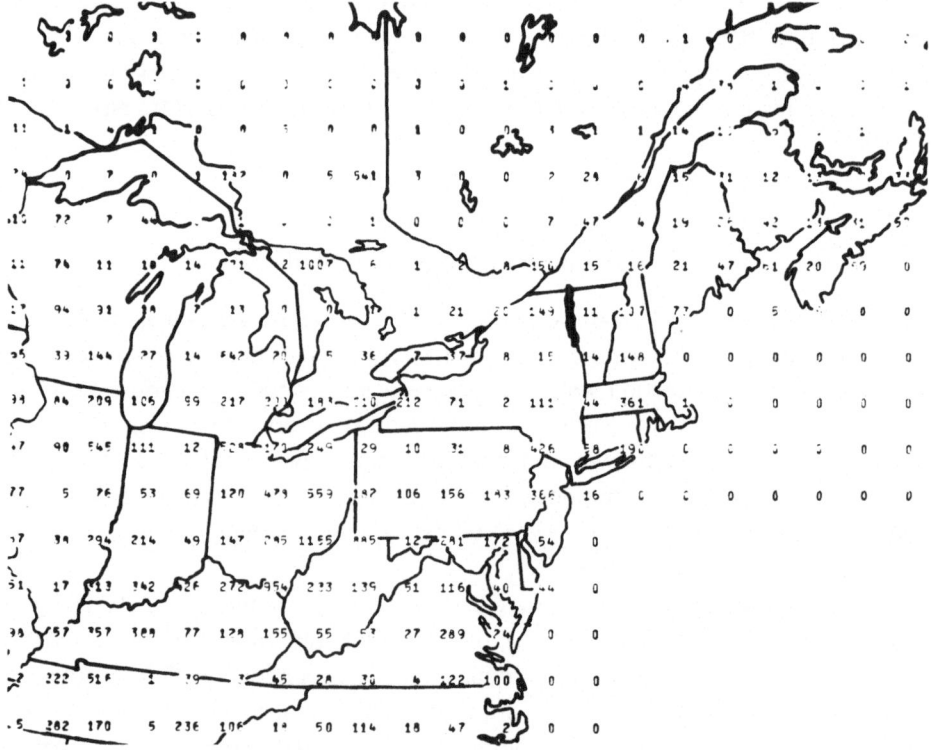

Fig. 2.   Annual SO_2 emissions inventory in kilotonnes.

Annual SO_2 emissions are gridded on the 127 km grid and are represented by the parameter Q. The value at each grid point represents a uniform emission over a grid square. Sub-grid scale removal of S or transformation of SO_2 to sulphate can be approximated by the factor 'f' applied to the inventory. The factors f_1 and f_2 are normally one and zero, respectively. An inventory subset is shown in Figure 2.

The SO_2 to sulphate transformation rate, k_t, has been given a constant value of 1% h^{-1}.

Dry deposition is parameterized in terms of a deposition velocity, V_{d1} for SO_2 and V_{d2} for SO_4. For this study V_{d1} had the value 1.0 cm s^{-1} and V_{d2} had the value 0.1 cm s^{-1}.

Wet deposition is parameterized by scavenging ratios for SO_2 (α_1) and SO_4(α_2) and by a gridded daily array of precipitation amount (P). For this study α_1 had the value 5×10^3 and α_2 had the value 8.5×10^5. The daily precipitation amounts from the North American synoptic networks were analyzed onto the 127 km grid for use by the AES-LRT concentration model. The precipitation fields are somewhat smoothed but still retain the essential precipitation intensity patterns and give realistic annual station totals of about 700 to 1100 mm yr^{-1}.

For this report, the full North American inventory and the eastern Canadian inventory only were used in order to estimate the deposition and S flux contribution to each provincial region and to the entire eastern Canadian region.

4. Budget Description

The spatial domain of eastern Canada is shown in Figure 1 wherein the provincial and international boundaries are represented by 35 line segments at the center of which a trajectory end-point is located. Similarly, the interior of the region is represented by 27 trajectory end-points. The total budget calculations are based on these 62 points. Each point represents an area and the sum of these areas approximates the total provincial or eastern Canadian area. Four day backward trajectories were computed from each end-point, four times a day, every day for the year 1978 using analyzed wind data from the Canadian Meteorological Centre data archives. These trajectories represent the atmospheric pathways which the air parcels have followed from their location four days back in time to their final designated end-point position. All trajectories started from the 925 mb level and concentrations and depositions are computed for each point as described in Section 3.

The annual average SO_2 and sulphate concentrations are shown in Figures 3 and 4 for the full emissions and in Figures 10 and 11 for the eastern Canadian emissions only. The sulphate concentration in precipitation is calculated by computing the precipitation-weighted sulphate concentration for the year multiplied by the sulphate scavenging ratio α_2 and the values are shown in Figures 5 and 12.

Dry deposition is computed from the product: $C_i \times VD \times \Delta t$. C_i is the concentration (SO_2 or SO_4) at the point 'i', VD is the deposition velocity and Δt is the time between trajectory starts, in this case 6 h. The dry depositions of SO_2 and sulphate are summed over the year at each point and molecularly combined to give the annual regional dry deposition of S per unit area based on a given emission scenario. Dry deposition patterns in kilograms of S per hectare are shown in Figures 7 and 14.

Wet deposition is similarly computed from the product:

$$C_i \times \alpha \times P_i \times \Delta t \,.$$

P_i is the precipitation at point 'i' and α is the scavenging ratio for either SO_2 or sulphate. The wet depositions of SO_2 and sulphate are also summed molecularly over the year at each point and the annual wet deposition of S is shown in Figures 6 and 13 in kilograms of S per hectare.

The wet and dry depositions at each point are combined to give the total annual S deposition which is shown in Figures 8 and 15.

The total annual S deposition at each point is multiplied by the area represented by that point and summed over the region to give the total S deposited within the region.

The mass transported across the boundaries is computed from the product:

$$\pm\, C_i \times L_i \times H_i \times V_{ni} \times \Delta t \,.$$

C is the concentration at the mid-point 'i' of a boundary segment with length L, mixing height H and normal wind speed V_n. The time between trajectory starts (Δt) is 6 h. The normal wind speed is determined by calculating the incident angle between the trajectory and the line segment at point 'i' and retrieving the computed wind speed from the last trajectory time-step. The $\pm$ sign shows whether the flow is directed out of the region or into the region. These mass transports are only calculated using the boundary trajectory endpoints.

All the input mass transports are summed over the regional boundary segments for a year to give the total input mass of S for each region. Similarly, the output mass transports are summed to give the total output mass of S from each region. These computations are done for each emission scenario to estimate relative mass transport contributions across all boundaries.

Schematic diagrams of the S budget for the eastern Canadian regions showing input and output mass transport and emissions and depositions within the regions are shown in Figures 9 and 16. The western Ontario boundary is a combination of three model boundary segments and the southwestern Ontario boundary (Upper Lakes) is a combination of seven boundary segments.

5. Budget Results

With these simplifications and approximations in mind, we would like to discuss some of the annual concentrations and depositions and the budget results as shown in Figures 3 to 16.

Figures 3 to 9 show results using the full North American emissions inventory and Figures 10 to 16 show results using only the eastern Canadian emissions. By subtracting

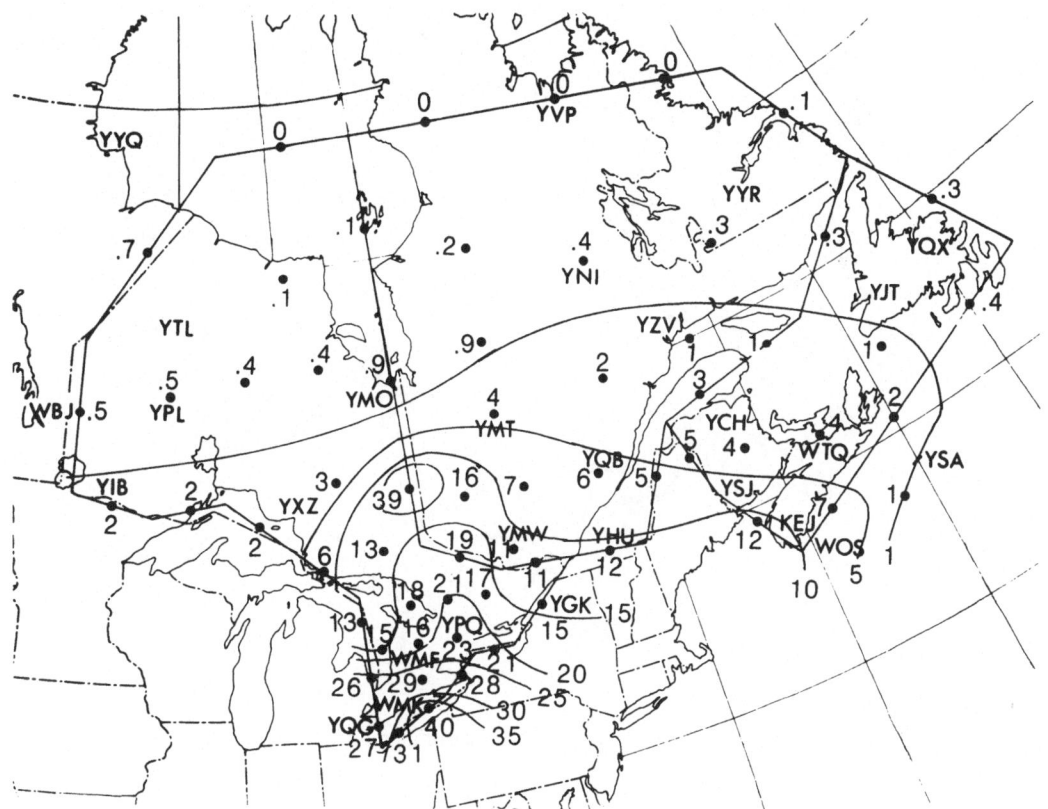

Fig. 3. 1978 annual average SO$_2$ concentration µg m^{-3} full emissions.

corresponding figures (3 and 10, 4 and 11 etc.), the effect of U.S. emissions on eastern Canada can be estimated.

Figure 3 shows the computed annual average SO$_2$ concentrations. An estimated value at Toronto (western Lake Ontario) would be between 20 to 25 µg m^{-3}. The annual average SO$_2$ concentration (1977–1978) at Toronto Island as determined from the Atmospheric Chemistry Division data (Ahmed *et al.*, 1980) is 22.9 µg m^{-3} which is in very close agreement with the model estimate. A receptor point located near the Noranda smelter (on the Ontario-Quebec border) gives a noticeably large concentration value.

Figure 4 shows the computed annual average SO$_4$ concentrations. An estimated value around Toronto would be between 7 to 8 µg m^{-3}. Data from the Toronto Island site give an annual average of 6.8 µg m^{-3} and data from the Ontario Hydro network in the Toronto-Niagara area indicate an average value of about 8.1 µg m^{-3}. Once again the agreement is very encouraging.

Figure 5 shows the computed annual average SO$_4$ concentrations in precipitation in mg l^{-1}. An estimated southern Ontario value would be between 5 to 6 mg l^{-1}. The Ontario Hydro network in southern Ontario indicates a value of about 5.7 mg l^{-1}.

In this area the model estimates agree very well with the measured data. Farther away from the major emissions regions, the model values are less than the CANSAP measure-

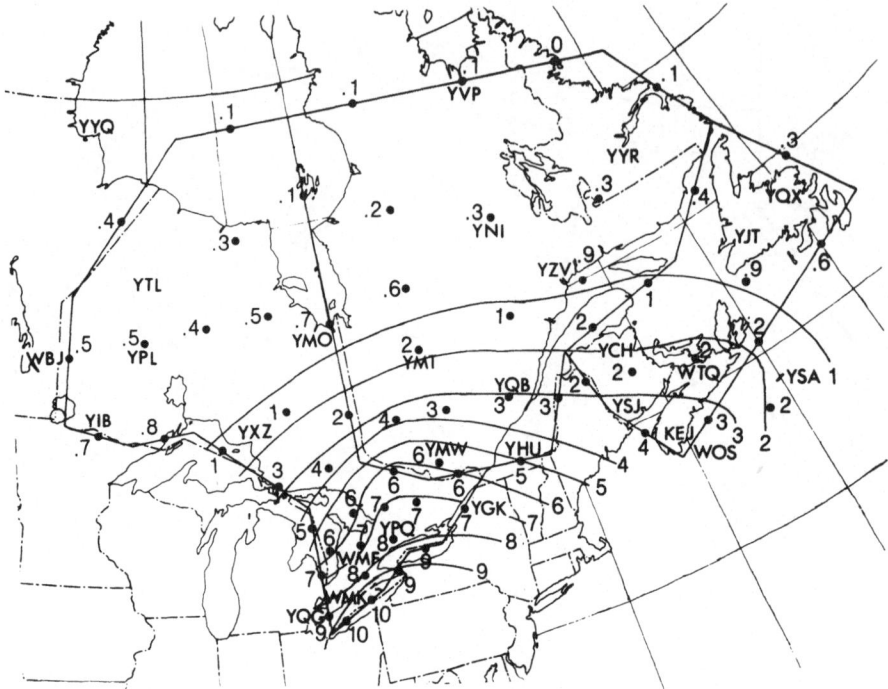

Fig. 4.   1978 annual average SO$_4$ concentration µg m^{-3} full emissins.

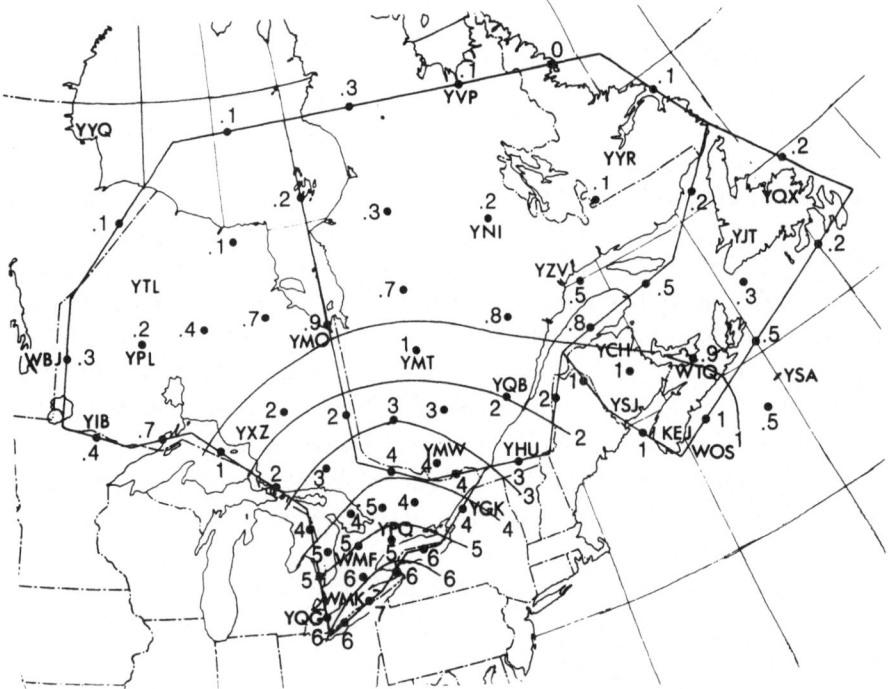

Fig. 5.   1978 annual precipitation-weighted SO$_4$ concentration in mg l^{-1} full emissions.

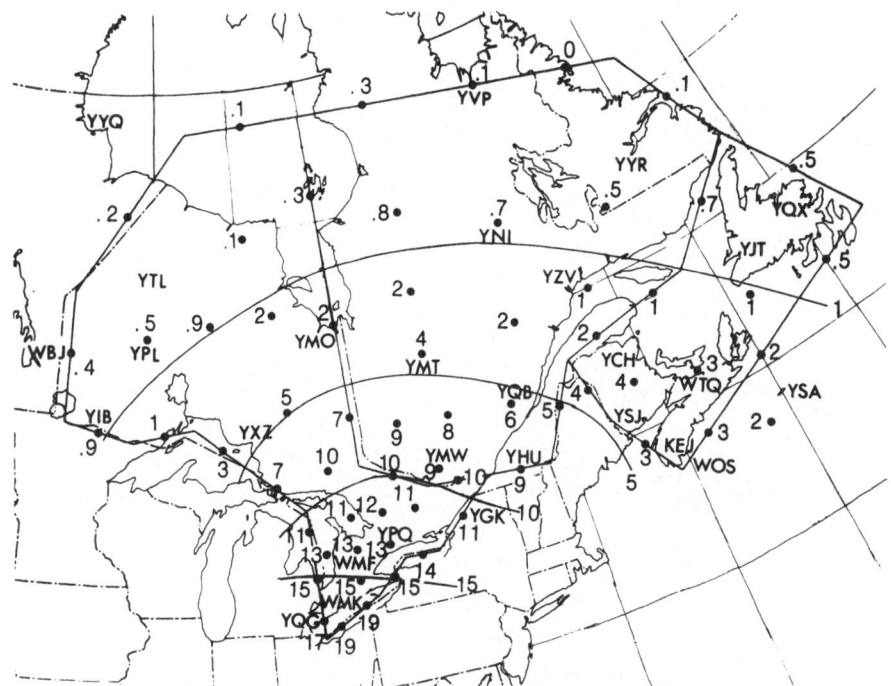

Fig. 6.   1978 annual wet S deposition in kg S ha^{-1} full emissions.

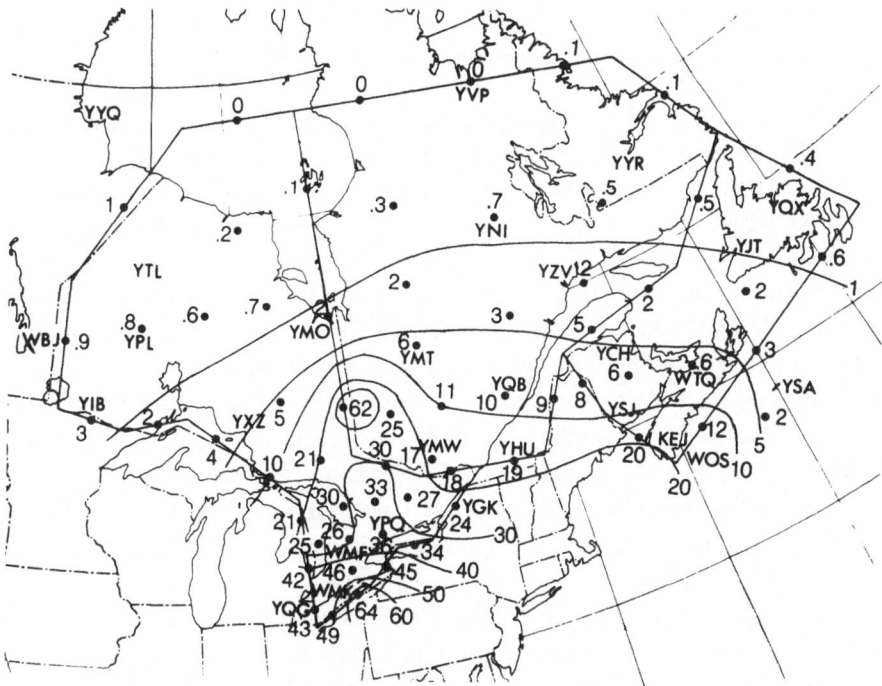

Fig. 7.   1978 annual average dry S deposition in kg S ha^{-1} full emissions.

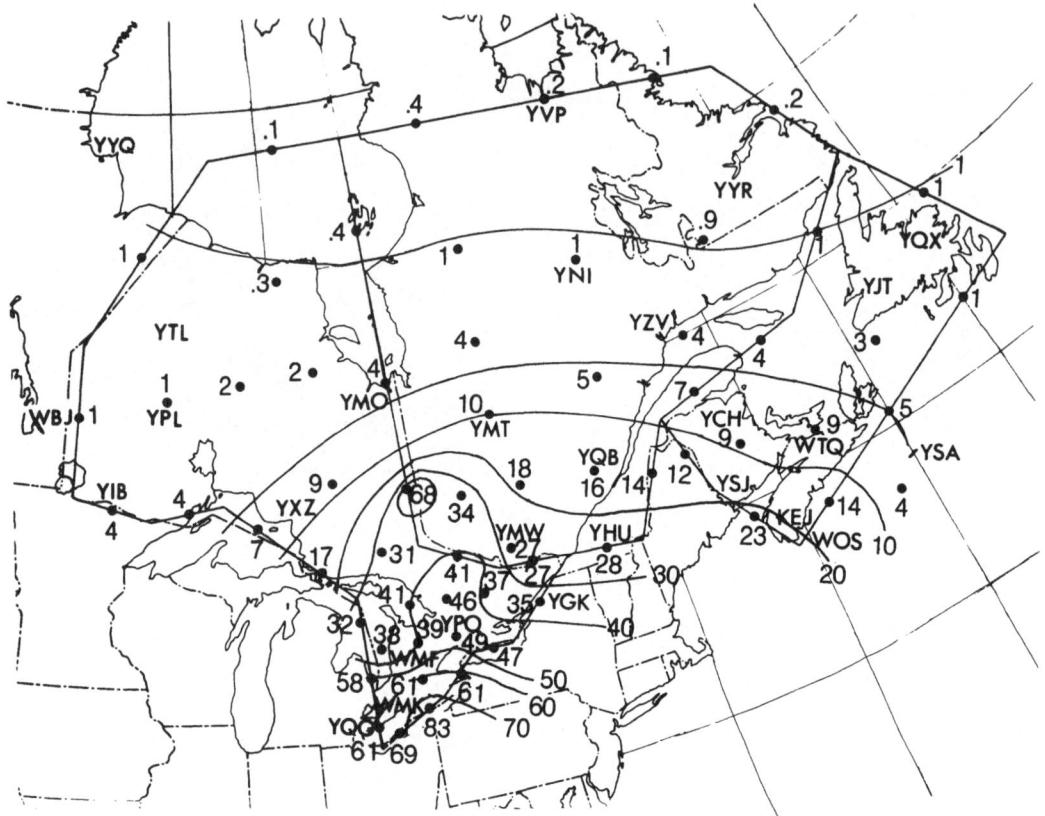

Fig. 8.   1978 annual S deposition (dry + wet) in kg S ha^{-1} full emissions.

ments by factors of 2 to 10. However, the measurements themselves also tend to be small making meaningful comparisons difficult and the model has no background values.

Figure 6 shows the total computed annual wet S deposition in southern Ontario to be about 15 kg S ha^{-1} decreasing to about 3 kg S ha^{-1} in the Nova Scotia and Lake Superior areas.

Figure 9 shows a schematic diagram of the computed 1978 Annual Sulphur Budget in teragrams (millions of tonnes) of S per year (Tg S). The three eastern Canadian regions are shown with emissions (E), depositions (D) and input-output transboundary mass transport indicated by arrows. The total United States S emissions are shown about 14 Tg yr^{-1} and the eastern Canadian S emissions are shown as 1.8 Tg. The total S deposition in eastern Canada is shown as 2.4 Tg annually and the total S mass transported into eastern Canada is shown as 2 Tg. The total S outflow to the U.S. from combined Canadian and United States sources is shown as 1.2 Tg with about 0.2 Tg flowing out to the Atlantic.

The model indicates that the largest amount of S flowing into eastern Canada from the U.S.A. comes through the western Ontario boundary which extends from Windsor to Kenora. Significant amounts of S flow across Ontario and into Quebec (0.6 Tg S) and also across Ontario and back into the U.S.A. (0.7 Tg S). More S is deposited in each

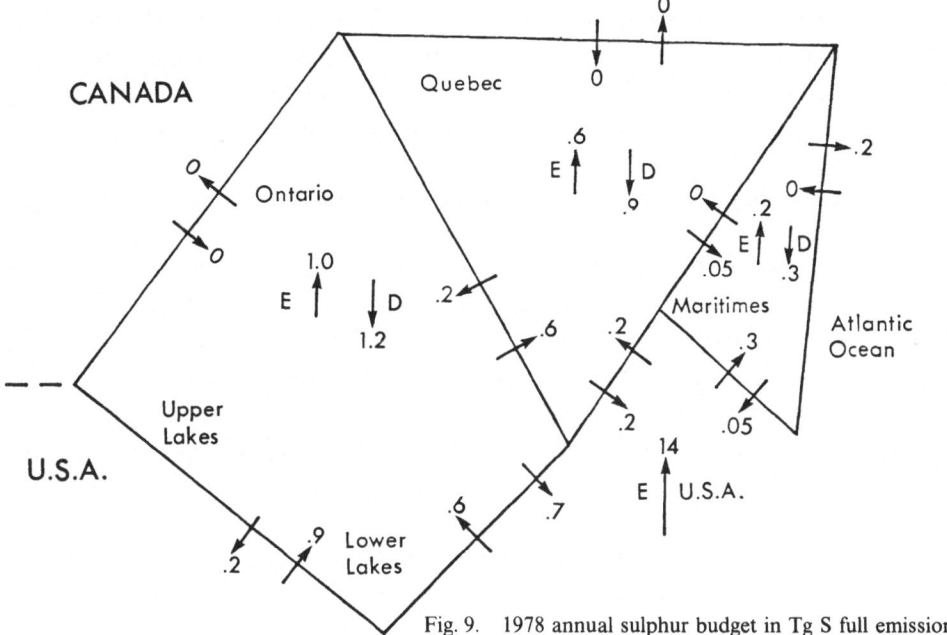

Fig. 9. 1978 annual sulphur budget in Tg S full emissions.

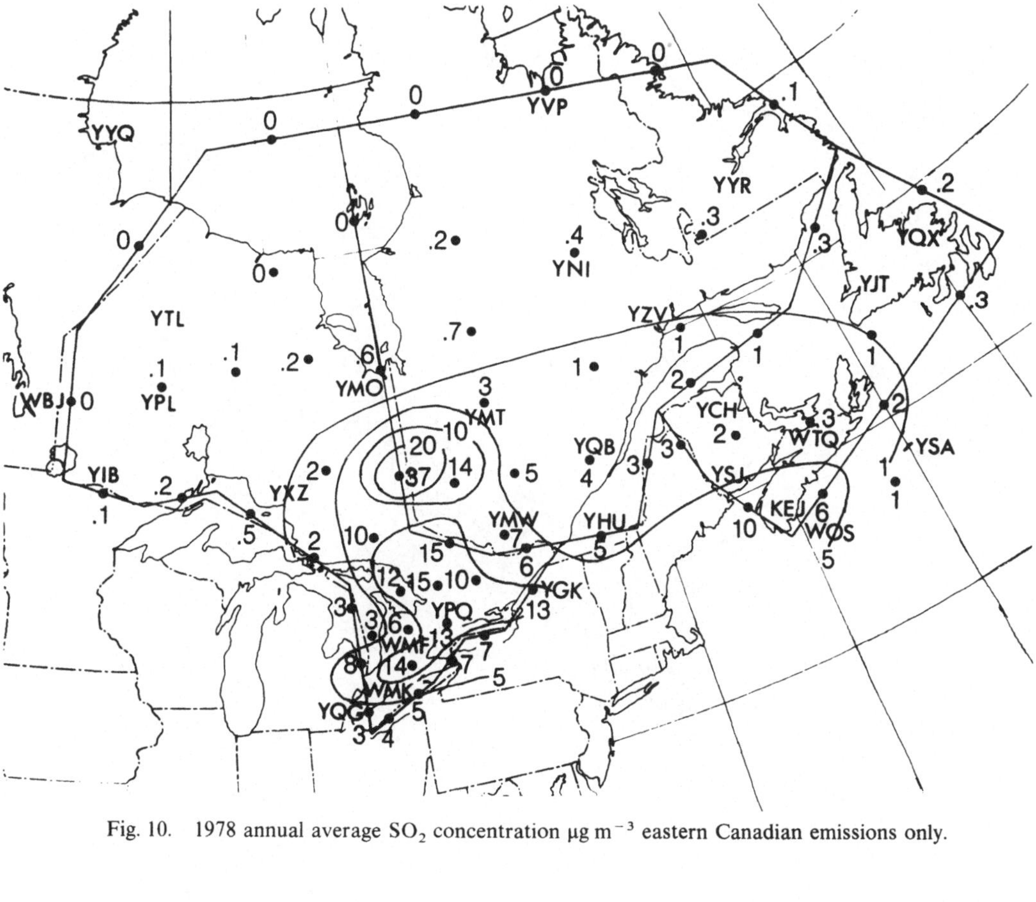

Fig. 10. 1978 annual average SO$_2$ concentration µg m^{-3} eastern Canadian emissions only.

of the three eastern Canadian regions than is emitted by them and Ontario receives the
highest annual S deposition although Quebec plus Labrador have a larger area.

This discussion only considers the anthropogenic S emissions from North America
as contained in the annual emissions inventory and does not consider any background
deposition.

The corresponding Figures 10–16 show similar quantities but are a result of using only
eastern Canadian emissions.

Figure 10 shows a SO_2 concentration maximum of about 10 to 15 µg m^{-3} in south-
central Ontario and a noticeably large concentration near the Noranda site on the
Ontario-Quebec border because of the receptor point location. The computed concentrations at many international boundary points are generally low showing that the model
indicates large contributions from U.S. sources when compared to Figure 3.

Figure 11 shows a maximum sulphate concentration of 3 µg m^{-3} and the location is
shifted somewhat eastward to the southern Ontario-Quebec border. Once again, the
computed international boundary point sulphate concentrations are quite low with
respect to the full emissions concentration scenario (Figure 4). The sulphate concentration pattern also shows an eastward extension to the Maritime Provinces as indicated
in Figure 4.

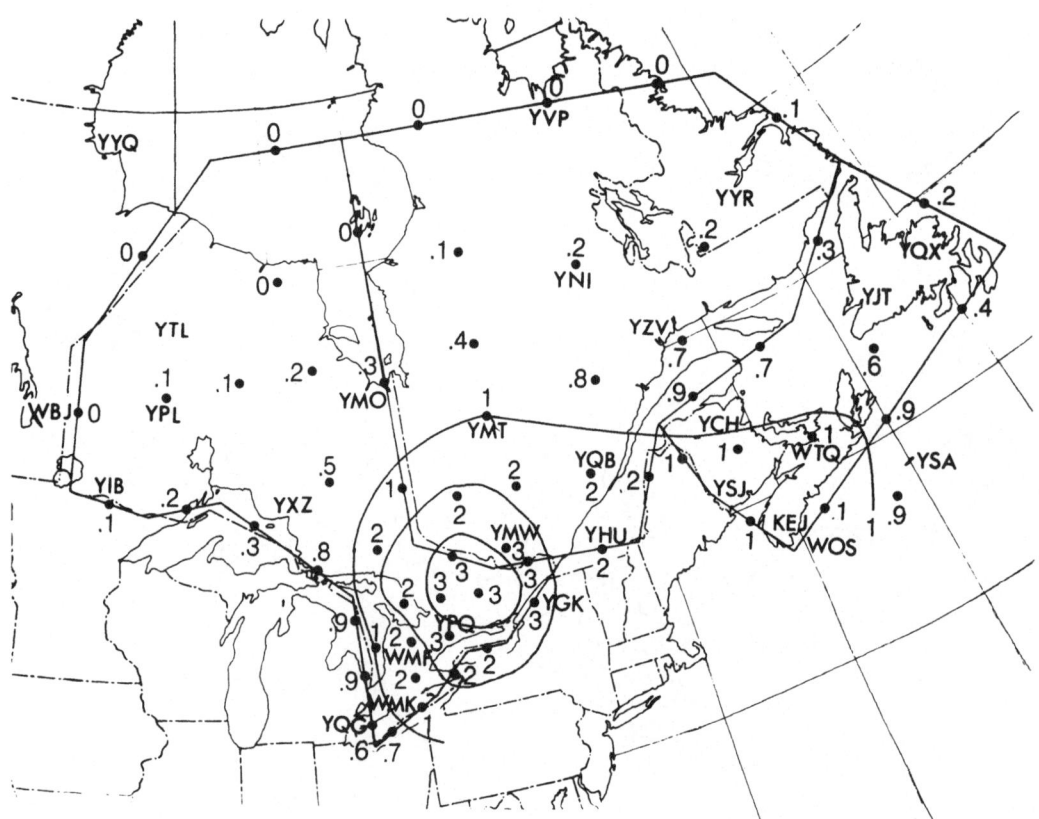

Fig. 11.   1978 annual average SO_4 concentration µg m^{-3} eastern Canadian emissions only.

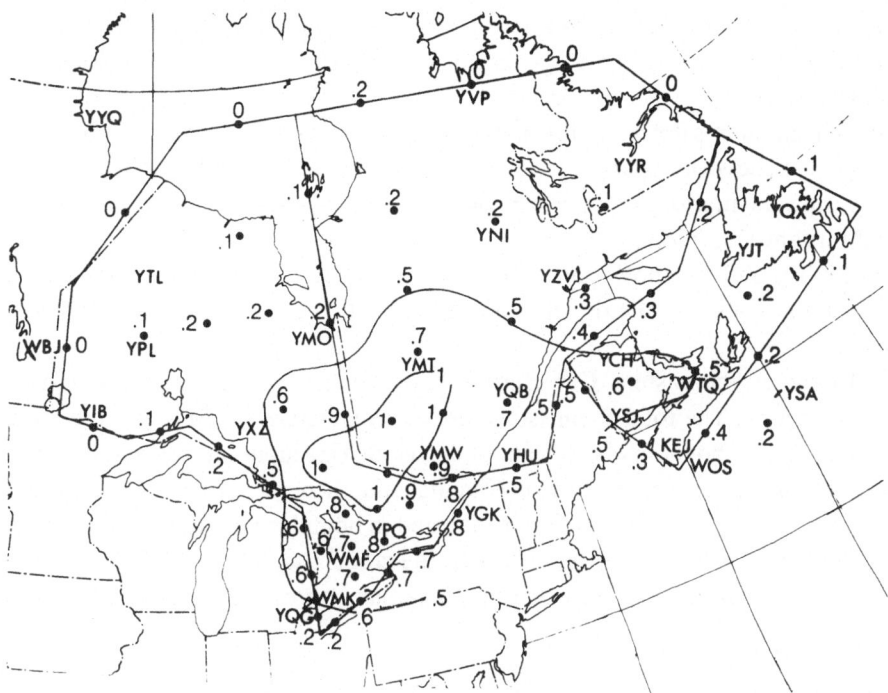

Fig. 12.   1978 annual precipitation-weighted SO_4 concentration in precipitation in mg l^{-1} eastern Canadian emissions only.

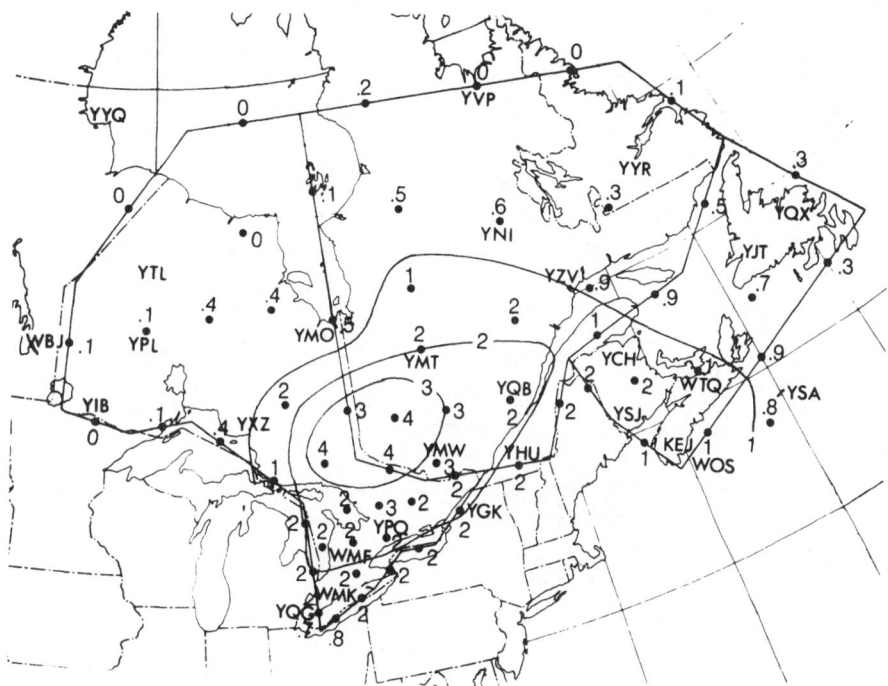

Fig. 13.   1978 annual wet S deposition in kg S ha^{-1} eastern Canadian emissions only.

Only in the extreme northern and eastern portions of the total region do the sulphate concentrations equal or exceed the SO_2 concentrations as shown by comparing Figures 3 with 4 and 10 with 11.

The concentration of sulphate in precipitation shows an annual average maximum of about 1 mg l^{-1} as shown in Figure 12 extending eastward into Quebec from the Georgian Bay area of central Ontario and decreasing down the St. Lawrende River valley into the southern Maritime Provinces.

Figures 15, 14, and 10 show very similar patterns because the total computed S deposition is dominated by the dry S deposition which is directly related to the SO_2 concentration.

Figure 16 shows the eastern Canadian S budget when considering only eastern Canadian emissions. The total S emissions are approximately 1.8 Tg S and the input mass of S through the boundary is only about 0.3 Tg S. This input represents the effects of trajectories that have passed out of Canada and then returned to Canada and represents any effect of Canadian emissions located in boundary emission grid squares. Of the total of 2 Tg S input into eastern Canada, the model suggests that 1.7 Tg S comes from U.S. sources and about 0.3 Tg S from Canadian sources re-entering Canada.

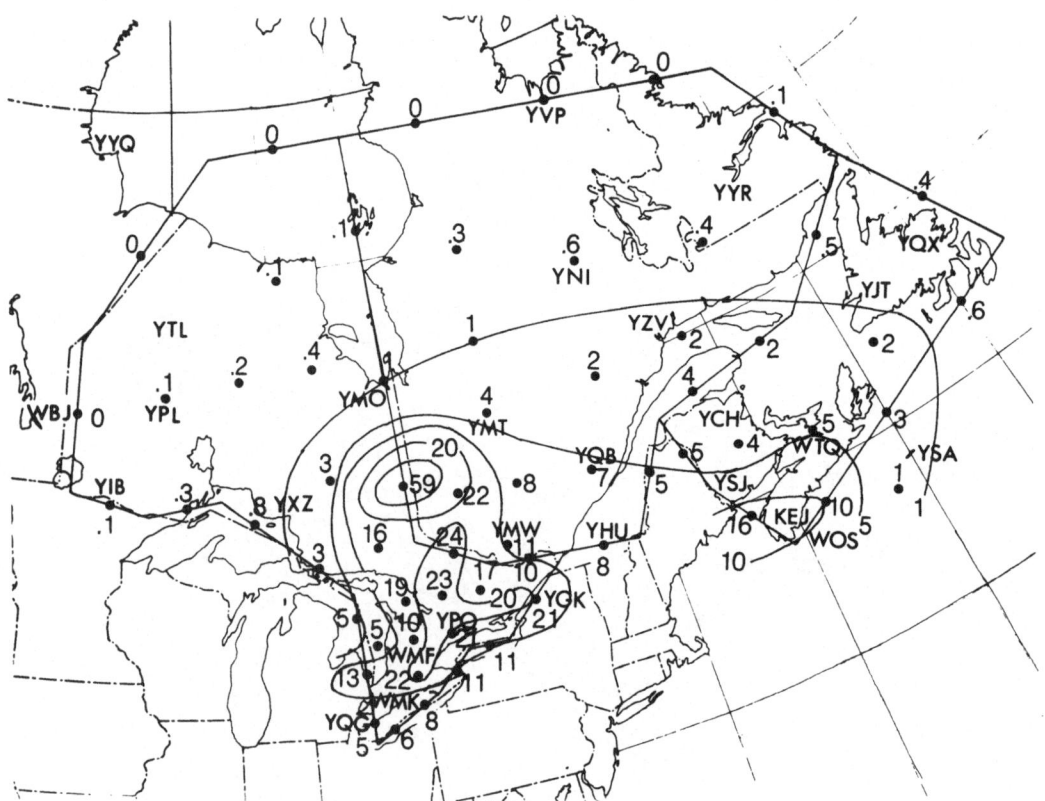

Fig. 14.   1978 annual average dry S deposition in kg S ha^{-1} eastern Canadian emissions only.

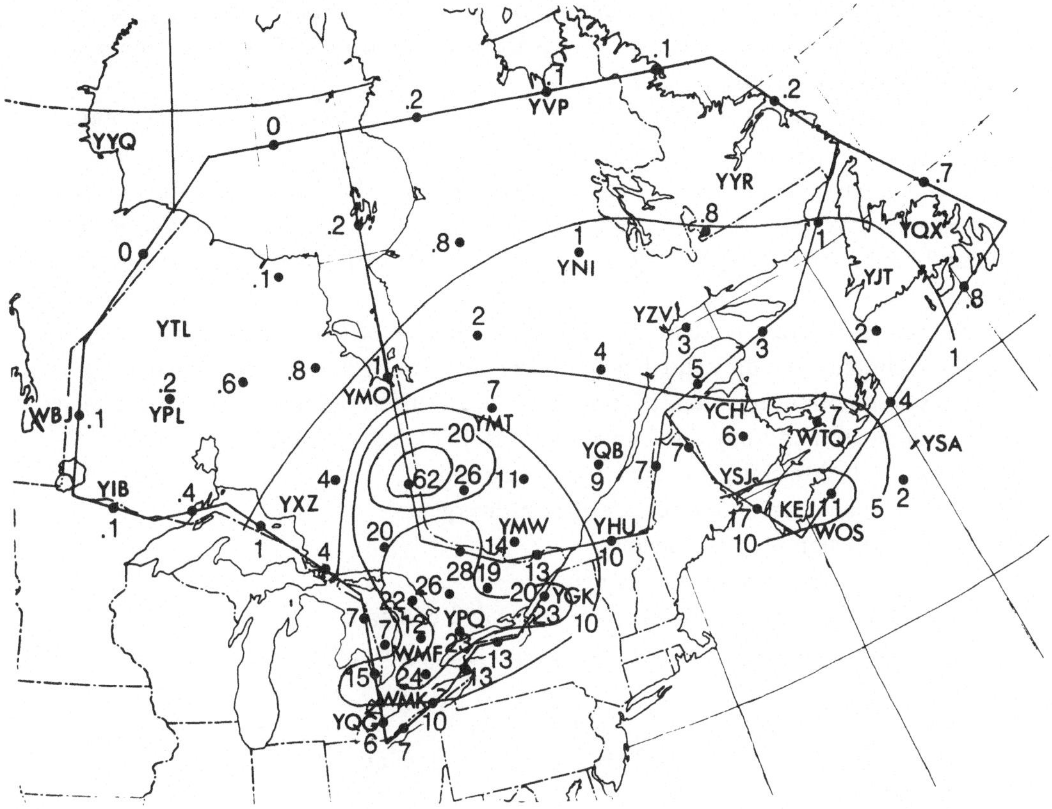

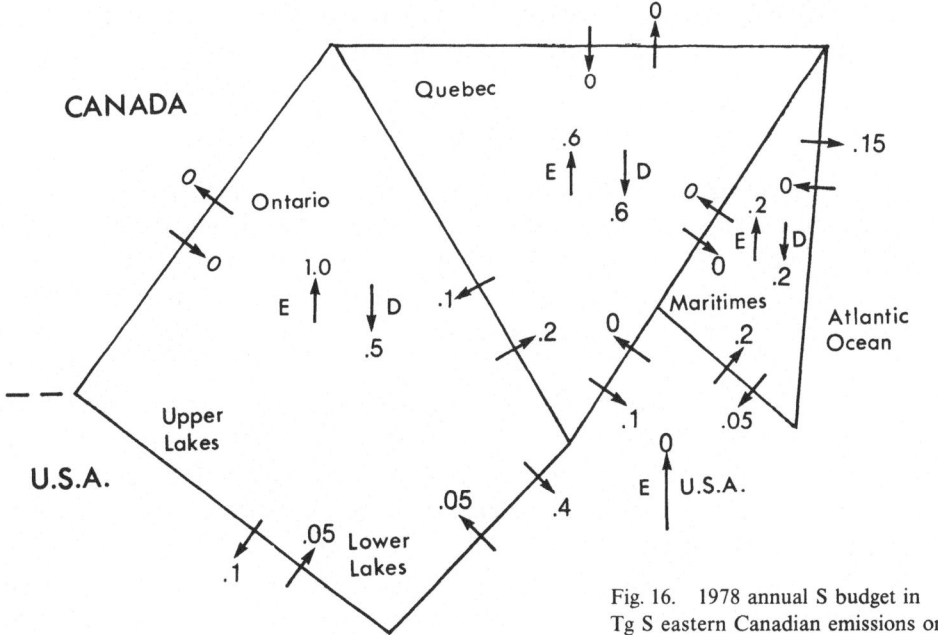

Fig. 15. 1978 annual total S deposition in kg S ha⁻¹ eastern Canadian emissions only.

Fig. 16. 1978 annual S budget in Tg S eastern Canadian emissions only.

The budget diagram shows a S deposition of 1.3 Tg S throughout the region from eastern Canadian sources and thus 1.1 Tg S are deposited from U.S. sources to complete the overall total S deposition of 2.4 Tg S.

Figure 16 indicates that 0.8 Tg S leave the region from eastern Canadian sources. Since Figure 9 indicates a total S outflow of 1.4 Tg S, then 0.6 Tg S are from U.S. sources that pass into and out of Canada. Of the 0.8 Tg S leaving Canada from eastern Canadian sources, 0.3 Tg S re-enter Canada, 0.15 Tg S flow out through the Atlantic Ocean boundary and 0.35 Tg S flow into the U.S. some of which will be deposited in the U.S.

The budget diagram indicates the largest S outflow of 0.4 Tg S occurs across the southern Ontario boundary and that Quebec receives the highest S deposition (0.6 Tg S).

Many assumptions regarding the transport and diffusion of S have been made namely:
– emissions are uniformly distributed over the 127 km grid squares,
– primary sulphate emissions have been neglected,
– the atmosphere within the regions is non-divergent,
– vertical diffusion is instantaneous,
– no mass flux through the mixing height,
– physical and chemical processes are linear and constant,
– wet deposition uses a fixed scavenging ratio and a precipitation rate that is constant for 24 h, and
– the model uses climatological mixing heights.
Therefore the budget results should be regarded as preliminary estimates and subject to revision.

The uncertainty in the emissions inventory is considered to be of the order of 20%, the uncertainty in the parameterization coefficients could be about 50% or higher in some cases and the receptor point distribution is quite arbitrary. The overall uncertainty in the S budget at this stage is estimated to be of the order of 40 to 50%.

6. Conclusions

The S budgets (Figures 9 and 16) indicate the following preliminary conclusions:
– Of the 1.7 Tg S entering eastern Canada from a U.S. emissions total of 14 Tg S, 0.55 Tg S return to the U.S., 0.05 Tg S flow out through the Atlantic Ocean boundary and 1.1 Tg S are deposited throughout eastern Canada. Ontario receives about 0.7 Tg S, Quebec (plus Labrador) receives about 0.3 Tg S and the Maritimes receive about 0.1 Tg S.
– Of the 1.8 Tg S emitted from eastern Canadian sources, approximately 0.3 Tg S leave and re-enter the region, about 0.35 Tg S flow across the border to the U.S., about 0.15 Tg S flow out to the Atlantic and 1.3 Tg S are deposited throughout the region. Ontario receives about 0.5 Tg S, Quebec receives about 0.6 Tg S and the Atlantic Provinces receive about 0.2 Tg S.
– The highest total S mass input from U.S. sources into eastern Canada occurs across the western Ontario boundary although the highest S input per unit boundary length

occurs through the southern Ontario boundary. The highest S mass output from eastern Canadian sources to the U.S. occurs across the southern Ontario boundary. A significant amount of S transport occurs from Ontario to Quebec across the southern portion of the border.

– Ontario receives the largest total S deposition (1.2 Tg S) and receives the largest S deposition from U.S. sources (0.7 Tg S). Quebec (plus Labrador) receives the largest S deposition (0.6 Tg S) from eastern Canadian sources.

– The total S input into eastern Canada is about equally divided between input from U.S. sources (1.7 Tg S) and regional emissions (1.8 Tg S). The total S deposition in eastern Canada is also approximately equally divided between deposition from regional emissions (1.3 Tg S) and deposition from U.S. sources (1.1 Tg S).

References

Ahmed, M. S. A., Anlauf, K. G., and Wiebe, H. A.: 1980, 'Data Report on Air Quality Measurements at Toronto Centre Island, July 1977–September 1978', ARQA-77-80, Atmospheric Environment Service, Downsview, Ontario, Canada.

Olson, M. P., Oikawa, K. K., and MacAfee, A. W.: 1978, 'A Trajectory Model Applied to the Long-Range Transport of Air Pollutants', LRTAP 78–4, Atmospheric Environment Service, Downsview, Ontario.

Olson, M. P., Voldner, E. C., Oikawa, K. K., and MacAfee, A. W.: 1979, 'A Concentration/Deposition Model Applied to the Canadian Long-Range Transport of Air Pollutants Project', LRTAP 79–5, AES, Downsview, Ontario.

Rutherford, I. D.: 1977, 'An Operational Three Dimensional Multivariate Statistical Objective Analysis Scheme', Issue No. 1, Notes Scientifiques et Techniques, RPN, Atmospheric Environment Service, Dorval, P.Q.

Voldner, E. C. and Shah, Y.: 1980, *Atmos. Env.* **14**, 419.

Voldner, E. C., Olson, M. P., Oikawa, K. K., and Loiselle, M.: 1980, 'Comparison Between Measured and Computed Concentrations of Sulphur Compounds in Eastern North America', AQRB-80-0003-T (LRTAP-02), Atmospheric Environment Service, Downsview, Ontario (to be published in Journal of Geophysical Research, 1981).

Walmsley, J. L., Mailhot, J., and Hopkinson, R.: 1981, 'Sensitivity Tests with a Trajectory Model', AQRB-81-027-L, Atmospheric Environment Service, Downsview, Ontario, Canada.

MODELING OF POLLUTANT TRANSPORT AND REMOVAL DURING A REGIONAL SULFATE EPISODE

HALÛK ÖZKAYNAK* and P. BARRY RYAN**

Atmospheric and Environmental Research. Inc. 840 Memorial Drive, Cambridge, MA 02139, U.S.A.

and

LANCE F. BOSART

State University of New York at Albany, Department of Atmospheric Sciences, 1400 Washington Avenue, Albany, NY 12222, U.S.A.

(Received July 31, 1981; Revised November 5, 1981)

Abstract. As part of a multidisciplinary study performed for the National Commission on Air Quality, the phenomena of atmospheric transport and the removal of SO_2 and $SO_4^=$ during a major regional sulfate episode (the period July 18–25, 1978 in the eastern U.S.) had been examined. The main objective of this study was the evaluation and the quantification of varying source/receptor relationships under atmospheric conditions conducive to long-range transport of fine particulate matter. In the case study presented here, air mass trajectories were obtained using the numerical NMC trajectory predictions and the results of the isobaric trajectory computations at the 850 mb level. The effects of alternative regional SO_2 emission reduction scenarios on the predicted ambient SO_2 and $SO_4^=$ concentrations were also investigated using the new modeling methods that were specifically developed for this purpose.

1. Introduction

Studies on atmospheric transport and transformation of pollutants have indicated the potential for significant long-range transport impacts under certain meteorological conditions for much of the United States. Recent regional studies focusing on the eastern United States have characterized the nature and extent of these events (see for example, Lavery *et al.*, 1980; AER, 1981). The key questions that still remain to be addressed, however, are: (a) is it possible to estimate the conversion and removal rates of precursor pollutants under a variety of atmospheric and meteorological conditions; and (b) can the effects of alternative regional emission control strategies on source/receptor relationships be predicted within a modeling framework.

In this work, we have attempted to address these key questions using information limited to a one week regional sulfate episode. In subsequent sections the meteorological and atmospheric chemistry related factors pertaining to the transport, transformation and removal of SO_2 and sulfates during the (northeast) regional episode period on July 18–24, 1978 are examined. (Although the results presented here may not be representative of other periods, the techniques used, however, could be readily generalized).

* Present Affiliation: Energy and Environmental Policy Center, Harvard University, 140 Mt. Auburn St., Cambridge, MA 02138, U.S.A.
** Present Affiliation: Department of Environmental Healt Sciences, Harvard School of Public Health, 665 Huntington Ave., Boston. MA 02115. U.S.A.

The case study to be presented here consisted of the following investigative steps: (1) processing of the SURE ground station data tape; (2) generation of both isobaric and numerical (NMC) 850 mb trajectories for the northeastern U.S.; (3) development of a regional scale Eulerian grid model and a Lagrangian trajectory model; (4) estimation of transformation and deposition rates by the use of actual measurements of ambient SO_2, $SO_4^=$ concentrations; and (5) studying the air quality impacts associated with the various geographical and regional emission scenarios in connection with the long-range transport phenomenon of SO_x.

In particular, we have tested the effects of alternative emission control strategies on the ambient SO_2 and $SO_4^=$ levels as influenced by long-range transport of precursor/pollutants using two different modeling approaches. Twenty four hour Lagrangian trajectory simulations (Section 4) have been specifically designed to investigate long-range transport impacts of pollutants on sites far downwind of source regions under a particular set of meteorological conditions. Eulerian Grid Modeling (Section 5), on the other hand, simulates grid-by-grid regional impacts on ambient air quality over a five-day impact period. Air quality impacts over longer study periods (monthly or annually) are discussed elswhere (cf. AER, 1981).

2. Background Climatology

Wagner (1978) has described the synoptic climatology of July 1978. The main point is that the month featured weaker than normal 700 mb westerlies across the United States with the latitude of the maximum westerlies displaced somewhat north of the climatological location. This set the stage for extensive above normal temperatures across much of the southern United States with comparatively little air movement. The episode period of 18–25 July coincided with a strong westward extension of the Bermuda anticyclone that allowed maritime tropical and continental polar air masses to become entrenched over the eastern and central United States.

3. Trajectory Methods

3.1. NUMERICAL PREDICTION MODEL

Trajectories for the July 1978 episode are documented in two ways. Figure 1 presents 24-h forecast trajectories terminating at 850 mb for Albany, NY and Williamsport, PA for the dates in question. The three dimensional trajectories were derived from the operational National Meteorological Center (NMC) primitive equation prediction model forecasts that are routinely available twice each day. (See Reap, 1978 for documentation.) The trajectories are only as good as the model forecast. Fortunately, however, there is ample evidence that existing NMC operational models provide excellent 24-h synoptic scale simulations (see e.g., Fawcett, 1977; Shuman, 1978; Reed, 1977). Subjective evaluation of the NMC trajectories for the 18–25 July period indicate that they are generally representative of the large scale flow patterns.

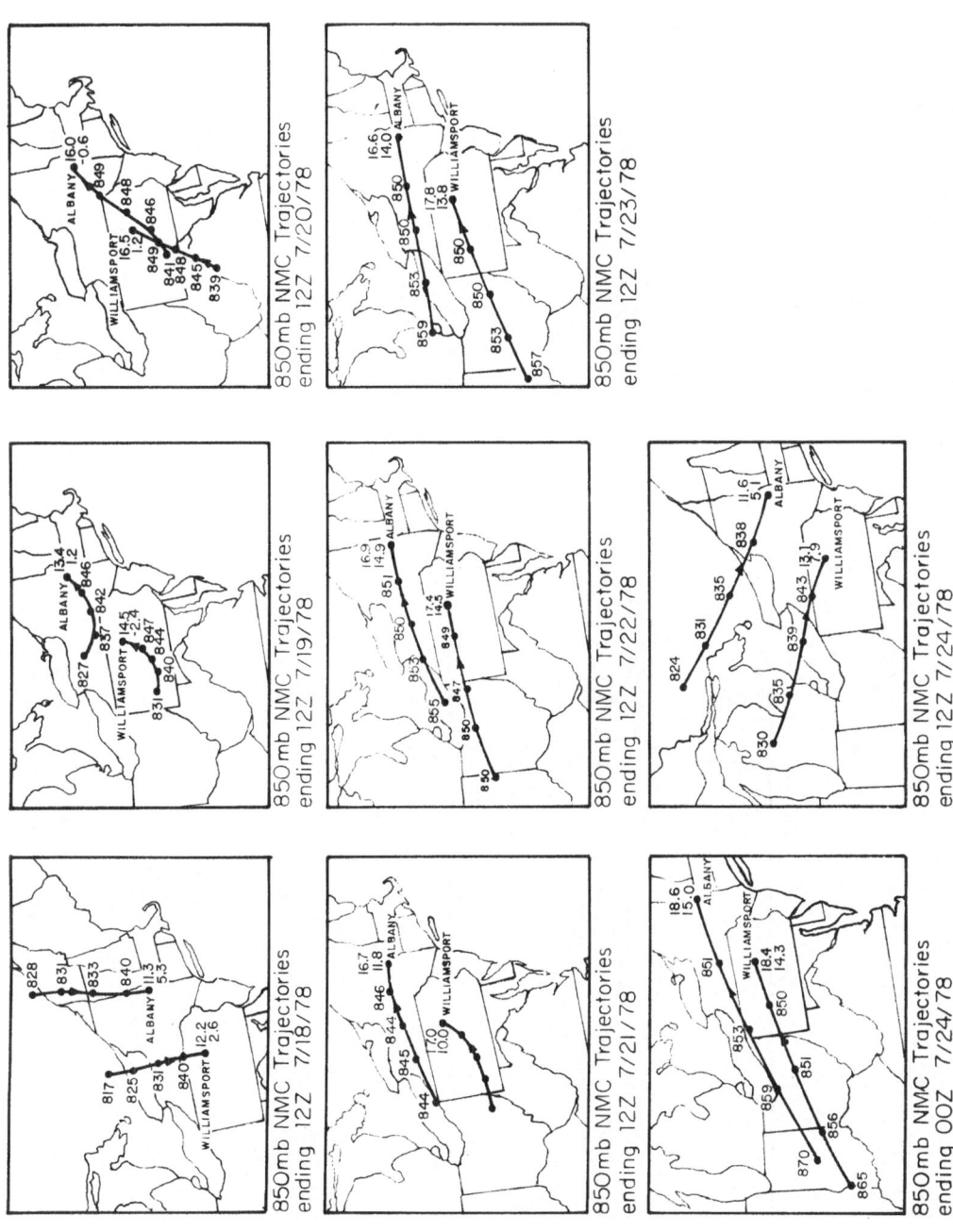

Fig. 1.   24-h 850 mb NMC Trajectories ending at Albany, NY and Williamsport, PA. (Each 6-h position along the trajectory is indicated by a solid circle. The adjacent three digit number indicates the forecast pressure level in whole millibars (mb). The numbers at the trajectory end points represent the forecast temperature and dew point at the 850 mb level in tenths of a °C.)

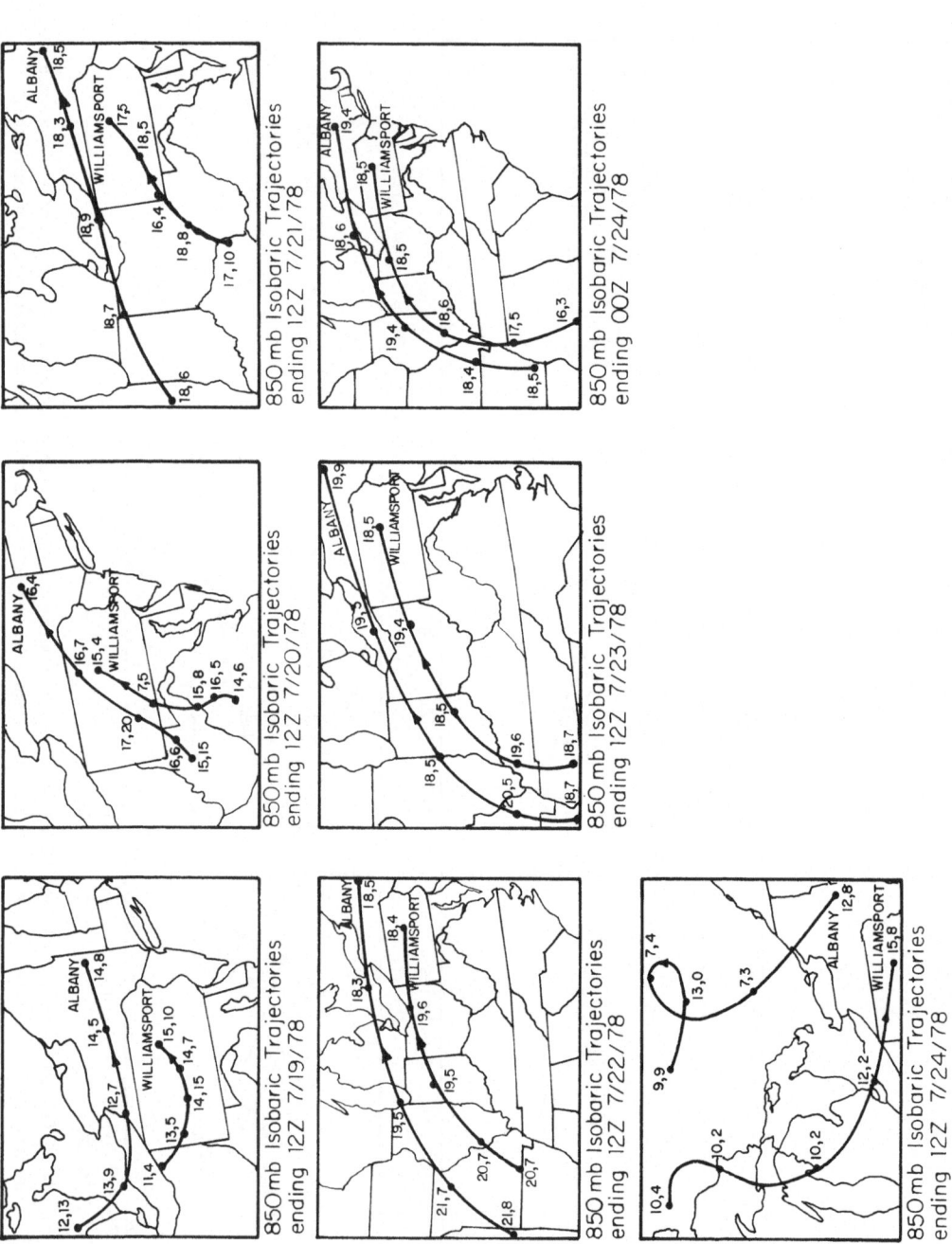

Fig. 2.   48-h 850 mb Isobaric Trajectories ending at Albany, NY and Williamsport, PA. (The solid circles represent the 12-h positions along the trajectory while the adjacent numbers give the observed temperature and dew point depression in tenths of °C.)

3.2. ISOBARIC TRAJECTORIES

Figure 2 provides 48-h isobaric trajectories for the 850 mb surface terminating at Albany, NY, and Williamsport, PA. These trajectories were computed subjectively using the observed winds at the 850 mb level. The computational procedure was to start at the desired trajectory end point and work backwards 48-h in 6-h increments. At the desired end point, a simple 6-h backwards extrapolation was made using the observed wind field (for details see Danielsen, 1961) and assuming the observed stream line pattern was essentially unchanged. The length of the trajectory was then adjusted by simple linear interpolation based upon the difference between the initial wind and the previous 6-h wind from the first guess trajectory. The resulting 6-h backwards extrapolation point was then transferred to the previous 850 mb map from 12-h earlier and the whole process repeated through two 6-h increments to obtain an 18-h trajectory, and so forth until the desired 48-h trajectory was obtained.

3.3. TRAJECTORY ERRORS

Individual trajectory errors may vary considerably. In a broad statistical sense the computational uncertainty may be anywhere from 10 to 20% of the length of the 24-h trajectory. A particularly woeful NMC 24-h synoptic circulation forecast would probably translate into a much greater error. The same could be said for the isobaric method if the trajectory crossed a region of active weather in which vertical motions were significant. The 48-h trajectory uncertainty may approach 25 to 40% of the trajectory length.

Another question about the trajectories is the appropriateness of using the 850 mb level as representative of the boundary layer air flow and to distinguish between days of long-range transport versus local stagnation. Terrain elevations from the Ohio Valley to the mid-Atlantic coast average 200 to 300 m above mean sea level (MSL) with 500 to 700 m more representative of the mountain region. The average height of the 850 mb surface is close to 1500 m above MSL so the 850 mb level is normally within the mixed layer during a summer afternoon. It would obviously be more desirable to use available wind information every 300 m in the vertical. (However, the processing and analysis of such data would involve a major research commitment and is beyond the scope of the trajectory analysis presented here.) Overall, the internationally determined data level of 850 mb is representative of the boundary layer under a variety of summer meteorological conditions with mixing heights often reaching 2000 to 2500 m above ground level (AGL).

3.4. TRAJECTORY ANALYSES FOR THE 18–25 JULY EPISODE

The NMC trajectories in Figure 1 establish that by 00Z (0000 GMT) 19 July a well defined anticyclone was centered over West Virginia. Air transport in the Ohio Valley was only 100 to 200 km for the previous 24-h period. Subsidence, characteristic of a developing anticyclone, was seen to the east of the anticyclone center. Quasi-stagnant conditions persisted for the next 48-h with little overall air transport from the Ohio Valley eastward to the middle Atlantic coast. During this period, vertical motions averaged less than 10 mb for 24-h (0.1 cm s^{-1}) which supports the use of 850 mb isobaric trajectories.

By 00Z 21 July increasing quasi-isobaric transport was seeen from the Ohio Valley eastward across New York, Pennsylvania, and New England. To the south of the east-west oriented ridge line a westward transport was seen. In the following 24-h a pattern of transport from the Tennessee Valley, across the Ohio Valley to New England was well established. Average ascent in the western Great Lakes region was less than 0.3 cm s^{-1}, which still supports the use of isobaric trajectories since little directional or speed shear was evident from the observed data between 850 and 700 mb.

By 00Z 23 July the main thrust of the transport across the Tennessee and Ohio Valleys was towards the middle Atlantic coast. Air flow into the New York-New England region was from the west-northwest. Quasi-isobaric conditions were again indicated except for the trajectories just ahead of a cold front approaching the western Great Lakes. In the ensuing 24-h a strong west-southwesterly flow from the Tennessee and Ohio Valleys was re-established across the northeastern quarter of the United States ahead of a moderate cold front. Continental tropical air from the southwestern United States was accelerated eastward on 23 July, culminating in the highest temperatures of the summer at most northeastern locations that day.

Cooler and drier conditions developed across the upper midwest and northeastern United States by 24 July with maritime tropical air confined farther south. Transport from the Ohio Valley now reached into North Carolina and Virginia.

3.5. SUMMARY OF EVENTS

In summary, the picture suggested by the NMC 24-h forecast trajectories is one of a local stagnation event early in the episode period as anticyclone conditions developed in the wake of a cold front which reached the Atlantic coast on 17 July. After 20 July the episode event became regional with long-range transport of polluted boundary layer air from the Tennessee and Ohio Valleys eastward and northeastward to the Atlantic coast. The 48-h isobaric trajectories shown in Figures 1 and 2 further strengthen this conclusion. This finding is consistent with that noted by Lavery *et al.*, 1980.

4. Modeling of Pollutant Transport and Removal

To study SO_2 and $SO_4^=$ patterns on regional scales and under real time transport conditions, the following mass balance equation was utilized:

$$\frac{\partial c}{\partial t} = -\nabla \cdot c\mathbf{v} + P - L \tag{1}$$

where the pollutant concentration (c), horizontal wind field ($\mathbf{v}$), production (P) and loss (L) terms are both time and space dependent. To infer pollutant behavior during episodic transport conditions and to evaluate source/receptor relationships, Equation (1) was solved by employing various numerical techniques. The solutions were obtained either by integrating Equation (1) along a trajectory in a Lagrangian framework or by explicitly integrating over a (640×320 km^2) two-dimensional Eulerian Grid centered around southwest Ohio. By incorporating the observed wind data and the recorded pollutant concentration profiles, both the Lagrangian and the Eulerian modeling methods were initially used to infer the effective transformation and deposition rates associated with the transported SO_x. Source/receptor relationships were later analyzed using these inferred oxidation and deposition rates in conjunction with the regional SO_2 emission inventory which was developed during the SURE program. The potential outcome of the regional emission reduction scenarios was similarly tested assuming a number of regulatory strategies that specified alternative allowable SO_2 emission levels for the utilities in the study areas.

4.1. Lagrangian Trajectory Modeling

A common technique used in solving Equation (1) involves integration along an air parcel trajectory within a Lagrangian framework rather than spatial and temporal referencing with respect to a fixed (Eulerian) coordinate system. Thus, for known or computed air mass trajectories one can readily calculate the time dependent changes for the pollutant concentrations using a representative set of transformation and deposition rates (this is also termed as the forward problem). Similarly, given information on the pollutant concentration values along the trajectory, one can in principal solve for, or infer, the estimated removal rates by inverting the linearized solutions to Equation (1) (mathematically this is defined as the inverse problem). In the work to be discussed here, both as a forward problem and also as an inverse problem, concentration histories associated with the pollutants or the precursors within a moving air parcel have been related to the regional emissions and the estimated removal rates of transported SO_2 and $SO_4^=$.

4.1.1. Description of Methods

Lagrangian formulation of Equation (1) assumes the following form for SO_2 and $SO_4^=$:

$$\frac{dg}{dt} \equiv \dot{g} = -k_t g - k_{d_1} g + \frac{Q_1}{h} \tag{2}$$

$$\frac{ds}{dt} \equiv \dot{s} = -k_{d_2} s + \frac{3}{2}\left(k_t g + \frac{Q_2}{h}\right) \tag{3}$$

where g and s are the concentrations of SO_2 and $SO_4^=$, respectively, Q_1 is the effective SO_2 emission rate, Q_2 is the corresponding effective sulfate emissions during the history of the parcel transport, and h is the mixing height.

In Equation (2) k_t denotes the (oxidation) conversion rate of SO_2 to $SO_4^=$ and k_{d_1} represents the rate of dry deposition of SO_2. Similarly k_{d_2} in Equation (3) represents the dry deposition rate of $SO_4^=$.

To solve the inverse problem [i.e., to estimate or infer the transformation (k_t) and deposition (k_{d_1}, k_{d_2}) parameters], one can either attempt to use the analytical form of the solutions to these Equations (by some type of a non-linear optimization scheme) or linearize the problem by substituting approximate but discretized differential operators in Equations (2) and (3). In the applications reported here, finite-difference techniques were used to linearize the Lagrangian form of the advection equations defined above. Lagrangian simulations typically employed 24-h (NMC) back trajectories ending at Albany, NY and Williamsport, PA. The trajectory simulations for the recovery and the estimation of the transformation and deposition parameters consisted of solving (in least squares sense) the following matrix equation:

$$[D] = [A][K] \qquad (4)$$

where

$$[D] = \begin{bmatrix} g \\ s \end{bmatrix} - \begin{bmatrix} Q_1/h \\ Q_2/h \end{bmatrix}$$

$$[K] = \begin{bmatrix} k_t \\ k_{d_1} \\ k_{d_2} \end{bmatrix} \quad \text{and} \quad [A] = \begin{bmatrix} -g & -g & 0 \\ 3/2g & 0 & -s \end{bmatrix}.$$

The source terms (SO_2 emission rates) were based on the SURE emission inventory. Primary sulfate emission were assumed to be equal to 3% of the SO_2 emissions (a representative figure based on the results of the SURE emission inventory).

Sulfur dioxide concentrations along the trajectory (at each hour) were estimated by using $1/r^2$ interpolation of data (since this approach is the one most commonly adopted when simple analytical approximation to atmospheric dispersion of pollutants are needed during spatial interpolation of data obtained from fixed monitor sites) from nearby SURE stations with a cutoff distance of approximately 170 km. Three-hour $SO_4^=$ concentrations from SURE Class I sites were interpolated to obtain hourly $SO_4^=$ data. Daily average $SO_4^=$ measurements from high volume samplers were also used to a limited extent to reproduce hourly $SO_4^=$ data from Class II sites (see AER, 1981 for further details). Using the interpolated SO_2, $SO_4^=$ concentrations along the NMC trajectories displayed in Figure 1, the removal rates, [K] were then inferred by inverting Equation (4). To control mathematical problems resulting from noise, data errors and the ill-conditioned nature of the associated Fredholm Equation of the first kind, we chose to implement a sophisticated optimization technique based on Lawson and Hanson's (1974) nonnegative least squares algorithm, along with the Tikhonov (1963) regularization formulation. Thus, SO_2 and $SO_4^=$ removal rates [K], averaged over 6-h increments were estimated by numerically solving the weighted constrained least-squares regularization problem:

$$\min_{K_i \geq 0} \{\| \tilde{D} - \tilde{A}K \|^2 + \lambda \| K \|^2\} \qquad (5)$$

where

$$\tilde{A} = WA, \quad \tilde{D} = WD, \quad \text{and} \quad W = \begin{pmatrix} 1 & 0 \\ 0 & \omega \end{pmatrix}.$$

Put differently, this optimization process (i.e., estimations using linear programming techniques) provides non-negative, thus physical removal rate estimates, that minimize the sum of square of differences between the observed and the predicted (concentration based) data values, while at the same time satisfying a number of pre-specified requirements that pertain to the overall distribution and uniformity of the magnitude of solutions, as well as to the accuracy by which $SO_4^=$ measurements are weighted relative to those of SO_2.

In the simulations performed the regularization parameter, λ, was varied between 0 and 100. To minimize the effect of possible systematic or even random errors associated with the interpolated concentration values, a representative relative weighting factor of 30% ($\omega = 0.3$) was selected (i.e., optimization equations containing the SO_2 data were given about 3 times more weighting than the equations containing the $SO_4^=$ information). Determination of optimum λ (in this case around $\lambda = 10$) resulted from several evaluations that included the study of both the residual and the solution norms (see AER, 1981 for details).

4.1.2. *Results of the Estimation Studies*

Detailed discussion of modeling results and the critique of uncertainties have been provided in AER (1981) and Özkaynak *et al.*, 1981b. Only the main results and the summary of essential findings will be presented in the following.

TABLE I

Estimated SO_2 removal rates ($\%$ h^{-1})[a]

Site	Date/time[b]	k_t	k_{d_1}
Albany, NY	7/19/12Z	0.8	4.4
Albany, NY	7/20/12Z	0.5	1.5
Albany, NY	7/21/12Z	1.0	2.6
Albany, NY	7/22/00Z	0.1	0.1
Albany, NY	7/23/12Z	1.4	2.7
Albany, NY	7/24/00Z	1.4	1.6
Albany, NY	7/24/12Z	0.0	0.2
Williamsport, PA	7/22/12Z	0.9	0.9
Williamsport, PA	7/23/12Z	2.3	3.2
Williamsport, PA	7/24/00Z	1.3	2.3
Williamsport, PA	7/25/00Z	0.7	2.1

[a] [1] Results shown are 24-h averages based on individual recoveries at 6-h intervals using $\lambda = 10$ and $\omega = 0.3$.
[2] Recoveries (expected accuracy within ± 10–40%) are based on hourly approximations to the equations of the Lagrangian model which simulates pseudo-first order $SO_2 \to SO_4^=$ chemistry.
[b] The dates and times shown here indicate the trajectory arrival times at either Albany, NY or Williamsport, PA.

Table I displays the estimated (recovered) 24-h averaged SO_2 oxidation and dry deposition rates for the events studied.

The regional characteristics of the episode are noted by the relatively higher conversion rates ($\sim 1\%$ h^{-1}) corresponding to events between the July 20th and July 24th. Albany 7/24/00Z and Williamsport, 7/23/12Z trajectories, for example, represent long range transport of precursor/pollutants over midwest and Ohio River Basin area into northeast U.S. (cf. Figures 1 and 2). In both of these instances a significant fraction of the 24-h average SO_2 oxidation rate has been determined to correspond to segments of the trajectory near or over high SO_2 emission regions in the vicinity of the Ohio River Valley. Albany 7/23/12Z trajectory corresponds to west to east transport mostly over urban areas with moderate emissions. Albany 7/20/12Z simulation on the other hand, represents relatively unpolluted transport from central Pennsylvania. Both the Albany 7/24/12Z and the Williamsport 7/25/00Z simulations are consistent with the cleaner Canadian air mass transported from the northwest and correspondingly therefore, with the low conversion and deposition rates that were recovered. In general, results of parameter estimations reveal stronger correlations between the recovered rates and the position of the air parcel rather than the time of day. Typically, near regions of high SO_2 emissions, higher removal rates were recovered (e.g., Williamsport 22/12Z, 23/12Z, 24/00Z, and Albany 21/12Z, 24/00Z trajectory simulations). In general, both the transport and conversion influences were noted (an exception to this observation was the Albany 22/00Z simulation). The recoveries for this date suggested transport rather than conversion effects. Diurnal patterns (although not always consistent) were also noted for air mass trajectories passing over relatively rural areas where daytime photochemistry influences were to be expected (for example, Albany 19/12Z simulation presented in AER 1981).

4.1.3. *Effect of* SO_2 *Emission Reductions on Source/Receptor Relationships*

Simulations performed in this section explore the effects of various regional emission reduction scenarios within the study area. The basic SO_2 control strategies tested for the period modeled, include: (1) plant or site/grid specific emission reductions corresponding to either 6.3 or 2.2 g of $SO_2/10^6$ cal heat input (3.5 or 1.2 lbs $SO_2/10^6$ BTU) limitation applied to 50 major power plant sources of uncontrolled SO_2 emissions; (2) statewide emission limitations to reduce the net SO_2 emissions from affected midwestern and eastern states (Illinois, Indiana, Kentucky, Ohio, Pennsylvania and Tennessee) to the same two levels of adjusted SO_2 emissions (i.e. aggregate emission reductions from major utilities in each of these states that belong to the same set of 50 major power plants).

The emission reduction scenarios were generated (by the staff at the National Commission on Air Quality, Washington, D.C.) using the basic SURE grid and the SURE SO_2 emissions inventory (see AER, 1981 for details).

State and grid emission reduction scenarios that allowed SO_2 emissions of 6.3 or 2.2 g/10^6 cal heat input were modeled using solutions to Equations 2 and 3 (in the forward simulation mode) and the previously recovered time dependent oxidation/deposition rates discussed above.

However, before proceeding with the results of the scenario testing, the following cautions must be made. It is assumed implicitly in this type of reduction scenario testing, that the effective chemistry involved in the transformation of SO_2 to $SO_4^=$ is unchanged by reduced emissions and concomitant reduced ambient concentrations. Indeed, this assumption of linearity of total rates on concentration precludes any effects of reaction saturation from being included in the scenario. For example, if the limiting species in a conversion reaction is not SO_2, then reducing the concentration of SO_2 may have little or no effect on the amount of $SO_4^=$ produced. Also, reduction of SO_2 emissions at their sources would conceivably affect the emissions of other pollutants, such as NO_x, TSP, hydrocarbons, etc., in a manner dependent on the technique used to reduce the emission. As these pollutants may take part in the overall reaction scheme, changes in their emissions may affect the overall, effective transformation rate constants.

Results presented here are based on either assuming a linear rollback in initial pollutant concentrations (using representative background values of $9\,\mu g\,m^{-3}$ of SO_2 and $3\,\mu g\,m^{-3}$ of $SO_4^=$ for this event period) or assuming no rollback in regional baseline values (i.e. initial SO_2, $SO_4^=$ concentration values were held fixed at the start of the air mass trajectory only allowing changes in the amount of SO_2 injected into the air parcel during its transport).

The general conclusion that can be reached from Lagrangian simulations is that, major emission reductions at upwind source regions appear to result in significant reductions in the predicted air quality impacts at downwind receptors under long-range transport conditions. More specifically, assuming the rollback hypothesis, corresponding to $\sim 15\%$ statewide SO_2 emission reductions in the upwind states that are located in the Ohio River Basin area, the predicted ambient levels of SO_2 and $SO_4^=$ at Albany, NY and Williamsport, PA were found to be reduced by 11%. However, more significant emission reductions (around 30 to 35%) were shown to result in approximately 26% reduction in the predicted SO_2 and $SO_4^=$ concentrations at the same receptor sites downwind of these source regions. On the other hand, a much more conservative approach (assuming fixed initial or baseline concentrations for SO_2 and $SO_4^=$ that are not influenced by regional or statewide emission reductions) resulted in predictions of more modest improvement in air quality. In this case for the two statewide scenarios, the predictions of SO_2 concentration reductions were around 8% and 24% and of $SO_4^=$ concentration reductions around 5% and 13% respectively.

Finally, it can be inferred from this and other simulations presented in AER 1981, that the $SO_4^=$ concentrations predicted at the receptor sites downwind of source regions located in the midwest U.S. are typically dependent on: (a) the initial levels of SO_2 and $SO_4^=$ in the air parcel that is transported; (b) the extent of SO_2 and $SO_4^=$ conversion rate experienced locally and regionally; and (c) the magnitude of regional or local SO_2 emissions injected into the air mass that is involved in the observed long-range transport phenomenon

4.2. Eulerian Grid Modeling

The Eulerian Grid (EG) approach to the solution of the time-dependent mass-balance equation proceeds by the simultaneous solution of the coupled differential equations (given in Equation (1)) describing the change in concentration of SO_2 and $SO_4^=$ with time. The procedure involves calculating the spatial derivatives of the concentrations at each point on the study and centered on Duncan Falls, Ohio extending approximately 160 km to the east and west and 80 km to the north and south from this central site.

The algorithm then steps forward in time for the period being modeled. One may note that the chemical and physical transformation rate 'constants' $(k_t, k_{d_1},$ and $k_{d_2})$ are parameters in this model. As the chemistry is not complete, these rate constants represent *effective* transformation rates and in no sense should be construed as rate constants for any direct transformation of SO_2 to $SO_4^=$. The estimation of these transformation rates is the essence of this procedure. To determine the best values of these rate constants, we have optimized these values by minimizing the mean square difference between our calculated value of $SO_2/SO_4^=$ concentrations and those given in the SURE data base for the period 17 July 78 – 22 July 78. The constants were optimized by an iterative search technique (see, for example, Adby and Dempster, 1974) based on sequential simplex techniques (Ryan *et al.*, 1980) until the change in 24-h average removal rate constants fell below .25% absolute. Further optimization was felt inappropriate due to the sparseness and noise of the data. After optimum values of the transformation constants were made, changes were then made in source terms to simulate the effects of emission reduction on various receptors in the region. A detailed model description is presented elsewhere (cf. AER, 1981; Özkaynak *et al.*, 1981a, b).

4.2.1. *Discussion of Results*

In Table II we present optimum transformation constants. Model validation statistics for the EG model is presented in Özkaynak *et al.*, 1981a, b.

In the scenario testing mode, the effect of reduced emissions on ambient concentrations was also included in the initial conditions and boundary conditions. We assumed

TABLE II

Optimum transformation and deposition rate constants ($\%\ h^{-1}$) for the EG model

Date	k_t	k_{d_1}	k_{d_2}[a]
7/18/78	1.75	1.25	1.00
7/19/78	0.75	1.75	1.00
7/20/78	0.60	5.90	1.00
7/21/78	1.50	4.50·	1.00
7/22/78	3.00	1.00	1.00

[a] $SO_4^=$ deposition rates were held fixed at $1\%\ h^{-1}$. (Results were shown to be insensitive to this assumption.)

that the initial concentrations were dependent on the emissions themselves. The initial concentrations were therefore scaled (by the previously discussed roll-back hypothesis) by a factor corresponding to the average emission reductions throughout the grid. The boundary values were scaled according to an exponentially-weighted average of the downwind emission reductions in an effort to simulate the reduced amount of SO_2 coming into the grid.

With these assumptions in mind, we now investigate the effects of reduced emissions on the predicted concentrations of SO_2 and $SO_4^=$. We report predictions for two different emission scenarios using state-wide and grid-by-grid reductions as discussed in Section 4.1 on Lagrangian modeling.

In Table III we present average daily reductions in $SO_4^=$ concentrations for the states in our study. For each date, the average absolute and relative reductions (shown in

TABLE III

EG model emission reduction scenarios simulating average daily reduction in $SO_4^=$ concentrations[a]

Date	Ohio		Pennsylvania		West Virginia	
	I_S	II_S	I_S	II_S	I_S	II_S
18 July 78	0.6	1.7	1.4	3.8	1.0	2.5
	(4.2)	(12.4)	(6.0)	(16.9)	(6.2)	(14.9)
19 July 78	1.9	4.4	2.9	8.6	1.3	3.3
	(8.8)	(21.8)	(8.8)	(26.2)	(5.1)	(18.3)
20 July 78	2.5	6.5	3.0	9.0	1.2	5.9
	(9.4)	(24.7)	(9.2)	(27.6)	(5.6)	(28.5)
21 July 78	1.8	5.4	4.2	12.5	1.9	7.8
	(7.1)	(21.9)	(9.7)	(29.2)	(6.2)	(28.1)
22 July 78	2.0	5.1	2.3	6.8	3.2	8.7
	(7.6)	(20.0)	(8.7)	(25.8)	(8.8)	(26.6)
Date	I_G	II_G	I_G	II_G	I_G	II_G
18 July 78	0.3	0.9	1.2	3.7	0.4	1.1
	(2.2)	(6.2)	(5.5)	(16.2)	(2.1)	(5.9)
19 July 78	1.3	3.2	3.5	9.4	1.7	4.5
	(6.9)	(17.9)	(10.6)	(28.0)	(9.2)	(24.5)
20 July 78	2.0	4.2	3.7	9.8	3.2	8.5
	(8.1)	(17.6)	(11.2)	(29.8)	(15.7)	(41.4)
21 July 78	2.1	4.7	4.8	13.0	4.7	11.4
	(8.6)	(20.0)	(11.1)	(30.1)	(18.1)	(42.6)
22 July 78	2.0	3.9	2.5	5.9	6.7	12.5
	(8.5)	(17.1)	(6.8)	(22.0)	(22.8)	(40.6)

[a] Figures represent Absolute Reductions ($\mu g\ m^{-3}$).
Relative Reductions are shown in parentheses (%).

I_S = 6.3 g/10^6 cal (3.5 lbs/MBTU) State Scenario.
II_S = 2.2 g/10^6 cal (1.2 lbs/MBTU) State Scenario.
I_G = 6.3 g/10^6 cal (3.5 lbs/MBTU) Grid Scenario.
II_G = 2.2 g/10^6 cal (1.2 lbs/MBTU) Grid Scenario.

parentheses) are given. It should be noted that within our simulation region Ohio is represented by 17 grids, Pennsylvania 7 grids and West Virginia 5 grids so that statistics are relatively better for Ohio, but trends should remain.

Study of the grid-by-grid predictions (not presented here) also reveals that overall reductions are approximately the same for state-wide or grid-by-grid reductions. The numbers are an 8 to 10% reduction in ambient $SO_4^=$ in the 6.3 g/10^6 cal SO_2 emission scenario and about 25% reduction for the 2.2 g/10^6 cal SO_2 emission scenario. The effect of the state-wide reduction scheme is to scale down all ambient concentrations. The grid-by-grid reduction scenarios, on the other hand, lowers certain grids extensively while leaving others basically unchanged. This process reduces the concentration gradients quite effectively.

The dominant feature of Table III is that, although the major reductions on both the state-wide and grid-by-grid reductions occur in Ohio, it is western Pennsylvania and northwestern West Virginia that see larger absolute and relative reductions in ambient $SO_4^=$ concentrations. This is most obvious in the grid-by-grid 2.2 g/10^6 cal scenario, but is apparent in all scenarios. This may be attributed to transport of $SO_2/SO_4^=$ from Ohio by the prevailing west and southwest winds of the period modeled.

We may summarize the results of scenario testing as follows. Reduction of both absolute and relative ambient $SO_4^=$ concentrations may be affected by any of the scenarios modeled. An increase in relative and absolute reductions comparable to the emission reductions is found in going from the selected SO_2 emission limitations of 6.3 g/10^6 cal to 2.2 g/10^6 cal. The differences between state-wide and grid-by-grid reduction scenarios is small when averaged over a large region but can be very significant on a local scale. We expect more sopisticated models to support the above conclusion though the quantitative results may differ.

5. Conclusions

The analysis of events during the July 18–25, 1978 sulfate episode showed that, after the 20th the episode became regional with long-range transport of polluted boundary layer air from the Tennessee and Ohio Valleys eastward and northeastward to the Atlantic coast. $SO_4^=$ measured during this period was found to be strongly associated with the transport and conversion of SO_2.

A number of optimization methods were used to infer the SO_2 oxidation and deposition rates which were representative of the observed phenomena. Typically, near areas of high SO_2 emissions, effective SO_2 removal rates were found to be high. In contrast, however, relatively smaller transformation and deposition rates were inferred for cleaner air masses transported from northwestern United States and Canada. Diurnal patterns suggesting daytime photochemistry influences were also noted for those air masses that were transported over certain rural areas. However, the proximity of the air parcels to high SO_2 emission zones and the specific nature of the episode meteorology were the key factors influencing the estimated $SO_2/SO_4^=$ removal rates.

Using the modeling tools developed, a number of regional emission reduction scenarios (based on emission limitations applied to 50 major power plant sources of uncontrolled SO_2 emissions) were also simulated in order to study source/receptor relationships. These scenarios utilized the meteorological conditions appropriate to July 18–25, 1978 and reduced SO_2 emissions to model the $SO_2/SO_4^=$ impacts on the region studied. According to the Lagrangian and the Eulerian grid simulations, a reduction in emissions from power plants in the six-state area of Illinois, Indiana, Kentucky, Ohio, Pennsylvania and West Virginia of about 15% is expected to result in 5 to 10% reduction in the predicted $SO_4^=$ concentrations within this region or at receptor sites as far downwind as Albany, NY. Similarly, a reduction of about 35% in SO_2 emissions in these 6 states would result in 20 to 40% reduction in the predicted $SO_4^=$ concentrations, again, within this study region or at receptor sites as far downwind as Albany, NY.

Acknowledgments

Much of this work was supported by the National Commission on Air Quality (NCAQ). We would like to acknowledge Mr. James E. Fairobent of the Nuclear Regulatory Commission (formerly with NCAQ) for his expert guidance and review during the course of the AER study. We are also grateful to Dr. Jack Shannon of Argonne National Laboratory for his assistance and modeling support. Finally, we acknowledge EPRI for the use of the SURE data base.

References

Adby, P. R. and Dempster, M. A. H.: 1974, *Introduction to Optimization Methods*, Chapman and Hall, London.
AER: 1981, 'Study of the Role of Transport in Fine and Total Suspended Particulate Air Quality', Report prepared by Atmospheric and Environmental Research, Inc., for the National Commission on Air Quality (NTIS PB81-168031).
Danielsen, E. F.: 1961, *J. Meteor.* **18**, 479.
Fawcett, E.B.: 1977, *Bull. Amer. Meteor. Soc.* **58**, 143.
Lavery, T. F., Hidy, G. M., Baskett, R. L., Mueller, P. K., and Warren, K. K.: 1980, Environmental Research and Technology, Inc., Internal Paper.
Lawson, C. L. and Hanson, R. J.: 1974, *Solving Least Squares Problems*, Prentice-Hall International, Inc., Englewood Cliffs, NJ.
Özkaynak, H., Ryan, P. B., and Bosart, L. F.: 1981a, *Proc. of the Environmetrics 81 Conference*, Washington, DC, April 8–10, 1981, SIAM Publ.
Özkaynak, H., Ryan, P. B., and Bosart, L. F.: 1981b, Atmospheric and Environmental Research, Inc., preprint.
Reap, R.: 1978, NOAA Technical Procedures Bulletin No. 225, The Trajectory (TRAJ) Model, March 6, 1978.
Reed, R. J.: 1977, *Bull. Amer. Meteor. Soc.* **58**, 390.
Ryan, P. B., Barr, R. L., and Todd, H. D.: 1980, *Anal. Chem.* **52**, 1460.
Shuman, F. G.: 1978, *Bull. Amer. Meteor. Soc.* **59**, 5.
Tikhonov, A. N.: 1963, *Sov. Math. Dokl.* **4**, 1624.
Wagner, A. J.: 1978, *Mon. Wea. Rev.* **106**, 1509.

EFFECTS OF VARYING AIR TRAJECTORIES ON SPATIAL AND TEMPORAL PRECIPITATION CHEMISTRY PATTERNS

GILBERT S. RAYNOR and JANET V. HAYES

Atmospheric Sciences Division, Brookhaven National Laboratory, Upton, NY 11973, U.S.A.

(Received May 22, 1981; Revised August 12, 1981)

Abstract. This study was designed to determine if judicious use of synoptic data and an operational trajectory model could identify probable source regions of anthropogenic pollutants in northeastern United States precipitation and thus relate receptor measurements to emissions data without consideration of the complex intervening meteorological and chemical processes. The storm event of April 8 to 10, 1979, was selected for intensive study. Precipitation chemistry data were obtained from event samples at six MAP3S sampling sites and from hourly samples at Brookhaven National Laboratory. Concentrations of hydrogen, sulfate, nitrate and ammonium ions were used as receptor data. Some emissions data for SO_x and NO_x were obtained from the MAP3S emissions inventory. Surface and upper air meteorological data were analyzed. Backward trajectories ending at each of the sampling sites during the precipitation period were computed with the Heffter Interactive-Terminal Transport Model using selected transport layers.

Results show that concentrations of pollutant species in event precipitation samples were much higher at stations at end points of trajectories passing through the Ohio River valley than at stations with other trajectories. Likewise, concentrations at Brookhaven were much higher during the end period of a trajectory through the same region than with more northerly and more southerly tracks. The model produced back trajectories consistent with synoptic flows. Concentrations of air pollutants in precipitation were roughly proportional to the number of major pollutant sources along the trajectory. These results suggest that a larger number of studies might identify more restricted source areas or even establish a quantitative relationship between source emissions along a trajectory and concentrations in precipitation at receptor sites.

1. Introduction

Much information has been obtained· in recent years on the occurrence and concentrations of the chemical constituents of precipitation in the northeastern United States and on their temporal and spatial variability. Reasonably complete inventories are also available which give locations and emissions rates of both point and area sources of important air pollutants. However, few attempts have been made to identify sources or even source regions of contaminants in precipitation. This is due, in part, to the complexities of the atmospheric motions, the cloud and rain formation mechanisms and the chemical reactions and transformations which occur during pollutant transport from source to receptor. Until recently, it was also due to the lack of precipitation chemistry data with adequate spatial coverage and temporal resolution. Although trajectory methods have been used in a number of air pollution studies (e.g., Hall *et al.*, 1973; Wolf *et al.*, 1977; Galvin *et al.*, 1978; Samson, 1981), a contributing factor to their infrequent use in wet deposition studies has been uncertainty concerning the adequacy of available trajectory methods for predicting or reconstructing accurate trajectories under frontal or storm conditions.

Despite these problems, several previous investigators have used trajectory analyses or related meteorological methods in precipitation chemistry studies. Munn and Rodhe

(1971) classified precipitation events in Sweden by surface geostrophic wind direction and found higher concentrations in precipitation with flow from the major source regions of central Europe and Great Britain. Førland (1973), Nordø (1976) and Ottar (1976) obtained essentially the same results for Norway. Smith and Hunt (1978) used back trajectories based on surface geostrophic winds and on wind profiles measured from aircraft to determine trajectories to sampling stations in England and Norway.

Wolff *et al.* (1979) analyzed surface layer air trajectories and pH measurements from a number of sampling sites in the New York City area and showed that precipitation was more acid with trajectories from the west and southwest. Miller *et al.* (1978) used the ARL trajectory model on 45 event samples at Ithaca, New York, and reported higher levels of acidity with transport from westerly sectors. Wilson *et al.* (1980) analyzed 1977–1979 data from the MAP3S precipitation chemistry sampling sites using the ARL-ATAD trajectory model to characterize individual events as to air mass origin. In a number of case studies, they found that concentrations in precipitation at the Whiteface Mountain sampling site were higher when the air passed over the midwest Ohio Valley region as contrasted to air coming through the Great Lakes region or from Canada.

Thus, except for Miller *et al.* (1978) and Wilson *et al.* (1980) these studies were based on single level trajectories which may or may not have been representative of the height intervals through which the materials of interest were transported or of the actual path of the pollutants. Their results were necessarily expressed in terms of direction sectors or very generalized probable source regions. This study was designed to determine if judicious use of synoptic data and an operational trajectory model could identify possible source regions with a higher degree of probability and thus relate receptor data to emissions without consideration of all the intervening processes. Thus, the study was planned primarily to develop and test analytical techniques rather than to obtain definitive results on source-receptor relationships.

2. Methods

Because of the effort required for development and testing of the analytical methods, a single precipitation event was selected for intensive study. The event of April 8 to 10, 1979, was chosen because it provided adequate amounts of precipitation with moderate to high concentrations of contaminants at nearly all northeastern stations. In addition, the frontal system was relatively uncomplicated.

2.1. DATA USED

Two sources of precipitation chemistry data were used. Event data were obtained from six of the eight MAP3S sites (Pacific Northwest Laboratory, 1980) and hourly sequential data from Brookhaven National Laboratory (BNL) (Raynor and Hayes, 1979). Chemical constituents considered in this study were sulfate (SO_4^{2-}), nitrogen in nitrate or in the sum of nitrate and nitrite ($NO_3^- + NO_2^-$ N), nitrogen in ammonium (NH_4^+ N) and hydrogen ion (H^+). Units used at MAP3S sites are μmol l^{-1}. Units at BNL are μeq l^{-1}. All of these species are believed to have largely anthropogenic or at least

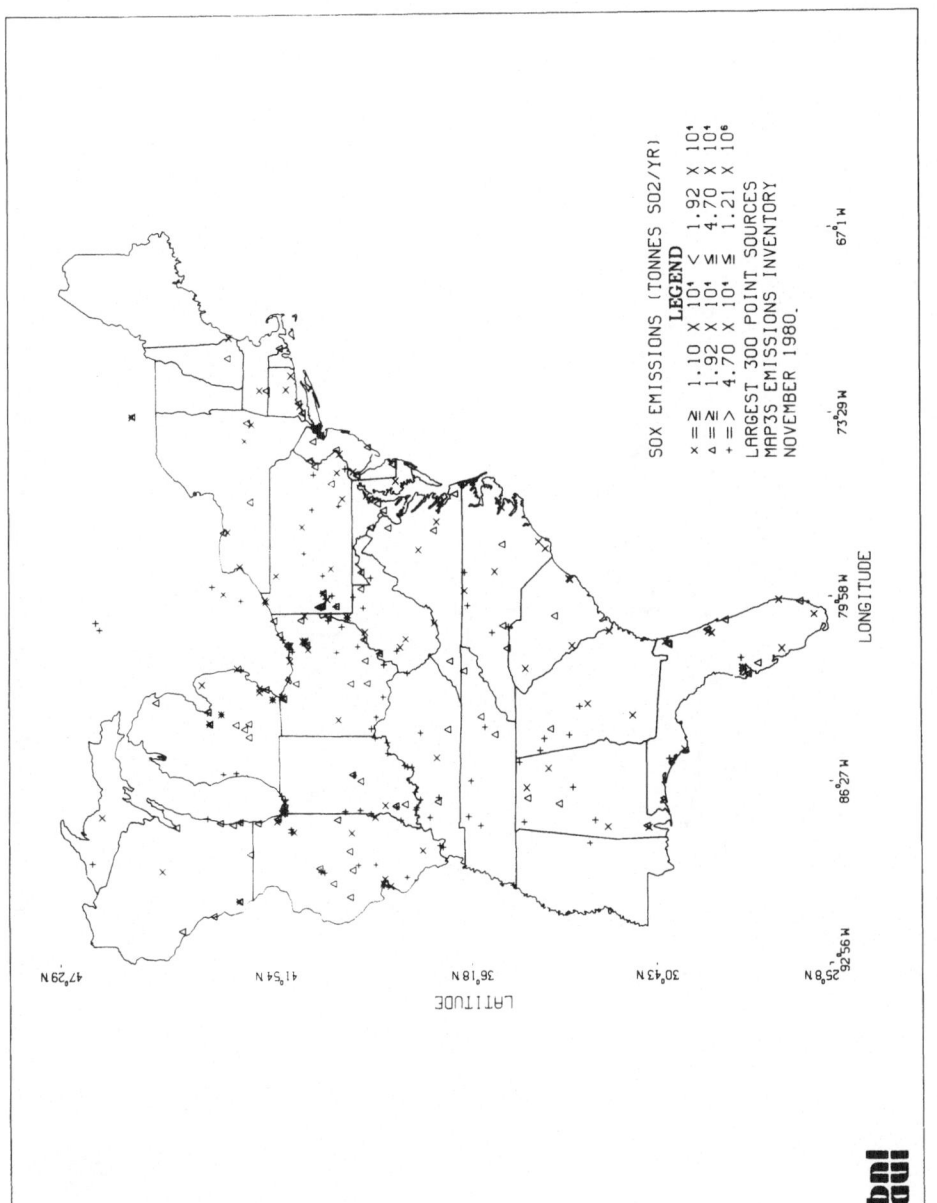

Fig. 1. Distribution of the 300 largest SO$_x$ point sources in the eastern United States and southern Canada.

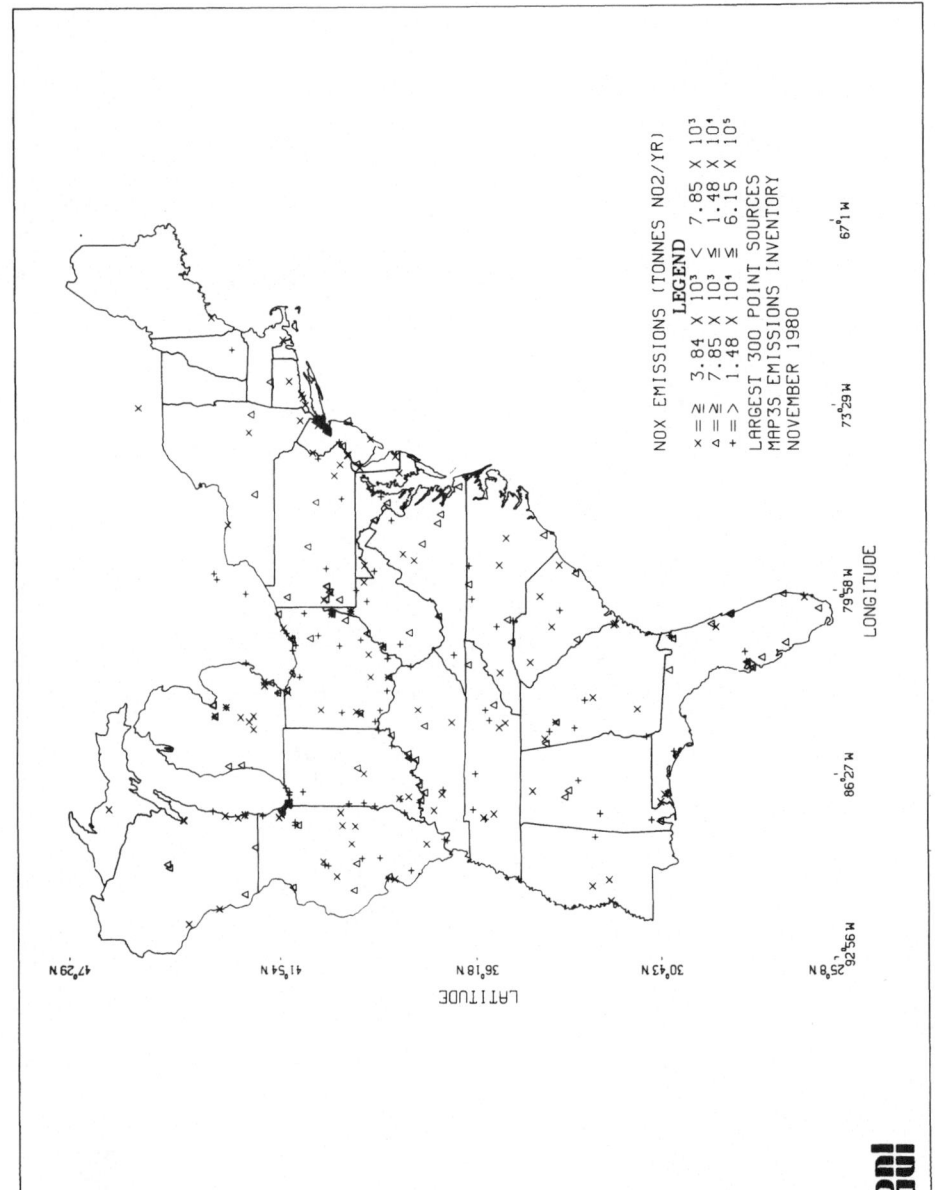

Fig. 2.   Distribution of the 300 largest NO$_x$ point sources in the eastern United States and southern Canada.

terrestrial sources in contrast to other major ions such as sodium and chloride which come largely from marine sources.

Emissions inventory data were obtained from the MAP3S data bank. The outputs of the 300 largest point sources of SO_x and NO_x in the eastern United States and southern Canada were classified into three annual emission groups and their locations plotted on maps of the region (Figures 1 and 2).

Surface and upper air meteorological data obtained earlier from the National Weather Service through a link to the BNL computer were available. Surface maps were plotted at 6-h intervals and 850,700 and 500 mb maps at 12-h intervals. Radiosonde data were also plotted and examined.

2.2. TRAJECTORY MODEL

The Heffter Interactive-Terminal Transport Model (Benkovitz and Heffter, 1980) was used to calculate back trajectories ending at each sampling site during precipitation periods. This model is a version of the Air Resources Laboratories Atmospheric Transport and Dispersion Model (ARL-ATAD) programmed to run on the BNL computer. The model calculates trajectories of three days duration either forward or backward in time from any selected location starting or ending every 6-h on selected days.

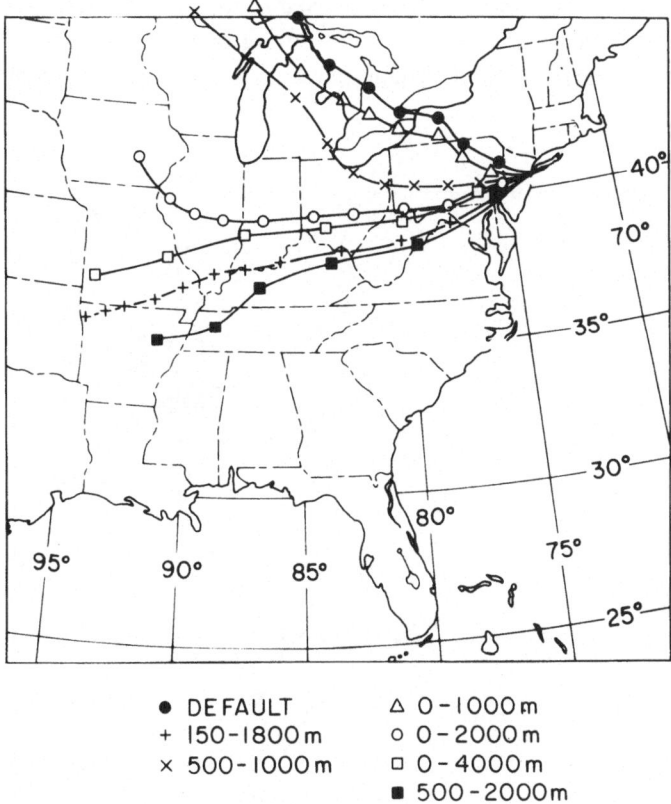

Fig. 3. Trajectories ending at BNL at 0600 GMT April 9 calculated by selection of the transport layer indicated.

Calculations are based on upper air observations stored in the computer data bank.

Each trajectory is calculated using transport winds averaged over a fixed or a variable vertical layer. Because winds change in both direction and speed with height as well as with location and time, results of model calculations are quite sensitive to the vertical layer selected (Figure 3). The model can be used in two modes. In the first (default mode), the height of the transport layer is calculated internally using the meteorological data to determine the height of the mixing layer at each 3-h time step. Since the mixing layer height is typically much greater during the day than at night, materials of low-level origin may be carried to appreciable altitudes during the day by vertical mixing and convective processes. Some may settle slowly during the night but most probably remain above the height of the nighttime mixing layer. Because these layers are not included in the nighttime trajectory calculations, use of the first mode can lead to erroneous results for this application.

In the second mode, the user selects a height interval which is used throughout the trajectory calculations for a whole day. However, care must be taken to select a layer representative of the actual transport layer. For this study, the model was run in the first mode for each case. The maximum height of the mixing layer calculated from the data was then used in the second mode as the top of the selected layer for that trajectory. In all cases a minimum altitude of 150 m was used to avoid local near-surface effects.

Each trajectory ending at a sampling site during or near the sampling period was plotted on a map. The position of each air parcel at 6-h intervals was indicated on each trajectory.

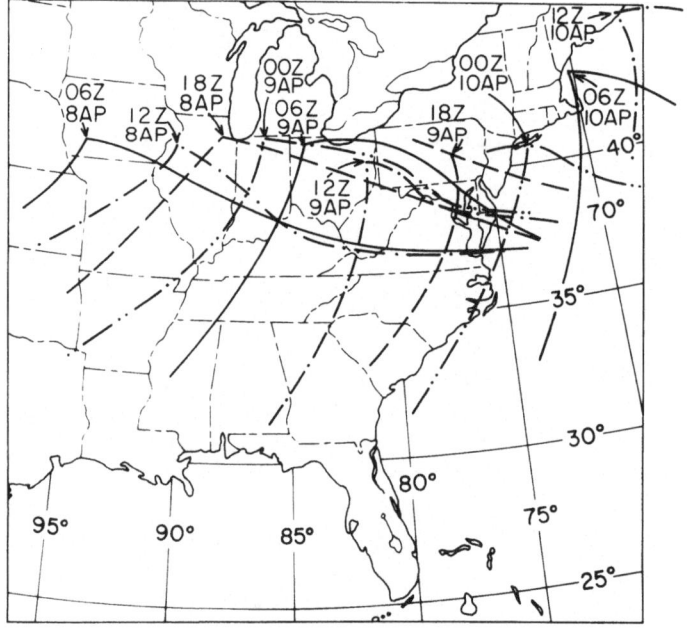

Fig. 4.   Positions of frontal systems at 6 h intervals. Z on figures = GMT.

3. Results

3.1. METEOROLOGICAL ANALYSIS

The low center of the frontal system selected for study moved nearly straight eastward from southwestern Iowa at 0600 GMT April 8 to the vicinity of Long Island, New York, at 0000 GMT April 10. It then moved northeastward along the New England coast (Figure 4). The system remained an open wave until it reached Pennsylvania on April 9 when partial occlusion occurred. The precipitation area covered all of the northeastern states at some time during the passage of the storm (Figure 5). Precipitation was primarily rain but fell as snow at the more northerly stations.

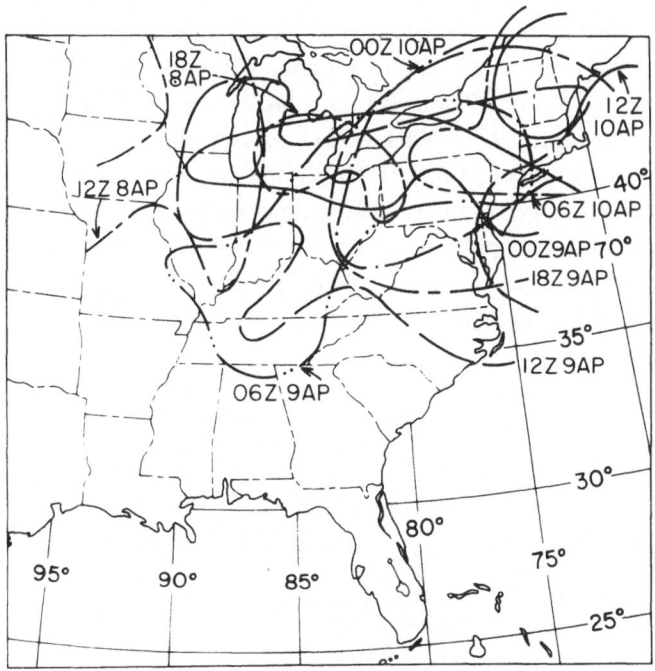

Fig. 5. Boundaries of precipitation areas at 6 h intervals.

The upper air patterns were typical of developing frontal systems. At 1200 GMT April 8, a ridge centered over Michigan was present at all three standard levels. West of the ridge, marked warm air advection was evident, particularly at the 850 mb level. By 0000 GMT April 9, a closed low had developed over lower Michigan at the 850 mb level and a trough was starting to develop in the same area at the higher levels. Meanwhile, the ridge had flattened and moved eastward.

At 1200 GMT April 9, an extensive 850 mb low was centered over northern Ohio and a closed 700 mb low over Lake Huron. The 500 mb trough had deepened and was present over the same region. By 0000 GMT April 10, the 850 mb low was located in southeastern New York and the 700 mb low over south central New York. The 500 mb trough was present over the same region. Cold air advection was evident west of the low centers. At 1200 GMT April 10, the 850 mb low was centered near the Maine-New Hampshire border, the 700 mb low was over the Vermont-New Hampshire line and the 500 mb trough was aligned over the New York-Vermont boundary.

Temperature and humidity profiles were examined on plots of radiosonde data from stations in the study area. Heights of cloud layers were inferred from the profiles. Consideration was given to using the top of the cloud layer as the top of the selected transport layer in the trajectory calculations. The maximum internally computed mixing layer height was used instead because cloud tops were generally well above that height and materials of surface origin could not be shown to mix to maximum cloud levels.

3.2. SPATIAL ANALYSIS

For the spatial analysis, precipitation chemistry data were used from Miami University, Oxford, Ohio (OXF); the University of Virginia, Charlottesville (VIR); the University of Delaware, Lewes (LEW); Pennsylvania State University, University Park (PEN); Cornell University, Ithaca, New York (ITH); and the State University of New York station at Whiteface Mountain (WHF). No sample was obtained from the Champaign, Illinois, station and the BNL sample was contaminated by bird droppings and discarded. As an example of trajectory patterns in relation to frontal positions, Figure 6 shows the frontal location and precipitation area at 0600 GMT April 9 and the trajectories ending at that hour at each of the seven stations.

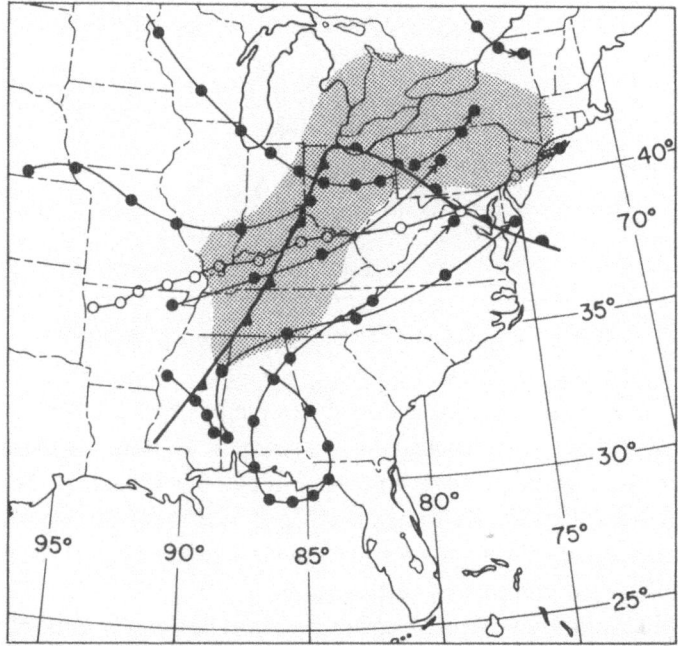

Fig. 6. Frontal positions (heavy lines), precipitation areas (stippled), and trajectories ending at seven sampling sites (OXF, VIR, LEW, PEN, ITH, WHF and BNL) at 0600 GMT April 9. See text for site names and locations. The dots and circles on the trajectories (light black lines) indicate positions of the air parcels at 6 h intervals.

Because plotting all trajectories ending at all stations during the entire precipitation period produces cluttered and confusing diagrams, selected stations only are illustrated in each figure. For clarity, WHF, PEN and VIR which are aligned in an approximate north-south direction are used as one set and the east-west OXF, PEN and LEW stations as a second set. Measurements at and trajectories to ITH were very similar to those for PEN and are not illustrated.

All trajectories ending at the three north-south stations during the period of precipitation are shown in Figure 7. Air parcel positions at 6 h intervals are indicated. H$^+$

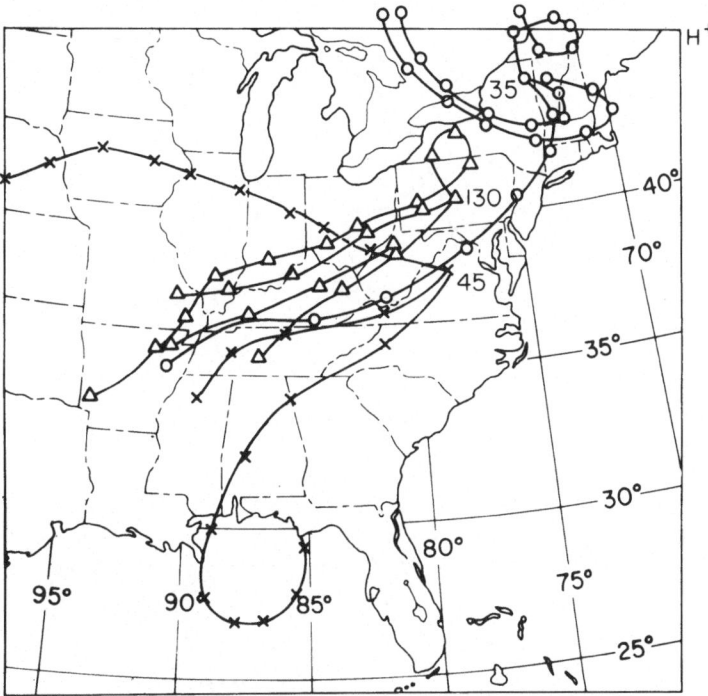

Fig. 7.   Trajectories ending at WHF, PEN and VIR during precipitation period with H$^+$ concentrations beside each station. Air parcel positions at 6 h intervals are indicated on each trajectory by symbols.

concentrations in the event precipitation samples are shown near each station. All trajectories to PEN passed through the highly industrialized Ohio River Valley and the H$^+$ concentration is about three times as great as those at the other two stations. With one possible exception each, their air arrived from regions with fewer major sources.

The SO$_4^{2-}$ results (Figure 8) are similar. The PEN concentration is nearly double that at VIR and three to four times greater than at WHF. The NO$_3^-$ pattern (Figure 9) is even more striking with PEN values about four times as great as those at WHF and VIR. The NH$_4^+$ distribution is similar with concentrations at WHF and VIR only a third of those at PEN (Figure 10).

The east-west differences are less extreme because trajectories to those stations are less divergent. As shown in Figure 11, H$^+$ values at PEN are still appreciably higher than those to the east and west. Only one trajectory to LEW passed through the midwest and

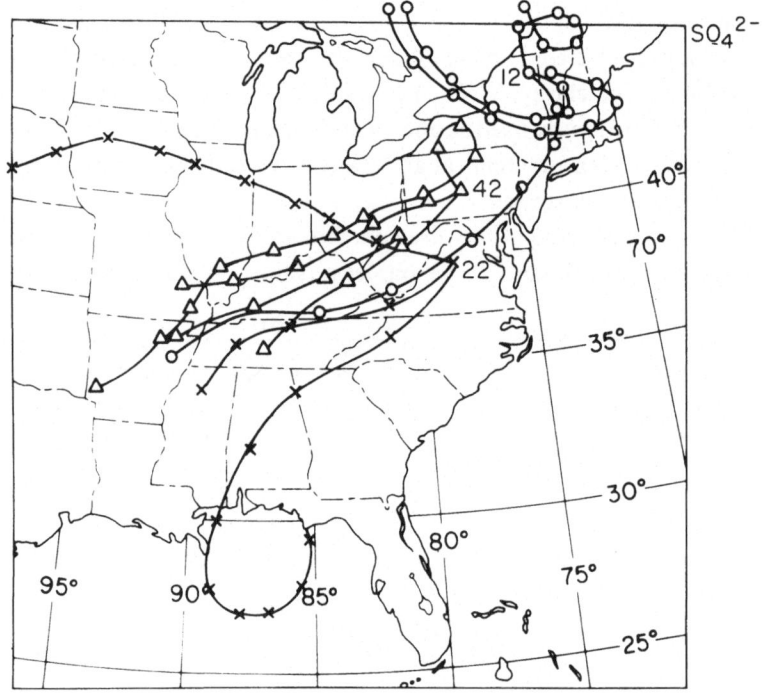

Fig. 8.   As Fig. 7 with SO_4^{2-} concentrations.

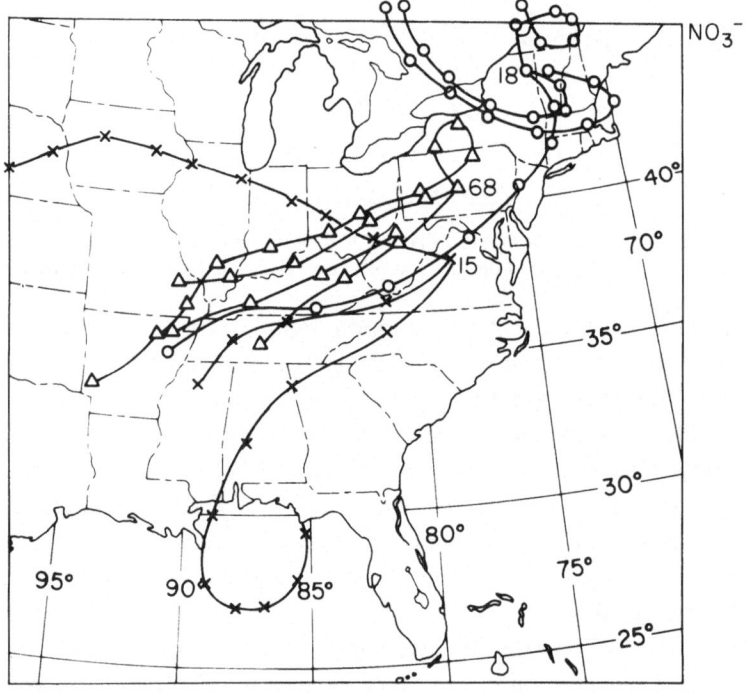

Fig. 9.   As Fig. 7 with NO_3^- concentrations.

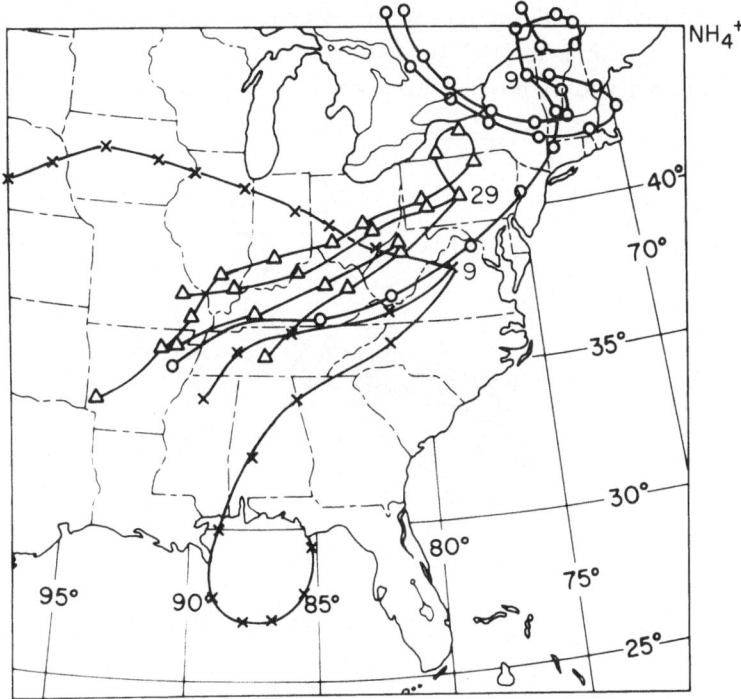

Fig. 10.   As Fig. 7 with NH_4^+ concentrations.

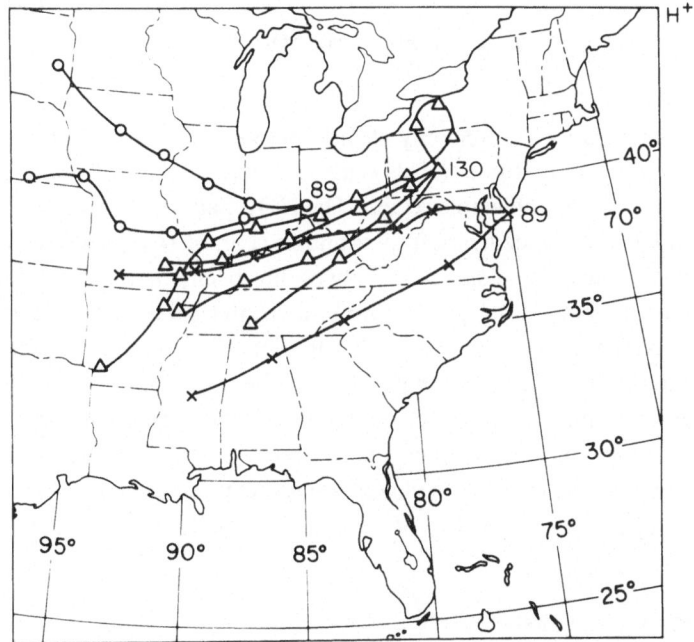

Fig. 11.   Trajectories ending at OXF, PEN and LEW during precipitation period with H^+ concentrations beside each station. Air parcel positions at 6 h intervals are indicated on each trajectory by symbols.

the distances from possible sources to the receptor site are greater permitting greater dilution. The trajectories to OXF passed over numerous large sources in Illinois and Indiana (Figures 1 and 2), possibly accounting for the relatively high values there.

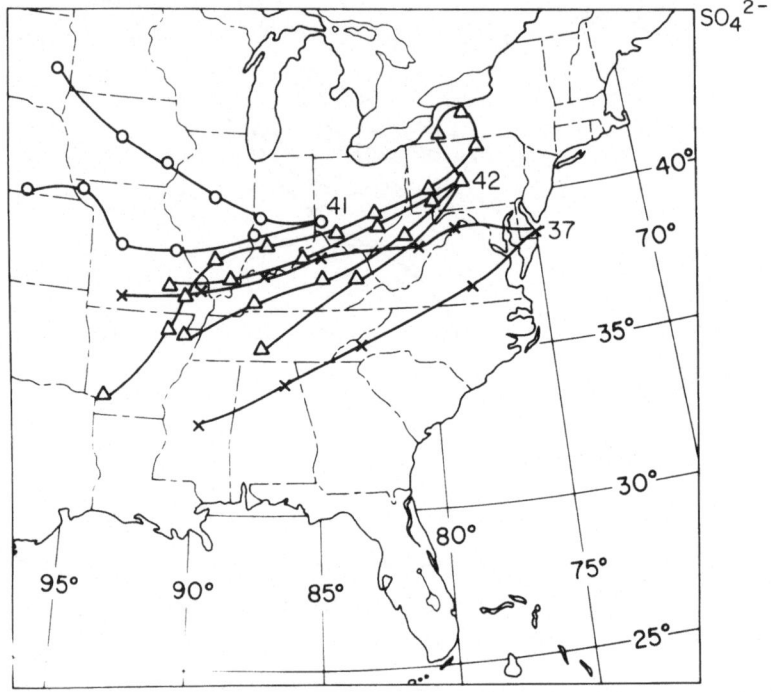

Fig. 12.   As Fig. 11 with SO_4^{2-} concentrations.

Sulfate concentrations differ only slightly between the three stations (Figure 12) in agreement with previous studies (Hidy *et al.*, 1978) which generally found rather uniform air concentrations over the whole region. The NO_3^- pattern (Figure 13) is similar to that of H^+ but the NH_4^+ concentration distribution (Figure 14) shows higher values to the west. This is not unexpected since the midwest has been identified previously as a major ammonia source region (Junge, 1958) primarily from terrestrial and agricultural origins.

Thus, the evidence from the spatial distribution of pollutant concentrations and the trajectories to the six stations strongly suggests the midwest, and the Ohio River Valley in particular, as an important source region of anthropogenic pollutants in northeastern U.S. precipitation. Obviously, many more events must be studied before a firm conclusion can be reached or more localized areas identified as specific sources. However, our analyses agree with those of Wolff *et al.* (1979), Miller *et al.* (1978) and Wilson *et al.* (1980).

3.3. TEMPORAL ANALYSIS

Precipitation occurred at BNL from about 0100 GMT April 9 to 0745 GMT April 10. Back trajectories were determined for five ending times near or during this period. The

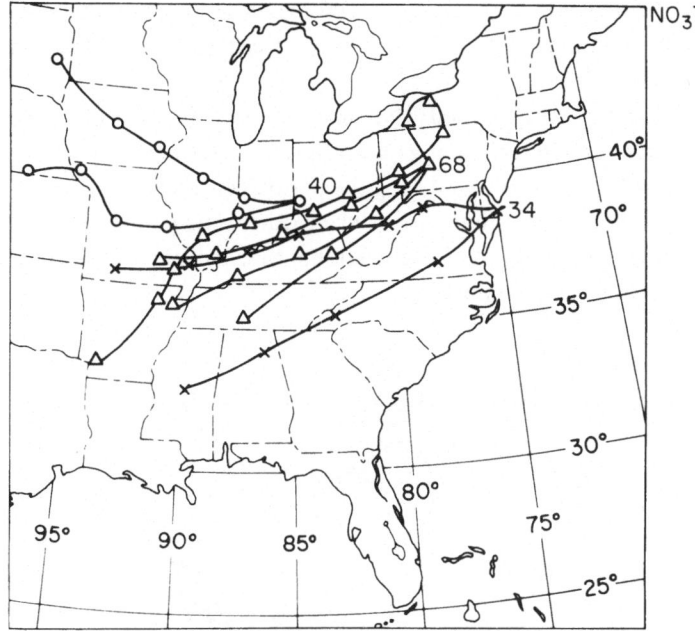

Fig. 13.   As Fig. 11 with NO_3^- concentrations.

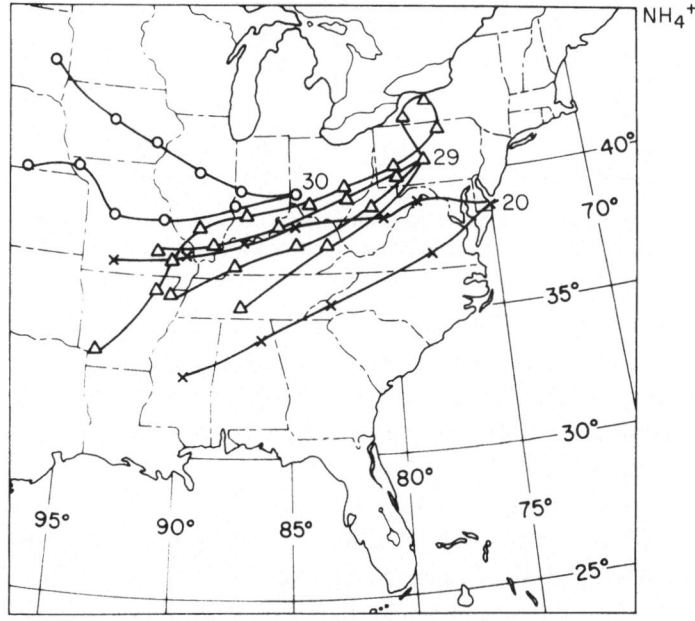

Fig. 14.   As Fig. 11 with NH_4^+ concentrations.

hourly sequential data were averaged in 6 h increments centered on the trajectory ending times. Thus, a mean concentration associated with each air parcel track was obtained for each chemical.

The earliest trajectory approached BNL from the northwest. The second paralleled the Ohio River Valley and the remaining three approached from the southwest. As shown in Figure 15, the H$^+$ concentration during the period centered on the second trajectory was three to seven times higher than those during other time periods. Two of the periods with southwest flow had identical concentrations. Measurements were not obtained for the third period due to small samples.

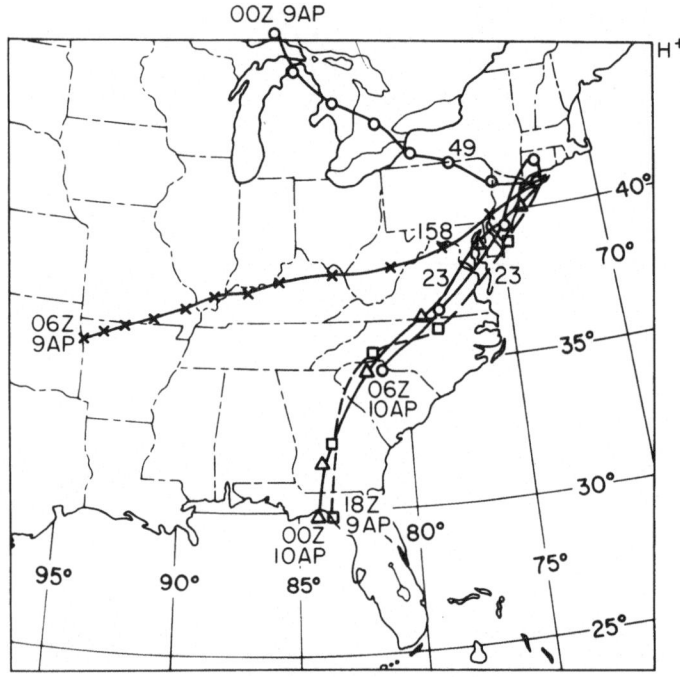

Fig. 15. Trajectories ending at BNL at indicated times and dates with H$^+$ concentrations for 6 h periods centered on trajectory ending times beside each trajectory. Z = GMT.

A similar pattern is evident for sulfate (Figure 16). The three periods with the almost identical southwest trajectories had rather similar concentrations. This suggests that the concentrations may be largely determined by the path of the air rather than time of day or some other factor. As with H$^+$, the period with northwest flow was somewhat higher and the period with westerly flow much higher.

With one exception, the nitrate pattern is similar (Figure 17). The trajectory ending at 0600 GMT April 10 passed just west of New York City and looped into Connecticut before reaching BNL and was associated with very high concentrations. This may be the result of local emissions since Figure 2 shows a concentration of large NO_x sources in and just west of the New York City area.

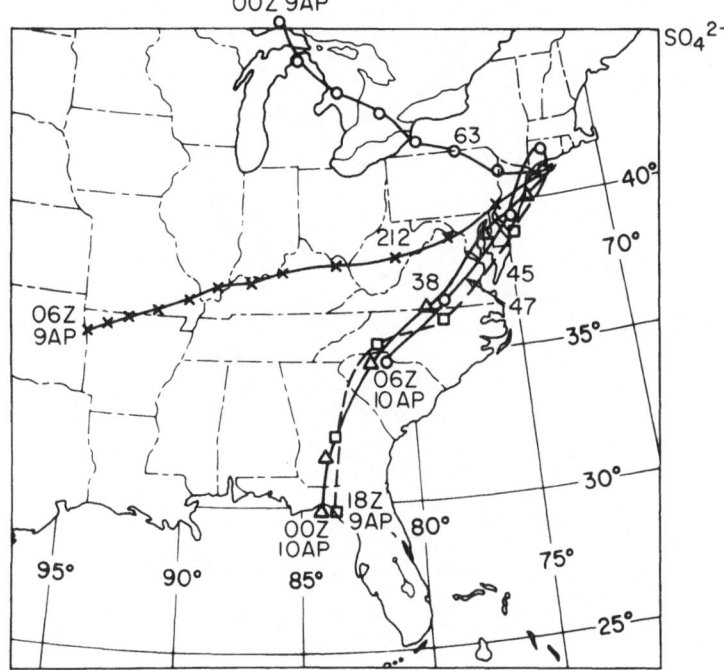

Fig. 16. As Fig. 15 with SO_4^{2-} concentrations.

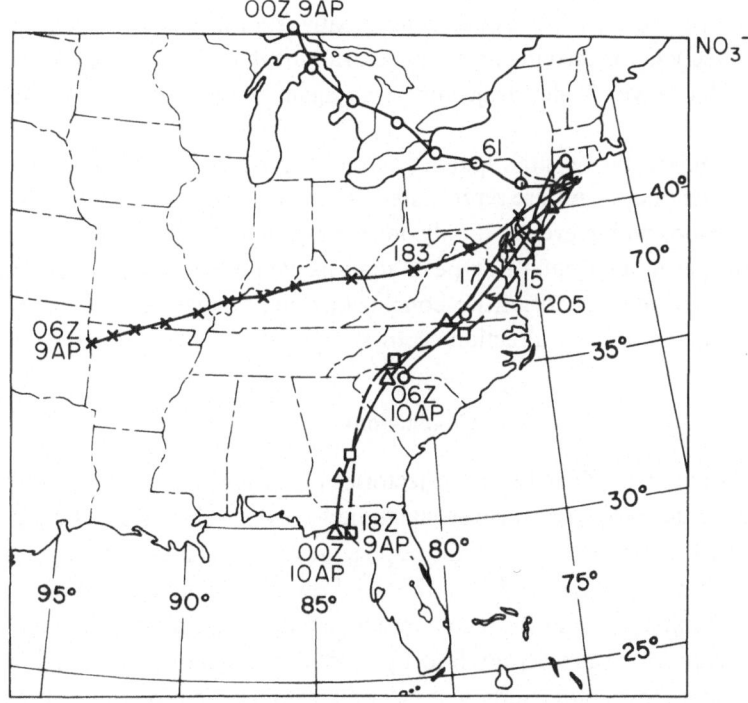

Fig. 17. As Fig. 15 with NO_3^- concentrations.

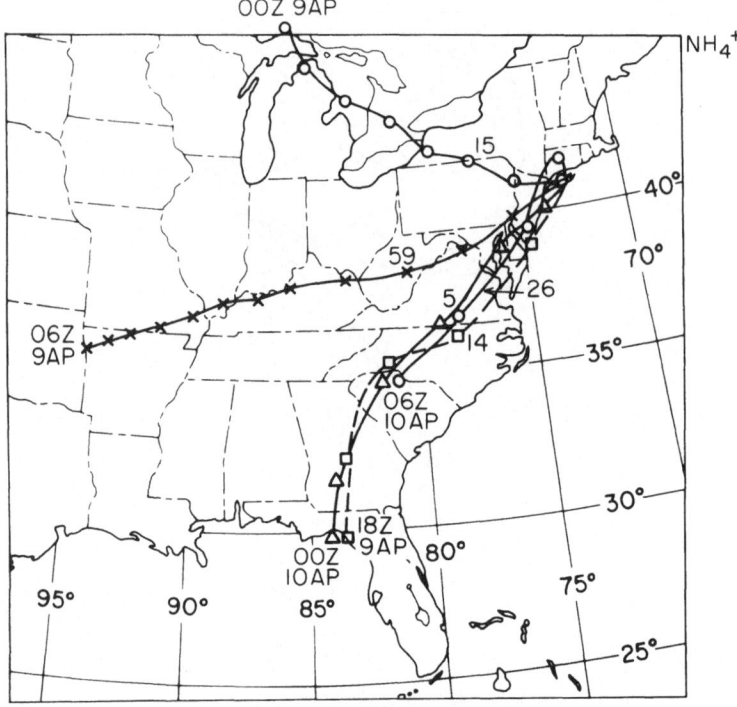

Fig. 18.   As Fig. 15 with NH_4^+ concentrations.

The NH_4^+ pattern (Figure 18) does not differ greatly from the others. The trajectory looping into Connecticut resulted in higher concentrations than the two other southwest tracks but the Ohio River Valley trajectory was again associated with the highest concentrations.

These results reinforce those of the spatial analysis and are even more indicative since three widely divergent pathways are represented. They also show the value of sampling over short time periods within events. If only one mean concentration measurement had been obtained for the entire event, this type of analysis would have been impossible and because of the divergent trajectories, no conclusions could have been reached about possible source regions or even direction sectors.

4. Conclusions

The ARL-ATAD model produced back trajectories consistent with surface and upper air synoptic flows in the particular storm event studied when physically realistic transport layers were selected. Concentrations of anthropogenic air pollutants in precipitation at the array of sampling stations included in this study were roughly proportional to the number of major point sources in the regions traversed by the air parcels arriving at the stations during the precipitation period. Short period samples at one station seem to be more useful for source-receptor studies than event samples at a number of stations because concentrations can be related to single trajectories rather than several trajectories which may have divergent paths. Maximum information, however, can best be obtained from both types of data. Results suggest that more restricted source regions might be identified by a larger number of these studies. It is even possible that quantitative relationships may be established between source emissions and concentrations in precipitation at receptor sites by means of studies of this type even though the physical and chemical processes occurring between sources and receptors are not measured.

5. Acknowledgments

This research was made possible by the participation of many individuals including the MAP3S site operators who obtained the event data. John McNeil was responsible for operation of the BNL sampler and Carmen Benkovitz assisted with application of the trajectory model. Kevin Buckley, Walter Jahnig and Rickey Petty assisted with the meteorological analyses. This research was performed under the auspices of the United States Department of Energy under Contract No. DE-AC02-76CH00016 and the United States Environmental Protection Agency under Contract No. 79-DX-0533.

References

Benkovitz, C. M. and Heffter, J. L.: 1980, *Users' Guide to the Heffter Interactive-terminal Transport Model (ARL-HITTM)*, Informal Report 27801, Brookhaven National Laboratory, Upton, New York. 29 pp.

Førland, E. J.: 1973, *Tellus* **25**, 291.

Galvin, P. J., Samson, P. J., Coffee, P. J. and Romano, D.: 1978, *Environ. Sci. and Tech.* **12**, 580.

Hall, F. P., Jr., Duchon, C. E., Lee, L. G. and Hagan, R. R.: 1973, *Monthly Weather Rev.* **101**, 404.

Hidy, G. M., Mueller, P. K. and Tong, E. Y.: 1978, *Atmos. Environ.* **12**, 735.

Junge, C. E.: 1958, *Trans. Amer. Geophys. Union* **39**, 241.

Miller, J. M., Galloway, J. N. and Likens, G. E.: 1978, *Geophys. Res. Letters* **5**, 757.

Munn, R. E. and Rodhe, H.: 1971, *Tellus* **23**, 1.

Nordø, J.: 1976, *Water, Air, and Soil Poll.* **6**, 199.

Ottar, B.: 1976, *Water, Air, and Soil Poll.* **6**, 219.

Pacific Northwest Laboratory: 1980, *The MAP3S Precipitation Chemistry Network: Third Periodic Summary Report (July 1978–December 1979)*, Report PNL 3400, UC–11, Pacific Northwest Laboratory, Richland, WA. 290 pp.

Raynor, G. S. and Hayes, J. V.: 1979: *Analytical Summary of Experimental Data from Two Years of Hourly Sequential Precipitation Samples at Brookhaven National Laboratory*, Report BNL 51058, Brookhaven National Laboratory, Upton, New York. 91 pp.

Samson, P. J.: 1981, *J. Appl. Meteor.* **19**, 1382.

Smith, F. B. and Hunt, R. D.: 1978, *Atmos. Environ.* **12**, 461.

Wilson, J., Mohnen, V. and Kadlecek, J.: 1980, *Wet Deposition in the Northeastern United States*, ASRC Publication 796, Atmospheric Sciences Research Center, State University of New York, Albany. 139 pp.

Wolff, G. T., Lioy, P. J., Wight, G. D., Meyers, R. E. and Cederwall, R. T.: 1977, *Atmos. Environ.* **11**, 797.

Wolff, G. T., Lioy, P. J., Golub, H. and Hawkins, J. S.: 1979, *Environ. Sci. and Tech.* **13**, 209.

BRIDGING THE GAP BETWEEN AIR QUALITY AND PRECIPITATION CHEMISTRY

G. M. HIDY

Environmental Research and Technology. Inc., 2625 Townsgate Road, Westlake, CA 91361, U.S.A.

(Received May 21, 1981; Revised November 16, 1981)

Abstract. Data sets recently have become available providing simultaneous, regional scale observations of ambient air quality and precipitation chemistry. The data cover parts of the greater northeastern United States. With certain key assumptions, the observations of ambient SO_x and NO_x concentrations can be linked with precipitation concentrations through Junge's concept of rainout efficiency, assumed to be qualitatively equivalent to the washout ratio. A preliminary comparison between data taken from the Sulfate Regional Experiment (SURE) and the parallel precipitation studies using Junge's approach reveals certain useful consequences. Apparent dramatic differences in SO_x and NO_x scavenging are found. Ratios between sulfate in the aerosol and in precipitation compared with trace elements suggest the importance of particulate scavenging processes. Such results show promise for simplified analysis of these data as approaches to differentiating mechanisms involved in cloud-precipitation chemistry.

1. Introduction

This note explores the use of a simple scheme for merging air quality and precipitation chemistry data taken at the same time and location with ground level sampling. The method applies the concept of a *rainout efficiency* defined many years ago, for example, by Junge (1963). In addition, the ratio of certain species to sulfate appears to be a useful adjunct to interpretation of the rainout efficiency.

The air quality data used come from the recent Sulfate Regional Experiment (SURE) discussed by Mueller and Hidy (1981), and summer 1977 data of Pierson *et al.*, (1980) taken near Pittsburgh, PA. The precipitation observations are taken mainly from nearby bulk collector samples (USGS, 1978), and from the Multistate Atmospheric Power Production Pollution Study (MAP3S) (e.g., Benkovitz, 1980). Interpretation of rainout efficiencies calculated from these data for different ions suggests differences in sulfate and nitrate scavenging, indicating possible routes for further study.

2. Scavenging Model and Hypotheses

The rainout efficiency is defined as follows (Junge, 1963):

$$c = \rho c_A \varepsilon_i / L \tag{1}$$

where

$c =$ concentration in cloud water (mg l^{-1});
$c_A =$ concentration of constituent in air (μg m^{-3} STP);
$\varepsilon_i =$ rainout efficiency;
$L =$ liquid water content (g m^{-3} STP);

Water, Air, and Soil Pollution **18** (1982) 191–198.   0049–6979/82/0182–0191$01.20.

$\rho =$ density of water $(g\ cm^{-3})$,
with the units adopted in Equation (1), $0 \le \varepsilon_i \le 1$ if the species concentration in the aerosol is consistently defined with water composition.

The rainout efficiency can be identified with the washout ratio (c/c_A) based on ground-level chemistry data with certain simplifying assumptions. These are:
− the scavenging of material below clouds by washout is small compared with in-cloud (rainout) processes[*];
− the concentration of scavenged species is constant with height;
− clouds have approximately constant liquid water content, taken as $1\ g\ m^{-3}$ (STP) for our purposes here[**]; and
− the precipitation composition collected at a given location is identified with cloud water composition, and is at least qualitatively related to ground-level ambient gas or particle concentration taken on the same day.

From a history of observations, these assumptions can only be valid in the coarsest sense, but they provide a framework to place the results of the interpretation in perspective.

For sulfate $(SO_4^=)$ and nitrate (NO_3^-), the appropriate concentration to be used is most likely a weighted concentration sum of the (soluble) gases, SO_2, NO, NO_2, or HNO_3, and the aerosol particle constituent $SO_4^=$ or NO_3^-. The weighting factor for species depends on the scavenging mechanism and presumably the gas solubility; it has not been determined, and is probably variable. A simple test that particles are the dominant factor in rainwater sulfate or nitrate is to use only the particle concentration of $SO_4^=$ and NO_3^- in the calculation. Thus, the coefficients are of interest for comparison, and are:

(1) ε_1, based on particulate sulfate concentration;

(2) ε_2, based on particulate $SO_4^=$ concentration and SO_2 concentration as $SO_4^=$; and

(3) ε_3, based on particulate nitrate concentration.

A fourth ratio could be defined using NO, NO_2, or HNO_3 concentrations, added in a weighted manner to the particulate nitrate concentration. Concentration of HNO_3 was not measured in available studies though it was estimated indirectly from SURE observations. The additional ratio has not been included in this analysis.

If ε_1 or $\varepsilon_3 > 1$, then the gas absorption must be important in scavenging since ε must be less than one. In practice, if ε_2 is greater than ~ 0.5 but less than 1, we interpret this to suggest that sulfate particle scavenging is probably dominant.

A second series of ratios is of interest in conjunction with the rainout efficiency. These are the ratios between aerosol particulate sulfate concentration and metals, Ca, K, and Al. The latter are components which derive directly from condensed material and do not involve production in the air. Although K is probably in a soluble form in aerosols, Ca and Al may be in only sparingly soluble form. If the scavenging efficiency and mechanism for particulate material is approximately the same, e.g., by collision of particles and

---

[*] See Hidy (1973) for discussion of this assumption.
[**] For justification, see Fletcher (1966).

droplets, and the particle composition is similar over the height to cloud level, then the ratios of the metals to sulfate will be similar in particles and rain. Such a result may be interpreted that sulfate is scavenged mainly as particulate matter. Otherwise the continuity in ratios with respect to sulfate would not be preserved.

Considerable variability in rainout efficiencies or excess cation levels is expected through differences in scavenging processes by event, by the nature of the precipitation (rain or snow), or by the synoptic character of the stormy weather. Furthermore, there may be variability in air mass aging. Such differences may be persistent and large enough to be seen in the combined aerosol and precipitation data. In this note, only data from two summer periods are considered. Thus, any tentative conclusions apply to summer convective showers or storm activity.

The air quality and precipitation data accessible to investigate the application of rainout efficiency interpretation are currently very limited. To explore the nature of the values of the rainout efficiency and the ratios of chemical components, data were chosen from a region in western Pennsylvania, which is known to be influenced by regionally uniform polluted air quality conditions, and is centered in a zone of large wet deposition levels for sulfate and nitrate. The choice of an area influenced in a regionally uniform way rationalizes an additional assumption that air quality and precipitation data are comparable even though they may be taken at stations as far apart as 300 km.

The comparisons use only 24-h averaged air quality data, or data representative of longer durations. Examples were chosen for August 1977 and July 1978, where simultaneous observations of air quality and precipitation chemistry were available. The event data for precipitation were taken at State College, PA; they are compared with SURE air quality observations at Scranton, PA, and neighboring stations within 300 km, if SO_2 data at Scranton were not available. Data from 1977 are compared from Allegheny Mountain (Pierson, *et al.*, 1980), August 1977, and from stations at Kane, PA (July, 1977) and Athens, PA (June 26 to July 3, 1979). These were chosen to look at several comparable ions of interest in addition to sulfate and nitrate, including H^+, NH_4^+, Na^+, Mg^{+2}, Al^{+3}, K^+, and Ca^{+2}.

The precipitation data used were obtained from event observations or monthly data reported for the same period as the air quality data where possible. These observations were chosen from a site nearest the air quality station in question. The comparison of such data relies on an extension of the assumption that regional scale precipitation and air quality chemistry are qualitatively comparable over distances of up to 300 km. An even more tenuous assumption that atmospheric aerosol and precipitation chemistry will be similar in properties during a month period or summer seasonal period is required to compare cation and anion ratios discussed below. These assumptions are probably far too simplistic for natural phenomena, but nevertheless offer a starting point for this type of analysis.

3. Results and Discussion

The calculations of the rainout efficiency for days in August 1977 and July 1978 when rain was recorded in Pennsylvania at Scranton and State College are shown in Table I.

TABLE I

Sulfate and nitrate composition in ambient particles and rainwater in rural Pennsylvania in summer 1977–1978

Date	Aerosol composition[a] (μmol m^{-3})			Rainwater composition[c] (μmol l^{-1})		Rain[c] volume	Sulfate rainout efficiency		Nitrate rainout efficiency
	SO_2	$SO_4^=$	NO_3^-	$SO_4^=$	NO_3^-	cm^3-h	ε_1	ε_2	ε_3
8/8/77	0.041	0.216	0.0032	35	25	222	0.16	0.14	7.8
8/9/77	0.041	0.080	0.0048	110	87	139	1.38	0.91[d]	18.0
8/11/77	0.123	0.213	0.0032	86	70	148	0.40	0.26	21.0
8/15/77	0.041	0.024	—	46	29	238	1.90	0.71[d]	—
8/18/77	0.041	0.021	0.0016	43	32	356	2.00	0.69[d]	20.0
8/25/77	0.082	0.047	0.0016	65	92	55	1.40	0.50[d]	57.0
8/27/77	1.68	0.134	0.0016	110	80	356	0.82	0.061	50.0
8/31/77	1.11	0.208	0.0016	64	65	255	0.31	0.048	41.0
7/4/78	0.082[b]	0.040	0	25	13	32	0.62	0.20	Large
7/11/78	0.123[b]	0.033	0.0064	69	77	196	2.09	0.43	12.0
7/15/78	0.082	0.224	0	95	68	46	0.42	0.31	Large
7/24/78	0.165[b]	0.027	0.0016	42	20	1010	1.56	0.22	12.0
7/28/78	0.123[b]	0.070	0.0016	81	47	270	1.20	0.42	29.0
7/30/78	0.123[b]	0.053	0.0016	51	50	71	0.96	0.29	31.0

[a] From Scranton, PA, SURE Site 02.
[b] Average of available data from SURE Sites, Roseton, PA (15), Wilmington, DE (19), Ithaca, NY (40), and Chester, PA (50), and York, PA (52).
[c] From MAP3S data tape for State College, PA (Benkovitz 1980).
[d] Hypothesized to be dominated by nucleation of water particle and particulate sulfate scavenging.

Concentrations of airborne SO_x are listed in μmol m^{-3} and the rainwater composition is listed as μmol l^{-1}. The rain intensity is included from State College to give an indication of the volume of water precipitated in (cm^3-h).

Calculated values of the sulfate rainout efficiency based on particulate sulfate alone (ε_1) show values of 2.1 or less, with most values approximately unity. Inclusion of both SO_2 and particulate sulfate in the air gives rainout efficiencies (ε_2) less than unity. Four of the fourteen cases cited have $\varepsilon_2 \geq 0.50$, but these cases all involve low ambient concentrations of SO_2 and $SO_4^=$ and involve large measurement uncertainly in SO_2 especially. These four examples are consistent with dominance of sulfate particle scavenging but are not in themselves proof of dominance because of the measurement uncertainties discussed below. In all cases however, the scavenging of particulate sulfate appears to account for a significant, but possibly minor part of the sulfate present. Ten of the cases cited can be rationalized by a contribution from SO_2 absorption into hydrometeors even in cases where the (ground level) SO_2 concentrations are similar to those of sulfate. Thus, by this analysis, the contribution of SO_2 absorption to rainwater sulfate cannot be ruled out as a significant factor in eastern, summer rain scavenging. For high ambient SO_2 concentrations relative to $SO_4^=$ in the particles, ε_2 values tend to be lower than for the cases with minimal SO_2 levels. This is consistent with the expectation that gaseous SO_2 absorption is less efficient as a scavenging process than scavenging of sulfate particles.

There appears to be no systematic difference in scavenging with rainfall volume in the summer cases explored.

The values of the nitrate rainout efficiency based on particulate nitrate levels (ε_3) also are listed in Table I. The particulate nitrate values obtained from the SURE data are much lower than sulfate. There is a dramatic difference between particulate nitrate to sulfate ratios and in rainwater. The latter contains more than ten times as much nitrate relative to sulfate than the particulate matter. We note further that the nitrate rainout efficiencies are all very large, much larger than those for sulfate. This is interpreted to mean that nitrate scavenging is dominated by gas absorption.

Both NO and NO_2 are present in the air over Pennsylvania. These gases are sparingly soluble in water as is SO_2 (NO_2 actually reacts directly with water to form HNO_2 and HNO_3). However, in summer, photochemical reactions will produce significant quantities of HNO_3, which is extremely soluble in water. Mueller (see Mueller and Hidy, 1981) has estimated indirectly from calculations that the HNO_3 vapor concentrations present in the SURE region could be in the parts-per-billion concentration range. This ambient concentration would be sufficient to account for the anomalously high apparent nitrate rainout efficiencies in Table I. Therefore, there is circumstantial evidence that the nitrate found in summer rainfall cases listed in Table I can be accounted for by HNO_3 vapor absorption in hydrometeors, with little or no contribution from particulate nitrate scavenging, or from NO and NO_2 absorption.

An estimate of the representativeness of the calculated ratios and their interpretation can be obtained from the cited measurements, and other available information. The particulate sulfate data reported from the SURE have a well defined measurement uncertainty of less than $\pm 10\%$; the error in SO_2 observations has a variable measurement error that increases dramatically near the instrument detection limit of 13 $\mu g\ m^{-3}$ (0.13 $\mu mol\ m^{-3}$). Thus, the low 24-h average values of SO_2 cited in Table I generally will have an uncertainty in excess of $\pm 100\%$. (See also Mueller and Hidy, 1981; Chapter 3.)

The particulate concentrations reported are based on sampling using a filter substrate which eliminates the artifact from SO_2 and NO_2 absorption, but has a positive HNO_3 absorption artifact. The resulting particulate nitrate artifact is offset by a possible negative artifact associated with a loss by particle acidification. The particulate nitrate values are believed to be representative of nonurban ambient conditions in the East, but may have a measurement uncertainty of 100%. The minimum uncertainty in precipitation chemistry reported for MAP3S and the NADP can be estimated from different investigations (Hansen and Hidy, 1981). The analytical uncertainty, excluding sampling and handling is estimated for rainwater sulfate and nitrate to be approximately $\pm 10\%$. Using the square root of the sum of squares of individual error estimates, these translate into fractional uncertainties in the ratios for ε_1 of $\pm 14\%$ and ε_3, approximately $\pm 100\%$.

The calculation for ε_2 is sensitive to SO_2 concentration uncertainties. Near the instrument detection limit for SO_2, and above, the uncertainty in ε_2 typically be $\pm 100\%$ or more; as the SO_2 concentration increases, the uncertainty decreases. However when both SO_2 and $SO_4^=$ concentrations are low and the same magnitude as in the August 1977 period (Table I), the uncertainties in ε_2 are large and not readily determined.

The SURE data analysis indicated only weak correlation between ground level concentration observations and observations between 160 and 1500 m (Mueller and Hidy, 1981). This raises questions about the representativeness of ground level data for conditions aloft. However, the apparent rainout efficiencies for very limited in-cloud data reported by Lazrus et al. (1981), based on $L = 1 \text{ g m}^{-3}$, and ratios to sulfate are similar to those in Table I and II for $SO_4^=$, NO_x, NH_4^+, Na^+, and Mg^{+2}. This suggests that the connection between rainout and ground level conditions is reasonable.

The uncertainty in liquid water content can be examined from Scott's (1978) theoretical relation between liquid water content and precipitation rate. The rainfall rates reported for data cited in Table I varied between 2 and 57 mm h^{-1}. Scott's relation indicates that the liquid water content for such precipitation rates would range between 1.5 ± 0.5 and $3.5 \pm 0.5 \text{ g m}^{-3}$. This range is relatively narrow, but suggests a bias in the ε values of perhaps a factor of two low. If this is the case, then the conclusions about the significance of gaseous NO_x species for rainwater nitrate are reinforced. The ε_2 ratios become larger, emphasizing the particulate sulfate scavenging as opposed to SO_2 absorption in accounting for rainwater sulfate.

The comparison between ambient aerosol chemistry and rainwater composition can be extended in another way, considering ion concentrations ratioed to sulfate concentration. These are illustrated in the data listed in Table II. Unfortunately, data of this kind are very limited from existing observations so that the comparison must be done in a qualitative way without direct comparability in space and time, as noted above.

The results listed give the major cations and anions measured in the samples. The anion sum, cation sum, and cation sum less ammonium and H$^+$ also are tabulated. Ratios of the ions to sulfate ion are calculated for comparison. The results of Pierson et al. (1980) for data taken in August 1977 at Allegheny Mountain, PA, suggest that roughly a 30% excess cation concentration over anions could be attributed to H$^+$ in the aerosol particles sampled. The rainwater composition determined for the month of July 1979 at Kane, PA, 100 km to the west of Allegheny Mountain, indicates a roughly similar excess in H$^+$ concentration, However, the rainwater also shows a much larger quantity of nitrate ion relative to sulfate ion than in the particles at Allegheny Mountain. The addition of ammonium ion and alkaline metals, particularly K, Ca, or Mg, relative to the particle sample appears to have compensated for the added nitrate so that the H$^+$ excess is similar between the particles and the rainwater.

The similarity in metal to sulfate ratio between the particles and the rainwater with increase in ratio for the latter suggests that moisture nucleation on aerosol particles could account for most of the sulfate present in the rain on the average. However, the dramatic increase in rainwater nitrate relative to the other ions reinforces the hypothesis that this species is dominated by gas absorption rather than particle scavenging processes. For additional comparison, rainwater data from Athens, PA, 250 km northwest of Allegheny, are also listed. These data were taken from the reports of the National Atmospheric Deposition Program (Anonymous, 1979). Here again, apparent nitrate enrichment

TABLE II

Composition of suspended particulate matter and rainwater in rural Pennsylvania in summer

Component	Particulate concentration[a] Allegheny mountain		Rainwater concentration			
			Kane, PA[c]		Athens, PA[e]	
	neq m^{-3}	Ratio to $SO_4^=$	μeq l^{-1}	Ratio to $SO_4^=$	μeq l^{-1}	Ratio to $SO_4^=$
H$^+$	155	0.53	128	0.90	120	1.7
NH$_4^+$	128	0.44	100	0.70	19	0.27
NO$_3^-$	(8.5)[b]	0.029	60[d]	0.42	54	0.76
SO$_4^=$	292	1.00	142	1.00	71	1.00
Na$^+$	9	0.031	4.3	0.030	16	0.23
Mg^{+2}	6	0.020	8.2	0.058	2	0.029
Al^{+3}	78	0.27	–		–	
K$^+$	2	0.0068	5.3	0.037	1.2	0.016
Ca^{+2}	16	0.055	26.4	0.19	5	0.070
Anion sum (A)	300	–	202	–	125	–
Cation sum (B)	394	–	272	–	163	–
Cation sum less H$^+$ and NH$_4^+$ (C)	111	–	44.2[f]	–	24.2[f]	–
Ratio A/C	2.7	–	6.2	–	5.1	–
Ratio A/(C + NH$_4^+$)	1.3	–	1.4	–	2.9	–

[a] August 1977 data reported by Pierson *et al.* (1980).
[b] Questionable data (Pierson *et al.*, 1980), but consistent with values at Scranton, PA and other nearby sites, as reported by the SURE.
[c] July 1977 data (USGS, 1978).
[d] NO$_3^-$ + NO$_2^-$.
[e] June 26 to July 3, 1979 data (NADP data, 1979).
[f] Aluminum unavailable.

relative to sulfate is found; there is apparent depletion with respect to ammonium compared with the Athens case. The ratio of cations and anions excluding ammonium is similar in the two rainwater cases, but the H$^+$ concentration is higher relative to sulfate in this set. If ammonium enters droplets as ammonium sulfate, then it is possible that the enhancement of H$^+$ in rainwater is linked more with the increase in nitrate relative to sulfate rather than sulfate enrichment from SO$_2$ gas absorption with particle scavenging in this case.

4. Summary and Recommendations

The results to date emphasize the strong differences between sulfate and nitrate scavenging, and give hints about the significance of particle removal processes relative to gas absorption in hydrometeors. Furthermore, the results suggest that expanded study of aerosol chemistry with precipitation chemistry offers promise for the elucidation of rainfall as a scavenger of air pollution.

Progress towards quantification of scavenging processes using the rainout efficiency-washout coefficient concepts will require carefully designed experiments which could include:

(a) simultaneous measurements of SO_x, NO_x and other aerosol constituents in ambient air entering clouds, cloud water concentrations, and precipitation water concentrations under conditions where the precipitation chemistry can reasonably be associated with the in-cloud conditions,

(b) vertical structure of pollutant concentrations from the surface below and through the cloud layer to isolate rainout and washout as opposed to clean air variations,

(c) seasonal or warm cloud/supercooled cloud sampling to establish variation in scavenging by hydrometeor condition, and

(d) analysis of in-cloud and precipitation soluble and insoluble fractions to check for differences in scavenging processes by constituent.

Soluble tracer scavenging experiments involving in-cloud and below cloud sampling also offer a potential for obtaining independent estimates of scavenging rates.

Acknowledgement

This work was sponsored by the Electric Power Research Institute under Contracts 862-1 and 862-2. The helpful criticism of Peter Mueller is gratefully acknowledged.

References

Anonymous: 1979, *National Atmospheric Deposition Program – Data Report First and Second Quarter 1979*, Natural Resource Ecology Laboratory, Colorado State University, Fort Collins, CO.

Benkovitz, C.: 1980, *MAP3S Precipitation Data Records*, Multistate Air Pollution Power Production Study Data Bank, Brookhaven National Laboratory, Upton, NY.

Fletcher, N. H.: 1966, *The Physics of Rain Clouds*, Cambridge, Tables 1.1 and 1.2.

Hansen, D. A. and Hidy, G. M.: 1981, *Atmospheric Environment*, submitted.

Hidy, G. M.: 1973, 'Removal Processes of Gaseous and Particulate Pollutants', in S. I. Rasool (ed.), *Chemistry of the Lower Atmosphere*, Plenum Press, NY, p. 121.

Junge, C. E.: 1963, *Air Chemistry and Radioactivity*, Academic Press, NY, p. 291.

Lazrus, A. L., Haagenson, P. L., Kok, G. L., Hulbert, B. J., Kreitzberg, C. W., Likens, G. E., Mohnen, V. A., Wilson, W. E., and Winchester, J. W.: 1981, *Acidity in Air and Water in a Case of Warm Frontal Precipitation*, Presented at the 74th Annual Meeting, Air Pollution Control Assn., Philadelphia, PA.

Mueller, P. K. and Hidy, G. M.: 1981, *The Sulfate Regional Experiment: Report of Findings*. Report EA-1901, Vol. I and II, Electric Power Research Institute, Palo Alto, CA, in preparation.

Pierson, W., Brachaczek, W., Truex, T., Butler, J., and Korniski, T.: 1980, *Ann. NY Acad. Sci.* **338**, 145.

Scott, B. C.: 1978, *J. Appl. Meteorol.* **17**, 1375.

U.S. Geological Survey: 1978, *Water Resources Data for New York Water Year 1977*, Report NY-77-1, Vol. I, USGS, Syosset, NY.

AN ANALYSIS OF SPATIAL VARIABILITY OF THE DOMINANT IONS IN PRECIPITATION IN THE EASTERN UNITED STATES

JERRE W. WILSON

Science Research Laboratory, United States Military Academy, West Point, 10996, U.S.A.

and

VOLKER A. MOHNEN

Atmospheric Sciences Research Center, State University of New York at Albany, Albany, NY 12222, U.S.A.

(Received 29 June, 1981; Revised 9 October, 1981)

Abstract. Data of the Multistate Atmospheric Power Production Pollution Study (MAP3S) and the National Atmospheric Deposition Program (NADP) were utilized to develop wet deposition spatial distribution patterns for the eastern United States for 1979. The ions of $SO_4^=$, NO_3^-, H^+, and NH_4^+ were selected for study since they are the most prominent ones found in precipitation.

Total wet deposition for 1979 was normalized to one centimeter of precipitation and objectively analyzed using the Synagraphic Mapping System (SYMAP) technique. Gradients of $SO_4^=$ and NO_3^- were found to be essentially uniform, both to the east and west of the major pollution regions. An increased gradient in normalized deposition for $SO_4^=$, NO_3^-, and H^+ was found in the Appalachian Mountain region.

Estimates of total wet deposition were obtained by using the normalized deposition values in conjunction with precipitation as reported by the National Climatic Center. SYMAP analyses of the estimated total wet deposition were localized in nature due to precipitation variations between sites.

1. Introduction

Regional patterns of air and precipitation quality have been developed by numerous networks for a variety of constituents. The Junge network (Junge and Werby, 1958; Junge, 1960) provided the first comprehensive spatial analysis for the majority of the dominant ions found in United States precipitation for the period July 1955 through July 1956. Later the National Precipitation Sampling Network (Lodge *et al.*, 1968) monitored the major ions as well as metals in precipitation from 1960 to 1966 for 33 United States stations.

The spatial distribution of pH within the eastern United States is often used to characterize regional precipitation quality. However, Granat (1978) points out that pH, or H^+ concentration, may be severely influenced by contamination from locally-produced soil dust. Therefore, knowledge of the regional distribution of the dominant ions in precipitation is essential for proper interpretation of wet deposition patterns. This paper will examine the spatial distribution of wet deposition for the four dominant ions of sulfate ($SO_4^=$), nitrate (NO_3^-), hydrogen (H^+), and ammonium (NH_4^+) for the year 1979.

2. Analysis Procedure

While several United States precipitation chemistry monitoring networks operate independently, it is possible to correlate their results through proper analysis techniques
(Pack, 1980). Two programs offering regional coverage, accessible data, and reliable
quality control are the Multistate Atmospheric Power Production Study (MAP3S) and
the National Atmospheric Deposition Program (NADP). Even though MAP3S is
event-oriented, while NADP collects weekly samples, their integrated ion deposition
measurements are compatible when compared over time scales on the order of one year.
Figure 1 and Table I show the location of the MAP3S and NADP sites of the eastern
United States used to develop regional wet deposition patterns for 1979 (not shown is
Mead, Nebraska). The combined network is composed of eight MAP3S sites and
27 NADP stations. There are no stations located in the lower Mississippi Valley,

Fig. 1.   MAP3S and NADP station network used for 1979 spatial deposition analysis.

TABLE I

MAP3S and NADP station locations

State	Site name	Latitude	Longitude	Elevation (m)
MAP3S Sites				
New York	Whiteface	44.40	73.88	604
New York	Ithaca	44.38	76.72	503
Pennsylvania	Penn State	40.78	77.95	396
Virginia	Virginia	38.05	78.55	171
Illinois	Illinois	40.05	88.37	213
New York	Brookhaven	40.88	72.88	24
Delaware	Lewes	38.76	75.00	0
Ohio	Oxford	39.53	84.73	283
NADP Sites				
Florida	Bradford Forest	29.97	82.20	44
Georgia	Georgia Station	33.18	84.40	268
Illinois	Bondville	40.05	88.37	213
Illinois	SIU	37.70	89.27	140
Illinois	Dixon Springs Ag. Ctr.	37.43	88.67	165
Michigan	U. Mich. Biological Sta.	45.57	84.68	235
Michigan	Kellogg Biological Sta.	42.48	85.38	289
Michigan	Wellston	44.22	85.85	291
Minnesota	Marcell Exp. Forest	47.50	93.47	430
Minnesota	Lamberton	44.25	95.32	342
Nebraska	Mead	41.09	96.50	354
New Hampshire	Hubbard Brook	43.95	71.70	252
New York	Huntington Wildlife	44.00	74.22	500
New York	Stilwell Lake–West Point	41.35	74.03	185
North Carolina	Lewiston	36.13	77.17	25
North Carolina	Coweeta	35.02	83.45	220
North Carolina	Piedmont Research Sta.	35.67	80.57	251
North Carolina	Clinton Crops Res. Sta.	35.01	78.28	55
North Carolina	Finley	35.73	78.68	128
Ohio	Delaware	40.28	83.06	262
Ohio	Caldwell	39.78	81.52	275
Ohio	Wooster	40.77	81.93	312
Pennsylvania	Kane Exp. Forest	41.55	78.77	610
Pennsylvania	Leading Ridge	40.55	77.93	274
South Carolina	Clemson	34.67	82.83	231
Virginia	Horton's Station	37.18	80.42	322
West Virginia	Parsons	39.10	79.65	508

therefore the objective analysis in this area must be considered preliminary at best. The station density of the western United States was considered insufficient for development of spatial analysis patterns. Even though some NADP stations did not operate throughout the entire year, the samples collected were from both winter and summer and are considered to be representative of the precipitation quality at these various locations.

Total wet deposition is governed primarily by the amount of precipitation. The amount of precipitation and calculated wet deposition will be influenced by the efficiency of the collector, the care in sample handling, and the percentage of events actually sampled.

These considerations are paramount when considering inter-site or regional total wet deposition comparisons. While weighted-concentration calculations provide essentially the same values as normalized deposition, the latter is more applicable from a meteorologists's point of view. A knowledge of the total amount of precipitation allows one to make quick estimates of the total wet deposition over extended periods of time. A simple normalization procedure was utilized where the integrated amount of pollutant material delivered per centimeter of precipitation was calculated for 1979 (Wilson *et al.*, 1980). The application of this technique at selected MAP3S and NADP sites represents overall characteristics of precipitation quality throughout the year and allows for meaningful inter-site comparisons.

Normalized wet deposition values were further utilized to obtain estimates of the total wet deposition for the various sites. Since the normalized deposition values are integrated over the entire year, they contain no seasonal bias. The actual precipitation was multiplied by the normalized deposition for each site to obtain estimates of the total wet deposition for the four dominant ions. The precipitation data of the National Climatic Center were used to calculate the estimated total wet deposition.

To insure complete objectivity, spatial distributions of both normalzed and estimated total wet deposition were analyzed using the Synagraphic Mapping System (SYMAP), developed by the Laboratory for Computer Graphics and Spatial Analysis, Harvard University (Dougenik and Sheehan, 1975).

3. Emissions for the Eastern United States

To evaluate adequately the chemical quality of precipitation and investigate possible source-receptor relationships, a detailed emission inventory is needed. The MAP3S emission inventory was chosen to sum selected emissions per square latitude and longitude for the eastern United States. The MAP3S emission inventory was compiled of data extracted from the National Emissions Data System and the Federal Power Commission. It was updated using corrections provided by MAP3S and by the Sulfate Regional Experiment. The five (5) pollutants routinely measured were:
– total particulates
– SO_2
– NO_x
– hydrocarbons
– CO.
Pollutant emissions were grouped into classifications of point or area sources. Point sources were defined as stationary sources with the potential of emitting at least 100 tonne day^{-1} of any of the five (5) criteria pollutants (Clark, 1980). Area sources were reported by county and included vehicular emissions, as well as industrial and residual heating. The geographical center of each county was used to locate the area emissions in a particular grid.

The total annual point source emissions per square latitude and longitude for SO_2 are depicted in Figure 2. Over 90% of the total annual SO_2 emissions are from point sources,

Fig. 2. Total annual SO_x emissions (10^6 kg yr^{-1}) per square latitude and longitude for the eastern United States.

this summation of stationary sources is representative of the actual annual total. Nearly 57% of the total point source emissions of SO_2 can be linked to bituminous coal-fired electric generating plants, 12% to primary metals industry, 7% to burning of bituminous coal for industrial fuel, and 4% to petroleum industrial sources (Clark, 1980).

The total NO_x emissions of Figure 3 are closely correlated with large population density centers. Total point source emissions and area source contributions were found to be of the same order of magnitude. Two-thirds of the point source emissions are from the burning of fuels at electric generating plants, while over 80% of area sources are the result of vehicular exhaust.

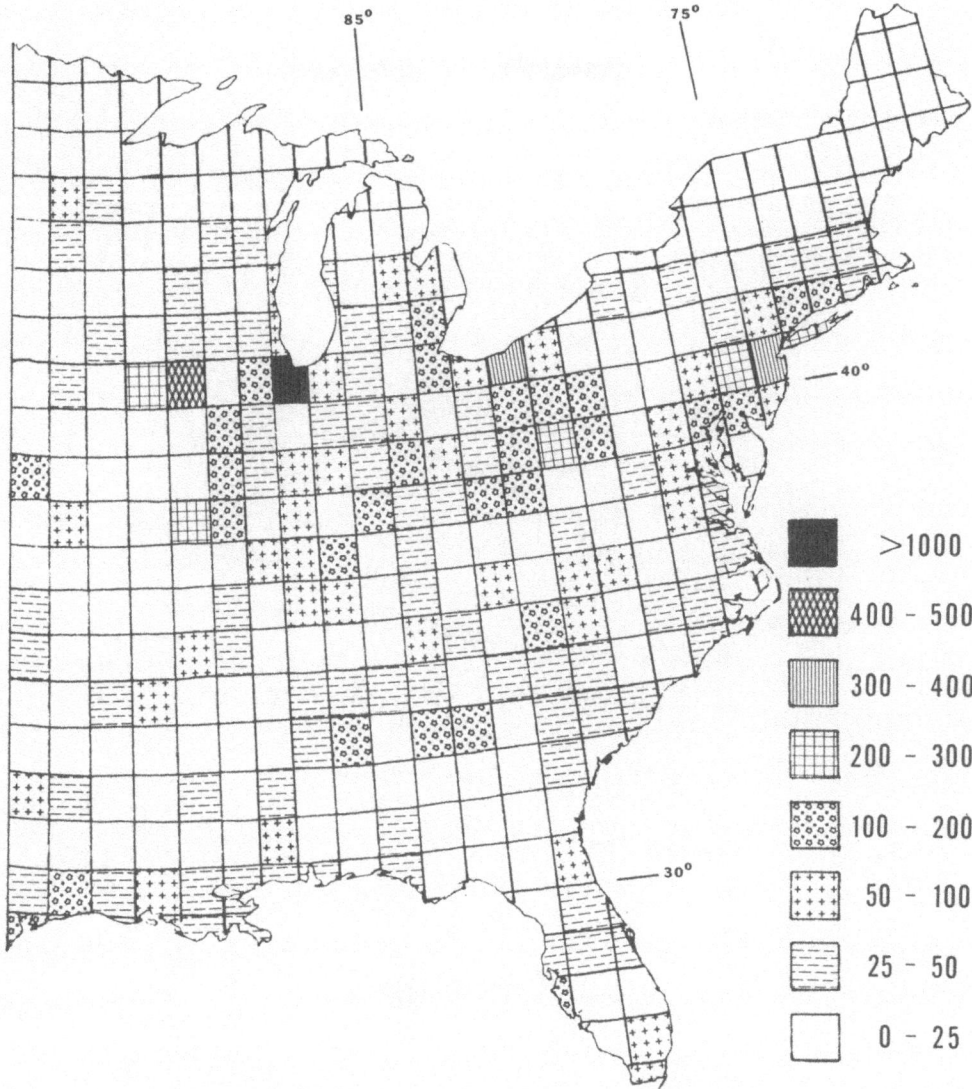

Fig. 3.   Total annual NO$_x$ emissions (10^6 kg yr^{-1}) per square latitude and longitude for the eastern United
States.

It should be noted that a variety of emission inventories exist, with no agreement as
to accepted emission values. Therefore, the emission distributions of Figures 2 and 3
should be regarded as somewhat general in nature and are essentially proportional to
the actual source strengths.

4. Normalized Wet Deposition – 1979

Figures 4 through 7 show the normalized wet deposition per centimeter of precipitation
for 1979 for SO$_4^=$, NO$_3^-$, H$^+$, and NH$_4^+$, respectively. These objective analyses, using

the procedure described above, present a rather uniform pattern for $SO_4^=$ and NO_3^-. A slightly more complex picture emerges for H^+ and NH_4^+.

The normalized $SO_4^=$ maximum of Figure 4 correlates well with the total point source emission patterns of SO_2 of the Ohio Valley, Figure 2. The secondary maxima in the Carolina's may be a result of the industrialized regions of the Southeast. Similar distribution characteristics are found for normalized NO_3^- wet deposition, Figure 5, when compared to the total NO_x emission of Figure 3. No direct source-receptor relationships are discernible from the data as a result of transport of pollutants over great distances.

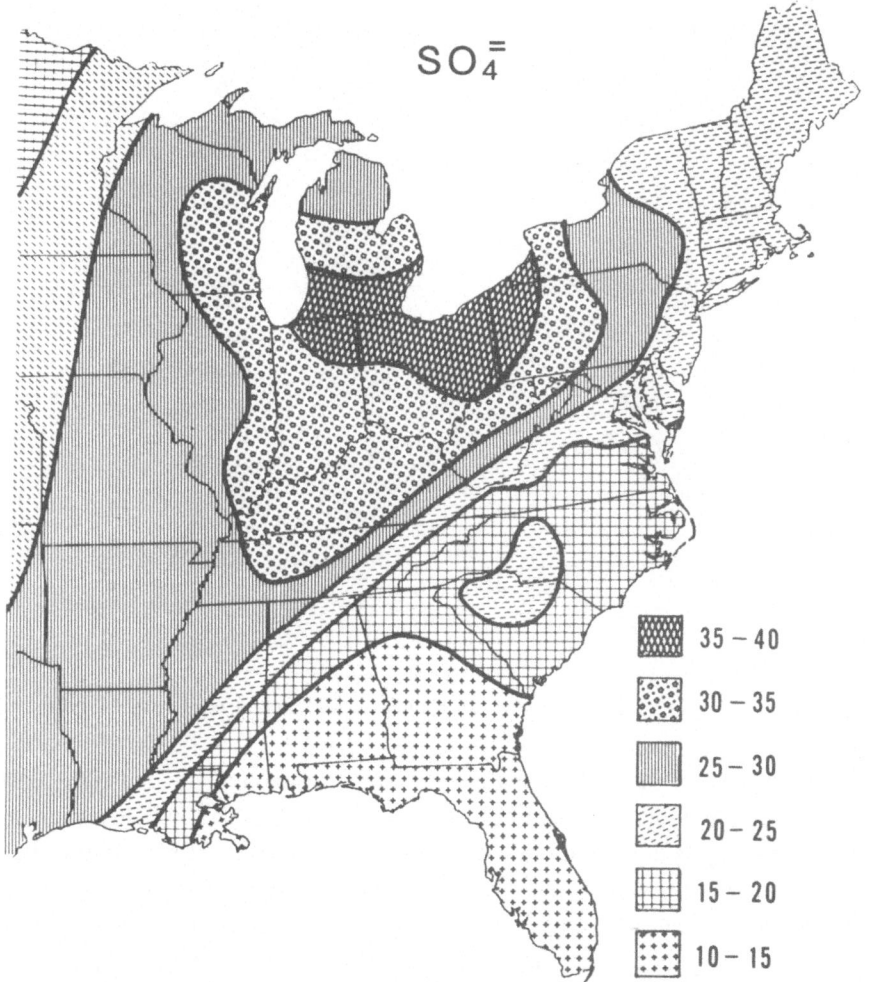

Fig. 4. $SO_4^=$ wet deposition (mg m^{-2}) normalized per cm of precipitation for 1979.

The integrated, normalized wet deposition of $SO_4^=$ and NO_3^- for 1979 shows a rather uniform gradient in almost every direction from the major emission area, except in the mountainous regions. The gradient for $SO_4^=$ deposition appears to be essentially the

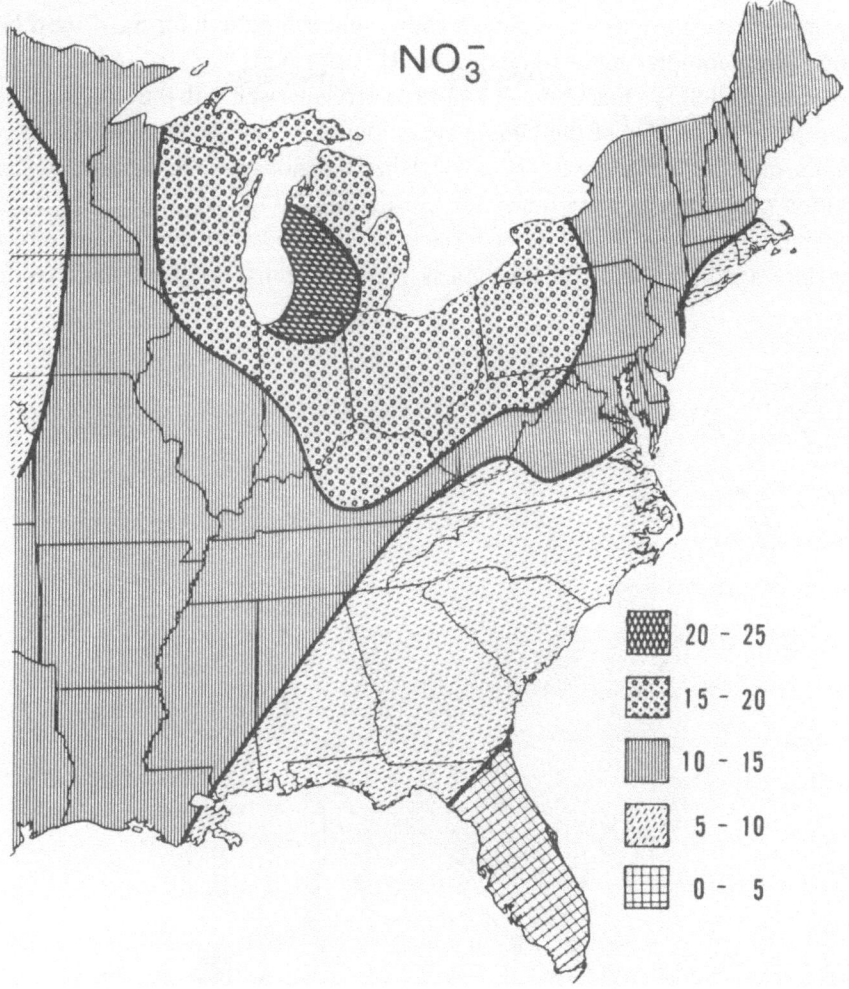

Fig. 5.   NO$_3^-$ wet deposition (mg m^{-2}) normalized per cm of precipitation for 1979.

same east and west of the Ohio Valley. Furthermore, no maximum located in the extreme
northeastern United States is seen in these deposition values that would be indicative
of increased wet deposition downwind of major emission areas. However, the effect of
the Appalachian Mountains is readily apparent as evidenced by the strong wet deposition
gradients shown in Figures 4 and 7. The mountainous regions of the eastern United
States above 600 m appear to have a pronounced effect on the quality of precipitation
in the Appalachian region. While acting to channel or divert low-level pollutants, the
higher elevations also act to produce varying types of precipitation in terms of moisture
output. Upslope precipitation on the eastern slopes of the Appalachian Mountains is
likely to have the Atlantic Ocean as its moisture source with little chance of continental
influence. On the other hand, upslope precipitation on the western slope will probably
be from a different source and have ample time for modification from continental sources.

· Since the free H^+ deposition results from the contribution of various constituents including sulfuric acid, nitric acid and soil-related components, such as Ca^{++}, Mg^{++}, and NH_4^+, the spatial distribution of normalized H^+ deposition can be expected to be more complicated than either $SO_4^=$ or NO_3^- alone. This is verified in Figure 6.

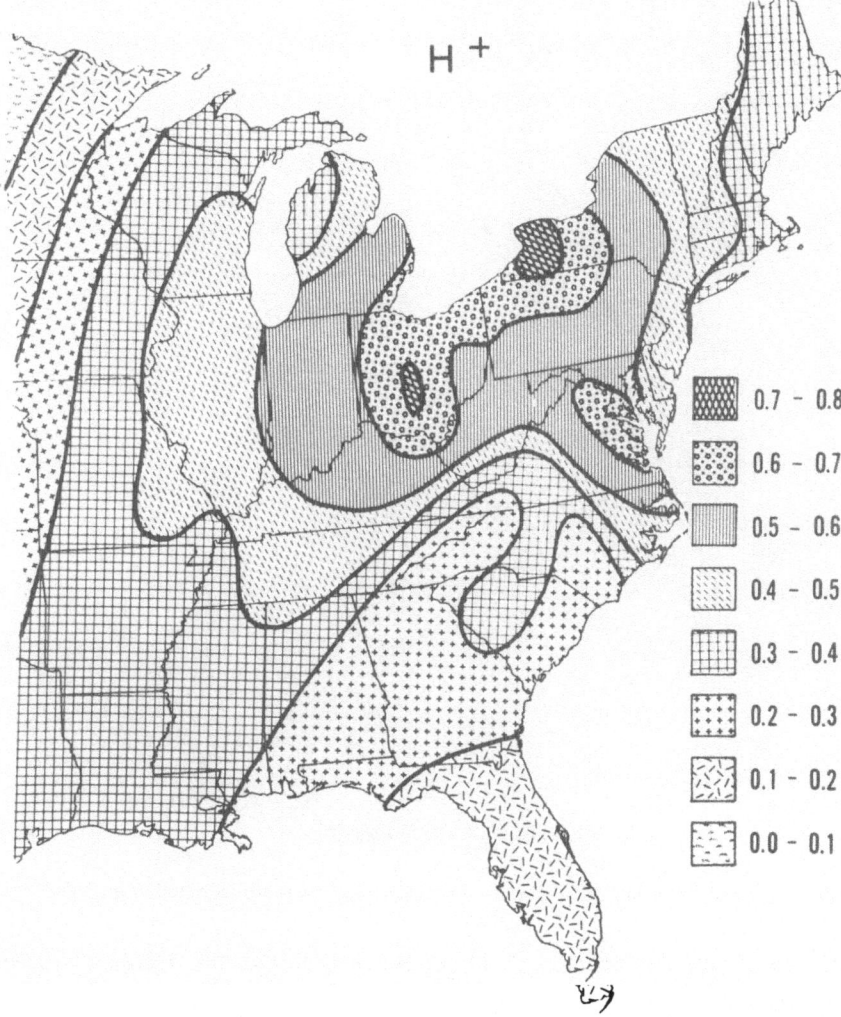

Fig. 6. H^+ wet deposition (mg m^{-2}) normalized per cm of precipitation for 1979.

Localized maxima in H^+ wet deposition are evident and a strong gradient emerges to the west of 85° longitude. One can speculate that soil-related material scavenged by precipitation might be responsible for the apparent change in free H^+ deposition. Topography effects and the Carolina maximum are again evident. This is not surprising, since a strong correlation exists between $SO_4^=$ and H^+ for the northeastern part of the United States.

The NH_4^+ distribution, Figure 7, is largely a function of natural sources and conse-quently displays no discernible source-receptor relationships. Increased NH_4^+ ion wet

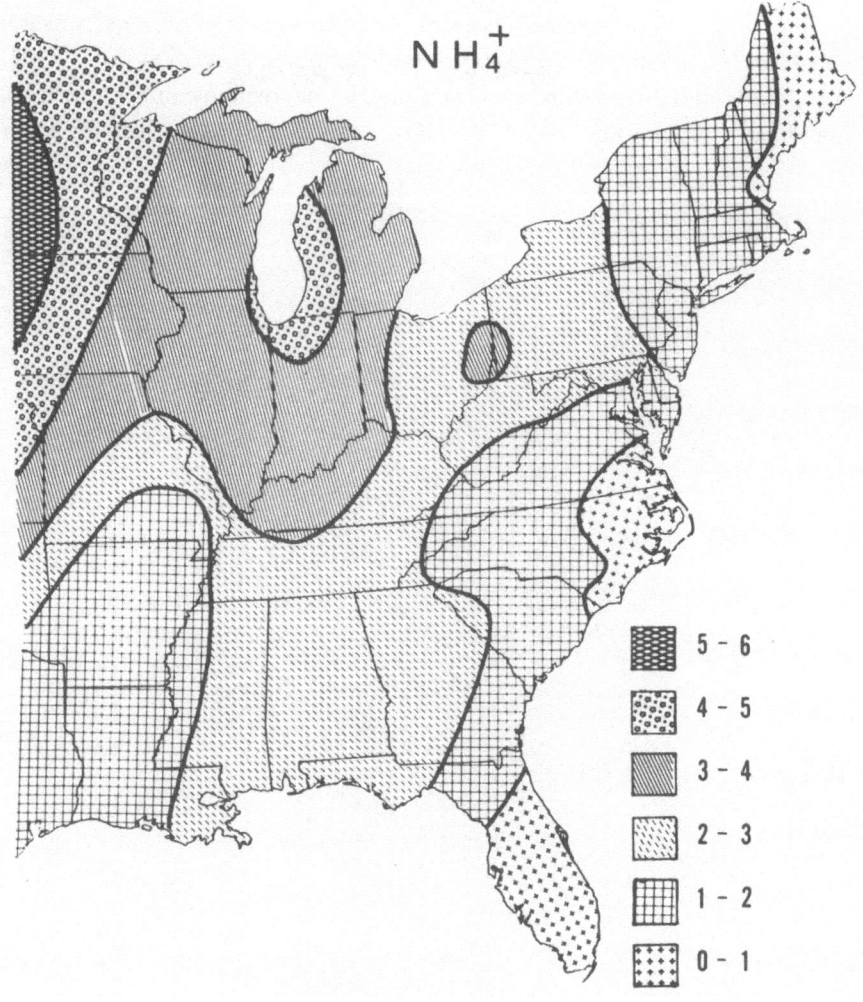

Fig. 7.   NH$_4$ ' wet deposition (mg m^{-2}) normalized per cm of precipitation for 1979.

deposition to the west of 85° longitude correlates well with the decrease in free H$^+$ deposition over the same geographic area.

5. Estimated Total Wet Deposition – 1979

While these normalized deposition, or weighted concentration, values provide an indicator of precipitation quality per unit amount of precipitation, localized total deposition maxima may result due to localized precipitation totals. Figures 8 through 11 depict estimates of total wet deposition for SO$_4^=$, NO$_3^-$, H$^+$, and NH$_4^+$, respectively. The estimates were derived from normalized deposition values measured at a particular MAP3S or NADP site and independent measurements of total rainfall was reported by the National Climatic Center.

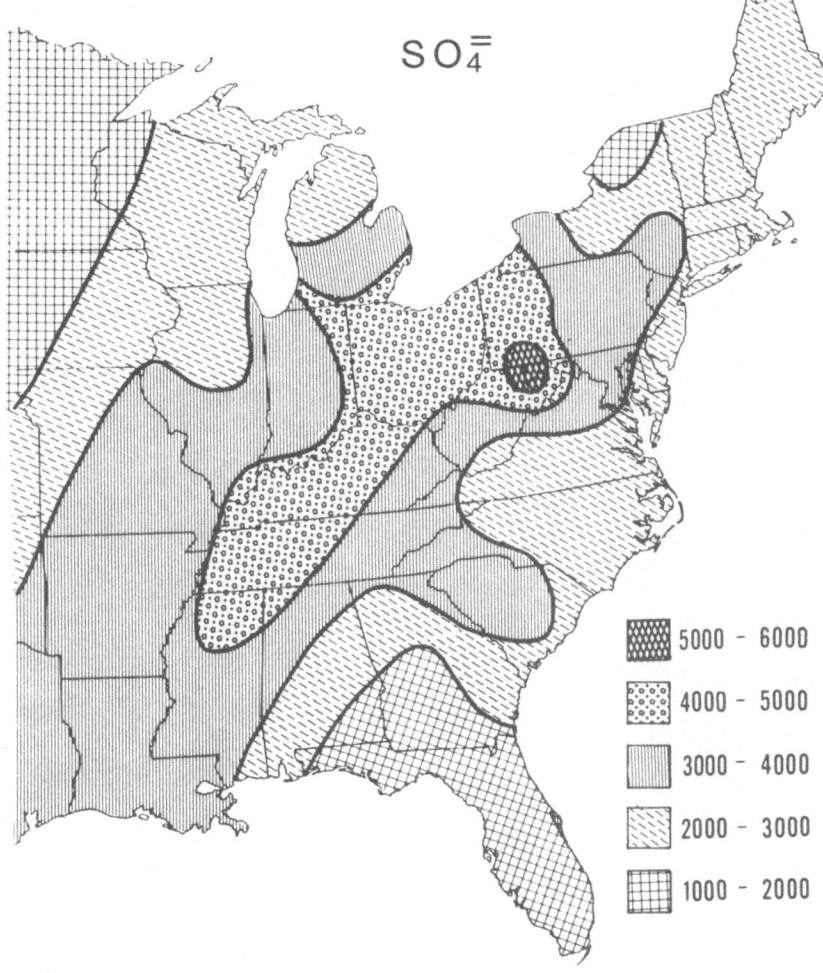

Fig. 8. Estimated total wet $SO_4^=$ deposition (mg m^{-2}) for 1979.

These computer-analyzed estimates reflect the localized nature of total rainfall over a time-span of a year. A precipitation maximum near the Pennsylvania–West Virginia border in excess of 186 cm is responsible for the increased estimated wet deposition of all ions experienced in that region.

With the exception of several localized maxima, the spatial analysis of estimated total wet deposition for $SO_4^=$, H^+, and NH_4^+ exhibits essentially the same basic pattern as that of normalized deposition. The estimated NO_3^- deposition pattern, Figure 9, was found to be more complex with regions of increased deposition in western Pennsylvania and the lower Hudson Valley of New York State.

While not an absolute measure of total ion deposition, these wet deposition estimates do provide reasonable approximations. A more dense network, operating over a longer time period, is needed to verify actual total wet deposition on a regional scale of less than 100 km grid space.

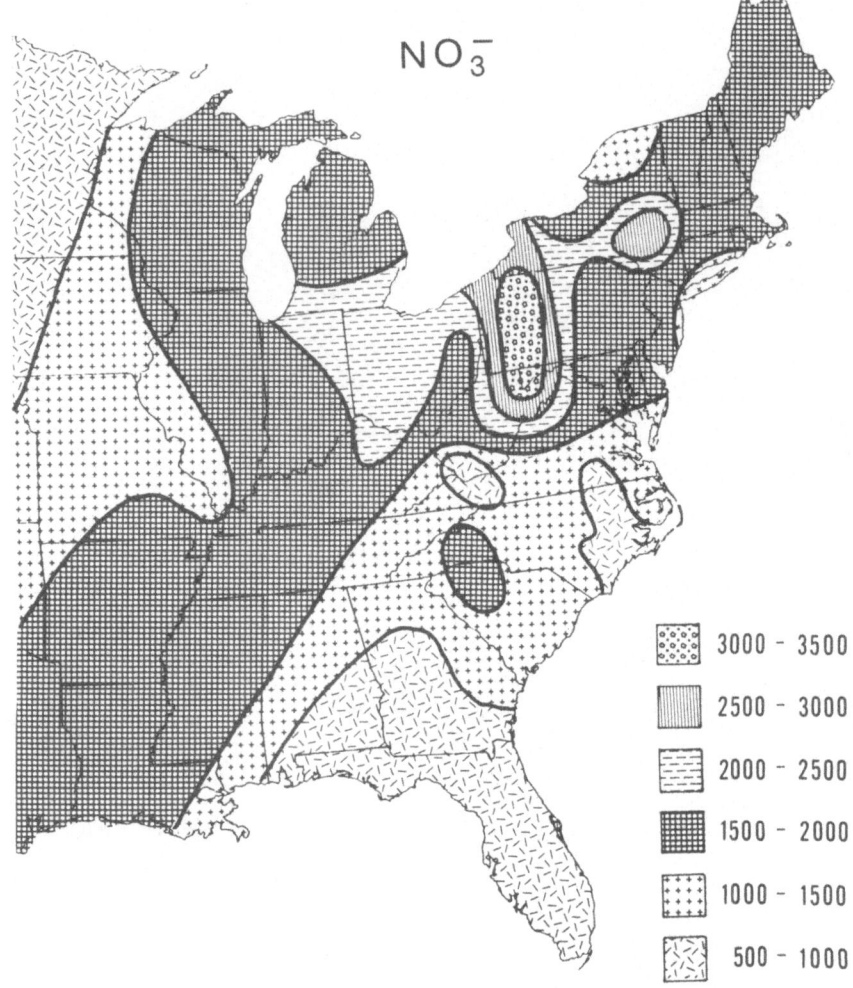

6. Summary

The concept of normalizing total measured wet deposition per centimeter of precipitation is introduced in an effort to characterize the amount of pollutant material delivered per unit amount of water. This technique allows for comparisons as well as long-term estimates.

Spatial variability in normalized wet deposition for sulfate and nitrate was found to be similar both to the east and west of the major emission region of the Ohio Valley. Maxima centered near the major emission regions suggest mixing processes on a variety of scales act to distribute pollutant material rather equally when the effects are integrated over an entire year. The gradient of normalized sulfate and nitrate wet deposition was found to decrease by approximately a factor of 1.5 from the Ohio Valley to Maine, as

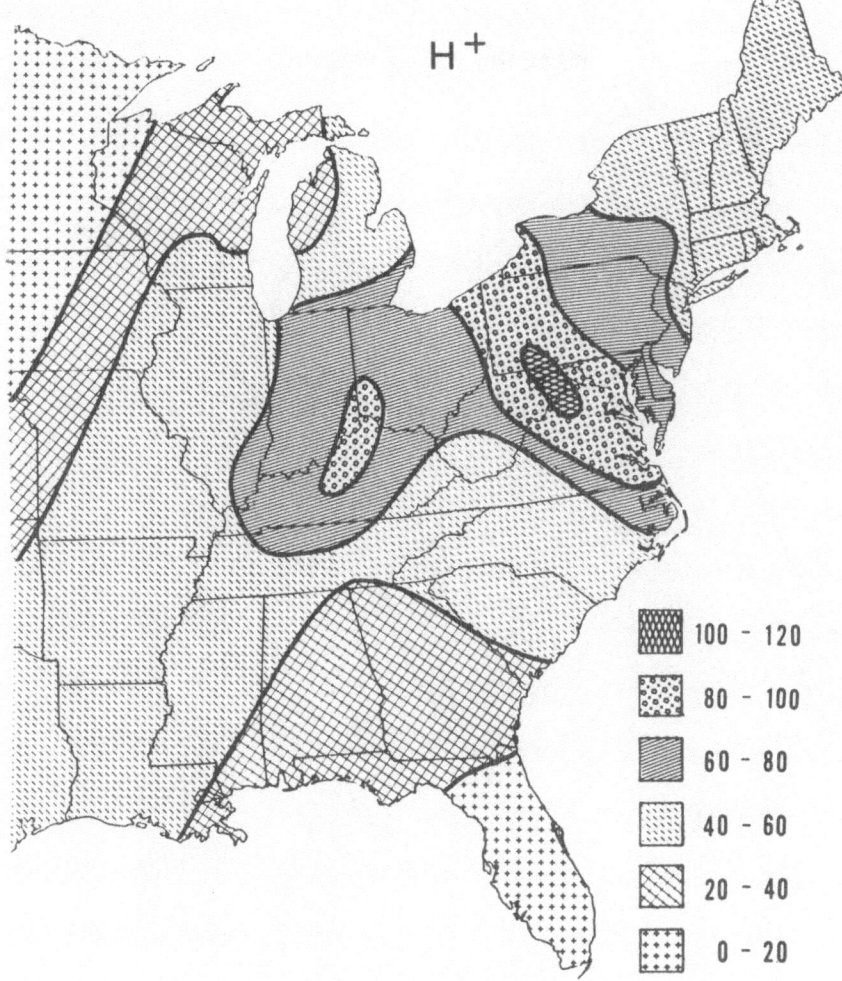

Fig. 10.   Estimated total wet H^+ deposition (mg m^{-2}) for 1979.

well as to the northwest of the Ohio Valley for the same distance. The findings are in agreement with those documented in the European Atmospheric Chemistry Network whereby sulfate concentrations and the large-scale concentration field is rather uniform (Granat, 1978).

The normalized H^+ deposition demonstrated a much stronger gradient west of 85° longitude than to the east. The H^+ wet deposition appears to be affected by locally-produced soil derivatives (Ca, Mg, ammonia, etc.). Interpretations of the H^+ deposition distribution may be erroneous if this fact is not taken into consideration.

Transport and distribution of the acidity of precipitation is therefore much more uncertain than that of the sulfate concentration. As mentioned earlier, it is likely that the H^+ wet deposition values are greatly influenced by the physical and chemical condition of the regional soil, as was hinted by Granat (1978).

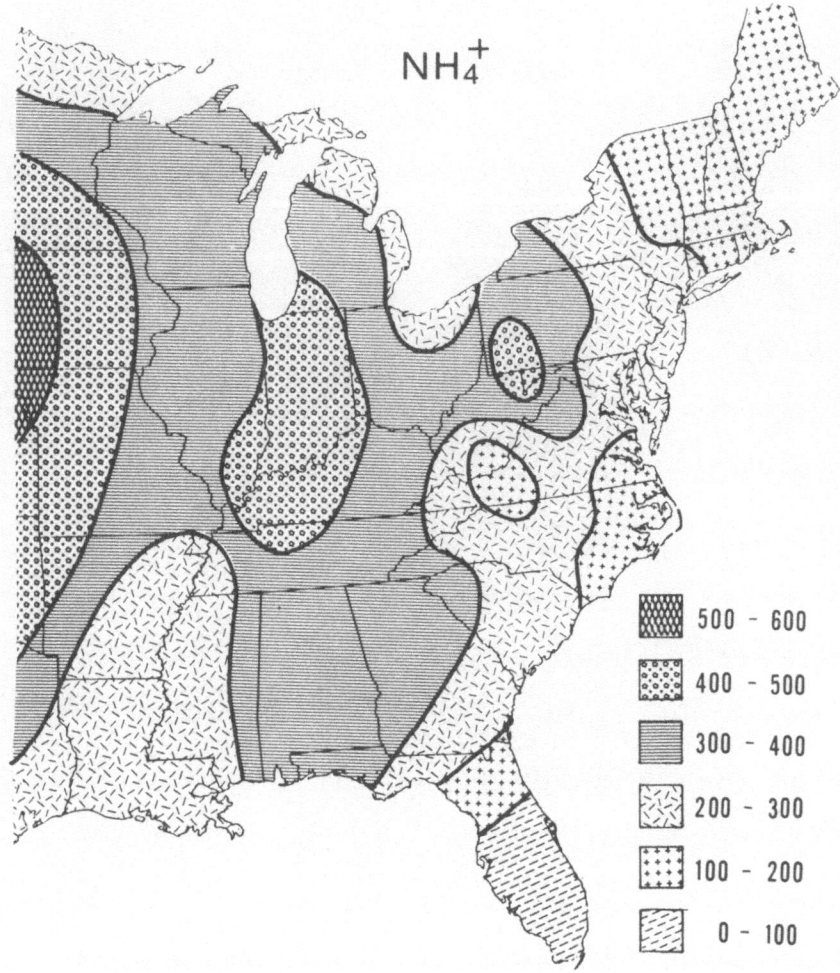

Fig. 11.   Estimated total wet NH_4^+ deposition (mg m^{-2}) for 1979.

Reasonable estimates of total wet deposition can be obtained using normalized deposition values and total measured precipitation. Spatial analyses of these estimated deposition fields were found to be somewhat localized in nature due to the nature of inter-site precipitation variation.

It is important to recognize that wet deposition events reported here do not include fog, dew, direct cloud droplet interception, etc., that might give rise to anomalous localized patterns. These 'special events', such as dry deposition, may play a significant role, particularly in mountainous regions.

Acknowledgments

The authors would like to thank LTC Frank Kolazar, United Military Academy, for his assistance in SYMAP techniques and Drs Austin Hogan, Lance Bosart and John

Kadlecek for their helpful comments. This research was supported by NSF grant ATM-7925435 and DOE grant DE-ACO2-76EVO2986.

References

Clark, T. L.: 1980, *Atm. Environ.* **14**, 961.

Dougenik, J. A. and Sheehan, D. C.: 1975, *SYMAP User's Reference Manual,* Laboratory for Computer Graphics and Spatial Analysis, Harvard University, 1–15.

Granat, L.: 1978, *Atm. Environ.* **12**, 413.

Junge, C. E.: 1960, *J. Geophys. Res.* **65**, 227.

Junge, C. E. and Werby, R. T.: 1958, *J. Meteorol.* **15**, 417.

Lodge, J. P., Jr., Hill, K. C., Pate, J. B., Lorange, E., Basbergill, W., Lazrus, A. L., and Swanson, G. S.: 1968, 'Chemistry of United States Precipitation', Final Report on the National Precipitation Sampling Network, Laboratory of Atmospheric Sciences, National Center for Atmospheric Research, Boulder, CO, 66 pp.

Pack, D. H.: 1980, *Science* **208**, 1143.

Wilson, J. W., Mohnen, V. A., and Kadlecek, J. A.: 1980, 'Wet Deposition in the Northeastern United States', ASRC Publication No. 796, State University of New York, Albany, NY, 139 pp.

THE CATION DENUDATION RATE AS A QUANTITATIVE INDEX OF SENSITIVITY OF EASTERN CANADIAN RIVERS TO ACIDIC ATMOSPHERIC PRECIPITATION

MARY E. THOMPSON

Aquatic Physics and Systems Division, National Water Research Institute, Canada Centre for Inland Waters, P.O. Box 5050, Burlington, Ontario, L7R 4A6 Canada

(Received 10 July, 1981; Revised 12 October, 1981)

Abstract. A model has been developed that relates the cation denudation rate (CDR) of a watershed (the rate that cations derived from chemical weathering are carried off by runoff), the atmospheric load of excess SO_4^{--}, and the pH of the river. Chemical and discharge data for rivers in Nova Scotia and Newfoundland were used to develop and test the model, which is based upon the common major ion chemistry of soft surface waters, and may be expressed by three statements:

(1) CDR (meq m^{-2} yr^{-1}) - Excess SO_4^{--} load (meq m^{-2} yr^{-1}) = HCO_3^- (meq m^{-2} yr^{-1}),

(2) HCO_3^- (meq m^{-2} yr^{-1})/Runoff (m^3 m^{-2} yr^{-1}) = HCO_3^- (meq m^{-3}),

(3) pH = pK + pP_{CO_2} - $pHCO_3^-$.

The model in concentration form applies well to lakes.

A detailed analysis of the data for the Isle aux Morts River, Newfoundland, is presented, showing that the CDR varies throughout the year, affected by both discharge and seasonal pattern.

1. Introduction

Chemical monitoring data for eleven rivers in Nova Scotia and ten rivers in Newfoundland have been analyzed in order to assess the impact of acid precipitation. The rivers were chosen on the basis of available data, water hardness, location, the evident lack of local pollution inputs, and the availability of discharge data. The earliest systematic monitoring data are for the water year 1954–55 in Nova Scotia and 1955–56 in Newfoundland (Thomas, 1960). Monthly sampling was resumed for some of the rivers in 1965 or 1966 and for others in 1969 or 1970. Regular monthly sampling was maintained for the Nova Scotia rivers through 1973, and for the Newfoundland rivers into 1979. The samples were analyzed at the Moncton, N.B. laboratories of the Water Quality Branch of Environment Canada; the data are stored in NAQUADAT, Canada's national water quality data storage system (Demayo, 1970). The rivers are gauged by Water Survey of Canada, and annual reports are published of daily, monthly and annual discharges. The sample date discharges are stored in NAQUADAT, as are also the early data of Thomas (1960).

Data for major ions, pH, color, conductance, and discharge were retrieved and subjected to several quality control tests: sums of cations and anions were compared to each other and to conductance; pH and alkalinity were tested for internal consistency. Obviously bad data were discarded; likely data entry errors were corrected. For example, if K^+ is reported as 0.4 mg l^{-1} in most samples for a river, but in one sample is reported 4.0 mg l^{-1}, and if the charge balance is off, a copy error is assumed to have been made.

Water, Air, and Soil Pollution **18** (1982) 215–226. 0049–6979/82/0182–0215$01.80.
Copyright © 1982 by D. Reidel Publishing Co., Dordrecht, Holland, and Boston, U.S.A.

In general, the quality of the data was good; a hierarchy of data quality from most to least reliable can be listed: cations, chloride, conductance, pH, sulfate, and alkalinity.

These coastal rivers receive varying amounts of seasalt. Because they are naturally soft waters, the seasalt is at times the major component of the sample. To correct data for seasalt, Cl^- was selected as the seasalt indicator species, and the other major ion components of seawater were assumed to be present in the same proportion to Cl^- as in normal seawater. A more detailed discussion of the arguments for this correction procedure is given in another paper (Thompson, 1982).

The assumption that the non-marine or excess SO_4^{--} is supplied to these rivers via long range transport is supported by the areal variations in loading that they display. The rivers at the southwest tip of Nova Scotia receive the highest loads of excess SO_4^{--}, and the loads decrease systematically to the northeast. The loads received by the Newfoundland rivers vary due to location and orographic effects. The Isle aux Morts River, at the southwest end of Newfoundland receives excess SO_4^{--} loads that vary from year to year in a pattern very similar to that of the Tusket River in southern Nova Scotia. Figure 1 shows the mean annual excess SO_4^{--} loads for those two rivers as well for the Medway River about 90 km to the north of the Tusket in Nova Scotia, and for the Upper Humber to the north and the Pipers Hole River to the east, respectively, of the Isle aux Morts River, Newfoundland. As these excess SO_4^{--} loads are calculated from 9 to 13 samples per river per year, the inter-river agreement is gratifyingly good.

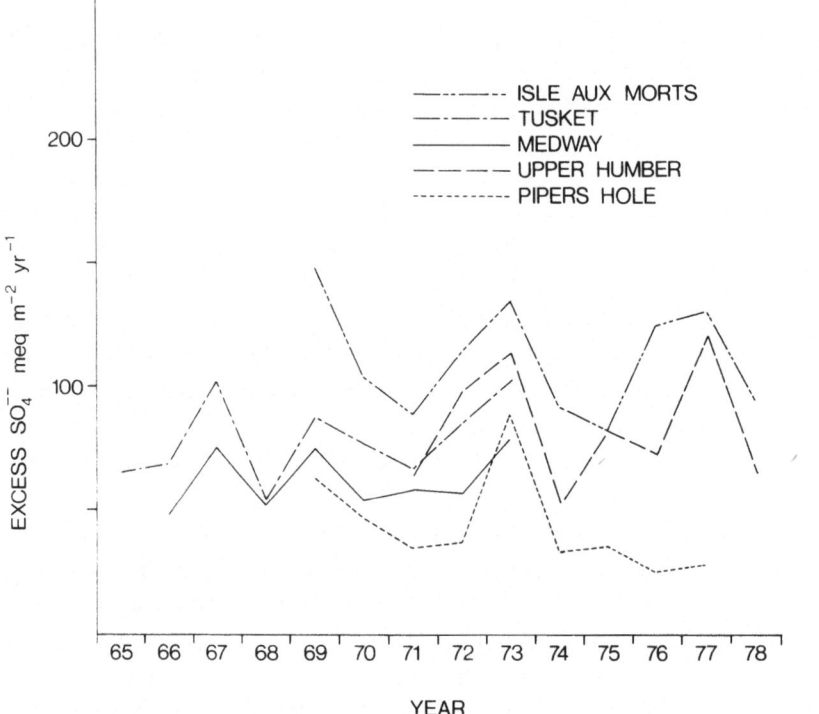

Fig. 1. Annual excess SO_4^{--} loads for the Isle aux Morts, Upper Humber, and Pipers Hole Rivers, Newfoundland, and for the Tusket and Medway Rivers, Nova Scotia.

The cation denudation rate (CDR) for each river was calculated using all available seasalt-corrected and discharge weighted sums of cations times mean sample-date discharge, divided by drainage area. The number of samples available per river ranged from 33 to 72 in Nova Scotia and from 53 to 119 in Newfoundland. The rivers, the number of samples, their CDRs, runoff, and 1973 excess SO_4^{--} loads are shown in Table I.

TABLE I

Rivers in Newfoundland and Nova Scotia, the Numbers of Samples from which CDR was calculated (n), CDR, Mean Runoff, and 1973 excess SO_4^{--} loads

Rivers	(n)	CDR meq m^{-2} yr^{-1}	Runoff m^3 m^{-2} yr^{-1}	Excess SO_4^{--} meq m^{-2} yr^{-1}
Rocky	(119)	133	1.30	101
Exploits	(93)	128	0.80	39
Isle aux Morts	(90)	145	2.14	134
Garnish	(83)	152	1.37	105
Pipers Hole	(89)	101	0.99	64
Gander	(53)	97	0.89	58
Terra Nova	(82)	76	0.91	46
Salmon	(81)	90	1.20	48
Upper Humber	(74)	243	1.29	115
Northeast Pond	(107)	71	1.1	65
Meteghan	(40)	129	0.93	123
Kelley	(48)	85	1.13	72
Wallace	(39)	203	0.95	69
Tusket	(49)	75	1.09	101
Roseway	(28)	56	1.13	128
Mersey	(51)	66	0.96	56
Medway	(66)	71	0.87	78
La Have	(33)	126	0.91	80
Liscomb	(37)	75	1.30	85
St. Mary's	(65)	100	1.08	60
Clam Harbour	(39)	182	1.33	90

2. The Model

The model is based upon simplifying assumptions about chemical weathering, as well as the common major ion chemistry of soft surface waters, and the relationships among pH, HCO_3^-, and CO_2.

The CDR, as used in the model, is the rate at which a watershed contributes cations to runoff as products of chemical weathering. The four conventional cations, Ca^{++}, Mg^{++}, Na^+, and K^+, are considered in the calculation of the CDR. These cations are considered to have been brought into solution during chemical weathering, by reactions with either H_2CO_3 or H_2SO_4. Because the discussion is restricted to watersheds underlain by resistant rocks and soils, reactions with silicate minerals are postulated:

$$\text{Ca silicate} + 2\,H_2CO_3 = Ca^{++} + 2\,HCO_3^- + H_2\ \text{silicate} \qquad (1)$$

$$\text{Ca silicate} + H_2SO_4 = Ca^{++} + SO_4^{--} + H_2\ \text{silicate} . \qquad (2)$$

In reaction (1), H_2CO_3, which is formed by solution of CO_2 in water, reacts with Ca-bearing silicate minerals to generate the soft $Ca^{++} - HCO_3^-$ waters typical of the Canadian Shield and other areas with resistant bedrock and overburden. The actual silicate minerals are not specified, and chemical weathering, when examined in detail, is a far more complex process, but for modelling purposes reactions (1) and (2) are sufficient. In reaction (2), the strong acid in acidic atmospheric precipitation is taken to be H_2SO_4. HNO_3 is not considered in this model because NO_3^- does not ordinarily appear in runoff in any abundance.

Therefore, in evaluating chemical data for surface waters, the cations accompanied by HCO_3^- are assumed to be products of normal chemical weathering, while those accompanied by SO_4^{--} are assumed to have been produced by reactions with strong acids in precipitation.

The common major ion chemistry of unpolluted soft surface waters is such that the four major cations, Ca^{++}, Mg^{++}, Na^+, and K^+ are balanced in solution by three major anions HCO_3^-, SO_4^{--}, and Cl^-. This balance is so generally found that the sum of these four cations compared to the sum of these three anions is used as a test of the quality of the analytical data. Cl^- in such waters is usually trivial, and if abundant, is generally attributed to inputs of neutral salts, e.g., roadsalt, seasalt. After correcting the data for chloride salt input, this relationship should hold:

$$\text{Sum of cations} = HCO_3^- + SO_4^{--}$$

or $\qquad$ $$\text{Sum of cations} - SO_4^{--} = HCO_3^- . \qquad (3)$$

The carbonic acid system operates such that there is a reasonably predictive relationship between HCO_3^- and pH. CO_2 dissolves in water to form H_2CO_3 which then dissociates to some extent, forming H^+ and HCO_3^-. Equations describing these reactions are usually written as

$$H_2O + CO_2 = H_2CO_3$$

and $\qquad$ $$H_2CO_3 = H^+ + HCO_3^-$$

but for this purpose it is simpler to combine them, obtaining

$$H_2O + CO_2 = H^+ + HCO_3^- . \qquad (4)$$

The equilibrium constant for Equation (4) has the form

$$K = (H^+)(HCO_3^-)/P_{CO_2}$$

in which the activity of H_2O is assumed to be unity, the pressure of CO_2 gas is equated to its activity, and parentheses are used to indicate active concentrations of H^+ and HCO_3^-.

Taking negative logarithms (using p to indicate negative log of) and rearranging, we obtain

$$pH = pK + pP_{CO_2} - pHCO_3^- . \tag{5}$$

pK varies only slightly over the temperature range encountered in Canadian surface waters, being 7.8 at 25° and 7.7 near 0° (Garrels and Christ, 1965). Therefore, if the CO_2 pressure is known, there is a direct, calculable relationship between pH and HCO_3^-. In the atmosphere pP_{CO_2} is about 3.5, but in surface waters it may vary considerably depending on the relative rates of solution or exsolution of CO_2, of biological production, and of respiration and decay. In the surface water data I have examined, for laboratory measurements of pH, pP_{CO_2} is commonly near 2.5, about ten times higher than atmospheric. For a first approximation, then, expression (5) may be reduced to:

$$pH = 10.3 - pHCO_3^- . \tag{6}$$

The model, in concentration form, is described by Equations (3) and (6). If the sum of cations is plotted against excess SO_4^{--} (which is the SO_4^{--} remaining after seasalt correction) in the same units and to the same scales (Figure 2), then the line from the origin at 45° describes the case where sum of cations is exactly equal to excess SO_4^{--}, and it divides the plot into two fields. Above that line, the sum of cations is greater than

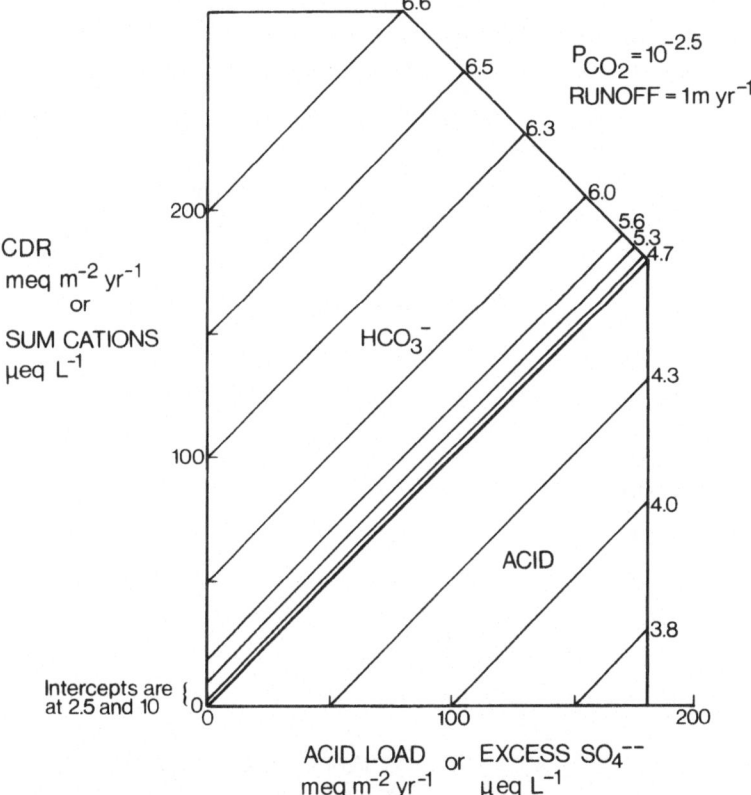

Fig. 2. A plot of the model that relates pH and sum of cations to excess SO_4^{--} in concentration units, or pH and CDR to rate of excess SO_4^{--} loading in rate units.

excess SO_4^{--}, and HCO_3^- is presumed to be present; below that line the sum of cations is less than excess SO_4^{--}, and excess, unreacted, acid may be assumed to be present. Lines drawn above the original one, and parallel to it, are lines of constant difference between sum of cations and excess SO_4^{--}, that is, of constant HCO_3^-, and, if the CO_2 pressure is the same, of constant pH. Lines drawn below the original, and parallel to it, may be considered to be lines of constant excess acid, and pHs may be assigned on that basis.

The concentration plot may be used as a way of displaying the present status of a group of lakes, or as a test of data quality, but for prediction purposes a rate model is probably more appropriate. For the rate model, the CDR, in meq m^{-2} yr^{-1} is plotted against excess SO_4^{--} in the same units, and the difference, HCO_3^-, is obtained in the same units. Before model pHs can be assigned, runoff must be considered. Runoff is usually expressed as m yr^{-1}, but is actually a volume per unit area, that is m^3 m^{-2}. Therefore, if HCO_3^-, in meq m^{-2} yr^{-1} is divided by m^3 m^{-2}, the mean HCO_3 concentration in the river in meq m^{-2} yr^{-1} is obtained, and the mean river pH can be calculated from that. Runoff in eastern Canada may vary from 0.4 to more than 2 m yr^{-1}, and neglecting the runoff factor will introduce an error in the calculated pH that is quite large at pH 6, but smaller at lower pH (Figure 3).

The plot shown in Figure 2 is similar to that of Henriksen (1980). The choice of axes is similar, except that this model uses sum of cations rather than just Ca^{++} + Mg^{++}, and calls for a series of parallel iso-pH lines, and can be used as a concentration or a rate model. As a rate model, it is a useful predictor of the mean pH to be expected for a given CDR, runoff, and excess SO_4^{--} load.

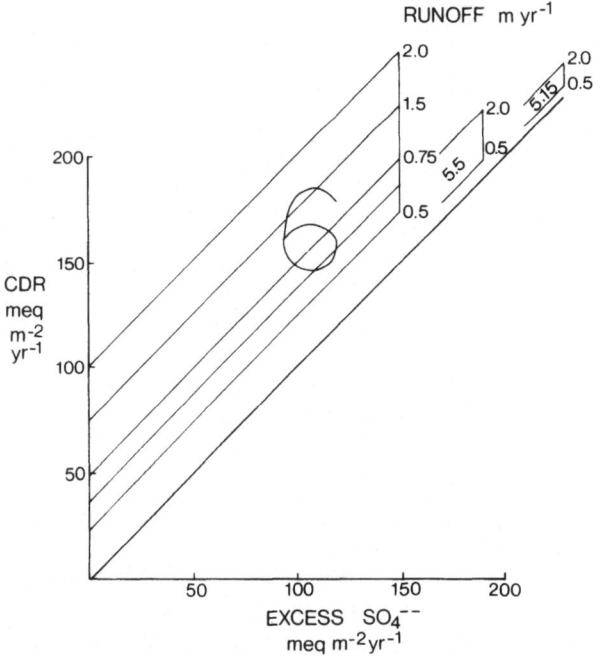

Fig. 3.   The effect of variations in runoff on model pH at pH 6, pH 5.5, and pH 5.15.

3. Examples

Several of the Nova Scotia and Newfoundland rivers have mean runoff near 1 m yr^{-1}, and can therefore be plotted on the same diagram. Figure 4 shows 1973 excess SO_4^{--} loads versus CDR, and pH for ten such rivers. The mean or median annual pHs agree reasonably well with the model pH. The three rivers with the lowest CDRs, the Roseway, the Mersey, and the Medway, are in southern Nova Scotia, and have strongly colored waters. Their pHs have been thought to be dominated by naturally occurring organic acids, but as can be seen on Figure 4, their low pHs can be explained quite well on the

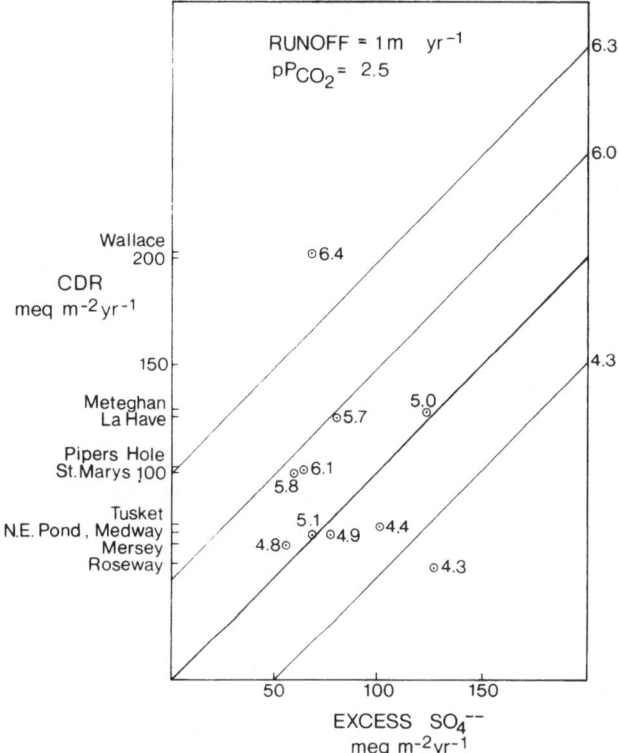

Fig. 4.   CDR plot for rivers with mean runopff near 1 m yr^{-1}, 1973 excess SO_4^{--} loads, and mean or median river pH.

basis of simple inorganic chemistry. The Mersey River data indicate a slightly smaller excess SO_4^{--} load than the Roseway to the south and the Medway to the north. The drainage basin of the Mersey includes the two largest lakes in Nova Scotia (Lakes Kejimkujik and Rossignol) and many smaller lakes and boggy areas, and the relatively light excess SO_4^{--} load may reflect SO_4^{--} reduction somewhere in the basin. SO_4^{--} reduction is known to occur in salt marshes (Howarth and Teal, 1979). As explained by Howarth and Teal, methanogenic and sulfate reducers working together produce this net effect:

$$CH_2COOH + SO_4^{--} + 2\,H^+ = 2\,CO_2 + H_2S + 2\,H_2O\,.$$

The reduction of SO_4^{--} consumes both SO_4^{--} and H^+, and if H_2S is lost as a volatile or precipitated as an insoluble sulfide, the net effect is beneficial to the runoff water. If, however, SO_4^{--} is reduced on a seasonal basis, and reoxidized at some stage in the annual cycle, there is no net benefit, but perhaps a delay in the relationship between the arrival of acid precipitation and the appearance of effects in runoff. Careful monitoring of both precipitation chemistry and volume and of runoff properties, as is now underway, will be necessary to clarify such effects.

The mean annual pH of a river may be calculated by summing the discharge-weighted antilogs of each pH value, dividing by the sum of the discharges, and taking the negative log of the quotient. But pH is not conservative when waters of different composition mix. Ordinarily one calculates the pH of a mixture by calculating the new HCO_3^- and the pH in equilibrium with it. Probably for this reason, the model assumptions work quite well for the Isle aux Morts River, Newfoundland, a river whose annual runoff ranges from 1.6 to 2.5 m yr^{-1}, when model pH is calculated assuming constant CDR and runoff, and varying excess SO_4^{--} loads (Figure 5). The change in annual CDR, if any, is compensated for by the change in annual runoff.

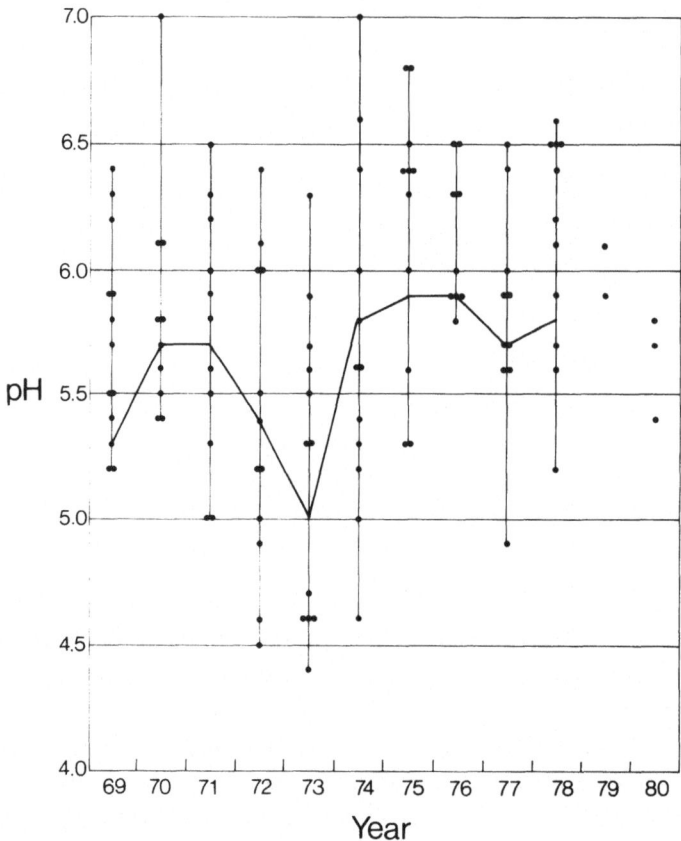

Fig. 5. Isle aux Morts River: all recorded pH data, by year, with the model pH drawn across the diagram.

Of more concern, probably, than the mean pH, is the lowest pH found in a river during the year. That question will be addressed in the next section.

4. The CDR – Dependence on Discharge

The CDR concept has been applied to annual means of data for rivers, because that seemed the most appropriate use of data based on monthly sampling. However, it is likely that a river whose pH is normally in the acceptable range might be subject to short term pH depressions because of seasonal changes in CDR and in acid load. The Isle aux Morts River, Newfoundland, has a fairly predictable seasonal discharge pattern, and data from a total of about 130 samples taken from 1966 to 1979 were available. Therefore, for each month, from eight to twelve SO_4^{--} values and from five to nine sum of cation measurements were available. (There are fewer sums of cations data because Mg^{++} was not reported until November, 1970.) The mean monthly sample-date discharge data

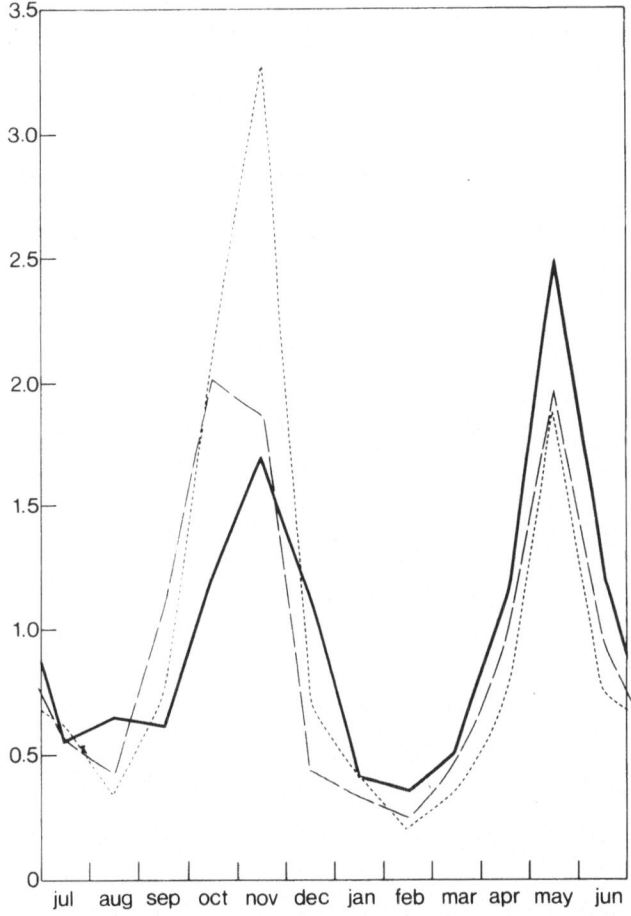

Fig. 6.   Isle aux Morts River: mean monthly discharge, excess SO_4^{--} load, and cation load (each data set divided by its mean). The solid line is discharge, the short dashed line is excess SO_4^{--}, the long dashed line is cations.

approximate quite closely the mean monthly daily discharge data. Mean monthly concentrations and loads (concentration times discharge) of excess SO_4^{--} and sum of cations were calculated. The mean monthly discharge, excess SO_4^{--} load, and cation loads are plotted on Figure 6. Each data set has been divided by its mean, so that only the pattern of monthly changes is seen. The plot runs from July to June so that the winter plots near the middle of the diagram. Discharge is low in the summer, rises sharply in October, peaks in November, decreases and remains low through the winter, during which the river is usually ice covered from December through March, then peaks sharply in May. The fall discharge peak is attributed to heavy autumnal rainfall, the spring peak to snowmelt. The major seasonal loading of excess SO_4^{--} apparently comes with the autumnal rains, although the river could also be carrying washed out summer dry fallout. The fall peak of discharge carries a similar peak of cation load, from which it can be concluded that, perhaps after a killing frost, the heavy autumnal rains are flushing out the year's major 'crop' of available cations. During the winter, under ice, when the flow is low and is presumably due to groundwater, the cation load is low. In the spring snowmelt, discharge, excess SO_4^{--}, and cation load all rise, in the relation: discharge > cation load > excess SO_4^{--}. From the regular annual pattern of events shown in Figure 6, it can be deduced that occurrences of low pH are most likely in the fall, and are also possible in the spring.

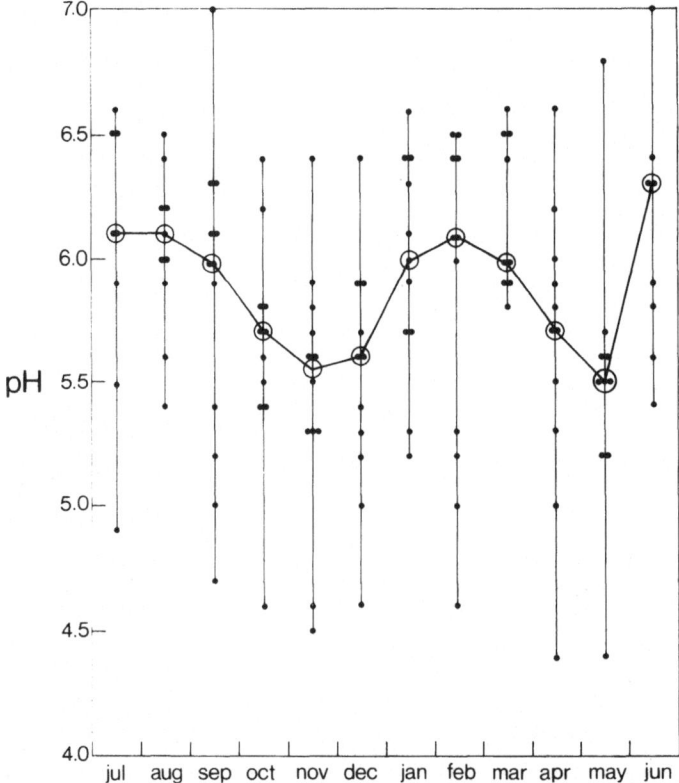

Fig. 7.   Isle aux Morts River: all recorded pH data, by month, with the trend of the median drawn across the diagram.

All recorded pH data for the river, by month, are shown on Figure 7, with the trend of the median pH drawn across the diagram. Though the scatter is large, the trend of the median is similar to that predicted from the hydrograph, the excess SO_4^{--}, and the cation loads of Figure 6. The lowest values of pH tend to occur in the fall or in the spring; the few low values of pH in February are associated with thaws and increased flow.

Depression of pH can be due to dilution as well as to acid load. For example, a water with sum of cations of 20 μeq l^{-1} and excess SO_4^{--} at 10 μeq l^{-1} (the dilute case) could have the same pH as a more concentrated sample with sum of cation at 100 and excess SO_4^{--} at 90 μeq l^{-1} (the acid case) that is, pH 5.3, at $pP_{CO_2} = 2.5$. The low median pH in the spring is probably due as much to dilution (by snowmelt water that has had relatively little contact with the ground) as to acid load, but the fall depression of pH must be chiefly due to acid load.

It is not known whether the CDR varies significantly from year to year. In this river, it certainly varies from month to month, and follows discharge. As the greater part of the acid load comes with rainfall which generates increased discharge, it appears that the natural seasonal pattern of cation load, which I think of as the river's self defence system, varies in such a way as to provide the river with the best defence available from limited resources.

The Isle aux Morts River was chosen for detailed study and presentation for a number of reasons:

– the long data record;
– the high mean runoff, and the high annual variations in runoff; and
– its unregulated flow.

Very few of the other rivers have as long a data record, unregulated flow, and a predictable discharge pattern, so that the conclusions reached with respect to the Isle aux Morts River as to its annual periods of vulnerability may not apply generally. However, because most chemical weathering, if related to plant growth, probably occurs during the summer, it would be logical to conclude that a river would be less vulnerable in the fall than during the spring snowmelt, unless, of course, the great bulk of the annual atmospheric acid loading is in summer and fall.

5. Summary and Conclusions

Chemical and discharge data for rivers in Atlantic Canada have been used to develop a model that relates the cation denudation rate of a basin, its excess SO_4^{--} load, and the pH of its runoff water. As a concentration model, it can be used as a test of data quality, or as a display of the current status of lakes. As a rate model, it can be used to establish acceptable rates of acid loading based upon minimum acceptable levels of pH that can be tolerated by the biota. The effect of seasonal patterns of rainfall, discharge and duration of snowpack on the annual pH cycle of one river suggests that permissible loadings ought to be defined by season as well as by amount.

References

Demayo, A.: 1970, 'A Storage and Retrieval System for Water Quality Data', Department of Energy, Mines and Resources, Inland Waters Branch, Report Series No. 9, Ottawa.

Garrels, R. M. and Christ, C. L.: 1965, *Solutions, Minerals, and Equilibria*, Harper and Row, New York, p. 89.

Henriksen, A.: 1980, 'Acidification of Freshwaters – A Large Scale Titration', Proceedings International Conference Ecological Impact of Acid Precipitation, Oslo, pp. 68–74.

Howarth, R. W. and Teal, J. M.: 1979, *Limnol. Oceanogr.* **24**, 999.

Thomas, J. F. J.: 1960, 'The Atlantic Provides, and the Saint John River Drainage Basin in Canada, 1954–56'. Water Survey Report No. 11, Industrial Water Resources of Canada, Department of Mines and Technical Surveys, No. 864.

Thompson, M. E.: 1981, *Geochem. et Cosmochem. Acta*, Vol. 46, No. 3 (in press).

MODERN AND PALEOLIMNOLOGICAL EVIDENCE FOR ACCELERATED LEACHING AND METAL ACCUMULATION IN SOILS IN NEW ENGLAND, CAUSED BY ATMOSPHERIC DEPOSITION

DENIS W. HANSON, STEPHEN A. NORTON, and JOHN S. WILLIAMS

Department of Geological Sciences, University of Maine at Orono, Orono, Maine 04469

(Received 10 July, 1981; Revised 2 October, 1981)

Abstract. Empirical field evidence for changing chemical processes in soils caused by atmospheric deposition of pollutants consists of: (1) Long-term water quality data including total dissolved solids, concentrations of specific metals (e.g. Ca), and conductivity; (2) Cation exchange capacity and base saturation values for soils located on precipitation pH gradients; (3) Lysimeter studies; and (4) Chemical analysis of organic soils on precipitation pH and metal gradients. For well-drained organic soils, as precipitation pH decreases, metals are differentially leached at an accelerated rate (Mn > Ca > Mg $\geq$ Zn > Cd and Na > Al). Experimental field and laboratory lysimeter studies on soil columns yield similar results, with increases in leaching rates for soil solutions with pH = 3 up to 100 × values for soil solutions with pH = 5. Nearly 100% of the Pb from precipitation is accumulating in the organic soil layer or sediments. Zn is accumulating in soils and sediments where the pH's of precipitation, soil solutions, and surface waters are generally above 5 to 5.5. At lower pH values Zn and other chemically similar elements are desorbed/leached (net) at an accelerated rate.

Chemical analyses of dated sediment cores from high and low altitude lakes, with drainage basins relatively undisturbed for the last 200 + yr, reveal that increased deposition of metals on a regional scale started in the northeastern United States as early as 1880, consistent with increased fossil fuel consumption. This suggests acidified precipitation as early as 1880. Cores from historically acidified lakes (pH < ~5.3 to 5.5) indicate that, as acidification of surface waters occurs (caused by acidic deposition), concentrations of Zn, Mn, and Ca decrease in the sediment. Apparently the metals are leached from the detritus prior to sedimentation. This conclusion results from data from experimental acidification of sediment cores and the general observation that precipitation pH is generally $\geq$ 0.5 pH units lower than lake water pH. Accelerated leaching of soil in New England dates to earlier than 1900.

1. Introduction

The United States and Canadian governments have identified acidic precipitation as one of the most serious environmental threats they will contend with in the 1980's (Galloway *et al.*, 1978). Although acidic precipitation is not a new phenomenon (TeBrake, 1975), the problem has intensified dramatically in the last 30 yr (Cogbill and Likens, 1974). Acidic precipitation is now a regional, even hemispheric, problem (Cragin *et al.*, 1975; Odén, 1976; Unpublished Internal reports of the National Atmospheric Deposition Program).

Acidic precipitation can deleteriously affect both terrestrial and aquatic ecosystems. Over large regions of North America and Europe, certain changes in aquatic ecosystems have been directly related to acidic precipitation (Likens *et al.*, 1979; Overrein *et al.*, 1980). The presence of pre-pollution data on aquatic ecosystems has allowed the identification of acidic precipitation-induced changes. Regional studies of acidic precipitation-induced changes in terrestrial ecosystems have not been undertaken due to the

Water, Air, and Soil Pollution **18** (1982) 227–239. 0049–6979/82/0182–0227$01.95.

relative complexity of terrestrial ecosystems, the lack of pre-acidic precipitation baseline data, and the long time scale of observable changes.

Indirect field evidence for increased leaching of forest soils and soil parent material due to acidic precipitation is abundant. Odén (1976) reported that podzol soils in southwestern Sweden have lost 55 to 70% of their normal content of cations. The amount of dissolved material discharged by Swedish rivers has increased considerably over a 50 yr period concurrent with an increase in precipitation acidity and a reduction in tilled acreage (Malmer, 1976). In small watershed areas on exposed granitic bedrock, the net output of Ca, Mg, and Al is directly related to H-ion input (Gjessing *et al.*, 1976). Studies in North America and Scandinavia indicate that water from acidified lakes has Al concentrations 10 to 50 times higher than non-acidified lakes (Wright *et al.*, 1980; Cronan and Schofield, 1979). The conductivity of lake water in oligotrophic lakes in southern Sweden has increased 80% since 1935 (Malmer, 1976). In Scandinavia, 80 to 95% of surface waters are a mixture of shallow soil water and and ground water (Odén, 1976). Consequently, the changes in surface water chemistry listed above are directly related to changes within the soil.

Most acidic precipitation related soil studies have been short-term projects. Controlled field plot and lysimeter studies indicate that leaching of the forest soil nutrients Ca, Mg, Mn, and probably K, and the non-nutrient elements Al and Na, is caused by increases in the acidity of precipitation (Overrein, 1972; Abrahamsen *et al.*, 1976; Cronan, 1980; Mayer and Ulrich, 1976; Wood and Bormann, 1976; Stuanes, 1980; Farrell *et al.*, 1980; Abrahamsen, 1980; Ulrich, 1980). Relative losses of K and Na are probably less than those of Ca, Mg, and Mn. Additionally, insoluble heavy metals associated with acidic precipitation accumulate in soil organic layers (Rühling and Tyler, 1968, 1969, 1971, 1973; Allen and Steinnes, 1980; Reiners *et al.*, 1975; Siccama and Smith, 1978; Siccama *et al.*, 1980).

The long-term loss or accumulation of plant nutrients and heavy metals in forest soils far removed from pollution sources has been determined in very few cases. Ulrich (1980) found that German forest soils exposed to natural acidic precipitation over an eight year period were depleted of nutrient cations, Na, and Al. Siccama *et al.* (1980) found that, since 1960, total Pb and Pb concentrations increased in soil organic layers in Massachusetts white pine forests. Zinc and Cu did not show this pattern. Smith and Siccama (1981) demonstrated lead accumulation in New Hampshire hardwood forest soils. Tyler (1981) studied accelerated leaching caused by increasing acidity in spruce forest soils.

2. Soil Transect Studies

Hanson (1980) used a regional soil-sampling transect to study acidic precipitation-induced changes in forest soils. In eastern North America, precipitation-pH isopleths (and probably heavy metals deposition isopleths) are expanding. At present, pH isopleths in New England and Atlantic Canada form a northeast trending gradient controlled by regional storm track patterns. A transect on this pH gradient is effectively a time study of precipitation-pH induced changes. For example, assume that the pH of precipitation in Vermont has decreased from 4.5 to 4.0 in the last 30 yr. Sampling and comparing a Vermont soil site with a *similar soil site* in Maine now receiving a precipitation of pH 4.5 will indicate the direction of changes which occurred in the Vermont soil due to increasingly acidic precipitation in that 30 yr period.

Hanson (1980) sampled subalpine conifer forest soil litter from 14 mountains in northern New England and eastern Canada (Figure 1). This region is downwind of the major North American industrial centers. Since 1940, precipitation has become increasingly acidic over the area (Cogbill and Likens, 1974). The pH of precipitation presently ranges from about 4.0 (site 1) to 4.6 (site 14) (Semonin *et al.*, 1981). (The sample sites are numbered so that increasing site number means increasing precipitation-pH (in a non-linear manner).) All sites were located on non-calcareous granitic or pelitic bedrock.

The subalpine conifer forest was selected for study because: (1) it is present on higher peaks throughout the region; (2) it is a structurally simple ecosystem dominated by balsam fir *(Abies balsamea)* (Reiners and Lang, 1979); (3) anthropogenic effects are restricted to airborne pollution; and (4) both wet and dry pollutant inputs are greater than at lower elevations due to increased wet deposition, and to cloud droplet and aerosol

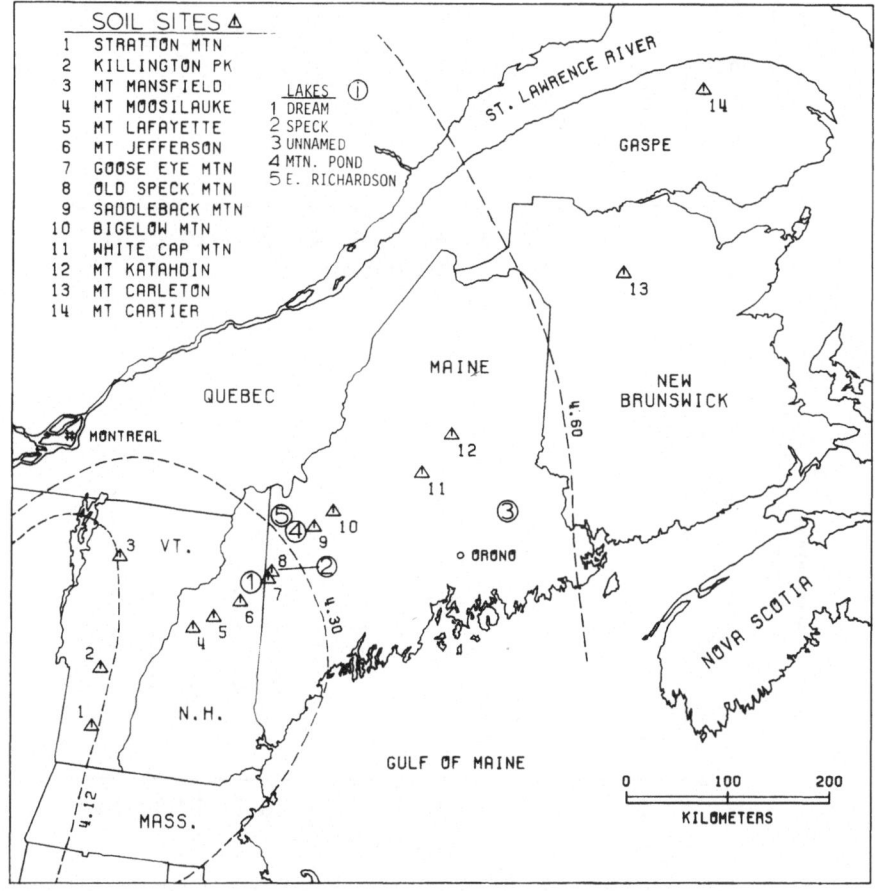

Fig. 1.   Location of soil transect sites and study lakes.

impaction on the dense foliage because of frequent cloud and fog cover and high winds (Schlesinger and Reiners, 1974; Reiners *et al.,* 1975). These factors suggested that soils of subalpine conifer forests might be the first component of the terrestrial ecosystem to exhibit acidic precipitation impact.

Litter samples were air dried at 50 °C, ground in a Wiley mill to 40 mesh, or ground in a mortar and pestle, and then digested in HNO_3 and $HClO_4$ (Hanson, 1980). The filtered solution was analyzed by atomic absorption spectrophotometry. Ca and Mn concentration gradients extend across the sampling transect (Figure 2). Concentrations increase to the northeast (higher-numbered sites). These gradients follow the regional precipitation-pH gradient, and probably mean that Ca and Mn are being leached more

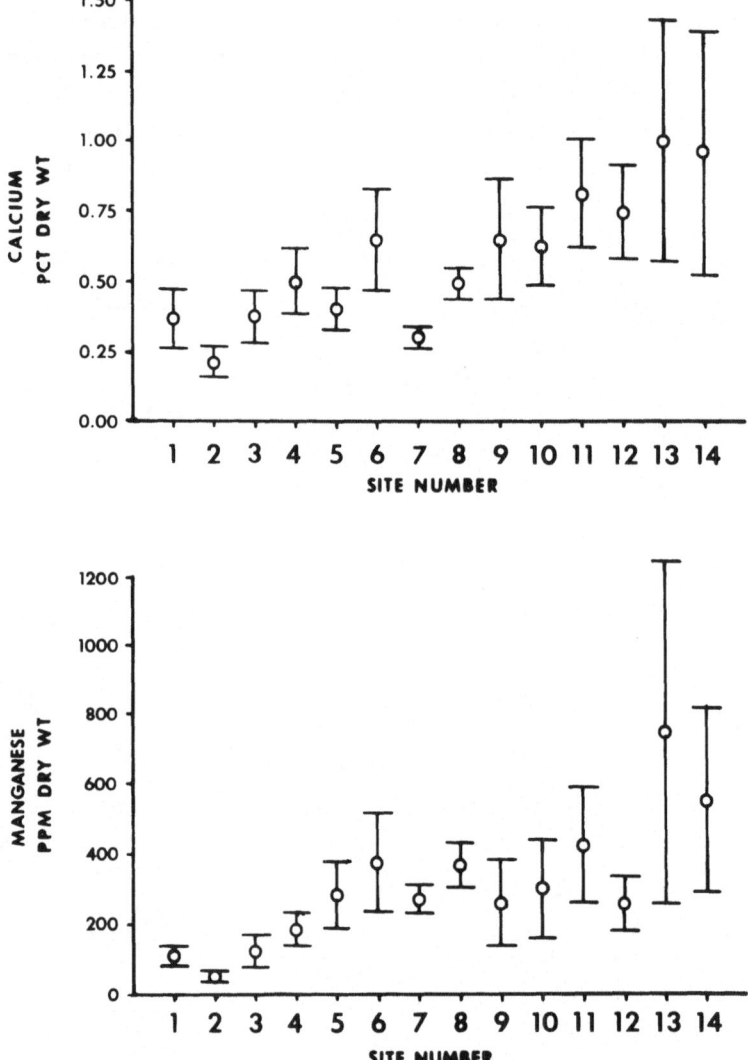

Fig. 2. Ca and Mn concentrations of organic soils from southern Vermont (1) to the Gaspé Peninsula, Quebec (14). The error bar is one standard deviation.

intensely in Vermont and New Hampshire (areas where the precipitation-pH is lower) than in Maine and Canada. The differences in values along the transect between Vermont and Canada probably indicate the changes which have occurred in Vermont subalpine conifer forest soil litter over the past 40 yr (post World War II) due to increasingly acidic precipitation during that period. The leaching gradients also indicate the direction of litter chemistry changes which may occur at presently less affected sites as precipitation becomes more acidic.

Potassium does not show a concentration gradient across the sampling transect (Figure 3a), but K concentrations at Vermont sites are lower than those at other sites, suggesting that the K leaching rate increases only when a threshold precipitation pH (ca. 4.0) is reached. Mg and possibly Na show similar patterns.

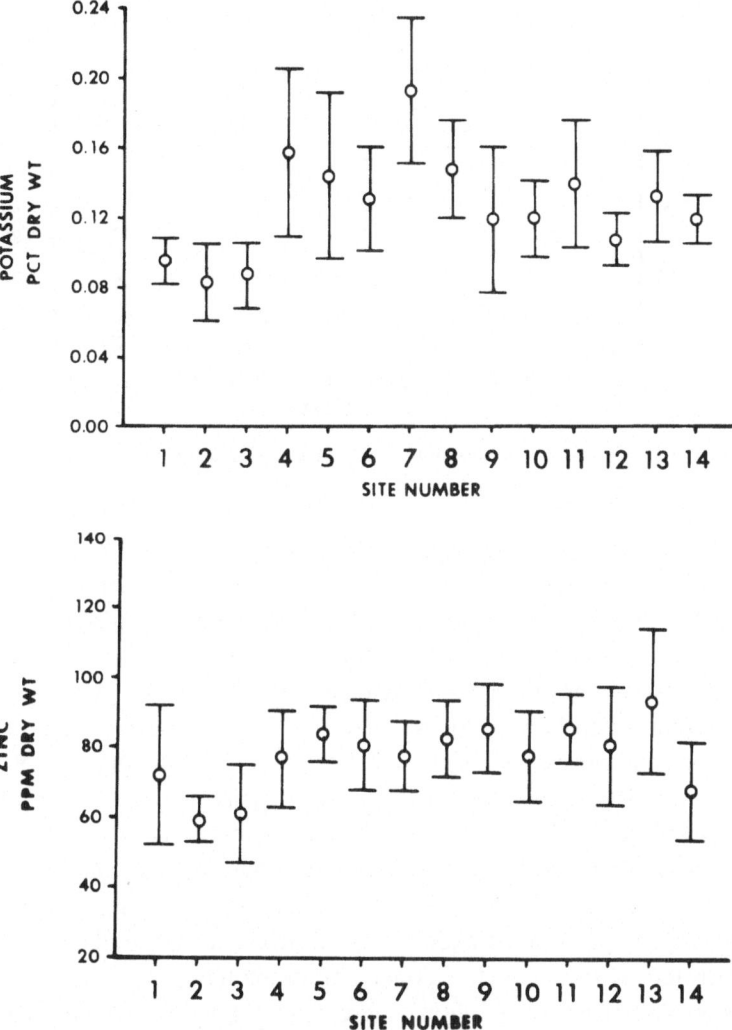

Fig. 3. K and Zn concentrations of organic soils from southern Vermont (1) to the Gaspé Peninsula, Quebec (14). The error bar is one standard deviation.

Atmospheric Zn deposition gradients probably exist across the sampling transect (Galloway *et al.*, 1981) because major Zn pollution sources are upwind of the region. If the deposition gradient exists and Zn accumulates in soils, Zn concentrations should be highest in Vermont, the part of the sampling transect closest to pollution sources. However, Zn concentrations are lower in Vermont than at other sites (Figure 3b). A possible explanation is that Zn leaching is similar to that of K; precipitation-pH in Vermont is low enough to cause Zn to be leached at a greater rate than at other sites.

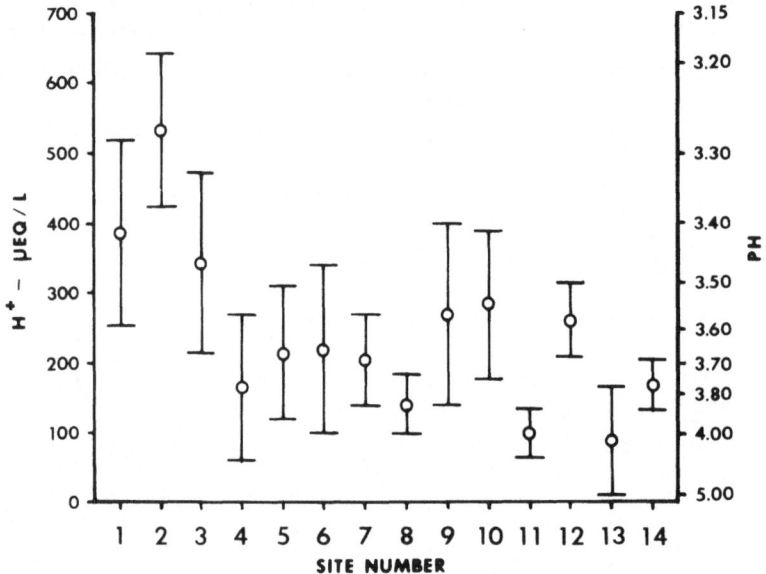

Fig. 4.   Soil litter pH from southern Vermont (1) to the Gaspé Peninsula, Quebec (14). The error bar is one standard deviation.

The soil litter pH gradient (Figure 4) supports the hypothesis that the rate of K and Zn leaching increases at pH's lower than those necessary to measurably increase the rate of Ca and Mn leaching. Increased K and Zn leaching rates appear to occur only at Vermont sites. Vermont soil litter is significantly more acidic than litter from other sites due probably to greater precipitation acidity in Vermont.

Lead concentrations (Figure 5) are highest at the southwestern end of the sampling transect, the area closest to major pollution sources. An atmospheric Pb gradient (Galloway *et al.*, 1981), combined with relative Pb enrichment at lower numbered sites due to greater leaching of other elements at those sites, causes the Pb concentration gradient across the transect. The high concentrations and accumulation pattern imply that much of the Pb deposited is not local in origin, even though population and motor vehicle densities in parts of the region are high. Iron shows a gradient similar to that of Pb. Copper and Al show no clear concentration pattern.

The order of relative susceptibility to increased leaching due to increased precipitation acidity is $Mn > Ca > Mg \geq K > Zn > (?)Na$ and Cd. The order of relative accumulation of the heavy metals is $Pb > Fe$.

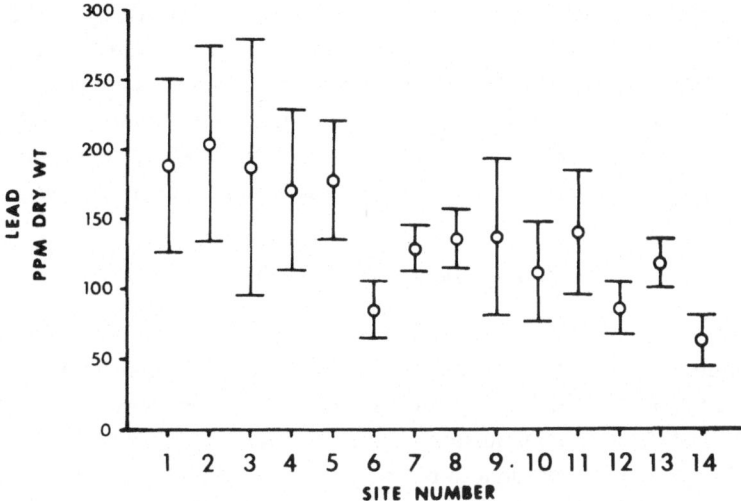

Fig. 5. Pb concentration (ppm) of organic soils from southern Vermont (1) to the Gaspé Peninsula, Quebec (14). The error bar is one standard deviation.

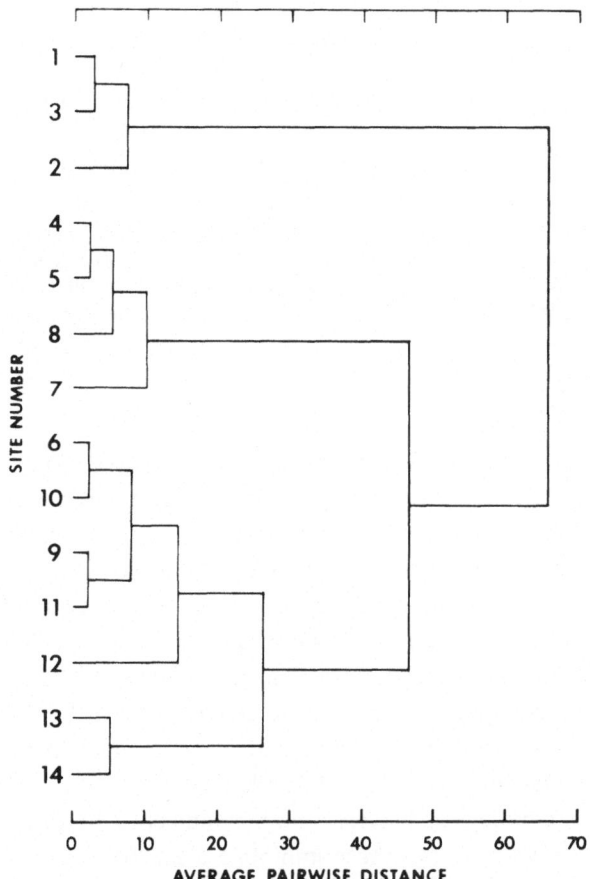

Fig. 6. Cluster analysis of 14 soil sites from southern Vermont (1) to the Gaspé Peninsula (14) using concentrations of Ca, Mg, K, Na, Al, Fe, Mn, Pb, Cu, Zn, and Cd, and ash content, sample elevation, and distance.

Cluster analysis was used to define groups of chemically similar sites by simultaneously considering all variables. Cluster analysis ordered the mountain sites, with one exception, according to their position along the regional precipitation-pH gradient (Figure 6). This implies that precipitation-pH, and associated heavy metal deposition, control soil litter chemistry in subalpine conifer forests in the region. Other possible sources of regional concentration gradients are probably not the cause of the concentration gradients.

3. Paleolimnological Studies

Sediment cores were obtained with a stationary piston corer (Davis and Doyle, 1969) from the profundal region of 5 lakes in New England (Table I) (Figure 1). The sediments have been dated by Cs^{137} and Pb^{210} (Johnston, 1980). Sediment was placed in solution by the method of Buckley and Cranston (1971). Chemical analyses were performed by flame atomic absorption spectrophotometry (TiO_2, CaO, MnO, FeO, and Zn) and by graphite furnace atomic absorption spectrophotometry for Pb.

TABLE I

Characteristics of study lakes

Lake	Summer pH	Area of lake	Altitude	Coring depth	Development of watershed	Dominant vegetation
Dream Lake, NH	4.5	3.7 ha	792 m	1.5 m	Pristine	Spruce/Fir
Speck Pond, ME	4.7	3.6 ha	1123 m	11.0 m	Pristine	Spruce/Fir
Unnamed Pond, ME	4.75	7.7 ha	141 m	6.7 m	Cut 30–40 yr ago (kettle lake)	Pine
Mountain Pond (R), ME	5.8	2.6 ha	837 m	3.5 m	Cut 40–50 yr ago	Spruce/Fir
E. Richardson Pond, ME	6.8	25.9 ha	537 m	5.2 m	Selective cutting 100 yr ago	Mixed hard and soft woods

These lakes were selected because of a history of relative mechanical stability in the watershed, particularly in approximately the last 50 yr. Accelerated erosion of soils and deposition of sediment due to logging, road construction, etc., confuses the interpretation of the chemical record preserved in the sediment (David and Norton, 1978). The relatively constant concentration of TiO_2 in the sediment (except in Unnamed) suggests a lack of serious disturbance in the watersheds over the last 200 + yr. None of the watersheds had undergone logging in recent time (30 to 50 yr at least). The drainage basins of Dream and Speck Ponds may never have been cut.

TiO_2 is a relatively immobile component against which other elements may be compared for relative rates of erosion. Other elements which are also relatively immobile, not concentrated in biological material, and not subject to diagenetic mobilization (e.g., Cr) should covary with Ti. All these immobile elements may vary reciprocally with variations in allochthonous organic matter or autochthonous organic matter (including diatom frustules) as well as autochthonous inorganic matter. Some elements (e.g., Na) will vary vertically in a core due to the changing proportions of allochthonous inorganic matter and other sediment components or due to variations in the degree of chemical weathering of the soil precursors of sediment (Davis and Norton, 1978). Other elements are partially controlled by organic content of the sediment. For example, Ca is readily leached from inorganic soils but may form as much as 25% of the ash from the organic matter. Thus, sediments high in organic matter may have relatively high Ca. Elements such as Fe and Mn, once deposited in the sediment, may be mobilized by diagenetic processes and translocated upward as the sediment accumulates.

With all these factors determining bulk composition of the sediment, unambiguous interpretation of time dependent changes in sediment chemistry is difficult. However, in lakes which have relatively undisturbed drainage basins, many of these confounding variables are constant and meaningful interpretation of chemical changes recorded in the sediment is possible.

Figure 7 shows the TiO_2, CaO, MnO, FeO (in weight percent of ignited sediment), and Pb and Zn (in parts per million of ignited sediment) for 5 lakes, arranged in order of increasing pH. Pb-210 dating has been carried out for the Speck, Unnamed, and Mountain Pond (Rangeley) cores. The increase in Pb and Zn in the cores (except Unnamed) occurs about 1870–1880. This date corresponds to other estimates of the increase in concentrations of Pb and Zn in sediments in other lakes in New England (Johnston et al., 1981). The sharpness of the increase in Pb concentration suggests minimal bioturbation of the sediment. The Pb profile in Unnamed increases sharply at about 10 cm, ca. 1880 ± 10 A.D. The Zn content of the lower 30 cm of the Unnamed core is controlled by (or at least covaries with) organic content of the sediment.

The five lakes have pH's spanning 4.5 to 6.8, altitudes ranging from 141 to 1123 m, have different vegetation, and are all in relatively remote areas. The synchronous and ubiquitous increase in sedimentation of Pb and Zn is attributed to increased atmospheric deposition of these metals directly into the lake and in the drainage basin with subsequent sedimentation of the metals by a variety of mechanisms including mechanical sedimentation of allochthonous particulates (especially Pb), adsorption onto sedimenting particulates (especially Zn), and scavenging by organisms and subsequent sedimentation of organic particulates (especially Zn). Vanadium and Cu show similar behavior.

The Zn and Pb in the atmosphere must be from the consumption of fossil fuels (largely coal, initially), supplemented later by the smelting of nonferrous metals. The emission of these metals to the atmosphere was obviously accompanied by the emission of SO_2 and NO_x gases in some unknown quantity. The presence of the excess metals thus suggests some acidification of atmospheric deposition at least 100 yr ago.

Pb concentrations in the sediment increase up to the present, with a slight leveling in the most recent sediments. The relative and absolute increases vary from lake to lake because of differing drainage basin and limnological factors. Data from Siccama et al. (1980), Reiners et al. (1975), and Hanson (1980) from soils also suggest that Pb loading has increased through time.

Zn concentration, on the other hand, increases upward starting at about 1880 and then, in the lower pH lakes (generally < 5.5), decreases toward the recent. Values for

Concentration vs Depth

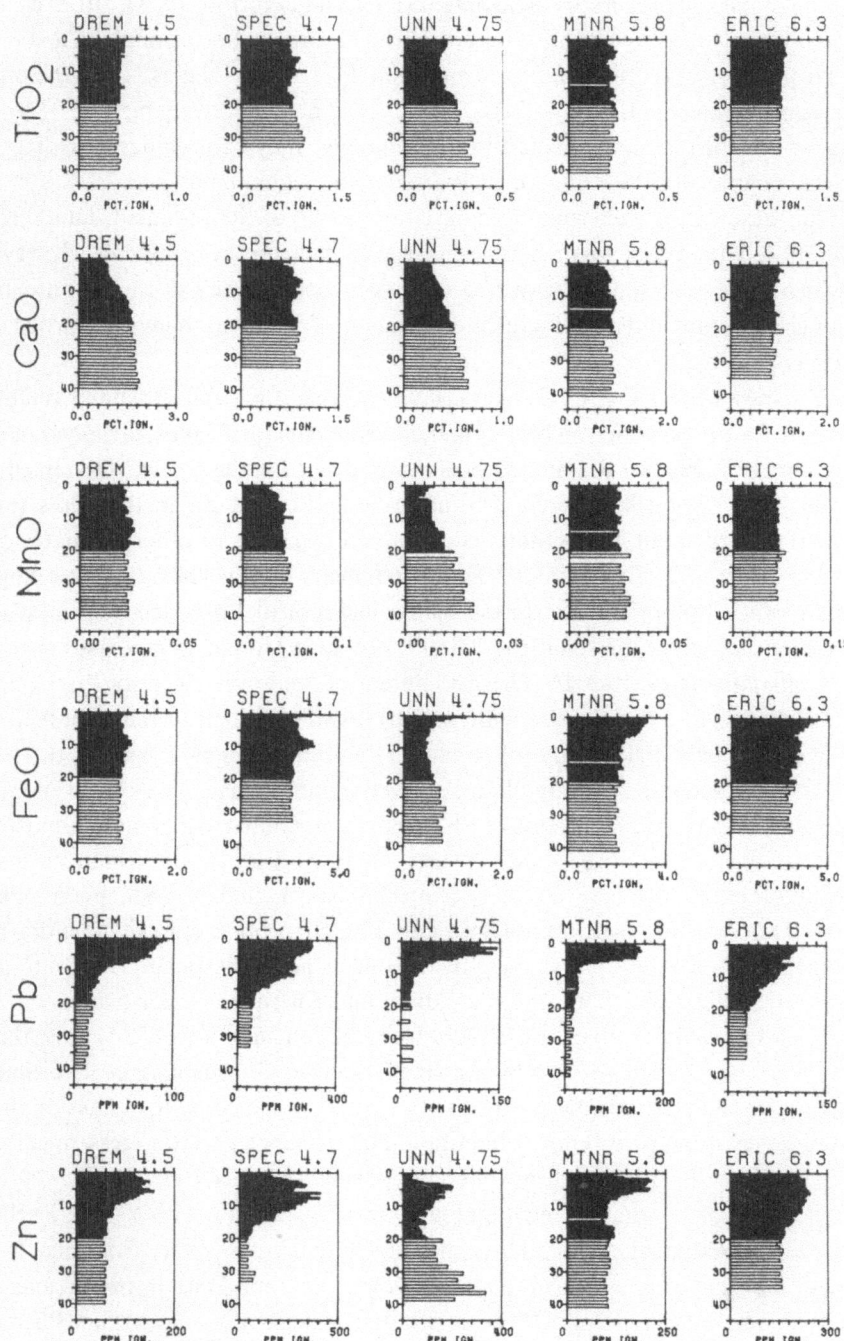

Fig. 7. Chemistry of lake sediments cores (weight percent or ppm of ignited sediment) from profundal sediment of 5 northern New England lakes.

recent sediment in Unnamed drop below pre-pollution levels. Lakes with high pH (> 7) have Zn concentrations which increase to the top of the core (unpublished data). Lakes with intermediate pH show intermediate behavior. We suggest that these Zn profiles are a result of two processes: (1) increased loading of Zn from the atmosphere; and (2) increased acidity of the precipitation which leaches Zn from the upper soil. Detritus from acidified and leached soil and in the lake would become impoverished in Zn with respect to pre-acid rain time. Lake sediment derived from these soils would be impoverished in Zn. Less acidic soils would be less leached or the Zn would be re-precipitated during in-soil neutralization.

Zinc concentrations in precipitation and surface waters are poorly known but do indicate that some of the Zn leached from the upper soil horizons (Hanson, 1980) does not enter the surface waters; rather it is translocated to deeper soil horizons where partial neutralization of descending soil waters occurs. However, there is a general relationship between pH and Zn concentration of lake waters in New England (Norton and Davis, 1981).

CaO (and MgO) content of the sediments in the four lower pH lakes (Figure 7) decreases upward relative to TiO_2. This decrease occurs in sediments deposited since the increase in Pb and Zn commenced, presumably accompanied by more acidic precipitation. We suggest that this is also caused by elevated leaching of Ca from the upper soils due to increasingly acidic precipitation. The more alkaline soils in the East Richardson drainage basin (as indicated by the higher pH of the lake) would be less well leached. The deciduous canopy of the East Richardson drainage basin would also contribute to the neutralization of precipitation.

MnO content of the sediments in the three lower pH lakes (Figures 7) decreases slightly relative to TiO_2, toward the top of the cores. We interpret these changes to be caused largely by increased leaching of soils from more acidic precipitation, possibly accompanied by increasing ion-exchange of H^+ for Mn^{++} in the sediment.

Although Fe is increasingly soluble at lower pH, its solubility, under oxidizing conditions, is less than that of Mn (Norton, 1973). Additionally the absolute amount of Fe in the sediment is large compared to the changes caused by acidic precipitation. Diagenetic upward migration and precipitation of Fe in the sediments is probably more important in determining the concentration of Fe in the sediment than are processes related to acid precipitation.

4. Summary

In the absence of historic data about the chemistry of soils it is possible to evaluate the long term effects of acidic precipitation by examining comparable organic soil litter along a pH gradient or by examining the chemistry of lake sediments deposited over a long period of time. Our studies indicate that:

(1) Pb has been accumulating at an accelerating rate in New England soils (and thus in lake sediments) for at least 100 yr.

(2) Zn has been accumulating at an accelerating rate in New England soils which are

circum-neutral. In acidic or acidified soils, there is a net loss of Zn due to solution by acidic precipitation. Zn concentration in lake sediments in acidifying drainage basins decreases as the ambient pH decreases.

(3) Accelerated leaching of K, Ca, Mg, and Mn has been occuring for at least 100 yr in areas with originally acidic soils, more recently in areas where soils have become acidified or are circumneutral.

Acknowledgments

Data from the soil transect studies are from Hanson (1980). That study and the paleolimnological data research were supported by the U.S. National Science Foundation (Grant DEB-78-10641 to R. B. Davis and S. A. Norton) and U.S. Fish and Wildlife Service (Grant 14-16-0009-79-040 to S. A. Norton and R. B. Davis). Chemical data for Dream Lake and East Richardson Pond are from Williams (1980) which was also supported by N.F.S. funds to Davis and Norton. Field and laboratory work were assisted by Elaine Hall and Marilyn Morrison.

References

Abrahamsen, G.: 1980, 'Leaching of Plant Nutrients', in *Ecol. Impact of Acid Precip. Proc.*, Sandefjord, Norway, 196 p.

Abrahamsen, G., Bjor, K., Horntvedt, R., and Tveite, B.: 1976, 'Effects of Acid Precipitation on Coniferous Forest', in F. H. Braekke (ed.), *Impact of Acid Precipitation on Forest and Freshwater Ecosystems in Norway*, SNFS-project, NISK, 1432 Aas-NLH, Norway, pp. 36–63.

Allen, R. O. and Steinnes, E.: 1980, 'Contribution from Long-range Atmospheric Transport to Heavy Metal Pollution of Surface Soil', in *Ecol. Impact of Acid Precip. Proc.*, Sandefjord, Norway, pp. 102–103.

Buckley, D. E. and Cranston, R. E.: 1971, *Chem. Geol.* **7**, 273.

Cogbill, C. V. and Likens, G. E.: 1974, *Water Resour. Res.* **10**, 1133.

Cragin, J. H., Herron, M. M., and Langway, C. C., Jr.: 1975, 'The Chemistry of 700 years of Precipitation at Dye 3, Greenland', Cold Regions Res. and Engineering Lab. Res. Rep. 341, 18 p.

Cronan, C. S.: 1980, *Oikos* **34**, 272

Cronan, C. S. and Schofield, C. L.: 1979, *Science* **204**, 304.

Davis, R. B. and Doyle, R. W.: 1969, *Limnol Oceanog.* **14**, 643.

Davis, R. B. and Norton, S. A.: 1978, *Pol. Arch. Hydrobiol.* **25**, 99.

Farell, E. P., Nilsson, I., Tamm, C. O., and Wiklander, G.: 1980, 'Effects of Artificial Acidification With Sulfuric Acid on Tree growth and Soil Chemistry in Scots Pine Forest', in *Ecol. Impact of Acid Precip. Proc.*, Sandefjord, Norway, pp. 186–187.

Galloway, J. N., Cowling, E. B., Gorham, E., and McFee, W. W.: 1978, 'A National Program for Assessing the Problem of Atmospheric Deposition: A Report to the Council on Environmental Quality', National Atmos. Dep. Prog., 97 p.

Galloway, J. N., Eisenreich, S. J., and Scott, B. C.: 1981, 'Toxic Substances in Atmospheric Deposition: A Review and Assessment', EPA 560 5-80-001: 1–148.

Gjessing, E. T., Henriksen, A., Johannessen, M., and Wright, R. F.: 1976, 'Effects of Acid Precipitation on Freshwater Chemistry', in F. H. Braekke (ed.), *Impact of Acid Precipitation on Forest and Freshwater Ecosystems in Norway*, SNSF-project, NISK, 1432 Aas NLH, Norway: 64–85.

Hanson, D. W.: 1980, 'Acidic Precipitation-induced Chemical Changes in Subalpine Fir Forest Organic Soil Layers', Unpub. M.S. thesis, Univ. of Maine at Orono, Maine, 90 p.

Johnston, S. E.: 1980, 'A Comparison of Dating Methods in Laminated Lake Sediments in Maine', Unpub. M.S. thesis, Univ. of Maine at Orono, Maine, 79 p.

Johnston, S. E., Norton, S. A., Hess, C. T., Davis, R. B., and Anderson, R. S.: 1981, 'Chronology of

Atmospheric Deposition of Acids and Metals in New England, Based on the Record in Lake Sediments', Proc. Am. Chem. Soc. Envir. Chem. Div., Ann Arbor Press Club.

Likens, G. E., Wright, R. F., Galloway, J. N., and Butler, T. J.: 1979, 'Acid Rain', Sci. Am. 241(4), 43–51.

Malmer, N.: 1976, Ambio 5, 231.

Mayer, R. and Ulrich, B.: 1976, 'Acidity of Precipitation as Influenced by the Filtering of Atmospheric Sulphur and Nitrogen Compounds – Its Role in the Element Balance and Effect on Soil', in L. S. Dochinger and T. A. Seliga (eds.), Proceedings of the First International Symposium on Acid Precipitation and the Forest Ecosystem, U.S. Dept. Agric. For. Serv. Gen. Tech. Rep. NE-23, Washington, D.C., pp. 737–744.

Norton, S. A.: 1973, Econ. Geol. 68, 353.

Norton, S. A. and Davis, R. B.: 1981, 'Responses of Maine Lakes to Atmospheric Inputs of Acids and Heavy Metals', Final Report, Office of Wat. Res. and Tech. U.S. Dept. Inter.

Odén, S.: 1976, 'The Acidity Problem – An Outline of Concepts', in L. S. Dochinger and T. A. Seliga (eds.), Proceedings of the First International Symposium on Acid Precipitation and the Forst Ecosystem, U.S. Dept. Agric. For. Serv. Gen. Tech. Rep. NE-23, Washington, D.C., pp. 1–36.

Overrein, L. N.: 1972, Ambio. 1, 145.

Overrein, L. N., Seip, H. M., and Tollan, Arne: 1980, 'Acid Precipitation – Effects on Forest and Fish', Final Report, SNFS Project: Oslo-As: 175 p.

Reiners, W. A. and Lang, G. E.: 1979, Ecology 60, 403.

Reiners, W. A., Marks, R. H., and Vitousek, P. M.: 1975, Oikos. 26, 264.

Rühling, A. and Tyler, G.: 1968, Bot. Not. 121, 321.

Rühling, A. and Tyler, G.: 1969, Bot. Not. 122, 248.

Rühling, A. and Tyler, G.: 1971, J. Appl. Ecol. 8, 497.

Rühling, A. and Tyler, G.: 1973, Water. Air. and Soil Pollut. 2, 445.

Schlesinger, W. H. and Reiners, W. A.: 1974, Ecology 55, 378.

Semonin, R. G., Bowersox, V. C., Gatz, D. F., Peden, M. E., and Stensland, G. J.: 1981, 'Study of Atmospheric Pollution Scavenging: Ill. Inst. Nat. Res.', 19th Progress Rept.

Siccama, T. G. and Smith, W. H.: 1978, Environ. Sci. Technol. 12, 593.

Siccama, T. G., Smith, W. H., and Mader. D. L.: 1980, Environ. Sci. Technol. 14, 54.

Smith, W. F. and Siccama, T. G.: 1981, J. Envir. Qual. 10, 323.

Stuanes, A. O.: 1980, 'Release and Loss of Nutrients, from a Norwegian Forest Soil Due to Artificial Rain of Varying Acidity', in Ecol. Impact of Acid Precip. Proc., Sandefjord, Norway, pp. 152–153.

TeBrake, W. H.: 1975, Tech. Cult. 16, 337.

Tyler, G.: 1981, Water. Air. and Soil Poll. 15, 353.

Ulrich, B.: 1980, 'Deposition, Production and Consumption of Hydrogen Ions in a Beech and a Spruce Ecosystem in the Solling District', in Ecol. Impact of Acid Precip. Proc., Sandefjord, Norway.

Wood, T. and Bormann, F. H.: 1976, 'Short-term Effects of a Simulated Acid Rain Upon the Growth and Nutrient Relationships of Pinus strobus l, in L. S. Dochinger and T. A. Seliga (eds.), Proceedings of the First International Symposium on Acid Precipitation and the Forest Ecosystem, Tech. Rep. NE-23, Washington, D.C., pp. 815–825.

Wright, R. F., Conroy, N., Dickson, W. T., Harriman, R., Henriksen, A., and Schofield, C. L.: 1980, 'Acidified Lake Districts of the World: A Comparison of Water Chemistry of Lakes in Southern Norway, Southern Sweden, Southwestern Scotland, the Adirondack Mountains of New York, and South-eastern Ontario', in Ecol. Impact of Acid Precip. Proc., Sandefjord, Norway, pp. 377–379.

THE USE OF CALIBRATED LAKES AND WATERSHEDS FOR ESTIMATING ATMOSPHERIC DEPOSITION NEAR A LARGE POINT SOURCE

P. J. DILLON, D. S. JEFFRIES, and W. A. SCHEIDER

Ontario Ministry of the Environment Box, 213, Rexdale, Ontario, M9W 5L1 Canada

(Received July 10, 1981; Revised November 9, 1981)

Abstract. We have measured the input and output rates of substances to and from both lakes and watersheds in the Sudbury and Muskoka-Haliburton areas of Ontario. At the former location, we have conducted mass balance studies on 5 lakes and their watersheds for $2\frac{1}{2}$ yrs. At the latter site, we have measured mass balances for 6 lakes and about 30 individual watersheds for the past 5 yrs. Substances studied included SO_4^{2-}, NO_3^-, NH_4^+, H^+, major cations (Ca^{2+}, Mg^{2+}, Na^+, K^+) and HCO_3^-.

During the course of the investigation at Sudbury we have made several observations that indicate that the inputs of some substances, specifically SO_4^{2-} or SO_4^{2-}-precursors and strong acids, to lakes and watersheds are underestimated when measured as bulk deposition (i.e. by collection in a continuously open container):

(a) The output of SO_4^{2-} from the calibrated watersheds was substantially greater than the input measured as bulk deposition.

(b) The SO_4^{2-} concentrations of the lakes could not be explained on the basis of the measured inputs. An additional input directly to the lake surface was needed to obtain a mass balance.

(c) The net input of acids measured as bulk deposition to the watersheds was much less than the acid consumed, which was estimated by the net output of Ca^{2+}, Mg^{2+}, Na^+, K^+, Al^{3+}, and the net retention of NO_3^-.

(d) The major cation content of the study lakes could be explained on the basis of weathering reactions in the lakes' watersheds only if the input of strong acid had been underestimated.

When these observations were quantified, they indicated a major portion of the total input of SO_4^{2-}-precursors and of strong acid was not included in our bulk deposition measurements. Deposition of SO_2 is the most likely explanation for these observations.

1. Introduction

Calibrated lakes and watersheds are those for which the input and output rates of substances are measured, and are an established research tool in environmental studies. For example, the development of strategies for the management of eutrophication of lakes by P control was based largely on mass balance studies and models (Vollenweider, 1975; Dillon and Rigler, 1975; Oglesby, 1977a,b; Reckhow, 1979).

Common reasons for the use of this approach include:

(a) the relative importance of the different components of the input of a pollutant can be assessed and abatement planned accordingly;

(b) the mass balances can be used with mathematical models to predict the chemical concentrations of the substance in the receiving body, either the stream draining the calibrated watershed, or the calibrated lake itself; and

(c) the quantitative accounting of the flow of substances in the watershed or lake may provide information concerning the processes and mechanisms occurring there.

Water, Air, and Soil Pollution **18** (1982) 241–258.   0049–6979/82/0182–0241$02.70.

The purposes of this paper are to describe mass balance studies carried out in southern Ontario, to use these studies to illustrate examples of basic principles relating to the processes occurring in lakes and watersheds, and to apply some of these principles to investigate the inputs of substances from the atmosphere to lakes and watersheds near a point source.

2. Study Sites

Calibrated lakes and watersheds have been studied in the Sudbury and Muskoka-Haliburton areas of Ontario (Figure 1). These areas, situated north and east of Georgian Bay, lie completely on the Canadian Shield. They are almost entirely underlain by Precambrian metamorphic, plutonic, and volcanic silicate rocks and are overlain by generally thin Pleistocene glacial deposits.

The Sudbury area is the site of major nickel production with 2 large smelters and associated operations emitting an estimated 1.3×10^6 mt yr^{-1} of SO$_2$ in the period 1972–78. Five lakes were studied at Sudbury (Figure 1). They are situated at varying

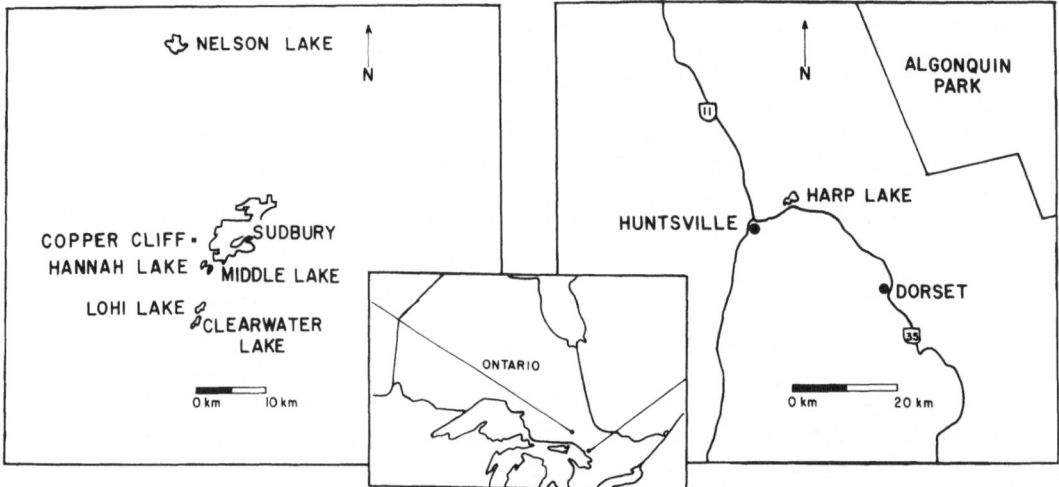

Fig. 1.   Location of the study lakes. The Sudbury area is shown on the left, Muskoka-Haliburton on the right.

distances from the central industrial activity. The 4 lakes south of the Inco Ltd. smelter at Copper Cliff are underlain by quartzite, gabbro, and gneiss (Table I) and are not contacted by the ore-bearing formation. Field reconnaissance of the watersheds confirmed that no obvious sulphide mineralization was present. Smelting practices employed in the early portion of this century denuded these watersheds of forest cover so that a large proportion of their surface area is now exposed bedrock (Table I).

The Nelson Lake basin lies ~ 30 km to the north of the Copper Cliff smelter. This area was not affected by SO$_2$ fumigation and thus is fully forested (mixed deciduous-

TABLE I

TABLE I

Location bedrock and surficial geology of the 5 study lakes in the Sudbury area

Lake	Distance (km) and direction from Copper Cliff	Bedrock geology	Surficial geology	
			% exposed bedrock	remainder (%)
Hannah	4, S	quartzite, gabbro	76	clay (14), thin till (7), peat (3)
Middle	5, S	quartzite, gabbro	72	thin till (13), clay (11), peat (4)
Lohi	11, S	quartzite, gabbro	65	thin till (29), peat (6)
Clearwater	13, S	quartzite, gabbro, gneiss	82	thin till (10), peat (5), pond (3)
Nelson	28, N	granophyre, volcanics	1	till plain (63), thin till (16), outwash (14), peat (4), pond (2)

coniferous) with minor exposure of bedrock. Surficial deposits are dominated by basal tills with lesser quantities of glacial outwash materials.

The Harp Lake watershed in Muskoka-Haliburton is underlain by gneiss and other metasedimentary bedrock. Basal tills cover the high areas of the basin and sand deposits, often covered by peat, occur in the valleys. The basin is forested with hemlock and balsam fir in the low-lying areas and sugar maple and yellow birch in the dry upland areas. The 90 dwellings around Harp Lake are used primarily for seasonal recreational purposes.

The Sudbury lakes and Harp Lake are all small (27 to 309 ha) with mean depths between 4 and 12 m (Table II). All but Hannah Lake are dimictic, with the latter being

TABLE II

Location, surface area (A_0), mean depth ($\bar{z}$), maximum depth (z_{max}), watershed area (A_d-excludes lake area), and water replenishment time (T_W) of the study lakes

Lake	A_0	Location		$\bar{z}$	z_{max}	A_d	T_W
	(ha)	Latitude (°N)	Longitude (°W)	(m)	(m)	(ha)	(yr)
Hannah	27.3	46° 27'	81° 02'	4.0	9	76	3.0
Middle	28.2	46° 26'	81° 02'	6.2	15	250	1.8
Lohi	40.5	46° 23'	81° 02'	6.2	20	460	1.3
Clearwater	76.5	46° 22'	81° 03'	8.3	22	340	4.4
Nelson	309	46° 44'	81° 05'	11.6	51	798	9.3
Harp	66.9	45° 23'	79° 07'	12.4	40	509	3.1

too shallow to have a well-defined hypolimnion. Four of the 5 Sudbury lakes have been manipulated by chemical additions (neutralizing agents – $CaCO_3$ and/or $Ca(OH)_2$), and fertilizers (P, N) in the period 1973–78. Details are reported by Dillon et al. (1979) and Yan et al. (1977). The treatments received and pH range during the course of the mass balance studies are summarized for each lake in Table III.

TABLE III

Lake manipulations carried out on study lakes and range of average annual
pH during course of mass balance studies

Lake	Treatment received	pH range
Hannah	neutralization (1975), fertilization (1975–78)	6.6 – 7.0
Middle	neutralization (1973), fertilization (1976–78)	6.4 – 6.7
Lohi	neutralization (1973, 1974, 1975)	4.7 – 5.3
Clearwater	none	4.1 – 4.4
Nelson	neutralization (1975, 1976)	6.3 – 6.5
Harp	none	6.1 – 6.3

3. Methods

Mass balance measurements at the Sudbury study lakes and watersheds were made from
June 1977 – December 1979 (31 months). Measurements at Harp Lake were intitiated
in June 1976 and are continuing. The data collected at Harp Lake for the period January
1977 – December 1978 are included in this report.

Input to the calibrated lakes and watersheds from the atmosphere was measured as
bulk deposition, defined as the material collected in a continuously open container. Three
stations consisting of a total of 7 collectors were used to estimate the bulk deposition
in the Sudbury area. Five collectors were used to estimate bulk deposition in the
Muskoka-Haliburton area. Details of the collector design, sampling frequency, analytical
methods, etc. are given in Scheider *et al.* (1979a) and Jeffries and Snyder (1981).

At Harp Lake, the 4 major inflowing streams and the outlet of the lake were hydro-
logically gauged. At the Sudbury site, 2 inflows of Clearwater Lake, and 3 inflows of
Nelson Lake were gauged, as were all study lakes' outflows. The Hannah Lake outflow
provides the major input to Middle Lake, while the Clearwater Lake outflow represents
the major input to Lohi Lake.

The output of substances from the calibrated watersheds was measured as the product
of the stream discharge and the concentration of the substances measured
20 to 100 times per stream over the 31-mo period or, in the case of Harp Lake, over the
24-mo period. Scheider *et al.* (1979b) reviewed methodologies of combining discharge
and concentration data to calculate output from watersheds.

Mass balances for 2 additional inflows of Clearwater Lake, 3 of Nelson Lake, and
1 of Middle Lake were also utilized. These were obtained by measuring the chemistry
and modelling the hydrology based on the calibrated watersheds. Details are given by
Scheider (1981).

The inputs to the study lakes from each lake's watershed were calculated as the sum
of the inputs from the gauged streams flowing into that lake, the inputs from the
secondary streams, the inputs from any upstream lakes, and the input from the portion
of the watershed of undefined drainage pattern. The latter was calculated by prorating
yields of substances from the nearest gauged tributaries i.e. the 4 inflows to Harp Lake
were used for the ungauged portion of the Harp Lake watershed, the 3 inflows of Nelson

Lake for the Nelson Lake watershed, and one of the 2 inflows of Clearwater Lake for the Lohi and Clearwater Lakes' watersheds. The calculated export (output per unit area) for Middle Lake watershed 1 was used to determine the inputs to Middle and Hannah Lakes from their ungauged watershed areas.

The mass balance model used to describe the relationship between the inputs and outputs of a substance and its concentration in the lake is given by:

$$\frac{d[X]}{dt} = \frac{J_A + J_T}{V} - \frac{Q[X]}{V} - S[X] \tag{1}$$

where

$[X]$ = the concentration of a substance X in the lake;
J_A = the input rate of X from the atmosphere;
J_T = the input rate of X from the watershed;
Q = the discharge rate of water at the lake outlet;
S = the sedimentation coefficient;
V = the lake volume.

If $S = 0$ (i.e. input rate = loss by outflow at steady-state), the time-dependent solution of the equation reduces to:

$$[X] = \frac{J_A + J_T}{Q}(1 - \exp(-t/T_W)) \tag{2}$$

where T_W is the water replenishment time of the lake ($= V/Q$).

4. General Principles

4.1. CONSERVATIVE SUBSTANCES

A conservative substance is defined as one which, at steady-state, has an input equal to its output; that is, the system does not act as a source or a sink for the substance. For example, the Cl^- ion is usually considered to be conservative with respect to both watersheds (Gjessing et al., 1976) and lakes (Sweers, 1969). Other substances may be conservative with respect to one system but non-conservative with respect to another. For example, Ca^{2+} is generally exported from watersheds in excess of the atmospheric deposition (Likens et al., 1977, Harvey et al., 1981), but behaves conservatively in soft-water lakes (Schindler et al., 1976). In contrast, total phosphorus (TP) is retained in both lakes (Vollenweider 1968; Vollenweider and Dillon 1974) and in watersheds (Schindler et al., 1976; Dillon and Rigler, 1975).

These different responses are illustrated for Harp Lake and its watershed (Table IV). The output and input of Cl^- to the terrestrial watershed were approximately equal while the watershed acted as a source for Ca^{2+} and a sink for TP. Harp Lake itself, however, was a sink for TP but neither a source nor a significant sink for Cl^- or Ca^{2+} (Table V).

TABLE IV

Input from the atmosphere (measured as bulk deposition) and output via stream-flow of Cl^- and Ca^{2+} (meq m^{-2} yr^{-1}) and total P (mg m^{-2} yr^{-1}) for the Harp Lake watershed (1977–78)

	Cl^-	Ca^{2+}	TP
Input	8.6	30.1	40.1
Output	10.9	82.8	6.8

The concentration of the latter substances in the lake can therefore be predicted using the mass balance model (Equation (2)). For example, the predicted Ca^{2+} concentration using this simple model was 3.2 mg L^{-1}, very close to the average measured concentration of 3.1 mg L^{-1}. Comparable models can, of course, be developed for TP; however, the sedimentation rate must be considered. For example 68% of the input of TP to Harp Lake was retained in the lake, presumably in the sediments.

TABLE V

Inputs and outputs of Cl^-, Ca^{2+} and total P to Harp Lake (kg yr^{-1}). Figures are averages for 1977–78

	Cl^-	Ca^{2+}	TP[a]
Input			
Atmospheric	200	400	26.8
Watershed	1980	8330	34.7
Total	2180	8730	61.5
Output	1940	8740[b]	19.6

[a] Based on 1977 only.
[b] Loss by sedimentation was measured as < 50 kg by combining sediment accumulation rate and sediment concentration.

4.2. INPUT AND CONSUMPTION OF ACIDS

Nitrate and NH_4^+, the principle inorganic N species in natural waters, are also often non-conservative. Almost all of the NO_3^- and NH_4^+ inputs to the Harp Lake watershed are retained and not exported via streamflow (Table VI). Therefore, the NH_4^+, which may be biologically utilized or microbially oxidized, has the same effect as a strong acid, since each of these processes results in generation of at least an equivalent amount of strong acid (Raven and Smith, 1976). That this occurs in lakes as well as watersheds is supported by results of whole-lake experiments, carried out in the Experimental Lakes Area (D. W. Schindler, pers. comm.) and in the Sudbury area (Yan and Lafrance, 1981). In these experiments, NH_4^+ salts, applied as nutrient source for phytoplankton, resulted

TABLE VI

Input (measured as bulk deposition) and output via streamflow of NH_4^+, NO_3^- and H^+ for the terrestrial watershed of Harp Lake. Figures are areally-weighted averages of 4 subwatersheds of Harp Lake (meq m^{-2} yr^{-1}) for 1977–78

	$NH_4^+ - N$	$NO_3^- - N$	H^+
Input	32.1	37.0	68.7
Output	1.9	4.7	1.9

in lake acidification. A corollary of this is that pH is an inadequate measure of the acidifying potential of atmospheric deposition.

On the other hand, the NO_3^- input is also almost entirely consumed in the Harp Lake watershed (Table VI). These reactions can generate a substantial amount of base (Raven and Smith, 1976; Reuss, 1975). Generation of alkalinity by NO_3^- uptake by phytoplankton in lakes is also a well-known phenomenon (Brewer and Goldman, 1976; Goldman and Brewer, 1980). For example, Harvey et al. (1981) reported concurrent pH increase and NO_3^- decrease in Crosson Lake in Muskoka-Haliburton. Complete utilization of nitric acid (HNO_3) should therefore result in no acidifying effect, as long as the utilized NO_3^- is stored in a sink, eg. as organic N in lake sediments.

The free H^+ is also retained or reacted in the Harp watersheds (Table VI) as it is in most other watersheds (reviewed in Harvey et al., 1981).

The effective net input of acid is equivalent to the sum of the H^+ and ammonium ion

TABLE VII

Input of acids and acids consumed in the Harp Lake watershed (1977–78). See text for detailed explanation

Net input of acid	meq m^{-2} yr^{-1}
H^+ retained	66.7
NH_4^+ retained	30.2
HCO_3^- exported	31.9
Total	129
Net acid consumed[a]	
Ca^{2+} export	54.5
Mg^{2+} export	34.0
Na^+ export	7.0
K^+ export	2.5
NO_3^- retained	32.3
Total	130

[a] Output of Al species may have contributed an additional 5–10 meq m^{-2} to the acids consumed.

retained and bicarbonate exported from the watershed, since the latter results from reaction of H_2CO_3 with the non-calcareous watershed substrate. This total can be compared to the net output of the products of the processes resulting in consumption of acids i.e. major ions (Ca^{2+}, Mg^{2+}, Na^+, K^+), and Al. The total can also be corrected for the NO_3^- retained in the watershed since this process may neutralize acid. This simplified balance between the input of acids and consumption of acids is justified only if (a) there is no other production of acids within the watershed, such as by sulphide oxidation; (b) organic acids do not contribute significantly, and, (c) there is no congruent dissolution of substances in the watershed.

Results of this kind of comparison for the Harp Lake watershed are shown in Table VII. This kind of analysis can be used to assess the quantitative importance of some of the processes occurring in the watershed; for example, major ion export is clearly more related to weathering from the atmospheric input of strong acids than to 'natural' weathering resulting from the dissociation of H_2CO_3.

TABLE VIII

Input (as bulk deposition) and output (via streamflow) of SO_4^{2-} for 11 watersheds in the Sudbury area (June 1977 – December 1979)

Watershed	Input	Output	Output
	(meq m^{-2} period^{-1})		Input
Middle 1	215	752	3.5
Clearwater 1	188	483	2.6[a]
Clearwater 2	188	519	2.8[a]
Clearwater 3	188	419	2.2[a]
Clearwater 4	188	413	2.2[a]
Nelson 1	186	448	2.4[b]
Nelson 2	186	454	2.4[b]
Nelson 3	186	408	2.2[b]
Nelson 4	186	567	3.1[b]
Nelson 5	186	533	2.9[b]
Nelson 6	186	454	2.4[b]

[a] Areally weighted difference between output and input for whole Clearwater Lake watershed is 281 meq m^{-2}.
[b] Areally weighted difference between output and input for whole Nelson Lake watershed is 271 meq m^{-2}.

5. Results and Discussion

5.1. SO_4^{2-} MASS BALANCES

5.1.1. *Watersheds*

In watersheds on Precambrian areas, remote from major SO_2 sources, the SO_4^{2-} ion behaviour approximates that of a conservative substance (Schindler *et al.*, 1975;

Gjessing *et al.*, 1976). On the other hand, watersheds in non-Precambrian regions such as Walker Branch watershed in eastern Tennessee may accumulate SO_4^{2-} (Shriner and Henderson, 1978; Henderson *et al.*, 1977). At Hubbard Brook in New Hampshire, the output of SO_4^{2-} exceeded the measured input by $\sim 40\%$, with the difference attributed to unmeasured inputs from atmospheric gaseous and aerosol sources (Likens *et al.*, 1977).

The measured inputs and outputs of SO_4^{2-} for the 11 headwater watersheds in the Sudbury area are shown in Table VIII. Bulk deposition of SO_4^{2-} to the watersheds of Middle, Clearwater, and Nelson Lakes was 215, 188, and 186 meq m^{-2} respectively. The measured loss of SO_4^{2-} from the watersheds by streamflow ranged from 408 meq m^{-2} (Nelson 3) to 752 meq m^{-2} (Middle 1). For each of the 11 watersheds, the loss of SO_4^{2-} from the watershed was greater than the input in bulk deposition. The ratio of loss : input ranged from 2.2 (Nelson 3 and Clearwater 4) to 3.5 (Middle 1), indicating that either a source of SO_4^{2-} existed in the watersheds or that bulk deposition underestimated the input of SO_4^{2-} and/or SO_4^{2-}-precursors to the watershed. As there is no evidence for the existence of S-bearing minerals in the watersheds, with the possible exception of the Onaping tuffs and breccia and peat bogs at Nelson Lake, it is reasonable to conclude that our bulk collectors underestimated the total input of SO_4^{2-} and/or SO_4^{2-}-precursors from the atmosphere. These results indicate that the extra deposition was highest at the Middle 1 watershed, presumably because it is closest to the smelting operations. However, extra depositions were approximately equal at the Nelson and Clearwater watersheds although they are ~ 12 km and 30 km distant from the Copper Cliff smelter, respectively.

5.1.2. *Lakes*

If a mass balance model, such as that employed for Ca^{2+} for Harp Lake, can be successfully used to predict the concentration of a conservative substance in a lake from the input rate and water budget, then conversely it may be used to predict the input rate from the lake concentration and the water budget. This approach can be used for SO_4^{2-} for the Sudbury lakes because the mass balance model was validated for these lakes for other conservative substances such as Mg^{2+}, K^+, and for some lakes Na^+, Ca^{2+} and Cl^- (Dillon, 1981). Sulphate almost certainly acts as a conservative substance in these lakes since the hypolimnia of all study lakes were well-oxygenated at all times; therefore, internal loss of SO_4^{2-} by reduction processes in the hypolimnia is not likely (Schindler *et al.*, 1980).

In order to determine if the steady-state version of the mass-balance model was applicable, it was necessary to examine the long-term trends in SO_4^{2-} concentration in the study lakes (Table IX). These data indicated that a small decline in SO_4^{2-} concentration in the study lakes may have occurred, possibly as a result of the short-term reduction in emissions from Inco Ltd. in Sudbury in 1978–79 (Scheider *et al.*, 1981). Although this would suggest use of a non-steady state model, this model has the disadvantage that small changes in lake concentrations over short time intervals can result in large variations in predicted input. It was therefore decided to use the steady-

TABLE IX

Average SO_4^{2-} concentration in Sudbury lakes during the period of mass balance studies. Standard error in brackets; sample number in a year ranged from 8 to 28.

Lake	1977	1978	1979
Hannah	49.8 (0.9)	46.7 (3.0)	39.8 (2.1)
Middle	39.2 (1.5)	36.8 (1.2)	33.8 (0.7)
Lohi	24.9 (0.6)	23.7 (1.7)	21.4 (0.8)
Clearwater	26.4 (0.9)	23.9 (0.7)	22.0 (0.4)
Nelson	15.2 (0.7)	14.9 (1.1)	14.4 (0.9)

state version of the model, realizing that the lag-time between SO_4^{2-} deposition on the watershed and its export from the watershed (Jeffries, 1981) would minimize the effect of the reduction in emissions on the model.

The mass balance data for Clearwater Lake are summarized in Table X for the period June 1977 – December 1979. For Clearwater Lake, the input of SO_4^{2-} in bulk deposition was 6.90 mt and the input from the watershed was 76.8 mt for a total input of 83.7 mt. However, the predicted input was 98.5 mt. If the terrestrial input was correct, 21.7 mt must have entered via the atmosphere (eg. bulk and unmeasured deposition). This implies that our bulk collector has undercollected SO_4^{2-} or SO_4^{2-}-precursor deposition to the lake surface by a factor of 3.1 and leads to an estimate of 404 meq m^{-2} as the unmeasured

TABLE X

Sulphate mass balances for Clearwater Lake and its subwatersheds. The input to the lake surface and each of the terrestrial components by bulk deposition (J_{BD}) was calculated as the area of each receptor multiplied by the bulk deposition rate, and the input to the lake from each receptor (J) was measured. The predicted total input to the lake was calculated using the mass balance model. The predicted atmospheric input to the lake surface was calculated as the predicted total input to the lake less the measured output from the terrestrial components. The ratio of the measured output from each receptor to the input measured as bulk deposition to the receptor (J/J_{BD}), and the ratio of the predicted atmospheric input to the measured bulk deposition for the lake surface and the watershed as a whole (J_{PR}/J_{BD}) are given. All inputs in mt (10^9 g) per 31-mo period.

Receptor	Measured bulk deposition to: (J_{BD})	Measured input to lake (J)	Predicted input to lake (J_{PR})	J/J_{BD}	J_{PR}/J_{BD}
Lake surface	6.90	6.90	21.7		3.1 (1.7–4.6)[a]
Subwatershed 1	8.16	21.0		2.6	
Subwatershed 2	2.42	6.68		2.8	
Subwatershed 3	2.00	4.47		2.2	
Subwatershed 4	6.00	13.1		2.2	
Ungauged	12.3	31.6		2.6	
Total	37.7	83.7	98.5		2.6 (2.4–2.9)[a]

[a] Range of ratio if accuracy of predicted total input is $\pm 10\%$ (i.e. 98.5 ± 9.9)

deposition to the lake surface. This is higher than the 281 meq m^{-2} estimated by analogous means for extra deposition to the terrestrial watershed (Table VIII).

The estimate of input directly to the lake surface is subject to substantial error, since it is in part derived as a difference between 2 large numbers. If the uncertainty in the model prediction of total input to the lake is estimated as $\pm 10\%$, then the ratio of actual direct input to measured direct input (i.e. bulk deposition) for the lake surface may be as low as 1.7 or as high as 4.6. However, if we consider the whole watershed, that is, the lake and its entire drainage area, the ratio can only range from 2.4 to 2.9. The ratios for the 4 subwatersheds varied from 2.2 to 2.8, and averaged 2.5. This is similar to and therefore gives credence to the result based on the lake mass balance which was calculated in an independent manner.

A summary of the results for the other 4 lakes is shown in Table XI. For Nelson Lake, the input of SO_4^{2-} in bulk deposition was 27.6 mt whereas 175.4 mt entered the lake

TABLE XI

Sulphate mass balances for 4 study lakes. Terminology as in Table X. All inputs in mt per 31-mo period.

Receptor	Measured bulk deposition to: (J_{BD})	Measured input to lake (J)	Predicted atmospheric input to lake (J_{PR})	J/J_{BD}	J_{PR}/J_{BD}
Hannah Lake Surface	2.81	2.81	26.1		9.3 (7.7–10.9)
Terrestrial Basin	7.84	18.5		2.4	
Total	10.7	21.3	44.6		4.2 (3.8–4.6)
Middle Lake Surface	2.90	2.90	13.8		4.8 (1.5–8.0)
Terrestrial Basin	25.4	80.2		3.1	
Total	28.3	83.1	94.0		3.3 (3.0–3.7)
Lohi Lake Surface	3.65	3.65	23.7		6.5 (2.5–10.5)
Terrestrial Basin	46.6	121.3		2.6	
Total	50.3	125	145		2.9 (2.6–3.2)
Nelson Lake Surface	27.6	27.6	71.6		2.6 (1.7–3.5)
Terrestrial Basin	71.4	175.4		2.5	
Total	99.0	203	247		2.5 (2.2–2.7)

via runoff from the terrestrial watershed for a total input of 203 mt. The predicted total input, however, was 247 mt. Once again, if it is assumed that the input of SO_4^{2-} from the terrestrial watershed is correct, then 71.6 mt (eg. 247 to 175.4) must have been the total input of SO_4^{2-} or SO_4^{2-}-precursors from the atmosphere. Using this approach, the total atmospheric input to the lake was greater by a factor of 2.6 than the measured input of SO_4^{2-} in bulk deposition. These data suggest that 296 meq m^{-2} of SO_4^{2-}-precursor is deposited on the surface of Nelson Lake (calculated from the difference between the total atmospheric input (71.6 mt) and the measured bulk input (27.6 mt) divided by the area of the lake surface). The extra deposition to the terrestrial basin was calculated to be 271 meq m^{-2} by an analogous method (Table VIII).

The SO_4^{2-} inputs to Lohi Lake from bulk deposition, from the terrestrial watershed and the total input were 3.65, 212.3, and 125 mt respectively. The total input predicted by the model and consequent total atmospheric input (as above) were 145 and 23.7 mt respectively, implying that SO_4^{2-} deposition to the lake surface from the atmosphere was underestimated by a factor of 6.5. The comparable ratio calculated for the terrestrial watershed was 2.9. However, in this case, the terrestrial ratio is not strictly independent information since SO_4^{2-} input from both the Clearwater outflow and the ungauged portion of the Lohi Lake watershed were calculated on the basis of Clearwater Lake and watershed data. The results suggest that the lake is a more efficient trap of the SO_4^{2-}--precursors than the terrestrial watershed.

Table XI shows that 2.9, 80.2, and 83.1 mt of SO_4^{2-} were input to Middle Lake from bulk deposition, the terrestrial watershed and in total respectively. To maintain the measured SO_4^{2-} concentration in the lake, the mass balance model predicts that 94 mt must have entered the lake from June 1977 – December 1979, 13.8 mt of that total from the atmosphere assuming the terrestrial inputs are correct. These data show that our bulk collectors undercaught the total SO_4^{2-} input to the lake surface by a factor of 4.8. Similarly, 3.1 times more SO_4^{2-} and SO_4^{2-}-precursors were deposited on the terrestrial watershed than collected in bulk samplers.

Finally, for Hannah Lake, 2.81, 18.5, and 21.3 mt of SO_4^{2-} were input from bulk deposition, from the terrestrial watershed and in total respectively. The predicted input was 44.6 mt with a total atmospheric component of 26.1 mt. The ratio of predicted total atmospheric SO_4^{2-}-precursor deposition to measured bulk deposition was 9.3 for the lake surface, compared to a value of 2.4 for the terrestrial watershed.

Although, in all cases, the ratio of calculated SO_4^{2-} input to measured bulk deposition is greater for the lake surface than the lakes' watersheds, the former figure has, in each case, much more uncertainty in it. The conclusion that the deposition of the SO_4^{2-}-precursors is greater to the lakes' surfaces than to the terrestrial portions of the watersheds must therefore be considered tentative.

The whole-watershed ratios (J_{PR}/J_{BD}; Tables X and XI) increased as the distance from Copper Cliff decreased, being 2.5 at Nelson Lake, 2.6 at Clearwater Lake, 2.9 at Lohi Lake, 3.3 at Middle Lake, and 4.2 at Hannah Lake. These results indicated that the unmeasured deposition of SO_4^{2-}-precursors was greatest closest to the major SO_2 source.

5.1.3. Summary

The most likely 'missing' SO_4^{2-}-precursor is gaseous SO_2 which would be under-collected by the bulk deposition samplers. Assuming that there are no S sources in the watersheds and that there is complete oxidation of SO_2 to SO_4^{2-}, the net loss of SO_4^{2-} from the watersheds (loss in streamflow – input by bulk deposition) can be used as an estimate of the unmeasured deposition of SO_2. Values estimated in this manner ranged from 222 meq m^{-2} (31 mo)$^{-1}$ for Nelson watershed 3 to 537 meq m^{-2} (31 mo)$^{-1}$ for

Middle watershed 1. The SO_4^{2-} mass balances calculated for the study lakes suggested that the lake's surfaces may be a more efficient sink for SO_2 than the terrestrial portion of the watersheds. This supposition is supported by the results of others (Whelpdale and Shaw, 1974) who have shown that water surfaces are better sinks for SO_2 deposition than terrestrial surfaces.

5.2. Input and Consumption of Acids

5.2.1. *Watersheds*

As discussed previously, the acid supplied to a watershed includes that portion of the bulk H^+ and NH_4^+ deposition which is retained by the watershed, plus H^+ derived from the dissociation of H_2CO_3 (probably in the soil environment). The latter is equivalent to the gross export of HCO_3^- when there is no carbonate rock in the watershed, since there is negligible atmospheric deposition of HCO_3^-. Measurements of these three components of the acid supply for 11 watersheds at Sudbury between June 1977 and December 1979 are reported in Table XII. An additional (unmeasured)

TABLE XII

Net acid supply and consumption (meq m^{-2} period^{-1}) in 11 watersheds at Sudbury from June 1977 – December 1979. Acid supply includes net watershed gain of H^+, NH_4^+ and loss of HCO_3^-. Acid consumption is reflected by net watershed loss of Ca^{2+}, Mg^{2+}, Na^+, K^+, and Al^{3+}, and net gain of NO_3^-.

Watershed	Net acid supply				Net acid consumption							Consumption
	H^+	NH_4^+	HCO_3^-	Sum	Ca^{2+}	Mg^{2+}	Na^+	K^+	Al^{3+}	NO_3^-	Sum	Supply
Clearwater 1	123	64.2	1.92	189	188	89.6	7.1	4.7	36.4	95.0	420	2.2
Clearwater 2	116	64.7	0.08	181	787[a]	195	172[a]	13.2	68.5	94.2	443[a]	2.4[a]
Clearwater 3	129	65.2	0.00	194	150	70.1	11.4	5.8	43.8	95.0	377	1.9
Clearwater 4	151	58.0	5.00	214	337[a]	123	49.6[a]	10.3	67.7	92.9	500[a]	2.3[a]
Middle 1	137	70.1	0.00	207	332	198	32.9	17.8	52.5	92.9	727	3.5
Nelson 1	127	60.4	0.91	188	219	73.4	-0.3	6.1	33.6	85.0	417	2.2
Nelson 2	128	62.6	0.00	191	226	67.6	6.4	1.8	41.6	90.0	434	2.3
Nelson 3	159	59.4	26.4	245	221	79.6	3.6	7.3	13.7	85.0	410	1.7
Nelson 4	157	62.1	33.4	253	301	127	17.8	12.2	16.8	89.3	564	2.2
Nelson 5	163	54.7	79.2	297	319	137	18.5	15.0	5.5	81.4	576	1.9
Nelson 6	159	62.4	48.6	270	255	191	26.7	12.6	20.2	85.7	591	2.2

[a] Ca^{2+} and Na^+ net export values for Clearwater 2 and 4 include contributions from road chemicals which are associated with an equal net export of Cl^-; therefore, the sum of acids consumed has been estimated by subtracting Cl^- net export.

component is the H^+ derived from the oxidation and dissociation of dry deposited SO_2. Of the measured components of the acid supply, $H^+ > NH_4^+ > HCO_3^-$ for all watersheds except Nelson 5 where the HCO_3^- component exceeded NH_4^+.

The total atmospheric supply of acid for the 31-mo period reported here is complicated by the fact that the largest local source (Inco Ltd.) was inoperative for approximately an 8-mo period (Oct. 1978 to May 1979). Scheider *et al.* (1981) have shown that bulk deposition of H^+ was not significantly reduced over the study area during the shutdown period; however, SO_4^{2-} deposition was significantly reduced at Middle, Hannah, Lohi, and Clearwater Lakes, the sites closest to the source. Nevertheless, because the shut-down period was relatively short compared to the total study period (8 to 31 mo), and the expected 'integrating' effect of the watersheds as the SO_4^{2-} passes through them, we do not believe that the temporary shutdown of Inco Ltd. significantly affected the results reported here. Balanced against the above input of acid are the substances produced or liberated by acid-consuming reactions (Table XII). These include the net export of Ca^{2+}, Mg^{2+}, Na^+, K^+, and other metals (noteably Al). Acid consumption resulting in the liberation of Al has been estimated (perhaps slightly over-estimated) by assuming that this metal is in the Al^{3+} form. In addition, NO_3^- can also be an acid-consuming substance (Section 4). The Ca^{2+} and Na^+ net export for Clearwater 2 and 4 were strongly affected by road maintenance activities which was reflected in an equivalent net Cl^- export. The total acid consumption for Clearwater 2 and 4 has therefore been 'corrected' by subtracting the value of the net Cl^- export.

Liberation of Ca^{2+} is the most important acid-consuming (weathering) process at Sudbury. Net release of Al^{3+}, Mg^{2+}, and retention of NO_3^- are intermediate and variable in importance while watershed release of Na^+ and K^+ are least important. The sequence in basic cation mobility implied by these results is consistent with observations for most granitic terrains (Holland, 1978).

The ratios of the measured net acid consumption to the measured net acid supply were always > 1, ranging from 1.7 (Nelson Lake watershed 3) to 3.5 (Middle Lake sub-watershed 1). These results are consistent with the hypothesis that bulk deposition underestimated the atmospheric deposition of strong acids in the Sudbury area. The consumption : supply ratios in Table XII are comparable in magnitude to the SO_4^{2-} mass balance ratios for the same watersheds (Table VIII); that is, the unmeasured input of SO_4 or SO_4^--precursors was similar to the unmeasured input of acid. These results again suggest that the principal unmeasured input is the dry deposition of SO_2.

5.2.2. *Lakes*

The steady-state concentrations of conservative cations (Ca^{2+}, Mg^{2+}, Na^+, and K^+) in lakes can be used to calculate the input rates of these substances. The concentrations of Mg^{2+} and K^+ were in steady state for Hannah and Middle Lakes. Although Ca^{2+} and Na^+ levels were not in steady state because of the addition of road treatment chemicals (NaCl, $CaCl_2$) and neutralizing agents ($CaCO_3$, $Ca(OH)_2$) concentrations measured in 1973–1974 prior to the influence of either of these factors are reasonable estimates of steady state values (Dillon, 1981). Using these data, it is possible to calculate the total input of each cation, and since we have measured all components of the major ion mass balance for Hannah and Middle Lakes except that contributed by ungauged portions of their watersheds, the yield of ions from the ungauged watersheds may be obtained by difference. It is assumed that the extra cations in the lake (i.e. not accounted for by the measured inputs) result from the deposition of strong acids or acidifying substances, followed by leaching and export from the ungauged watershed. As in the previous section, the release of ions from the watershed may then be compared to the measured acid supply.

The calculated or predicted total inputs of Ca^{2+}, Mg^{2+}, Na^+, and K^+ (J_{PR}) to Hannah and Middle Lakes from June 1977 to December 1979 are given in Table XIII.

TABLE XIII

Calculation of net cation export rates (June 1977 to December 1979) from the Hannah and Middle Lake watersheds using mass balance models. Predicted input (J_{pr}) was calculated from steady-state lake concentrations. Terrestrial input from the ungauged portion of the lakes' watersheds (J_{tr}) is J_{pr} − sum of all measured inputs. Input values are in mt. Gross export and net (gross-bulk deposition) export (meq m^{-2}) are calculated. The net export sum (Ca^{2+} + Mg^{2+} + Na^+ + K^+) is also presented.

Ion		Hannah Lake[a]	Middle Lake[b]
Ca^{2+}	J_{PR}	12.6	26.2
	J_{tr}	12.1	12.9[c]
	Gross export	794	462.2
	Net export	710	379
Mg^{2+}	J_{PR}	4.19	8.54
	J_{tr}	4.13	4.19
	Gross export	446	248
	Net export	428	230
Na^+	J_{PR}	3.46	6.41
	J_{tr}	2.98	2.36[c]
	Gross export	169	73.4
	Net export	92.6	− 3.1
K^+	J_{PR}	1.97	4.15
	J_{tr}	1.90	2.07
	Gross export	62.4	37.3
	Net export	55.9	30.8
	Net Export Sum	1290	637

[a] Area of ungauged watershed for Hannah Lake
= 76.3 × 10^4 m^2.
[b] Area of ungauged watershed for Middle Lake
= 139.2 × 10^4 m^2.
[c] Terrestrial input corrected for contributions from road treatment and lake neutralization chemicals (see text).

The contribution of the ungauged terrestrial input (J_{tr}) to the total input was obtained by subtracting all the other measured inputs (bulk deposition to lakes' surfaces for both Hannah and Middle, plus the Hannah Lake output and Middle watershed 1 output for Middle Lake only). Since a smaller portion of the total Middle Lake input was from the ungauged terrestrial watershed (compared to Hannah Lake), the J_{tr} for Middle Lake is probably less precise.

After 1974, road construction and maintenance contributed significant quantities of $NaCl/CaCl_2$ to the lakes. The effects of this input and also the application of $Ca(OH)_2$

and $CaCO_3$ for lake neutralization must be considered so that the J_{tr} for Ca^{2+} and Na^+ for the Middle Lake watershed are not over-estimated. Some of the Na^+ and Ca^{2+} distributed in the Hannah Lake watershed was supplied to Middle Lake via the Hannah Lake outflow. Using lake concentration data collected prior to road treatments or neutralization (i.e. from 1973 and 1974), an input to (and output from) Hannah Lake may be calculated using the steady-state model. If it is assumed that this 'background' level of input to Hannah Lake remained constant over the study period and was passed on to Middle Lake via the outflow, then a correction can be calculated to account for latter road and neutralization inputs. Elevated Ca^{2+} and Cl^- export (but not Na^+) from Middle watershed 1 also indicate the use of $CaCl_2$ on roads at this location. Export of Ca^{2+} was corrected by subtracting the measured net Cl^- export as in the previous section.

In agreement with observations in the previous section, steady state ion concentrations in the lakes suggest that the net export of Ca^{2+} and Mg^{2+} from the ungauged watersheds was much greater than that for Na^+ and K^+. Moreover, the total net export of basic cations for the ungauged watershed of Hannah Lake was 2-fold greater than that for Middle Lake. The predicted net export sum for the ungauged Middle watershed compares favorably (within 9% over the 31-mo period) with that measured for the Middle Lake watershed 1 which provides additional support for the validity of the prediction.

The bulk deposition of strong acid (H^+) plus acidifying substances (primarily NH_4^+) over the 31-mo period at Hannah and Middle Lakes was 169 and 74.8 meq m^{-2} respectively. The cation export alone for both watersheds thus greatly exceeded this bulk acid input which again indicates that there is a major, unmeasured input of strong acid. Consideration of Al and NO_3^- increases the difference. The unmeasured input must also be greater for the Hannah watershed than the Middle watershed.

6. Summary

The SO_4^{2-} mass balances for the 11 watersheds and 5 lakes in the Sudbury area demonstrated that the measured inputs of SO_4^{2-} as bulk deposition could not account

TABLE XIV

Proportion ($\%$) of the input of SO_4^{2-}-precursors or acid not measured by bulk deposition.

Method	Clearwater	Lohi	Hannah	Middle	Nelson
Watershed SO_4^2 budgets	59	–	–	72	61
Lake SO_4^2 budgets	62	66	76	70	60
Acid input-Acid consumed	56	–	–	72	50
Lake ion budgets	–	–	86	73	–

for the export of SO_4^{2-} from the watersheds or the SO_4^{2-} concentrations in the lakes. The balance between the measured input of acids and the measured consumption of acids in the 11 watersheds, and the calculated consumption of acids in the Hannah and Middle Lake watersheds indicated that a major portion of the input of acid was not included in the bulk deposition measurements. The fact that the SO_4^{2-} or SO_4^{2-}-precursors were underestimated by the same amount as the acids indicates that deposition of SO_2 is the most probable unmeasured acid input.

The results of the previous calculations are summarized in Table XIV. The proportion of the input of SO_4^{2-}-precursors/acid which was not measured by bulk deposition collectors was calculated by 4 independent methodologies which gave extremely consistent results within a site. For example, at Middle Lake where all 4 methodologies were used, the predicted unmeasured input ranged from only 70 to 73% of the calculated total input. Nelson Lake had the broadest range (50 to 61%), but even these results were remarkably consistent.

The importance of the unmeasured inputs decreased with distance from Copper Cliff, the unmeasured fraction averaging 81, 72, 66, 59, and 57% of the total input at Hannah, Middle, Lohi, Clearwater, and Nelson Lakes respectively. This provides additional evidence that Copper Cliff is the source of the SO_2.

Acknowledgements

The authors wish to thank Claude Lafrance, Bruce Cave, Jim Jones, Lem Scott, Bob Girard, Ron Reid, and Bev Clark for assistance with the field work. Jane Smith, Chuck Cox, and Warren Snyder helped with the data analyses while Al Nicolls undertook all of the computer programming necessary.

References

Brewer, P. G. and Goldman, J. C.: 1976, *Limnol. Oceanogr.* **21**, 108.

Dillon, P. J.: 1981, 'Mass Balance Models: An Explanation for the Observed Chemistry of Lakes in the Sudbury Study Area', in *Studies of Lakes and Watersheds Near Sudbury, Ontario*, Ontario Ministry of the Environment, Tech. Rep.

Dillon, P. J. and Rigler, R. H.: 1975, *J. Fish. Res. Board Can.* **32**, 1519.

Dillon, P. J., Yan, N. D., Scheider, W. A., and Conroy, N.: 1979, *Arch. Hydrobiol. Beih, Ergebn. Limnol.* **13**, 317.

Gjessing, E. T., Henriksen, A., Johannessen, M., and Wright, R. F.: 1976, 'Effects of Acid Precipitation on Freshwater Chemistry', in Braekke, F. (ed.), *Impact of Acid Precipitation on Forest and Freshwater Ecosystems in Norway*, Sur Nedbors Virkuing Pa Skog og Fisk Res. Rep. 6/76.

Goldman, J. C. and Brewer, P. G.: 1980, *Limnol. Oceanogr.* **25**, 352.

Harvey, H. H., Pierce, R. C., Dillon, P. J., Kramer, J. R., and Whelpdale, D. M.: 1981, 'Acidification in the Canadian Aquatic Environment', *Nat. Res. Coun. Can. Report* No. 18475.

Henderson, G. S., Hunley, A., and Selvidge, W.: 1977, 'Nutrient Discharge from Walker Branch Watershed', in Correll, D. (ed.), *Watershed Research in Eastern North America: A Workshop to Compare Results*, Vol. 1. 2/28-3/3/77. Chesapeake Bay Center for Environmental Studies, Smithsonian Institution, Edgewater, M.D.

Holland, H. D.: 1978, *The Chemistry of the Atmosphere and Oceans*, John Wiley and Sons, New York, 351.

Jeffries, D. S.: 1981, 'Atmospheric Deposition of Major Ions, Nutrients, and Metals in the Sudbury Area', in *Studies of Lakes and Watersheds Near Sudbury, Ontario*, Ontario Ministry of Environment, Tech. Rep.

Jeffries, D. S. and Snyder, W. R.: 1981, *Water, Air, and Soil Pollut.* **15**, 127.

Likens, G. E., Bormann, F. H., Pierce, R. S., Eaton, J. S., and Johnson, N. M.: 1977, *Biogeochemistry of a Forested Ecosystem*, Springer-Verlag, New York.

Oglesby, R. T.: 1977a, *J. Fish. Res. Board Can.* **34**, 2255.

Oglesby, R. T.: 1977b, *J. Fish. Res. Board Can.* **34**, 2271.

Raven, J. A. and Smith, F. A.: 1976, *New Phytol.* **76**, 415.

Reckhow, K. H.: 1979, in Scavia, D. and Robertson, A. (eds.), *Perspectives on Lake Ecosystem Modeling*, Ann Arbor Sci., Ann Arbor.

Reuss, J. S.: 1975, 'Chemical and Biological Relationships Relevant to Ecological Effects of Acid Rain', EPA-660 375 032.

Scheider, W. A., Snyder, W. R., and Clark, B.: 1979a, *Water, Air, and Soil Pollut.* **12**, 171.

Scheider, W. A., Moss, J. J., and Dillon, P. J.: 1979b, 'Measurement and Uses of Hydraulic and Nutrient Budgets in Lake Restoration', EPA 440/5-79-001, 77.

Scheider, W. A.: 1981, 'Hydrology of Lakes and Watersheds in the Sudbury Area', in *Studies of Lakes and Watersheds Near Sudbury, Ontario*, Ontario Ministry of Environment, Tech. Rep.

Scheider, W. A., Jeffries, D. S., and Dillon, P. J.: 1981, *Atmos. Environ.* **15**, 945.

Schindler, D. W., Newbury, R. W., Beaty, K. G., and Campbell, P.: 1976, *J. Fish. Res. Board Can.* **33**, 2526.

Schindler, D. W., Wagemann, R., Cook, R. B., Ruszczynski, T., and Prokopowich, J.: 1980, *Can. J. Fish. Aquat. Sci.* **37**, 342.

Shriner, D. S. and Henderson, G. S.: 1978, *J. Environ. Qual.* **7**, 392.

Sweers, H. E.: 1969, *Proc. 12th Conf. Great Lakes Res.* 1969, 734.

Vollenweider, R. A.: 1975, *Schweiz. Z. Hydrol.* **37**, 53.

Vollenweider, R. A.: 1968, Water Management Research. OECD Paris. DAS/CSI/68.27.

Vollenweider, R. A. and Dillon, P. J.: 1974, 'The Application of the Phosphorus Loading Concept to Eutrophication Research', Nat. Res. Coun. Can. Report No. 13690.

Whelpdale, D. M. and Shaw, P. W.: 1974, *Tellus* **26**, 196.

Yan, N. D. and Lafrance, C.: 1981, 'Experimental Fertilization of Lakes in the Sudbury Area', in *Studies of Lakes and Watersheds Near Sudbury, Ontario*, Ontario Ministry of Environment, Tech. Rep.

Yan, N. D., Scheider, W. A., and Dillon, P. J.: 1977, *Proc. 12th Can. Symp. Wat. Poll. Res. Can.*, 213.

BIOLOGICAL, CHEMICAL AND PHYSICAL RESPONSES OF LAKES TO EXPERIMENTAL ACIDIFICATION

D. W. SCHINDLER and M. A. TURNER

Department of Fisheries and Oceans, Freshwater Institute, 501 University Crescent, Winnipeg, Manitoba R3T 2N6 Canada

(Received 10 July, 1981; Revised 2 November, 1981)

Abstract. Changes in physical, chemical and biological factors were observed during a 5-yr experimental acidification study in Lake 223 of the Experimental Lakes Area, and compared to a 2 yr pre-acidification period. Significant changes included increased transparency, rates of hypolimnion heating and rates of thermocline deepening; increased concentrations of Mn, Na, Zn, Al, and chlorophyll; decreased concentrations of suspended C, total dissolved N, Fe and chloride; increases in Chlorophyta but decreases in Chrysophyta; the disappearance of the opossum shrimp *Mysis relicta* and the fathead minnow *Pimephales promelas*; the appearance of epidemics of the filamentous alga *Mougeotea*; decreased fitness and decline in numbers of *Orconectes virilis*; and increased embryonic mortality of the lake trout *Salvelinus namaycush*.

Sulfur budgets for two lakes experimentally acidified with sulfuric acid reveal that an average of 1/4 to 1/3 of added sulfate is sedimented, presumably as FeS, reducing the efficiency of acidification. The sedimentation occurs under both oxic and anoxic conditions. The utility of whole-ecosystem mass balance studies of S in 'experimental' and 'observational' mass balance studies is discussed.

1. Introduction

Documentation of the ecological effects of acidification of freshwater lakes has been largely based upon observational studies in lakes where pre-acidification data are few. As a result, only the most obvious changes, such as disappearance of important fish stocks or major changes in water chemistry, can usually be stated with confidence. Because such changes are detectable only relatively late in the acidification process, it is often difficult to determine whether effects on fishes are due to direct toxicity of H^+, the analogous decrease in bicarbonate, or to secondary causes such as disappearance of critical food organisms or increased concentrations of toxic trace metals. This dilemma has resulted to a large extent from the low priority given in the past to monitoring programs by investigators and by funding agencies. Because we are without the detailed 'baseline' data which would be necessary to document changes caused by acid precipitation, we must resort to whole-lake experiments in order to quantify changes which occur early in the acidification process, and to eclucidate the ecological mechanisms which cause the observed biological changes. This paper summarizes the work of several investigators on biological, physical and chemical changes which occurred early in the experimental acidification of a small lake, and discusses some examples of how they interact. As we shall demonstrate, in many cases effects are difficult to assess, even when acid inputs are exactly known and two years' data are available prior to acidification. In other cases, some surprisingly dramatic effects occur early in the acidification process, before toxic metals or food-chain effects are likely to be important.

Water, Air, and Soil Pollution **18** (1982) 259–271. 0049–6979/82/0182–0259$01.95.

2. History of the Lake Acidification Experiments

Lake 223 was studied for 2 yr, 1974–75, prior to acidification with 'electrolyte grade' sulfuric acid in 1976–1980. More detailed information on initial conditions, acid additions and subsequent analyses is provided by Schindler (1980a, b) and Schindler et al. (1980). After depletion of the lake's alkalinity in 1976, the objective has been to decrease the pH of the epilimnion by approximately 0.25 pH units per year. The pH in 1980 averaged 5.37, down from an initial average of 6.6. Values for other years are given by Schindler (1980b).

Lake 114, a non-stratified lake, was also acidified with sulfuric acid. The additions were made once a month in 1979 and 1980, and were equivalent to adjusting the natural pH of rain on the lake's surface downward by one pH unit with H_2SO_4 (Schindler 1980a). This lake is one of the most sensitive in the ELA area, with an initial alkalinity of 10 to 30 $\mu eq\, l^{-1}$ and a natural pH of 5.8 to 6.2.

3. Results

3.1. EARLY BIOLOGICAL CHANGES IN LAKE 223

A number of important biological changes occurred at pH values of 5.8 to 6.0. Most notable were the disappearance of the benthic crustacean *Mysis relicta*, which had been an important item in the diet of lake trout, and of the minnow *Pimephales promelas* (Nero 1981 and K. Mills, unpublished data).

The sensitivity of *Pimephales promelas* was a complete surprise. The ease with which this species can be kept in the laboratory suggested that it might be very hardy. Recent embryological work with the effects of acid on the species agree with the sensitivity which we have seen in Lake 223 (McCormick *et al.*, 1980 and pers. comm.). *Salvelinus namaycush* also exhibited a high incidence of embryological deformity and abnormality at pH values of 5.8 to 6.0 (Kennedy, 1980), but through 1980 population structure has been stable and total year-class failures have not been detected in either lake trout, or the white sucker *Catostomus commersoni* (K. Mills, unpublished data).

Earlier physiological studies of the crayfish *Orconectes virilis* predicted that the physiological stresses imposed by acidification would inhibit carapace hardening after moulting when pH values decreased below 5.6 (Malley, 1980). This has occurred as predicted. Furthermore, egg mortality and infestation with the protozoan parasites *Thelohania* sp. have increased. In 1980 and 1981, females with only a few eggs, or with armorphous masses of degenerating eggs, have been observed (R. France, unpublished data). Recruitment of young of the year crayfish and population numbers diminished in 1980, and demise of the population seems imminent (I. Davies, pers. comm.).

Also at about pH 5.6, mats of filamentous algae, chiefly of the genus *Mougeotea*, invaded the littoral zone of the lake. These results are similar to those observed in Scandinavia, and to the results in experimentally acidified microcosms (Müller, 1980). Likewise, as pH decreased, there was a continous decrease in Chrysophyceae, with a corresponding increase in Chlorophyceae. Overall, a substantial increase in algal biomass and chlorophyll occurred (Findlay and Saesura, 1980). The increase was most pronounced in the hypolimnion, presumably due to the increased clarity of the lake, as discussed below.

Biological effects, and the approximate pH at which they were detected, are summarized in Table I. Other information is given by Schindler (1980b).

TABLE I

Biological changes observed early in the acidification of Lake 223 with sulfuric acid and the approximate pH at which they occurred

pH		
below 6.5	Increased sulfate reduction	Schindler *et al.*, 1980.
	Increased abundance of Chlorophyta,	Findlay and Saesura, 1980.
	decreased abundance of Chrysophyta	Müller, 1980.
5.8–6.0	Disappearance of *Mysis relicta*	Nero, 1981.
	Disappearance of *Pimephales promelas*	K. Mills, unpubl. data.
	Increased embryonic mortality in	Kennedy, 1980.
	Salvelinus namaycush	
5.6	Reduced calcification of *Orconectes*	
	virilis exoskeleton begins	Malley, 1980.
	Increased infestations of *Orconectes*	
	virilis with Thelohania sp.	R. France, unpubl. data.
	Epidemics of *Mougeotea* in littoral zone	
5.3–5.4	Decline in *Orconectes* populations	I. Davies, unpubl. data.

3.2. PHYSICAL CHANGES IN LAKE 223

Scandinavian investigators have observed that acid lakes have more transparent waters (Grahn *et al.*, 1974). It has been suggested that this should allow more rapid heating of waters deep in the lake, causing thermoclines to deepen (Canadian NRC, 1981), although the only data published are for an acidified lake before and after liming. This hypothesis has been explored using the 7 yr of data available for Lake 223.

Extinction coefficients indicated that Lake 223 has become more transparent after acidification, although results were by no means clearcut. In 1976, the first year of acidification, when pH did not decrease, and in 1978, a very cool year, extinction coefficients were significantly higher than 1977, 1979 and 1980 (Figure 1). As a result of increased light penetration, there has also been a substantial increase in the rate of summer warming in waters just beneath the thermocline (Figure 2). Also as a result of the increase in penetration of radiant energy, the thermocline of the lake has deepened more rapidly (Figure 3). The effect on the lake's total heat budget has been insignificant, however, because the latter is dominated by the interaction of the large mass of epilimnetic water and prevailing air temperature. A comparison of thermocline depth in Lake 223 with a nearby control lake, No. 239, in the same years leads to similar conclusions (Malley *et al.* in press).

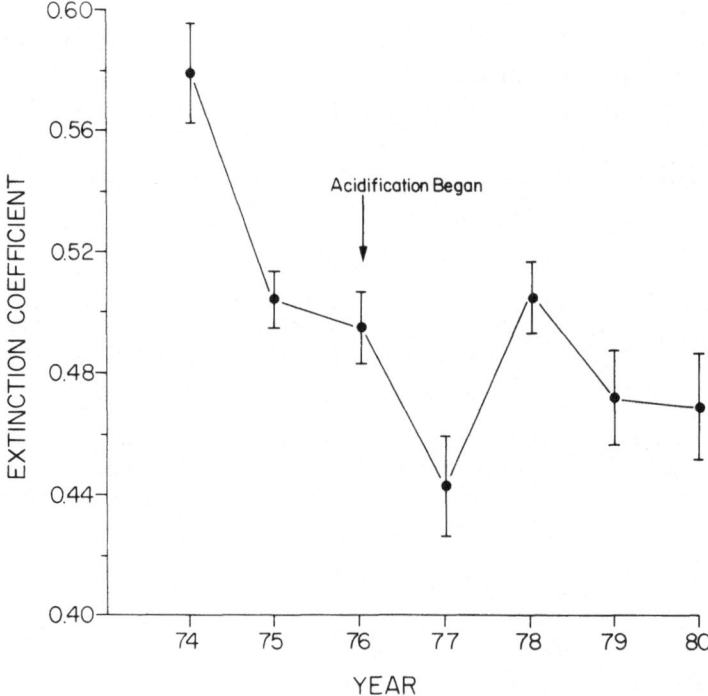

Fig. 1. Mean annual extinction coefficients for Lake 223. Bars represent one standard error either side of the mean, where se equals the standard deviation–n, where n is the number of dates on which measurements were made (11 dates each year during the ice-free season). On each date, the measurement is an average of extinction at 12 to 14 points, usually between the surface and a depth of 10 to 12 m (for example, Shearer and DeClercq, 1980). At $P = 0.05$, 1974 > 75, 76, 78 > 79, 80 > 77.

TABLE II

Substances for which an *increase* ($P < 0.2$) in lake 'mass' was observed from 1976 through 1980. Degrees of freedom ranged from 45 to 54.

	Estimated % change from 1976 to 1980 ($\pm$ SE)	P	r
Nutrient elements none observed to increase			
Other substances			
H^+	+ 1090 $\pm$ 230[a]	< 0.001	0.844
Na	+ 26 $\pm$ 13	< 0.001	0.500
Mn	+ 980 $\pm$ 139	< 0.001	0.900
Al	+ 155 $\pm$ 147[b]	< 0.05	0.315
Zn	+ 550 $\pm$ 299[b]	< 0.001	0.499
SO_4^{2-}	+ 141 $\pm$ 20	< 0.001	0.899
conductivity	+ 37 $\pm$ 7	< 0.001	0.855
chlorophyll	+ 63 $\pm$ 60	< 0.05	0.304

[a] Expressed as % of preacidification (H^+).
[b] Estimated changes for Al and Zn are for 1976 through 1979.

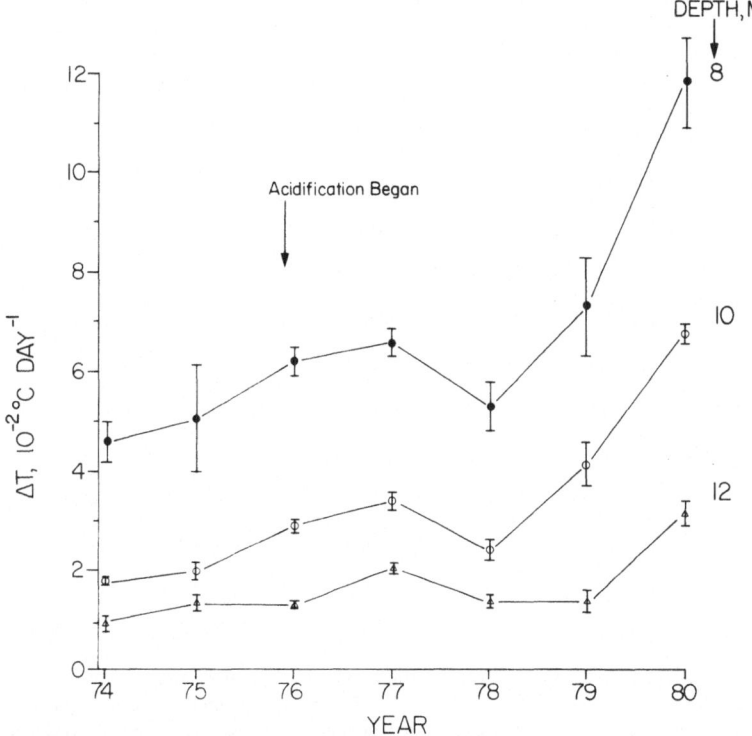

Fig. 2. The average rate of change of temperature (ΔT) at four subthermocline depths in Lake 223, before and during acidification. The period of calculation for each depth and year began with the first sampling after the onset of thermal stratification in spring, and ended when the thermocline deepened to within one meter of the depth considered. Analysis of covariance of ΔT vs time reveals that the probability that increased heating is due to chance alone is < 0.05 for 8 m; < 0.025 for 10 m; and $\simeq 0.07$ for 12 m.

TABLE III

Substances for which a *decrease* ($P < 0.2$) in lake mass was observed from 1976 to 1980. Degrees of freedom ranged from 47 to 54.

	Estimated % change from 1976 to 1980 ($\pm$ SE)	P	r
Nutrient elements			
DIC	$- 57 \pm 29$	< 0.001	$- 0.506$
Suspended C	$- 18 \pm 27$	< 0.2	$- 0.195$
Total C	$- 18 \pm 14$	< 0.01	0.357
TDN	$- 13 \pm 19$	< 0.2	$- 0.193$
Total N	$- 13 \pm 14$	< 0.05	$- 0.281$
Other substances			
Fe	$- 60 \pm 75$	< 0.1	$- 0.215$
Cl$^-$	$- 66 \pm 34$	< 0.001	$- 0.504$

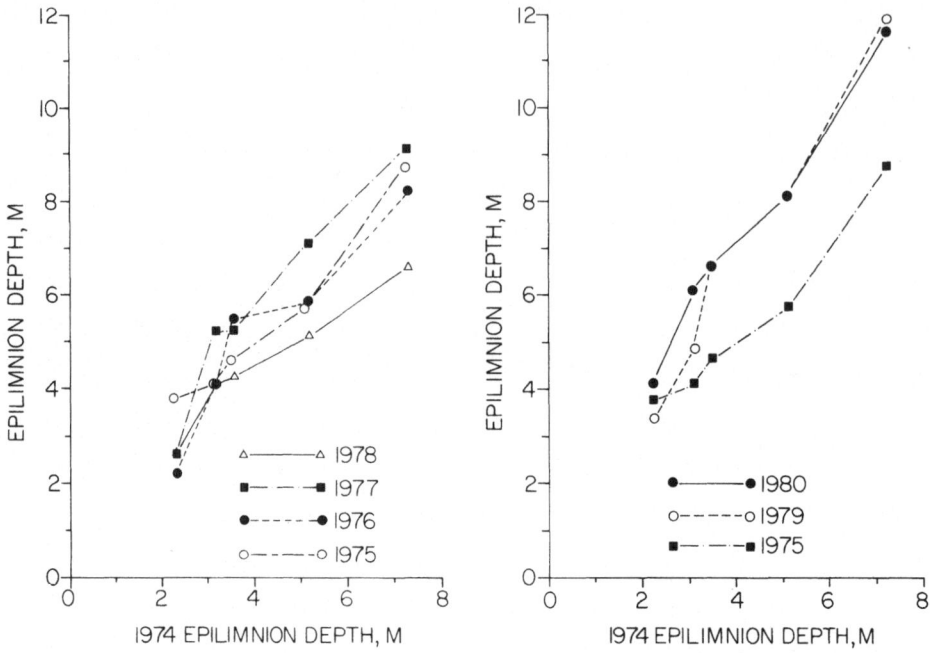

Fig. 3. A test of the hypothesis that the rate of thermocline deepening (ΔZ_e) increases as the result of acid-induced increases in the transprency of Lake 223. The rate of deepening in the control year 1974 is compared to a second control year, 1975, and to 5 yr while the lake was progressively acidified. The probability that a slope different from 1.0 is due to chance is as follows: '74 vs '75 $P \simeq 0.5$; '74 vs '76 $P \simeq 0.5$; '74 vs '77 $P \simeq 0.4$; '74 vs '78 $P \simeq 0.07$; '74 vs '79 $P < 0.025$; '74 vs '80 $P < 0.05$.

3.3. CHEMICAL CHANGES IN LAKE 223

Linear regression analysis revealed that only a few chemical analyses showed any significant trends over 5 yr in acidification. Some of these are obvious ones – for example, increases in H^+, conductivity and SO_4^{2-} and decreases in dissolved inorganic carbon (DIC) (Table II and III). Other results were no surprise – for instance, the increase in Mn and Al (Table III) and the lack of change in many parameters (Table IV). More surprisingly, chlorophyll and Na increased very significantly, chloride and total N decreased significantly, and TDN, suspended C and Fe exhibited decreases of 'border-line' significance. The increase in chlorophyll appears to be largely due to a hypolimnetic bloom which has developed in response to the increased water clarity (Schindler, 1980b). Other changes cannot be so readily explained at present.

3.4. SULFUR BUDGETS

Precise S budgets for lakes in polluted areas are difficult to calculate, due to the poor precision with which dry deposition can be estimated. Because dry deposition at ELA is low, and most of the Lake 223's S income is from acid additions, we can perform such calculations with somewhat more precision, allowing some insights into the S cycle.

TABLE IV

Substances for which *no change* $(P > 0.2)$ in lake mass was observed from 1976 to 1980

Nutrient elements
DOC–C

NO_3–N
NH_3–N

Suspended N

Suspended *P*
TDP
Total *P*

Si (soluble reactive)

Other substances
Ca
Mg
Fe (total dissolved)
K
Cd[a,b]
Cu[b]

H_2S–S

Oxygen
Color (abs. @ 425)

---

[a] Cd concentrations were frequently below detection limits.
[b] Period of observation from 1976 to 1979.

As described earlier (Schindler *et al.*, 1980), the reduction of sulfate to sulfide, followed by permanent sedimentation, causes a gain in alkalinity which partially counteracts the acidification of the lake. The amount of alkalinity generated by sulfate reduction is a function of several factors, including thermocline depth, concentrations of O_2 and sulfate, and light penetration. A deeper thermocline causes a decreased hypolimnion volume. As a result, less sulfate is entrapped for reduction in the hypolimnion.

The concentrations of O_2 and sulfate depend on the completeness of the overturn. An incomplete overturn leaves little O_2 in the hypolimnion so that anoxic processes, including sulfate reduction, begin quickly once the lake has restratified. On the other hand, hypolimnetic sulfate has usually been depleted since the last overturn, at least at depths greater than 10 m, and this is not effectively replenished if circulation is incomplete. If this is the case, sulfate reduction will be substrate limited during the following stratified period.

More complete overturns, which result from prolonged cool weather during periods when thermal stratification is absent, occur infrequently in small ELA lakes and at other north continental locations. When complete overturn occurs, high concentrations of O_2 and sulfate are present in the hypolimnion at the onset of stratification. As a result, sulfate reduction begins later in the summer, because it does not begin until anoxia develops. Once initiated, however, all of the sulfate below 10 m is usually reduced before the onset of fall turnover, so that the total amount of alkalinity generated tends to be greater following more complete circulation. Likewise, in wet years when water renewal is faster, more of the added sulfate is washed out of the outflow, causing the proportion of input which is sedimented to be lower in wet years.

As mentioned earlier, greater light penetration has also caused greater photosynthesis in the upper part of the Lake 223 hypolimnion. By generating O_2, this reaction may delay or prevent development of the anoxic conditions necessary for sulfate reduction.

In anoxic hypolimnions, co-diffusion from sediments of bicarbonate with ferrous iron, and reaction of ammonium ion with carbonic acid consume additional H^+. Unfortunately, the alkalinity generated by such mechanisms is only of a seasonal nature, i.e. Fe^{2+} and HCO_3^- re-sediment when overturn renders the hypolimnion oxic once again (Cook, 1981).

Altogether, these internal processes combine to make prediction of the amount of sulfate reduced in any year highly unpredictable (Table V). The problem is further complicated by the fact that ΔSO_4^{2-}, the change in concentration or mass of sulfate during a period of time, is likely to be very small relative to the precision of sulfate analyses. The total error in ΔSO_4^{2-} in a given year is estimated to be about 25%. It can be stated, however, with some certainty that about 1/3 of the sulfate added to the lake during the past 5 yr has been sedimented. Only a small proportion of this was organic S; most was lost as FeS (Cook, 1981).

TABLE V

Annual S budget for Lake 223. All values are in 10^3 moles.

	1976	1977	1978	1979	1980
Natural Inputs					
Bulk precipitation	2.6	1.7	2.0	2.0	2.0[a]
Direct runoff	8.9	12.8	16.6	10.9	11.8
Streamflow	0.8	3.8	6.5	4.6	0.8
Total (I_{nat})	12.3	18.3	23.1	17.5	14.6
Acid Addition	101.8	53.5	62.0	51.7	56.9
Total Input (I)	114.1	71.8	87.1	69.2	71.5
Outflow (O)	8.4	23.3	43.5	37.9	9.5
Change in Mass (ΔM)	78.3	24.0	20.8	41.7	−7.3[a]
Sedimentation (S)	27.4	24.5	22.8	−10.4	69.3
S/I, %	24	34	26	−15	97	$\bar{x}$ = 33

[a] Approximate values. Neither January 1981 lake chemistry nor precipitation values for 1980 are available.

Analysis of the Lake 114 data reveals that sulfate reduction and sedimentation take place even though the lake does not have an anoxic hypolimnion. In contrast to Lake 223, sediments in Lake 114 are highly organic, and methanogenesis occurs within a few centimeters beneath the sediment surface, even though sediments are overlain by oxic water (C. Kelly, pers. comm.). While calculated rates of sulfate reduction per unit area of sediment are lower than in the hypolimnion of Lake 223, presumably because reduction is controlled by the rate at which sulfate diffuses into sediments, the large area of epilimnion sediments renders them an important sink for S in the lake. Although highly variable, overall, sedimentation of S in Lake 114 is almost as important a component of the S budget as in Lake 223 (Table VI). The high variability in sulfate reduction is not

TABLE VI

Annual S budget for Lake 114. All values are in 10^3 moles.

	1978	1979	1980	Total 1978–1980
Natural Inputs				
Precipitation	0.88	0.88	0.89	2.65
Direct runoff	5.62	3.66	3.98	13.26
Total (I_{nat})	6.50	4.54	4.87	1591
Acid addition	0	2.52	4.41	6.93
Total Input (I)	6.50	7.06	9.28	22.84
Outflow (O)	4.74	4.04	1.62	10.40
Change in Mass (ΔM)				
January to January	−0.69	6.18	−1.20	4.29
Sedimentation (S)	2.45	−3.16	8.86	8.15
S/I, %	38	−45	95	36

totally explicable at present, and is probably a complex function of the availability of sulfate and organic matter, temperature, and oxygen. It is certain that the high variance is not due to errors in budget calculations, because inputs are known exactly, and throughout much of the ice-free season of 1980, there was no outflow from either lake.

4. Discussion

A number of organisms disappear quickly once pH values decrease to below 6.0. In this pH range, toxic metals are present only in very low concentrations. In 1977, Al concentrations averaged less than 20 $\mu g \, l^{-1}$, and Zn less than 3 $\mu g \, l^{-1}$. Other metals were below limits of detection. As a result, the observed biological effects are attributable either to increased H^+ or the corresponding decrease in bicarbonate.

Benthic Crustacea as a group appear to be extremely sensitive to acidification. *Mysis* in our study disappeared at pH values very similar to those which Scandinavian studies have shown to limit the distribution of Lepidurus, Gammarus and Asellus (Almer *et al.*,

1978; Økland and Økland, 1980). Crayfish appear to be only slightly less sensitive than the above organisms, as indicated by our results and by Scandinavian distribution studies (Almer *et al.*, 1978; Abrahamsson, 1972). The effects of pH on the amphipod *Pontoporeia affinis* has not been studied, but its taxonomy and habits indicate that it may be as sensitive as *Mysis* or *Gammarus*. A number of molluscs appear to be equally sensitive (Økland and Økland, 1980).

The disappearance or absence of *Mysis*, or perhaps of *Pontoporeia*, may be an important early indicator of acid stress in North American lakes. Both are glacial relicts, whose natural distribution in North America is reasonably predictable from glacial patterns and ecological considerations (Pennak, 1953; Dadswell, 1974). Both are important items in the food chains of northern lakes (Rawson, 1961; Johnson, 1975; Juday and Birge, 1927; Larkin, 1948; Parker, 1980) and their disappearance would be expected to exert some stress on carnivorous fish populations.

Among fishes, the minnow *Pimephales promelas* appeared to be as sensitive as benthic crustaceans. This species is ubiquitous in northern lakes, forming an important part of aquatic food chains, and providing part of the basis for a multi-million dollar bait fishery (Beamish *et al.*, 1976). Furthermore, the species normally flourish in small shallow lakes of the sort likely to be most vulnerable to acidification.

In should be noted that the physical parameters significantly affected by acid-induced increases in transparency are the *rate of hypolimnion heating* and the *rate of thermocline deepening*, rather than the hypolimnion temperature and the thermocline depth *per se*. The latter two parameters are functions not only of increased light penetration, but also of the hypolimnion temperature and thermocline depth at the onset of summer stratification, which are dependent on spring weather as well as on lake transparency.

Sulfate reduction is not the only process which can affect the rate of lake acidification. Denitrification, nitrification, reduction of nitrate to ammonium, and biological uptake of nitrate or ammonium followed by permanent sedimentation will generate or consume H^+ depending on whether electrons are consumed or yielded. These will vary widely from lake to lake, depending on the trophic status and amount and ionic change of N input (Kelly *et al.*, in press). The possible reactions which can consume or produce alkalinity are set out in some detail by the Canadian NRC (1981).

Because the terrestrial watersheds of Lake 223 and 114 were not acidified, the observed chemical changes indicate changes in sediment-water or atmosphere-water exchange. The analyses presented here must be considered to be only a first, rough attempt, which can be refined in the future to increase the sensitivity with which significant changes may be detected. For example, a very high proportion of the variance in many parameters is of a highly predictable, cyclic nature, due to biological metabolism or oxidation-reduction phenomena (Figure 4). Such variance should be reducible by simple techniques such as Fourier transformation. Techniques of this sort would also be of considerable utility in determining significant long-term trends from monitoring data.

Despite the fact that sulfate inputs are large and precisely known in our studies, mass-balance techniques have revealed only the gross features of the S cycle. In non-

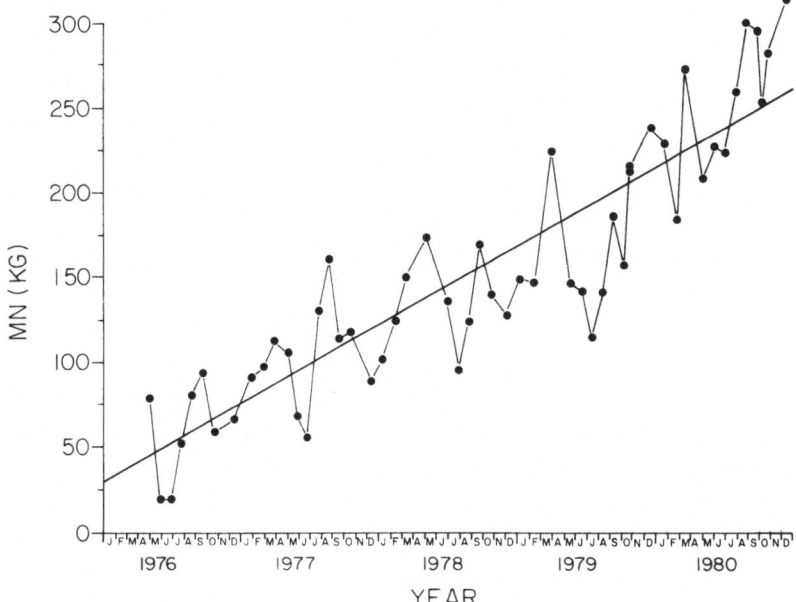

Fig. 4.   The total mass of Mn in the water column of Lake 223, demonstrating the increase in mass due to acidification, and the highly predictable, cyclic changes in mass due to seasonal changes in redox conditions. The probability that a linear regression yields a slope significantly different from zero due to chance is < 0.001.

experimental studies where dry deposition is an important component of input, S mass balances will be too imprecise to allow calculation of sedimentation rates as we have done. Whereas our limitation is chiefly one of analytical precision, which will be improved from 1981 onward by acquisition of a liquid chromatograph, measurement of dry deposition presents a more complicated problem, which cannot be offset by the most precise analyses of S in 'wet' components of the budget. As Horntvedt et al. (1980) have revealed, different moisture conditions or receptor species alone may cause dry deposition to vary by an order of magnitude or more, so that calculated S inputs must be very imprecise.

We predict that a more useful approach in non-experimental calibrated watersheds will be to measure sedimentation of S directly by comprehensive coring, chemical analysis and dating programs, then to combine these with measurement of change in mass (ΔM), outflow (O) and the 'wet' proportion of input (I) to back-calculate dry deposition. Sedimentation measured in this way is certainly less accurate than as calculated in our experimental systems, but we believe that it should be accurate enough to provide a considerable improvement in estimates of (I) over those done from models based on deposition velocity and atmospheric concentration.

Investigators may be deluded by the fact that mass-balance techniques have been informative in studies of whole-lake nutrient cycling (Schindler et al., 1973; Vollenweider, 1976). However, in lakes where full-scale mass balances have been successfully used to interpret whole-ecosystem nutrient balances, (I) has been known rather precisely, because direct input of fertilizer or sewage is a major component. Furthermore, sedimentation (S) of P or N is typically a high proportion of the difference between (I) and (O), and in eutrophic systems P concentrations can be measured with a precision of 2 to 5%. None of these advantages apply to similar budgets for S. Hence, there appears to be a major difference in the way in which 'experimental' and 'observational' 'calibrated watersheds' can be employed to the best avantage in the elucidation of biogeochemical questions about the effect of acid precipitation on the S cycle. The latter are unlikely to yield S balance sheets precise enough to elucidate even the gross features of S dynamics in lakes. Instead, they may provide a means of improving our knowledge of atmosphere-biosphere exchanges.

Acknowledgments

I. Davies and K. Mills allowed inclusion of their unpublished findings for this summary. P. Campbell, D. Malley, and R. Hesslein made helpful criticisms of the manuscript.

References

Abrahemsson, S.: 1972, *Rep. Inst. Freshwater Res. Drottningholm*, **52**, 23.

Almer, B., Dickson, W., Ekström, C., and Hörnström, E.: 1978, 'Sulfur Pollution and the Aquatic Ecosystem', in J. Nriagu (ed.), *Sulfur in the Environment*, part 2, John Wiley and Sons, New York, Chapter 7, 464 pp.

Beamish, R. J., Blouw, L. M., and McFarlane, G. A.: 1976, 'A Fish and Chemical Study of 109 Lakes in the Experimental Lakes Area (ELA), Northwestern Ontario, With Appended Reports on Lake Whitefish Ageing Errors and the Northwestern Ontario Baitfish Industry', *Can. Fish. Mar. Serv. Tech. Rep.* **607**, 116 pp.

Canadian National Research Council: 1981, *Acidification in the Canadian Aquatic Environment*, NRCC No. 18475, 369 pp.

Cook, R. B.: 1981, 'The Biogeochemistry of Sulfur in Two Small Lakes', Ph.D. Thesis, Columbia University, New York, 354 pp.

Dadswell, M. J.: 1974, *Distribution Ecology and Postglacial Dispersal of Certain Crustaceans and Fishes in Eastern North America*, Publications in Zoology, National Museum of Natural Sciences (Ottawa), **11**, 110 pp.

Findlay, D. L. and Saesura, G.: 1980, *Effects on Phytoplankton Biomass, Succession and Composition in Lake 223 as a Result of Lowering pH Levels from 7.0 to 5.6. Data from 1974 to 1979*, Can. MS Rep. No. 1585, Fisheries and Aquatic Sciences, Western Region, Dept. of Fisheries and Oceans, Winnipeg, 16 pp.

Horntvedt, R., Dollard, G. J., and Joranga, E.: 1980, 'Effects of Acid Precipitation on Soil and Forest. 2. Atmosphere-Vegetation Interactions', in Drabløs, D. and Tollan, A. (eds), *Ecological Impact of Acid Precipitation*, SNSF Project, Oslo, Norway, p. 192–193.

Grahn, O., Hultberg, H., and Landner, L.: 1974, *Ambio*. **3**, 93.

Johnson, J.: 1975, *J. Fish. Res. Board Can.* **32**, 1981.

Juday, C. and Birge, E. A.: 1927, *Ecology* **8**, 445.

Kelly, C. A., Rudd, J. W. M., Cook, R. B., and Schindler, D. W.: (in press), *Limnol. Oceanogr.*

Kennedy, L. A.: 1980, *Can. J. Fish. Aquat. Sci.* **37**, 2355.

Larkin, P. A.: 1948, *Bull. Fish. Res. Board Can.* **78**, 3.

Malley, D. F.: 1980, *Can. J. Fish. Aquat. Sci.* **37**, 364.

Malley, D. F., Findlay, C. L., and Chang, P. S. S.: (in press), *Ecological Effects of Acid precipitation on Zooplankton*, Proc. Conf. Acid Precip. Mich. State University, Apr. 1981, Mich. State Univ. Press.

McCormick, J. H., Stokes, G. N., and Portele, G. J.: 1980, 'Environmental Acidification Impact Detection by Examination of Mature Fish Ovaries', in Klaverkamp, J. H., Leonhard, S. L., and Marshall, K. E. (eds.), *Proceedings of the 6th Annual Aquatic Toxicity Workshop*, Can. Tech. Rep. No 975, Fisheries and Aquatic Science, Western Region, Dept. Fisheries and Oceans, Winnipeg, p. 41.

Müller, P.: 1980, *Can. J. Fish. Aquat. Sci.* **37**, 355.

Nero, R. W.: 1981, 'The Decline of *Mysis relicta* Loven in response to Experimental Acidification of a Whole Lake', M.Sc. Thesis, University of Manitoba, Winnipeg.

Økland, J. and Økland K. A.: 1980, 'pH Level and Food Organisms for Fish: Studies of 1,000 Lakes in Norway', in Drabløs, D. and Tollan, A. (eds.), *Ecological Impact of Acid Precipitation*, SNSF Project, Oslo, Norway, p. 326–327.

Parker, J. I.: 1980, *J. Great Lakes Res.* **6**, 164.

Pennak, R. W.: 1953, *Fresh-water Invertebrates of the United States*, Ronald Press Co., New York, 769 pp.

Rawson, D. S.: 1961, *J. Fish. Res. Board Can.* **18**, 423.

Schindler, D. W.: 1980a, 'Ecological Effects of Experimental Whole-Lake Acidification', in Shriner, D., Richmond, C., and Lindberg, C. (eds.) *Atmospheric Sulfur Deposition*, Ann Arbor, Mich., Science, p. 453–464.

Schindler, D. W.: 1980b, 'Experimental Acidification of a Whole Lake: A Test of the Oligotrophication Hypothesis', in Drabløs, D. and Tollan, A. (eds.) *Ecological Impact of Acid Precipitation*, SNSF Project, Oslo, Norway, p. 370–374.

Schindler, D. W., Hesslein, R. H., Wagemann, R., and Broecker, W. S.: 1980, *Can. J. Fish. Aquat. Sci.* **37**, 373.

Schindler, D. W., Kling, H., Schmidt, R. W., Prokopowich, J., Frost, V. E., Reid, R. A., and Capel, M.: 1973, *J. Fish. Res. Board Can.* **30**, 1414.

Shearer, J. A. and DeClercq, D. R.: 1980, *Light Extinction in the Experimental Lakes Area – 1979 Data*, Can. Data Report of Fisheries and Aquatic Sciences, No. 189, Western Region, Dept. of Fisheries and Oceans, Winnipeg, 63 pp.

Vollenweider, R. A.: 1976, *Schweiz. Zeits. Hydrol.* **38**, 29.

EFFECTS OF EXPERIMENTAL ACIDIFICATION ON MACROINVERTEBRATE DRIFT DIVERSITY IN A MOUNTAIN STREAM

RONALD J. HALL*, J. MICHAEL PRATT, and GENE E. LIKENS

Section of Ecology and Systematics, Division of Biological Sciences, Cornell University, Ithaca, New York 1485, U.S.A.

(Received July 10, 1981; Revised December 3, 1981)

Abstract. A small stream (Norris Brook) within the Hubbard Brook Experimental Forest was acidified to determine what effect elevated H^+ stress may have on the ecology of a mountain stream. The experiment was designed to simulate a pH level (4.0) that can occur during initial snowmelt (acute period) and during longer term (chronic period) acidification. Daily macroinvertebrate drift samples were collected from treatment and reference areas of Norris Brook. Drift diversity at the generic level was calculated using Brillouin's formula and partitioned hierarchically following macroinvertebrate classifications based on taxonomy (orders) and feeding strategies (functional groups or guilds).

The rate of movement of individuals and genera was significantly greater for those organisms leaving the acid-stressed area during the first five days than for those entering, whereas no difference between the rate of macroinvertebrates entering or leaving the acid-stressed area was apparent for either numbers or genera over the remaining 25-day study period. For the acute period (first five days), the increased macroinvertebrate drift leaving the acidified area was significantly more diverse at the levels of aquatic insect orders and functional groups but less diverse at the generic level than the drift entering. For the chronic period (25-day period) no significant differences were detected in either major taxa, functional group (with the exception of collectors), or generic diversity between the drift entering and leaving the treatment reach. Mayflies and probably chironomids leaving the acid-stressed area during the acute period were generically more diverse than those entering, whereas stoneflies drifting out of the acidified reach were generically less diverse than those drifting in. The overall change in the normal pattern of spatial and temporal variation in drift rate and diversity provides quantitative evidence that H^+ stress significantly altered the structure and function of the macrobenthic community.

1. Introduction

The deposition of acidic atmospheric pollutants has been linked to the gradual acidification of an increasing number of streams and lakes in southern Scandanavia (Oden, 1976; Braekke, 1976), in the Canadian Shield (Beamish, 1976; Conroy et al., 1976), particularly in the Sudbury area, and in northeastern United States (Likens, 1976; Schofield, 1976). Four well-documented observations, in particular, argue persuasively for this causal link. First, airborne emissions of SO_x and NO_x released during the combustion of fossil fuels have increased markedly in amount from the mid-1940's to the present (Likens et al., 1979). Secondly, the H^+ concentration and geographical incidence of acid precipitation also have increased during this same period (Cogbill and Likens, 1974; Likens and Butler, 1981). Thirdly, sulfuric and nitric acids are the dominant acids in precipitation (Galloway et al., 1976). Lastly, rain and snow deposit

---

* Present address: Ministry of the Environment, Limnology and Taxonomy Section, Water Resources Branch, Rexdale, P.O. 213, Ontario, Canada M9W 5L1

Water, Air, and Soil Pollution **18** (1982) 273–287. 0049–6979/82/0183–0273$02.25.

these acids preferentially in areas (e.g. S. E. Canada and N. E. U.S.A.) downwind from significant sources of air pollution (Likens *et al.*, 1979). Numerous investigators have linked acidification of lakes and streams to acidic wet (rain and snow) and dry deposition as the major factor contributing to ecological changes of poorly buffered aquatic ecosystems (Almer *et al.*, 1974; Grahn *et al.*, 1974; Beamish *et al.*, 1975; Hendrey *et al.*, 1977; Dickson, 1978).

Regions most affected by acid deposition are located at higher elevations in northern latitudes where soils underlain by granitic bedrock are poorly developed (e.g. Braekke,

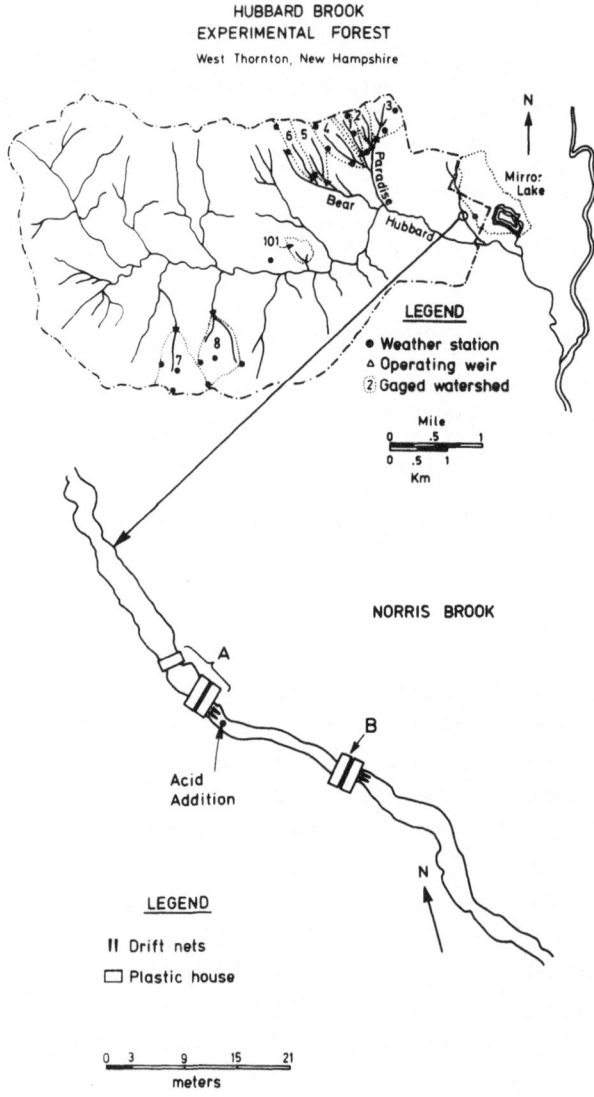

Fig. 1.   Norris Brook study area located within the Hubbard Brook Experimental Forest. Reference area (site A) was located 1 m above acid addition point. Treatment area was located 15 m below acid dripping point.

1976) and snow accumulates during winter. The initial meltwater from the acid-contaminated snowpack can be very acidic (Schofield, 1976; Leivestad and Muniz, 1976; Jonannessen and Henriksen, 1978; Overrein et al., 1980). As a result streams and lakes often have pH values that are particularly low (e.g. pH 4.0) during initial snowmelt (Leivestad and Muniz, 1976). This sudden increase in H^+ in spring can deleteriously affect fish (Leivestad and Muniz, 1976) and invertebrates (Hall et al., 1980; Hall and Likens, 1980a, b; Hall and Likens, 1981) in just a few days.

This paper discusses the response of macrobenthic invertebrates to elevated H^+ stress as reflected by changes in the rate (numbers/day) at which these organisms move downstream (drift) and by corresponding changes in the diversity of such drift. Monitoring macroinvertebrate drift as a behavioral response to increased H^+ stress can provide quantitative ecological data that helps elucidate what potential impact acid precipitation may have on the nutrient flux (Hall and Likens, 1980b; Hall and Likens, 1981) and community structure of poorly buffered lotic ecosystems at high elevations (Hall et al., 1980; Pratt and Hall, 1981). Diversity as a measure of community structure is a frequently used index for describing and gauging biological effects of environmental stress in freshwater ecosystems (Wilhm, 1967, 1972; Wilhm and Dorris, 1960; Cairns and Dickson, 1971; Cairns et al., 1971).

2. Study Area

The characteristics of the Norris Brook watershed are described in detail elsewhere (Hall et al., 1980; Figure 1). The watershed is 87.2 ha in area and is south-facing. The stream flows through a deciduous forest underlain by granitic bedrock. The pH of Norris Brook stream water in the study area (third-order reach, Strahler, 1957) ranged from 5.4 to 6.4. Mean width and depth in the study area were 1.8 and 0.2 m, respectively. The substrate varied in composition from fine sand to boulders 1 m in diameter. During the six-month study period (March – September 1977) water temperature ranged from 0.0 to 4.5 °C in March to 11 to 20 °C in July. Electrical conductivity varied from 21 to 30.5 µS cm^{-1}, and dissolved oxygen was near saturation throughout the study period.

3. Methods and Materials

3.1. ACIDIFICATION EXPERIMENT

The pH of the stream water in the treatment section was maintained near 4.0 (relative to about 6.4 in the reference area) by continuously adding dilute concentrations of sulfuric acid (0.05 to 1N) for approximately five months (18 April – 22 September, 1977). Stream pH was measured approximately every 5 min during spates and acid additions adjusted to maintain the treatment section at pH 4; during constant discharge pH readings were taken every 6 to 8 hr. Stream water pH was measured routinely 12 m below and 5 m above the acid addition point. Only the drift data collected 1 m above (site A) and 15 m below point of acid addition (Figure 1) will be presented in this report, although there were additional drift sampling stations (described elsewhere in detail; Hall *et al.*, 1980) further downstream.

3.2. INVERTEBRATE DRIFT

Drift nets were positioned upstream (site A) and downstream (site B) of the acid addition to sample macroinvertebrates moving into and out of the acidified reach. Two nets with 253 μm mesh Nitex netting filtered the stream water at both sites (A and B, Figure 1). These nets, constructed of a polyvinyl chloride frame (15 × 30 cm), were anchored in mid-channel of each sampling location by wooden dowels driven to in the substrate (Waters, 1962). Drift samples for the first week of acid addition were collected at 24-hr intervals to determine differences in sample size and the macrobiota moving into and out of the acidified area. Thereafter, drift samples were collected for 24-hr periods every week until 31 May. We report here results for the first 30 days (18 April – 31 May) of continuous acid addition divided into the first 5 days (acute response) and the subsequent 25 days (chronic response) of elevated stream water acidity. The initial 5 days of H^+ stress simulated an abrupt change in pH regime that can occur in small mountain streams during initial snowmelt episodes (Hall *et al.*, 1980; Hall and Likens, 1981); the subsequent 25-day period may simulate long-term acidification (Hall and Likens, 1981) at pH levels (4.0) commonly found in ambient rain in the northeastern U. S. (Hall *et al.*, 1980; Likens *et al.*, 1979).

Organisms collected in the drift samples were identified to genera except for individuals within the family Chironomidae (midges), which were identified to subfamily (e.g., Tanypodinae, Orthocladinae, and Chironominae; Merritt and Cummins, 1979).

The two drift samples collected daily per station were combined to obtain a larger sample size for determining drift diversity above (site A) and below (site B) the H^+ stress. The larger sample from combining the two nets provides a more reliable and accurate indication of what genera and how many organisms drifted in 24 hr. Generic diversity was calculated for ten samples – five samples for the acute period (5 days) and five samples for the chronic period (25 days) from each site (A and B). A total of 7850 individuals were examined.

3.3. DIVERSITY CALCULATIONS

Brillouin's formula (1) defines the total diversity (B) of a fully sorted and enumerated collection of N organisms distributed among g groups (e.g., orders, genera, species) in the amounts $N_1, N_2, \ldots, N_g$ (Pielou, 1975):

$$B = \log \frac{N!}{N_1! N_2! \ldots N_g!} \tag{1}$$

Dividing total diversity (B) by the number of individuals (N) in the collection yields average diversity per individual (H), which permits comparison of collections of different sizes. The resulting equation or diversity index can be rewritten in a form that simplifies computation when used in conjunction with a programmable calculator and a table of $\log_{10} N!$ values (Lloyd *et al.*, 1968):

$$H = \frac{c}{N} \left(\log_{10} N! - \sum_{i=1}^{g} \log_{10} N_i! \right) \tag{2}$$

where c is the scale factor for converting logarithms from base 10 to whatever base is chosen; base 2 is commonly used, $c = 3.3219$. Generic diversity computed with the Brillouin diversity index reflects the average degree of uncertainty in correctly predicting which of g genera an organism belongs to if it is randomly selected from a fully sorted and counted collection of N individuals. Generic diversity, therefore, is a function of both number (richness) and equitability of abundance (evenness) of the genera comprising a collection. The frequency distribution of organisms among the genera in a sample can range between two extremes – (1) minimum evenness when all genera but one are each represented by a single organism with the remaining organisms belonging to just one genus and (2) maximum evenness when the distribution of organisms among genera is as equitable as possible. Similarly, the generic diversity of a sample consisting of N individuals and g genera also can fall between minimum and maximum values (i.e., H min and H max) depending upon the distribution of organisms among genera. Thus, the ratio of the actual diversity (H) to the maximum diversity (H max) of a sample served as a measure of the degree of evenness (J) N organisms were distributed among g genera:

$$J = H/H_{max} \tag{3}$$

where $H_{max} = (c/N) \{ \log_{10} N! - g \cdot (\log_{10} k!) - r \cdot \log_{10} (k + 1) \}$, k equals the integer part of N/g, and $r = N - (g \cdot k)$ (Hamilton, 1975).

Most diversity indices conveniently summarize large amounts of data. However, these indices provide information that is ecologically vague when characterizing and comparing samples of organisms from different or even the same biotic communities. The limitations of the ecological reality of diversity indices arise primarily from a failure to reflect qualitative or biological differences among organisms. A diversity value such as H provides, for example, little information about the taxonomic affinities and trophic relationships of the organisms in a sample corresponding to the N_i's substituted into Equation (2). The Brillouin diversity index, however, can be partitioned to reflect specific biological and ecological properties of samples when diversity in Equation (2) is divided into hierarchical components according to some meaningful classification scheme. For example, it is possible to express in a systematic way the degree to which the diversity of a sample is determined by different taxonomic or functional groups. Therefore, valuable biological information can be taken into account that otherwise would have been lost if the census data (i.e., numbers/taxon) of different samples were merely condensed mathematically into a set of diversity values.

Generic diversities of drift samples were divided into two catergories of components that represent different ways of classifying macroinvertebrates above the generic level (Table I):

  (1) taxonomic (e.g., mayflies, stoneflies, caddisflies) and
  (2) functional according to feeding methods (e.g., collectors, predators, shredders; Merritt and Cummins, 1979).

TABLE I

Schematic diagram of the partitioning of generic diversity according to classifications based on major macroinvertebrate taxa (i.e., insect orders) and functional groups reflecting modes of feeding (e.g., collectors, predators, and shredders).

Generalized scheme for partitioning generic diversity	Partitioning generic diversity according to taxonomic and functional classifications	

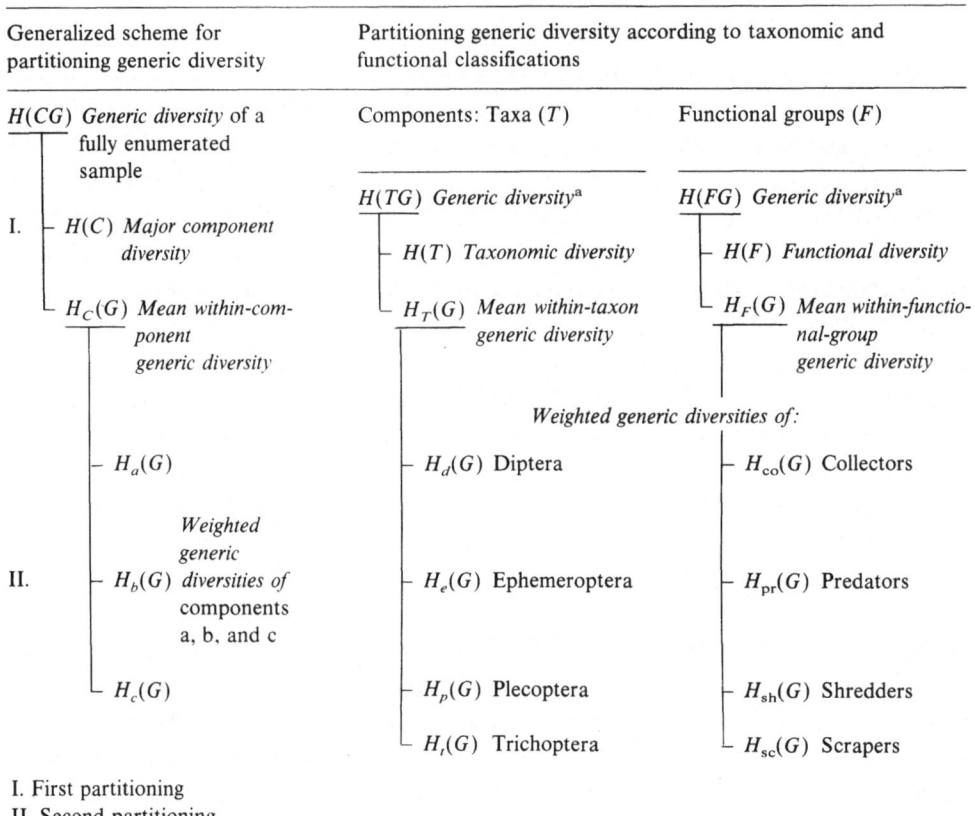

I. First partitioning
II. Second partitioning

[a] $H(TG)$ and $H(FG)$ are always equal for a particular sample with the different notation merely indicating whether the generic diversity of a sample is partitioned taxonomically (T) or functionally (F). However, the various products of partitioning diversity, such as $H(T)$ and $H(F)$, and $H_d(G)$ and $H_{co}(G)$, may or may not be equal in magnitude.

We summarize in Equation (4), then, the first hierarchical partitioning of generic diversity:

$$H(GC) = H(C) + H_C(G) \qquad (4)$$

where $H(GC)$ is the generic diversity of a sample, $H(C)$ is the diversity among components or simply the component diversity of a sample (e.g., diversity at the level of major macroinvertebrate taxa such as insect orders), and $H_C(G)$ is the mean within-component generic diversity averaged over all components (e.g., all stream insect orders in a sample). We partition $H_C(G)$ in Equation (5) into the generic diversities of the individual components (e.g., taxa or functional groups) comprising a certain classification:

$$H_C(G) = \sum_{j=1}^{c} (N_j/N) \cdot H_j(G)' \tag{5}$$

where $H_j(G)'$ is the generic diversity of the jth component (e.g., Plecoptera) weighted in proportion to its contribution (N_j) to sample size (N)* (Pielou, 1967, 1975). $H(CG)$ and H are quantitatively equivalent in that they both equal the generic diversity of a sample as calculated by Equation (3). The notation $H(CG)$, however, reflects the hierarchical nature of the Brillouin's diversity index and that the generic diversity of a sample can be partitioned systematically as described by Equations (4) and (5).

Statistical comparisons were made on pooled daily drift data. The drift collections from sites A and B were related since they were collected simultaneously. Thus comparisons were made using a student's t-test on the mean differences between A and B.

4. Results

4.1. DRIFT RATE

Macroinvertebrates drifted into the acidified reach (site A) at a mean rate of 270/day ($n = 5$) and drifted out (site B) at 1300/day ($n = 5$) during the acute period (first 5 days, Table II). During the subsequent 25 days (chronic period), the mean daily numbers of individuals entering and leaving the treatment area were approximately equal (site A – 380, site B – 370, $N = 5$). The net rate of flux ($A - B$) of individuals from the experimental area was -378% and significant (student's t, one-tailed, $p < 0.05$) during the first five days, but the small net rate of gain for the experimental reach during the 25-day period was not significant ($p > 0.5$).

4.2. GENERIC RICHNESS OF DRIFT

The mean daily number of macroinvertebrate genera (24.2) moving into the experimental (site A) area was less than the total number of genera (33.8) leaving the treatment area during the acidification for the acute period (Table II). The mean drift rate of genera entering (22.4) and leaving (23.4) the acid stressed area during the chronic stress period was essentially the same. The net flux of genera during the 5-day H^+ increase was highly significant ($p < 0.005$, -40%), while a net flux of only one genus occurred during the chronic stress period.

4.3. GENERIC DIVERSITY OF DRIFT

The mean daily generic diversities of the drift at sites A and B during the acute and chronic periods were comparable (Table II). The net flux in generic diversity of drift samples was positive during the first five days and negative during the subsequent 25 days. These net fluxes, however, were not statistically significant.

* $H_j(G) = (N_j/N)H_j(G)' = (N_j/N) (\log N_j! - \sum_{i=1}^{gj} \log_{10} N_{ij}!)$.

TABLE II

Mean daily number of individuals, number of genera (generic richness), generic diversity (H), and evenness (J), and net flux ($A - B$) of aquatic insects drifting during 5-day (acute, 18–22 April) and 25-day (chronic, 23 April – 17 May) periods of increased acidification. Negative flux values equal a loss from the experimental reach ($A < B$) and positive values equal a gain to the experimental reach ($A > B$).

	Mean daily drift ± SE ($N = 5$)					
	Acute response			Chronic responce		
	A	B	Net flux (loss or gain)	A	B	Net flux (loss or gain)
Number of individuals	270 ± 36	1300 ± 390	−1030 ± 370[a]	380 ± 103	370 ± 109	10 ± 45.9
Number of genera	24.2 ± 1.83	33.8 ± 1.91	−9.6 ± 1.83[c]	22.4 ± 2.96	23.4 ± 3.7	−1.0 ± 2.17
Diversity (H)	3.27 ± 0.088	3.24 ± 0.037	0.03 ± 0.086	2.32 ± 0.236	2.53 ± 0.210	−0.21 ± 0.112
Evenness (J)	0.749 ± 0.0156	0.655 ± 0.0117	0.094 ± 0.0211[b]	0.55 ± 0.050	0.61 ± 0.063	−0.06 ± 0.036

[a] $p < 0.025$
[b] $p < 0.01$
[c] $p < 0.005$

4.4. Evenness of Drift

Macroinvertebrate drift entering the acidified reach at site A for the first five days of lowered pH was on the average more evenly distributed among genera ($J = 0.75$) than the drift leaving at site B ($J = 0.66$) (Table II). During the subsequent 25-day period of acid addition, however, mean evenness of drift at A was slightly less than at B. The percent difference $[(A - B)A^{-1} \cdot 100\%]$ in mean evenness of drift at A and B was significant ($12.6\%, p < 0.01$) for the first five days of elevated H^+ concentration but insignificant ($10.9\%, p < 0.10$) over the following 25 days of acidification.

4.5 Generic Diversity partitioned taxonomically and functionally

We diagram schematically the partitioning of generic diversity according to classifications based on insect orders and functional groups reflecting modes of feeding (e.g., collectors, predators, and shredders) in Table I. The generic diversities of individual components such as mayflies or scrapers are weighted in proportion to their relative abundance in a sample.

The macrofauna moving into the experimental area was on the average 11% more diverse taxonomically (i.e., $H(T)$ = diversity at the level of orders) than that moving out during the acute acidification period [$H(T)$ at A and B = 1.94 and 1.73, respectively] (Table III). This difference in mean $H(T)$, however, was not significant ($p > 0.4$). Over the next 25 days, the chronic acidification period, mean $H(T)$ was 7% less at site A than at site B, a difference that was also insignificant ($p < 0.5$).

TABLE III

Mean diversity of daily drift and net flux ($A - B$) of aquatic insects during 5-day (acute) and 25-day (chronic) periods of increased acidification. Taxonomic diversity at the level of insect orders is represented by $H(T)$, and mean generic diversity within taxa by $H_T(G)$. The weighted generic diversities[a] of Diptera [$H_d(G)$], Ephemeroptera [$H_e(G)$], Plecoptera [$H_p(G)$], and Trichoptera [$H_t(G)$] also are included. A net flux in generic drift diversity or in one of its various taxonomic components is positive when the macrofauna entering the acidified reach is more diverse than that leaving ($A > B$) and negative when it is less diverse ($A < B$).

Mean diversity of daily drift $\pm$ SE ($N = 5$)						
Acute response				Chronic response		
A	B	Net flux (loss or gain)		A	B	Net flux (loss or gain)
$H(T)$ 1.94 $\pm$ 0.036	1.73 $\pm$ 0.126	0.21 $\pm$ 0.144		1.30 $\pm$ 0.097	1.39 $\pm$ 0.429	-0.09 $\pm$ 0.026
$H_T(G)$ 1.33 $\pm$ 0.053	1.51 $\pm$ 0.13	-0.18 $\pm$ 0.118		1.01 $\pm$ 0.146	0.98 $\pm$ 0.121	0.03 $\pm$ 0.122
$H_d(G)$ 0.60 $\pm$ 0.0275	0.51 $\pm$ 0.039	0.09 $\pm$ 0.031[b]		0.67 $\pm$ 0.092	0.64 $\pm$ 0.098	0.03 $\pm$ 0.102
$H_e(G)$ 0.40 $\pm$ 0.064	0.75 $\pm$ 0.183	-0.35 $\pm$ 0.168		0.16 $\pm$ 0.044	0.14 $\pm$ 0.039	0.02 $\pm$ 0.0280
$H_p(G)$ 0.232 $\pm$ 0.0135	0.138 $\pm$ 0.0208	0.094 $\pm$ 0.0128[c]		0.128 $\pm$ 0.0292	0.176 $\pm$ 0.0193	-0.048 $\pm$ 0.0101[c]
$H_t(G)$ 0.13 $\pm$ 0.047	0.068 $\pm$ 0.0101	0.06 $\pm$ 0.048		0.028 $\pm$ 0.0159	0.050 $\pm$ 0.0216	-0.022 $\pm$ 0.0120

[a] Weighted generic diversity of the ith component (e.g. Ephemeroptera) = $H_i(G) = H_i(G)' \cdot (N_i/N)$, where $H_i(G)'$ is the unweighted or absolute value of the generic diversity of the ith component comprising a sample.
[b] $p < 0.025$.
[c] $p < 0.005$.

The drift entering the acidified reach at A during the acute and chronic periods was 14 and 3% less diverse generically within taxa [(mean $H_T(G)$] than was the drift leaving at B (Table III). The difference in mean $H_T(G)$ was not significant for either the acute ($p > 0.2$) or chronic ($p > 0.5$) acidification periods.

Dividing $H_T(G)$ into the weighted generic diversities of the dominant insect orders found (Table III) shows that Ephemeroptera (mayflies) moving out of the experimental area at B were more diverse generically [$H_e(G) = 0.75$] than those moving in at A[$H_e(G) = 0.40$] during the acute period, whereas Diptera ($H_d(G)$, true flies), Plecoptera ($H_p(G)$, stoneflies), and Trichoptera ($H_t(G)$, caddisflies) entering the experimental reach were each more diverse generically than the corresponding taxa leaving. Because the diversity values for Coleoptera (beetles) at site A and B are left out of this analysis ($H_c(G) = 0.03$, and $H_c(G) = 0.04$, respectively), the $H_j(G)$ values (i.e., $H_d(G) + H_e(G)$, etc.) in Table III do not sum to $H_T(G)$.

Net flux ($A - B$) in weighted generic diversity [$H_j(G)$] during the acute period was significantly positive ($A > B$, a net gain) for true flies ($p < 0.025$) and stoneflies ($p < 0.005$), nonsignificantly positive for caddisflies ($p > 0.1$), and almost significantly negative ($A < B$, a net loss) for mayflies ($p < 0.1$) (Table III). Over the chronic period of stream acidification, net flux in weighted generic diversity [$H_j(G)$] for the same four orders was insignificantly positive for true flies ($p > 0.25$) and mayflies ($p > 0.25$), insignificantly negative for caddisflies ($p > 0.1$), and significantly negative for stoneflies ($p < 0.005$).

TABLE IV

Mean diversity of daily drift and net flux ($A - B$) of aquatic insects during 5-day (acute) and 25-day (chronic) periods of increased acidification. Functional diversity at the level of major feeding groups is represented by $H(F)$, mean generic diversity within functional groups by $H_F(G)$. The weighted generic diversities of collectors [$H_{co}(G)$], predators [$H_{pr}(G)$], shredders [$H_{sh}(G)$], and scrapers [$H_{sc}(G)$] also are included. A net flux in generic drift diversity or in one of its various functional components is positive when the macrofauna entering the acidified reach is more diverse than that leaving ($A > B$) and negative when it is less diverse ($A < B$).

	Mean diversity of daily drift ± SE ($N = 5$)					
	Acute response			Chronic response		
	A	B	Net flux (loss or gain)	A	B	Net flux (loss or gain)
$H(F)$	2.03 ± 0.030	1.67 ± 0.096	0.36 ± 0.087[c]	1.48 ± 0.14	1.53 ± 0.121	−0.050 ± 0.104
$H_F(G)$	1.24 ± 0.059	1.57 ± 0.093	−0.33 ± 0.1147[b]	0.83 ± 0.098	1.0 ± 0.10	−0.170 ± 0.0242[a]
$H_{co}(G)$	0.74 ± 0.037	1.10 ± 0.136	−0.36 ± 0.135[a]	0.55 ± 0.062	0.64 ± 0.074	−0.090 ± 0.0268[b]
$H_{pr}(G)$	0.134 ± 0.0112	0.174 ± 0.0297	−0.04 ± 0.035	0.086 ± 0.0147	0.13 ± 0.041	−0.044 ± 0.036
$H_{sh}(G)$	0.286 ± 0.0160	0.14 ± 0.033	0.15 ± 0.040[c]	0.164 ± 0.0183	0.192 ± 0.0299	−0.028 ± 0.0265
$H_{sc}(G)$	0.088 ± 0.0281	0.17 ± 0.044	−0.08 ± 0.0357[a]	0.038 ± 0.0193	0.032 ± 0.0120	0.006 ± 0.0143

[a] $p < 0.05$.
[b] $p < 0.025$.
[c] $p < 0.01$.
[d] $p < 0.005$.

Mean daily diversity among functional groups, or functional diversity $[H(F)]$, was greater for the drift entering the acid-stressed reach (site A, $H(F) = 2.03$) than that leaving (site B, $H(F) = 1.67$) for the first five days (Table IV). In contrast, mean functional diversity was greater for drift moving out than moving in for the chronic period. The net change in mean functional diversity of drift was significantly negative ($A < B$, $p < 0.01$) during the acute period but not during the chronic period ($p < 0.3$).

Mean generic diversity within functional groups $[H_F(G)]$ was significantly greater on a daily average for drift leaving the acid-stressed reached than entering during both the acute $[(A - B)A^{-1} \times 100\% = 27\%, p < 0.025]$ and chronic periods (20% $p < 0.005$).

Dissecting $H_F(G)$ into its subcomponents (Table IV provides ecological insight into the functional make-up of the macroinvertebrate community that drifted in response to elevated acidity. The mean daily generic diversities of collectors $[H_{co}(G)]$, predators $[H_{pr}(G)]$, and scrapers $[H_{sc}(G)]$, were, respectively, 49, 30, and 93% greater at site B than at site A for the acute period, while the generic diversity of shredders $H_{sh}(G)$ was 104% greater at site A than B for the same time period (Table IV). For the 25-day period collectors, predators, and shredders were 16, 51, and 17%, respectively, more diverse generically at site B than at site A. In contrast, scrapers in the drift were more diverse at site A than site B.

The net changes in the generic diversities of collectors, scrapers, and shredders drifting during the initial five days of acid addition were significantly negative $[H_{co}(G), H_{sc}(G),$ and $H_{sh}(G)$ greater at B than at A, p's < 0.05]. Only the generic diversity of collectors continued to be significantly greater ($p < 0.025$) at B than at A over the subsequent 25 days of stream acidification.

5. Discussion

The introduction of a strong acid into a soft-water mountain stream for five days altered the magnitude and the taxonomic and functional diversity of the macroinvertebrate drift. The rates at which individuals and genera drifted were significantly greater below the point of acid addition than above during the first five days (acute period) of stream acidification. An additional 25 days of acid addition, however, apparently had no significant effect on either the rate or the generic diversity, evenness, and richness of macroinvertebrate drift in Norris Brook. Therefore, the mass exodus of organisms from the bottom of the stream, as reflected by increased drift rate, in the acid-stressed area represented a sizeable depletion of acid-sensitive, macrobenthic populations within just the first few days. Apparently the more tolerant organisms drifted into the acid-stressed area (e.g. some Diptera, Plecoptera and Trichoptera, Table III) and remained there, while the more sensitive macroinvertebrates (e.g. mayflies, Table III) drifted through to avoid the elevated acidity stress.

Genera that were collected either infrequently or not at all above the acid stress at site A were responsible for the greater generic richness (number of genera) of the macrofauna leaving the experimental area (site B) compared to that entering for the acute period. Many invertebrate genera not prone to drift, therefore, entered the water-column and moved downstream to avoid the acid stress. Although the number of genera drifting at B was greater than at A, the generic diversities of macroinvertebrates leaving and entering the acidified area were not significantly different for the first five-days of stream acidification (Table II). This insensitivity of generic diversity to increased generic richness is due to the mathematical properties of the Brillouin diversity index, which condenses the number of genera in a sample with their frequencies of occurrence into a single number. Although richer generically, the drift leaving the acidified reach was dominated to a greater degree by a few genera (e.g. *Epeorus*, *Ephemerella*, *Prosimulium* and chirononid genera within the subfamily Orthocladiinae) than was the drift entering. Consequently, the expected increase in generic drift diversity that would have been attributable to increased generic richness was counterbalanced by a corresponding decrease in evenness. (J). The net effect was no statistical difference in diversity between reference (fewer genera but greater evenness) and treatment drift (more genera but lower evenness) for teh acute acidification period. Thus, the effect of H^+ stress on drift in Norris Brook was not conveyed by calculating and comparing only the generic diversities of drift samples from above and below a point of acid addition. A better interpretation of the effects of elevated acidity may be gained by analyzing diversity qualitatively as well as quantitatively with respect to phylogenetic (e.g., insect orders: mayflies, stoneflies, etc.) and functional classifications (e.g., feeding guilds: collectors, predators, etc.; Merritt and Cummins 1979, Table I).

Macroinvertebrates entering the acidified section daily were on average taxonomically and functionally more diverse than those leaving for the acute period (Tables III and IV). These differences in $H(T)$ and $H(F)$ are a consequence of one or two taxa and functional groups dominating to a greater degree the drift at site B relative to site A. These shifts in the relative abundance of certain taxa and feeding guilds, which decreased evenness (Table II), and equality in the taxonomic (Table III) and functional (Table IV) richness of macroinvertebrates (i.e., the same number of taxa and functional groups drifted into as drifted out of the experimental reach in a day) collected in the drift at sites A and B combined to decrease $H(T)$ and $H(F)$ at site B.

The mean generic diversity within taxa $[H_T(G)]$ and generic diversity within functional groups $[H_F(G)]$ were generally higher at site B relative to site A and tended to balance the decreases in $H(T)$ and $H(F)$ at site B during the acute period (Table III and IV. The increase in treatment $H_T(G)$ resulted from the large number and high diversity of mayflies leaving the acidified area. Also, the increase in $H_F(G)$ at site B occurred because of the increase in numbers and diversity of collectors, scrapers, and predators leaving the experimental area relative to those entering. The individuals within the family Chironomidae were not identified to the generic level and, thus, the diversity of Diptera [true flies, $H_d(G)$] at site B relative to site A may have been higher because large numbers of individual chironomids responded to elevated acidity by drifting downstream.

Thus, all of the diversity components outlined in Table I give a fuller description of the effect of increased H^+ stress on macroinvertebrates. The most useful measure, however, in terms of showing statistically significant changes in macrobenthic community structure and function were total numbers of individuals, generic richness, evenness (Table II), and the generic diversities of individual functional groups (Table IV).

A shift in macroinvertebrate generic drift diversity may have important consequences for the competitive interactions for space on the stream bottom. For example, the rapid exodus of mayflies (Ephemeroptera) and chironomids (Diptera) drifting from the experimental area relative to those entering, together with their corresponding increase in drift diversity at site B, suggests that stream acidification greatly reduced overall benthic diversity as well as the diversity of these two taxa. In contrast, stonefly diversity (Plecoptera) was greater for those entering than those leaving the experimental area, suggesting that their diversity on the stream bottom increased. With the removal of mayflies and chironomids and the subsequent increased availability of physical space on the substrate, reduced competitive pressure may have resulted such that stoneflies could increase in abundance for the first five days (acute period). However, given enough time for the stonefly densities to increase on the stream bottom, intra- and interspecific competitive interactions may have intensified and possibly caused more stoneflies to drift out of the experimental area, as was evident for the chronic stress period (Table III). However, the physiological differences among genera as the primary factor causing the change in distribution and abundance of organisms (Hall et al., 1980) within the experimental area relative to the upstream reference area can not be ruled out, while the competition between macroinvertebrate taxa in the study areas are probably of secondary importance.

Similarly, the acid-induced change in the functional group diversity of drifting macrofauna may alter, in turn, the biological and physical-chemical quality of organic matter. For example, the generic diversities of collectors and scrapers were significantly greater for those leaving the acidified area, while the diversity of shredder organisms was greater for those entering during the acute period. Collectors, and shredders play an important role in changing the particle size of allochthonous organic matter (leaves from riparian vegetation) while scrapers forage on periphytic algae and fungi growing on substrates within the stream (Cummins, 1974). The abandonment of the treatment area by macroinvertebrate collectors may result in the transport of organic particles to greater distances downstream, at least during the acute period of acidification (Hall and Likens, 1981). Similarly, the increase in shredders in the treatment area could cause a greater particle size reduction of organic matter and therefore a greater propensity of the smaller particles to be exported downstream from the experimental area. The reduction in numbers and diversity of scrapers feeding on periphyton could cause an increase in attached microflora. An increase in attached microalgae under prolonged acidic conditions (pH 4), with a concomittant decrease in grazers (scrapers), was observed in our study (Hall et al., 1980) and in artificially acidified lakes and streams in the Canadian Shield (Muller, 1980) and in Norway (Hendrey, 1976).

If our experimental results are extrapolated to headwater (first- through third-order), poorly buffered mountain streams, acidic snowmelt during high discharge in spring may significantly alter the biotic structure and function in these stream ecosystems. If the snowmelt proceeds from stream origin to some distance downstream (i.e., from first- to third-order streams; Leivestad and Muniz, 1976), the potential long-term impact of a short-term acid pulse lasting a few days (Schofield and Trojnar, 1980) may be of great ecological significance, particularly if upstream macroinvertebrate drift is the major mechanism for macrobenthic recovery in the acid-stressed area.

Summary

The rate of movement of individuals and genera was significantly greater for those organisms leaving the acid-stressed area during the first five days than for those entering, whereas no differences between the rate of macroinvertebrates entering or leaving the acid-stressed area was apparent for either numbers or genera over the remaining 25-day study period. The effect of H^+ stress was not apparent by calculating and comparing only the generic diversities of drift samples from above and below the experimental acidification. A clearer interpretation of the effects of elevated acidity was conveyed by analyzing diversity in relation to phylogenetic (e.g., insect orders) and functional classifications (e.g., feeding guilds). The most useful measures concerning the significant effects of stream acidification on macroinvertebrate community structure and function were changes in total numbers of individuals, generic richness, evenness and the generic diversities of individual functional groups.

Acknowledgments

This is a contribution to the Hubbard Brook Ecosystem Study. The National Science Foundation, the International Paper Company Foundation, and the Environmental Protection Agency, Duluth, provided financial support for the field study at Hubbard Brook. The Northeastern Forest Experiment Station, U.S. Department of Agriculture, Forest Service, Broomall, Pennsylvania, furnished the facilities. We thank B. Wesley and R. Haddock for technical assistance.

References

Almer, B., Dickson, W., Ekstrom, C., Hornstrom, E., and Miller, U.: 1974, *Ambio* **3**, 30.
Beamish, R. J.: 1976, *Water, Air, and Soil Pollut.* **6**, 501.
Beamish, R.J., Lockhart, W. L., van Loon, J. C., Harvey, H. H.: 1975, *Ambio* **4**, 98.
Braekke, F. W.: 1976, *Impact of Acid Preciptiation on Forest and Freshwater Ecosystems in Norway*, Research Report 6/76, SNSF Project, Oslo, Norway.
Cairns, J., Jr. and Dickson, K. L.: 1971, *J. Water Pollut. Contr. Fed.* **43**, 755.
Cairns, J., Jr., Crossman, J. S., Dickson, K. L., and Herricks, E. E.: 1971, *Biologists Bulletin* **18**, 79.
Cogbill, C. V. and Likens, G. E.: 1974, *Water Resour. Res.* **10**, 1133.
Conroy, N., Hawley, K., Keller, W., and Lafrance, C.: 1976, *Journ. of Great Lakes Research* **2** (Supplement 1), 146.
Cummins, K. W.: 1974, *Bioscience* **24**, 631.
Dickson, W.: 1978, *Verh. Internat. Verein. Limnol.* **20**, 851.
Galloway, J. N., Likens, G. E., and Edgerton, E. S.: 1976, 'Hydrogen Ion Speciation in the Acid Precipitation of the Northeastern United States', in: *Proceedings of the First International Symposium on Acid Precipitation and the Forest Ecosystem*, Forest Service General Technical Report NE–23, Northeastern Forest Experimental Station, Upper Derby, Pennsylvania, pp. 383–396.
Grahn, O., Hultberg, H., and Landner, L.: 1974, *Ambio* **3**, 93.
Hall, R. J., and Likens, G. E.: 1980a, 'Ecological Effects of Whole-stream Acidification', in: Shriner, D. S.,

C. R. Richmond, and S. E. Lindberg (eds.), *Atmospheric Sulfur Deposition: Environmental Impact and Health Effects*, Ann Arbor Science, Ann Arbor, Michigan. pp. 443–462.

Hall, R. J. and Likens, G. E.: 1980b, 'Ecological Effects of Experimental Acidification on a Stream Ecosystem', in: Drablos, D. and A. Tollan (eds.), *Proceedings of an International Conference on the Ecological Impact of Acid Precipitation*, SNSF Project, Oslo, Norway, p. 375–376.

Hall, R. J. and Likens, G. E.: 1981, Chemical Flux in an acid-stressed stream. *Nature* **292**, 329.

Hall, R. J., Likens, G. E., Fiance, S. B., and Hendrey, G. R.: 1980, *Ecology* **61**, 976.

Hamilton, M. A.: 1975, *J. Water Pollu. Contr. Fed.* **47**, 630.

Hendrey, G. W.: 1976, *Effects of pH on the Growth of Periphytic Algae in Artificial Stream Channels*, Research Report 25/76, SNSF Project, Oslo, Norway.

Hendrey, G. R., Baalsrud, K., Traaen, T., Laake, M., and Raddum, G.: 1977, *Ambio* **5**, 224.

Johannessen, M. and Henriksen, A.: 1978, *Water Resour. Res.* **14**, 615.

Leivestad, H. and Muniz, I. P.: 1976. *Nature* **259**, 391.

Likens, G. E.: 1976, *Chemical and Engineering News* **54**, 29.

Likens, G. E. and Butler, T. J.: 1981, *Atmos. Environ.* **15**, 1103.

Likens, G. E., Wright, R. F., Galloway, J., and Butler, T.: 1979, *Sci. Am.* **241**, 39.

Lloyd, M., Zar, J. H., and Karr J. R.: 1968, *The Am. Midl. Natur.* **79**, 257.

Merritt, R. C. and Cummins, K. W. (eds.): 1979, *An Introduction to the Aquatic Insect of North America.* Kendal/Hunt, Dubuque, Iowa.

Muller, P.: 1980, *Can. J. Fish. Aq. Aci.* **37**, 355.

Oden, S.: 1976, *Water, Air and Soil Pollut.* **6**, 137.

Overrein, L. N., Seip, H. M., and Tollan, A.: 1980, *Acid Precipitation – Effects on Forest and Fish*, Final report of the SNSF project 1972–1980 pp. 1–175.

Pielou, E. C.: 1967, 'The Use of Information Theory in the Study of the Diversity of Biological Populations', *Proc. 5th Berkeley Symp. Math., Stat., and Prob.* **4**: 163–177.

Pielou, E. C.: 1975, *Ecological Diversity*, John Wiley and Sons, New York.

Pratt, J. M. and Hall, R. J.: 1981, 'Acute Effects of Stream Acidification on the Diversity of Macroinvertebrate Drift', in: Singer, R. (ed.), *Effects of Acid Precipitation on Benthos*, Proceedings of a Regional Symposium on Benthic Biology, North American Benthological Society, Hamilton, New York.

Schofield, C. L.: 1976, *Ambio* **5**, 228.

Schofield, C. L. and Trojnar, J. R.: 1980, 'Aluminum Toxicity to Brook Trout in Acidified waters', in: *Proceedings of a Conference on Polluted Rain*, Plenum Publishing Corporation, Rochester, New York, pp. 341–366.

Strahler, A. N.: 1957, *Transactions of the American Geophysical Union* **38**, 913.

Waters, T. F.: 1962, *Ecology* **43**, 316.

Wilhm, J. L.: 1967 *J. Water Pollu. Contr. Fed.* **39**, 1673.

Wilhm, J. L.: 1972, *Annual Rev. Ent.* **17**, 223.

Wilhm, J. L. and Dorris, T. C.: 1968, *Bioscience* **18**, 477.

ALUMINUM TOXICITY TO FISH IN ACIDIC WATERS*

JOAN P. BAKER

School of Forestry and Environmental Studies, Duke University, Durham, NC 27706, U.S.A.

and

CARL L. SCHOFIELD

Department of Natural Resources, Cornell University, Ithaca, NY 14853, U.S.A.

(Received July 10, 1981; Revised December 10, 1981)

Abstract. An important consequence of acidification is the mobilization of Al from the edaphic to the aquatic environment. Elevated Al levels in acidic waters may be toxic to fish. Eggs, larvae, and postlarvae of white suckers (*Catostomus commersoni*) and brook trout (*Salvelinus fontinalis*) were exposed in laboratory bioassays to pH levels 4.2 to 5.6 and inorganic Al concentrations of 0 to 0.5 mg l^{-1}. Aluminum toxicity varied with both pH and life history stage. At low pH levels (4.2 to 4.8), the presence of Al (up 0.2 mg l^{-1} for white suckers; 0.5 mg l^{-1} for brook trout) was beneficial to egg survival through the eyed stage. In contrast, Al concentrations of 0.1 mg l^{-1} (for white suckers) or 0.2 mg l^{-1} (for brook trout) and greater resulted in measurable reductions in survival and growth of larvae and postlarvae at all pH levels (4.2 to 5.6). Aluminum was most toxic in over-saturated solutions at pH levels 5.2 to 5.4. The simultaneous increase in Al concentration with elevated acidity must be considered to accurately assess the potential effect of acidification of surface waters on survival of fish populations.

1. Introduction

Acidification of surface waters is a major problem in areas of the northeastern United States (Schofield, 1976; Davis *et al.*, 1978), eastern Canada (Beamish and Harvey, 1972; Dillon *et al.*, 1978), and southern Scandinavia (Oden, 1968; Almer *et al.*, 1974; Gjessing *et al.*, 1976). An important consequence of acidification is the mobilization of Al from the edaphic to the aquatic environment (Cronin and Schofield, 1979; Driscoll *et al.*, 1980). Elevated metal concentrations (e.g., Al, Zn, Mn) are often associated with low pH levels in lakes and streams (Schofield and Trojnar, 1980; Driscoll *et al.*, 1980; Dickson, 1975; Wright and Gjessing, 1976; Johnson, 1979; Hutchinson *et al.*, 1978). An inverse correlation between Al concentration and pH level has been identified for surface waters in the Adirondack Region of New York State, southern Norway, the west coast of Sweden, and Scotland (Wright *et al.*, 1980). Inorganic Al concentrations of 0 to 0.6 mg l^{-1} were measured in Adirondack lakes and streams with pH levels ranging from 7.2 to 4.2 (Driscoll *et al.*, 1980). These elevated Al levels in acidic waters may be toxic to aquatic biota (Schofield and Trojnar, 1980; Dickson, 1978).

Field and laboratory studies in the Adirondack Mountains, New York State and in southern Norway have demonstrated that inorganic Al concentrations in acidified waters

---

* Based in part on a thesis submitted by the senior author to the Graduate School of Cornell University in partial fulfillment of requirements for the degree of Doctor of Philosophy.

are potentially toxic to fish. Schofield and Trojnar (1980) analyzed survival of brook trout (*Salvelinus fontinalis*) stocked into 53 Adirondack lakes as a function of 12 water quality parameters. Levels of pH and concentrations of Ca and Mg were significantly higher and concentrations of total Al were significantly lower in lakes with trout survival than in lakes in which few or no trout survived. Covariate analysis indicated that the concentration of total Al was the primary chemical factor controlling fish stocking success. In laboratory bioassays at pH levels 4.4 to 5.2 with Al levels 0 to 0.5 mg l^{-1}, survival of brook trout postlarvae (fry after initiation of feeding) decreased at Al concentrations above 0.2 mg l^{-1} (Schofield and Trojnar, 1980; Cronan and Schofield, 1979). Some minor gill damage and reduced fish growth occurred with 0.1 mg Al l^{-1}. Mortality appeared to result from severe necrosis of the gill epithelium. Muniz and Leivestad (1980) also report significant fish mortality at Al concentrations 0.2 mg l^{-1} and above in laboratory bioassays. In experiments with fingerlings of brown trout (*Salmo trutta*) at pH levels 4.0 to 6.0 with Al levels up to 0.9 mg l^{-1}, maximum Al toxicity occurred at pH 5.0. The toxic mode appeared to be a combined effect of impaired ion exchange and respiratory stress caused by gill damage and mucous clogging of gills.

Aluminum occurs in acidic surface waters primarily as the free Al ion and complexed with hydroxide, fluoride, and organic ligands. Driscoll *et al.* (1980) concluded that only inorganic forms of Al were toxic to postlarvae of brook trout and white sucker (*Catostomus commersoni*) in laboratory bioassays with synthetic acidic Al solutions and natural Adirondack surface waters.

Decreased recruitment of young fish has been cited as the primary factor leading to the gradual extinction of fish populations in acidic lakes (Jensen and Snekvik, 1972; Leivestad *et al.*, 1976; Beamish and Harvey, 1972; Wright and Snekvik, 1978). Field observations indicated changes in population structure over time. Declining fish communities consist primarily of older and larger fish with a decrease in total population density. Recruitment failure may result from inhibition of adult fish spawning and/or increased mortality of eggs and larvae. Leivestad *et al.* (1976) concluded that egg and larval mortality is the main cause for fish reproductive failure in acidic Norwegian lakes.

The objective of our investigation was to evaluate one aspect of acidification effects on fish populations: the sensitivity of early life history stages of fish to acidic conditions with respect to elevated concentrations of H$^+$ and Al. Eggs, larvae (from hatching to resorption of the yolk sac), and postlarvae (after initiation of feeding) of two common Adirondack lake fish species – brook trout (*S. fontinalis*, Mitchill) and white sucker (*C. commersoni*, Lacepede) were exposed in laboratory bioassays to pH levels 4.2 to 5.6 and Al concentrations 0 to 0.5 mg l^{-1}. These chemical conditions cover the range of pH and inorganic Al levels typically measured in acidic Adirondack lakes and streams (Driscoll *et al.*, 1980). The purpose was two-fold: (1) determine the relative sensitivities of early life history stages of fish to acidic conditions; and, (2) quantify the relative importance and interactions of H$^+$ and inorganic Al as toxic agents to fish eggs, larvae, and postlarvae.

2. Experimental Procedures

Laboratory bioassays were conducted in 1978 and 1979 at the Cornell University Fisheries Research Laboratory, Ithaca, New York. The water supply for all experiments was softened, dechlorinated Ithaca City tap water. Tap water was softened with a Culligan Aqua-Cleer reverse osmosis system and dechlorinated by a Barnstead Organic Carbon Removal Column. Aliquots of dechlorinated tap water were remixed with the softened water to adjust Ca levels to around $2 \, \text{mg} \, l^{-1}$. Concentrations of ions in the experimental water and in Adirondack surface waters were similar with the exception of slightly higher Na and lower SO_4 concentrations in the experimental water (Table I).

TABLE I

Mean concentrations $(\text{mg} \, l^{-1})$ ($\pm$ standard error) of selected ions in softened, dechlorinated water utilized in all laboratory experiments and in Adirondack lakes and streams sampled during 1977–1978

Ion	Water supply for experiments	Adirondack waters[a]
Ca	2.2 $\pm$ 0.1	2.0 $\pm$ 0.6
Mg	0.5 $\pm$ 0.02	0.4 $\pm$ 0.1
Na	4.5 $\pm$ 0.4	0.7 $\pm$ 0.3
K	0.3 $\pm$ 0.1	0.4 $\pm$ 0.1
free F	0.019 $\pm$ 0.005	0.025 $\pm$ 0.060
SO_4	< 2.0 $\pm$ 0.5	6.2 $\pm$ 0.9

[a] Data from Driscoll (1980).

Stock solutions of treatment waters were prepared in 100 l polyethylene containers. Waters were acidified with concentrated hydrochloric acid (12 N, reagent grade). Aluminum was added as reagent grade anhydrous $AlCl_3$. Initial stock solutions at appropriate pH and Al levels were aged at least three days prior to the start of each experiment. Thereafter, stock solutions were prepared and aerated at least overnight, and pH levels readjusted prior to use.

Fish larvae (20 to 30 per experimental unit) were held in 5 l polyethylene containers set in a recirculating water bath. Eggs were held in a nylon mesh basket suspended in polyethylene units. Each basket contained 50 eggs. Water temperatures were controlled at an average of 9.5 °C (range 7 to 13 °C) for experiments with brook trout; 17.9 °C (range 16 to 21 °C) for experiments with white suckers. Photoperiod was not controlled. All containers were aerated continuously. Fresh treatment water was introduced into experimental units via a gravity-feed head flow system (Baker, 1981). Five liters of fresh water were added daily to each unit.

The pH level in each container was measured and adjusted (if necessary) once per day; Al level measured one to two times per week. For the most part, pH levels were maintained within ± 0.1 units of nominal pH levels. Measured total Al concentrations were generally within $0.03 \, \text{mg} \, l^{-1}$ of levels of Al added except in treatments in which levels of Al added were over-saturated with regard to formation of $Al(OH)_{3(s)}$ (see Stumm and Morgan, 1970 or Driscoll, 1980). In these treatments, portions of the Al added precipitated out of solution presumably as $Al(OH)_{3(s)}$, and measured concentrations of total Al in the water were consistently less than total quantities of Al added.

Fish utilized in bioassays were obtained from Adirondack fish stocks. Eggs were collected from adult brook trout from Cranberry Lake on October 27 to 28, 1978, and from adult white suckers from North Lake on June 8 and 12, 1979. Eggs and larvae were reared in softwaters (pH > 6) until utilized in laboratory experiments. For bioassays with fish eggs, eggs were stripped in the laboratory from mature fish, dry fertilized, and water hardened in treatment waters. Postlarvae, after swimming-up, were fed a combination diet of brine shrimp nauplii (*Artemia*) and commercially prepared dry trout diets. Postlarvae were fed dry food in excess once in the early morning and *Artemia* three to four times during the remainder of the day. Units were siphoned clean immediately after each feeding with dry food and again at the end of the day. In experiments with larvae, units were siphoned clean only once per day; with eggs, at least every other day.

Experiments with larvae and postlarvae were conducted for 13 to 14 days. Bioassays with eggs were maintained through hatching of embryos and until initiation of feeding by fry (postlarval stage) in all but extreme treatments (low pH and/or high levels of Al). Fish in treatments experiencing high levels of stress were allowed twice the amount of time necessary for larval development and initiation of feeding compared to control treatments. After this period of time, the experiment was terminated and numbers of postlarvae swimming and feeding vs larvae still inactive and resting on the bottom were enumerated for each experimental unit. Dead eggs and larvae were removed from units one or two times per day. Fish were weighed and measured at the end of each experiment.

Total and monomeric concentrations of Al were measured by the ferron-orthophenanthroline techniqiue as described by Baker (1981) and Driscoll (1980). Concentrations of aqueous inorganic Al species (free Al ion and Al complexed with hydroxide and fluoride ligands) were calculated assuming thermodynamic equilibrium and based on measured levels of pH and free fluoride. Sulfate concentrations in treatment waters (See Table I) were below limits of detection and were not included in equilibrium calculations.

Fish survival was assumed to be a logistic function of H^+ and Al levels in treatment waters. The logit transformations of percent of larvae and postlarvae surviving to the end of the experiment and percent of embryos surviving through the developmental stage identified by the appearance of eye pigmentation, through hatching, and through initiation of feeding (swim-up) were analyzed as a function of pH level and Al level and form. Independent variables in least squares regression analyses were measured average concentrations of H^+, total Al, and monomeric Al, estimated concentrations of the free Al ion and Al-hydroxide complexes, and levels of Al added in preparation of treatment waters. Polynomial terms and interactions between these parameters were also included in regression analyses. The logit transformation is undefined at 0 and 100%. Thus, in experimental units with total mortality of larvae or postlarvae, the logit transformation of P (proportion surviving) was approximated based on an estimate of mean survival time and a maximum likelihood iterative solution technique for the logistic function. In all other cases, values of $1/(2n)$ and $1 - [1/(2n)]$ were utilized in place of observations $P = 0$ and $P = 1$, respectively, as suggested by Ashton (1972) and Robson (1980).

Beginning with the full regression model, parameters were eliminated if the sequential sum of squares associated with the independent variable, when fit last in the model sequence, resulted in an insignificant ($p > 0.05$) reduction in the residual mean square. Based on this criterion, the full regression model with all independent variables and their interactions was reduced to yield the simplest model which adequately described variations in proportion of fish surviving. Isopleth lines of percent survival in Figures 1 to 8 are based on these models. For each level of Al added and pH, expected concentrations of total Al, monomeric Al, free Al ion and Al-OH complexes are derived from data in Tables II to VII.

3. Results

Fish response to elevated Al levels varied markedly at different pH levels and for different fish life history stages and species. As a result, simple generalizations concerning expected effects of increasing Al concentrations with acidification are not possible. A detailed examination of each separate experiment is necessary. Several trends are, however, noteworthy. White suckers (Figures 3, 4, 7, and 8) were much more sensitive

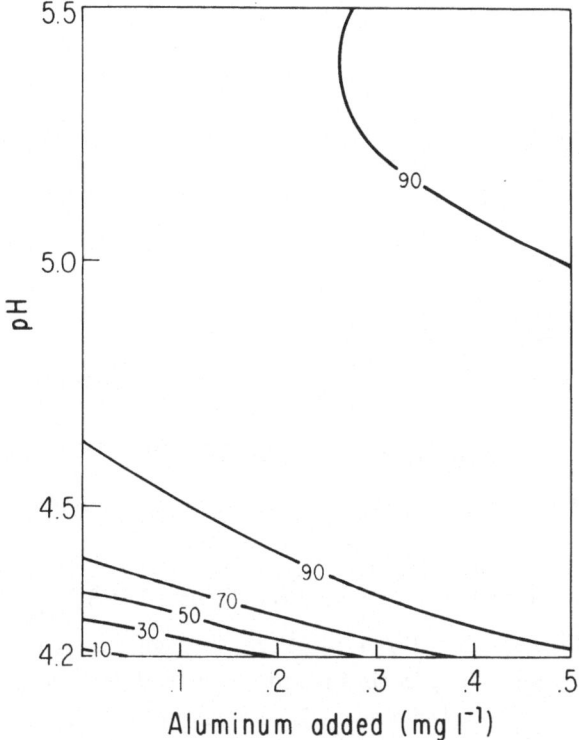

Fig 1.   Isopleths of % survival of brook trout embryos through the eyed stage of development in acidic waters with Al.

than were brook trout (Figures 1, 2, 5, and 6). The presence of Al was beneficial to survival of both white sucker and brook trout embryos prior to hatching at low pH levels (Figures 1 and 3). For brook trout eggs, more embryos hatched successfully and eventually initiated feeding at low pH levels (pH < 4.6) with the presence of Al (0.1 to 0.5 mg l^{-1}) than without Al (Figure 2). However, in bioassays with brook trout and white sucker larvae and postlarvae, Al was highly toxic even at low concentrations (0.1 to 0.2 mg l^{-1}) (Figures 5 to 8). The toxicity of a given Al concentration varied at different pH levels, perhaps reflecting shifts in the relative importance of Al forms with pH. Aluminum was found to be most toxic in solutions over-saturated with Al at pH levels 5.2 and 5.4. Elevated levels of Al (0.3 to 0.5 mg l^{-1}) were also highly toxic to larvae and postlarvae at low pH levels (4.2 to 4.6) (Figures 5 and 6). In general, older life history stages were less sensitive to low pH levels than were younger fish (Figure 9).

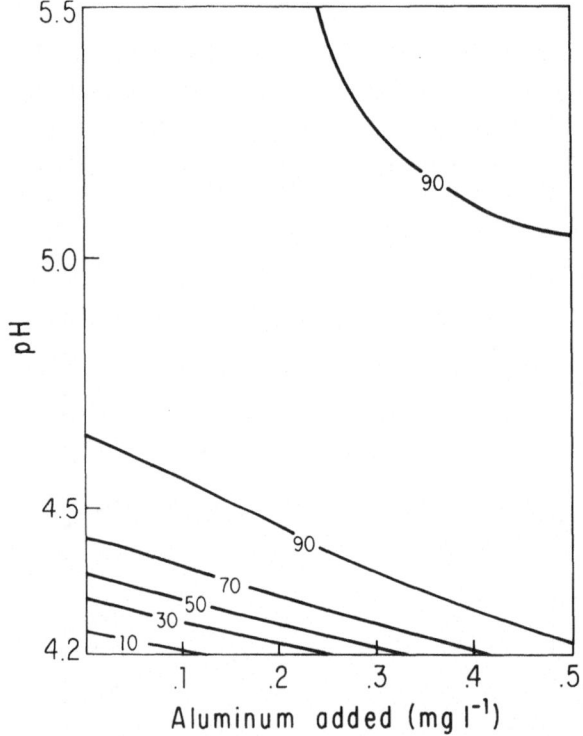

Fig. 2.   Isopleths of % hatch of brook trout embryos in acidic waters with Al.

3.1. SENSITIVITY OF BROOK TROUT EMBRYOS (Figures 1 and 2; Table II)

In the absence of Al, survival of brook trout eggs through the eyed stage and through hatching was fairly stable at pH levels 4.6 and above, with 90% or greater survival in all units (Table II). At pH levels 4.4 and 4.2, survival dropped off sharply. Addition of Al to these low pH waters appeared to mitigate the effect of low pH and substantially increased embryo survival. At pH 4.2, Al increased the percent hatch from 0 up to 52%;

TABLE II

Survival of brook trout embryos exposed to acidic waters with Al, 28 December, 1978 to 24 February, 1979

Al added (mg l^{-1})	pH			Average measured Al concentrations (mg l^{-1})		Percent survival to		Percent Swim-up
	Mean	Range				Eyed Stage	Hatching	
				Total	Monomeric			
0.0	4.18	4.09–4.23		0.00	–	2.0	0.0	0.0
	4.19	4.12–4.23		0.00	–	0.0	0.0	0.0
0.2	4.17	4.06–4.23		0.20	0.20	54.0	50.0	–
	4.17	4.09–4.24		0.20	0.20	56.0	42.0	32.0
0.5	4.17	4.09–4.24		0.51	0.50	78.0	76.0	–
	4.18	4.08–4.25		0.51	0.51	84.0	62.0	0.0
0.0	4.39	4.31–4.45		0.00	–	54.0	48.0	42.0
	4.38	4.28–4.46		0.00	–	60.0	60.0	58.0
0.1	4.40	4.29–4.47		0.10	0.10	90.0	86.0	72.0
	4.40	4.26–6.43		0.12	0.10	80.0	76.0	76.0
0.3	4.40	4.26–4.49		0.30	0.29	90.0	90.0	–
	4.40	4.30–4.47		0.30	0.31	96.0	90.0	68.0
0.0	4.62	4.50–4.74		0.01	–	90.0	90.0	50.0
	4.60	4.50–4.67		0.01	–	94.0	94.0	84.0
0.2	4.61	4.46–4.71		0.20	0.19	88.0	88.0	56.0
	4.61	4.48–4.69		0.20	0.19	93.9	93.9	69.4
0.5	4.60	4.43–4.69		0.50	0.49	100.0	100.0	72.0
	4.62	4.40–6.26		0.50	0.49	100.0	100.0	92.0
0.0	4.90	4.81–5.05		0.00	–	96.0	94.0	74.0
	4.91	4.81–5.12		0.00	–	95.9	93.9	91.8
0.1	4.90	4.76–5.07		0.10	0.09	94.0	94.0	86.0
	4.90	4.82–5.07		0.09	0.09	95.9	95.9	85.7
0.3	4.90	4.76–5.05		0.29	0.28	94.0	94.0	58.0
	4.91	4.79–5.29		0.28	0.27	93.9	91.8	87.8
0.0	5.21	5.06–5.51		0.00	–	98.0	98.0	86.0
	5.20	5.03–5.40		0.00	–	98.0	98.0	88.0
0.2	5.19	4.95–5.34		0.18	0.16	98.0	96.0	82.0
	5.19	4.99–5.33		0.17	0.16	90.0	88.0	88.0
0.5	5.19	5.04–5.32		0.35	0.23	88.2	88.2	3.9
	5.19	5.04–5.32		0.34	0.21	92.0	88.0	16.0
0.0	5.49	5.35–5.64		0.00	–	94.0	94.0	88.0
	5.48	5.32–5.65		0.00	–	95.9	91.8	83.7
0.1	5.48	4.76–5.60		0.05	0.05	94.0	94.0	94.0
	5.51	5.39–5.64		0.05	0.04	92.0	92.0	92.0
0.3	5.48	5.35–5.60		0.16	0.05	64.0	64.0	10.0
	5.49	5.41–5.60		0.14	0.05	96.0	96.0	14.0
0.0	7.01	6.84–7.15		0.00	–	94.0	94.0	82.0
	7.01	–		0.00	–	93.9	93.9	93.9
	7.01	–		0.00	–	96.0	96.0	86.0

at pH 4.4 from 58 up to 83%. In higher pH waters (5.2 and 5.5), high levels of Al (0.3 and 0.5 mg l^{-1}) had a slight adverse effect on egg survival. At pH 5.2, percent hatch decreased from 98% with no Al to 88% with 0.5 mg l^{-1}; at pH 5.5, from 93 to 80% with 0.3 mg l^{-1}. However, high levels of Al at these pH levels had no substantial impact on embryo survival and development until the larval stage. Development of larvae hatched in treatments with over-saturated Al concentrations (pH levels 5.2 and 5.5 with 0.3 and 0.5 mg Al added l^{-1}) was very slow and mortality levels gradually increased. The majority of larvae in these treatments were small and still had large quantities of yolk left when the experiment was terminated (about 1 mo after larvae in unstressed treatments had initiated feeding and swim-up).

3.2. SENSITIVITY OF WHITE SUCKER EMBRYOS (Figures 3 and 4, Table III)

White sucker eggs were very sensitive to low pH levels. In treatments without Al, no embryos survived to the eyed state at pH levels 5.0 and below. At pH levels 5.4 and 5.6, survival to the eyed stage ranged from 38 to 69%; in control water from 74 to 81%. Ninety-nine percent of the embryos which reached the eyed stage in treatments without Al successfully hatched and initiated feeding. Addition of Al to low pH waters (< 5.0) increased embryo survival through the eyed stage but not through hatching. Out of all embryos in all treatments with Al added (0.2 mg l^{-1} or greater), only 4 larvae successfully hatched. At pH levels 5.2 and above, the presence of Al resulted in embryo deformities, and only 1% hatched successfully. At lower pH levels, it is not known whether embryos died during hatching as a result of the low pH levels or the elevated Al concentrations.

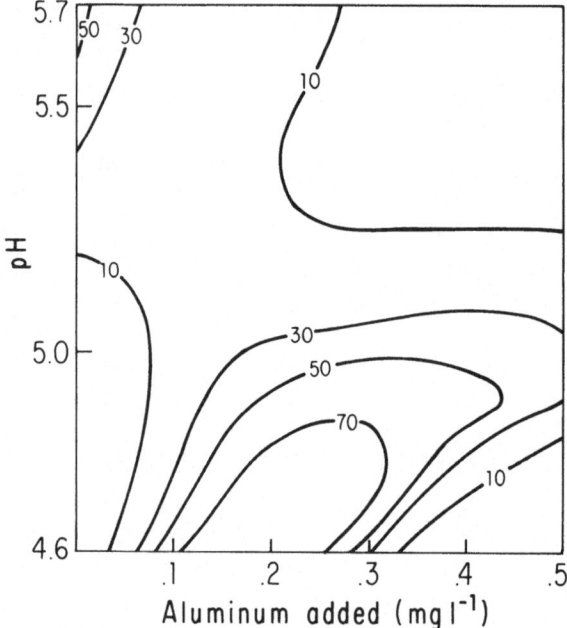

Fig. 3.   Isopleths of % survival of white sucker embryos through the eyed stage of development in acidic waters with Al.

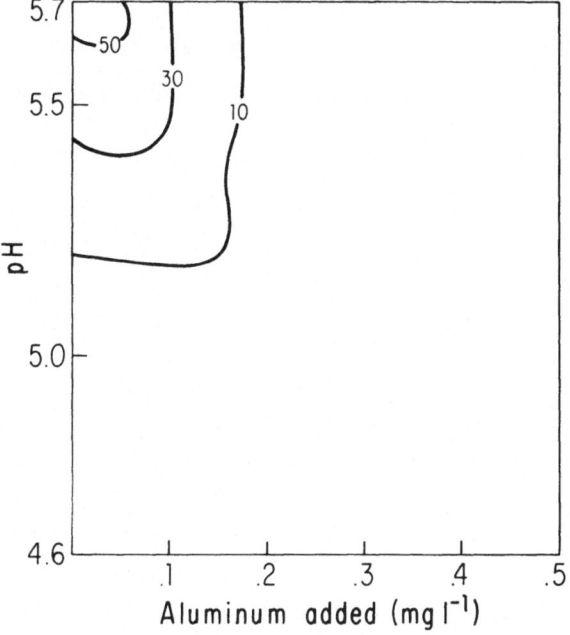

Fig. 4. Isopleths of % hatch of white sucker embryos in acidic waters with Al.

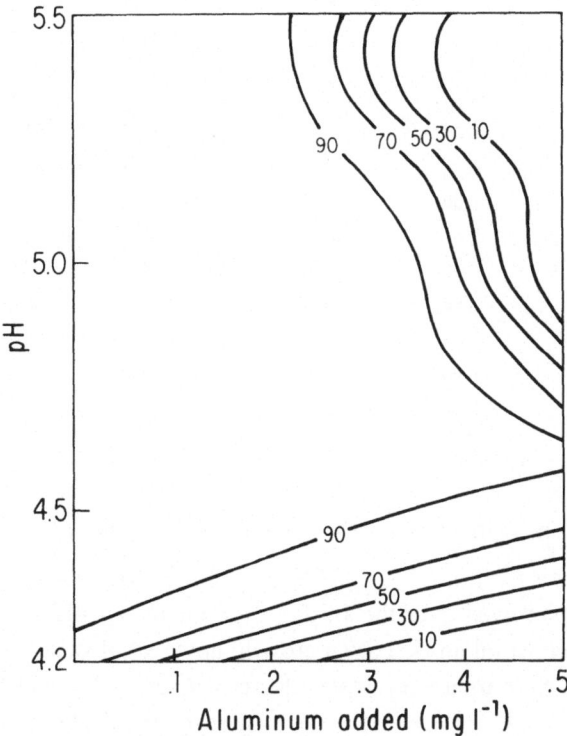

Fig. 5. Isopleths of % survival of brook trout larvae after 14 days in acidic waters with Al.

TABLE III

Survival of white sucker embryos exposed to acidic waters with Al, 12 June to 5 July, 1979

Al added (mg l^{-1})	pH		Average measured Al concentrations (mg l^{-1})		Percent survival to		Percent Swim-up
	Mean	Range	Total	Monomeric	Eyed Stage	Hatching	
0.0	4.63	4.63–4.64	0.00	–	0.0	0.0	0.0
	4.63	4.62–4.64	0.00	–	0.0	0.0	0.0
0.3	4.62	4.57–4.71	0.29	0.27	28.6	0.0	0.0
	4.60	4.56–4.68	0.30	0.27	54.0	0.0	0.0
0.0	4.84	4.84–4.84	0.00	–	0.0	0.0	0.0
	4.84	4.84–4.84	0.00	–	0.0	0.0	0.0
0.2	4.77	4.68–4.87	0.19	0.17	70.6	0.0	0.0
	4.77	4.70–4.86	0.19	0.18	80.0	0.0	0.0
0.5	4.76	4.69–4.84	0.48	0.47	0.0	0.0	0.0
	4.77	4.69–4.89	0.47	0.45	0.0	0.0	0.0
0.0	5.03	5.03–5.04	0.00	–	0.0	0.0	0.0
	5.03	5.03–5.03	0.00	–	0.0	0.0	0.0
0.0	5.15	5.08–5.19	0.00	–	0.0	0.0	0.0
	5.19	5.07–5.29	0.00	–	61.7	59.6	59.6
0.2	5.19	5.10–5.26	0.16	0.14	65.3	0.0	0.0
	5.17	5.12–5.25	0.15	0.13	0.0	0.0	0.0
0.5	5.16	5.04–5.32	0.34	0.15	16.0	0.0	0.0
	5.17	5.05–5.33	0.36	0.20	26.0	0.0	0.0
0.0	5.41	5.33–5.49	0.00	–	56.2	56.2	56.2
	5.41	5.30–5.49	0.00	–	69.4	69.4	69.4
0.0	5.58	5.50–5.75	0.00	–	65.3	65.3	65.3
	5.57	5.53–5.70	0.00	–	37.5	37.5	37.5
0.2	5.58	5.50–5.69	0.08	0.03	85.7	8.2	0.0
	5.60	5.53–5.70	0.06	0.02	0.0	0.0	0.0
0.5	5.60	5.53–5.67	0.28	0.05	0.0	0.0	0.0
	5.60	5.54–5.67	0.29	0.05	0.0	0.0	0.0
0.0	7.10	–	0.00	–	74.0	74.0	74.0
	7.10	–	0.00	–	79.6	79.6	79.6
	7.10	–	0.00	–	81.2	77.1	75.0

3.3. SENSITIVITY OF BROOK TROUT LARVAE (Figure 5, Table IV)

Brook trout larvae were tolerant of low pH levels. At pH levels 4.4 and above, 97% or greater of the larvae survived over the 14-day experiment. At pH 4.2, 56.7 and 90% survived in two experimental units. Larval growth, however, declined at pH levels less than 5.2. Addition of Al to treatment waters decreased survival at all pH levels (except for pH 5.5 with 0.1 mg Al l^{-1}). Larvae were most severely affected in treatments with both low pH levels and high Al concentrations or in treatments with over-saturated Al

TABLE IV

Survival and growth of brook trout larvae exposed to acidic waters with Al, 18 January to 1 February, 1979

Al added (mg l^{-1})	pH		Average measured Al concentrations (mg l^{-1})		Percent survival	Mean weight (mg)
	Mean	Range	Total	Monomeric		
0.0	4.20	4.16–4.24	0.00	–	90.0	50
	4.19	4.15–4.23	0.00	–	56.7	52
0.5	4.17	4.15–4.22	0.51	0.15	0.0	–
	4.17	4.15–4.22	0.50	0.51	0.0	–
0.0	4.39	4.32–4.46	0.00	–	96.6	50
	4.40	4.35–4.48	0.00	–	100.0	50
0.3	4.40	4.35–4.49	0.31	0.31	90.0	52
	4.41	4.35–4.48	0.31	0.31	63.3	54
0.0	4.60	4.54–4.75	0.00	–	100.0	51
	4.61	4.53–4.74	0.00	–	100.0	56
0.2	4.57	4.53–4.62	0.21	0.20	86.7	51
	4.57	4.53–4.62	0.20	0.20	90.0	50
0.5	4.58	4.53–4.63	0.51	0.50	69.0	54
	4.58	4.54–5.63	0.52	0.50	53.3	53
0.0	4.90	4.72–5.02	0.00	–	100.0	66
	4.90	4.74–5.07	0.00	–	100.0	61
0.1	4.91	4.78–5.01	0.10	0.09	100.0	53
	4.91	4.78–5.03	0.10	0.09	93.3	56
0.3	4.88	4.73–4.98	0.30	0.29	96.7	48
	4.87	4.71–4.97	0.30	0.29	96.7	53
0.0	5.16	4.92–5.37	0.00	–	100.0	69
	5.16	4.91–5.36	0.00	–	100.0	67
0.2	5.21	5.04–5.34	0.17	0.15	93.3	51
	5.22	5.11–5.33	0.17	0.15	83.3	50
0.5	5.12	5.00–5.19	0.40	0.25	0.0	–
	5.11	4.99–5.22	0.41	0.27	0.0	–
0.0	5.45	5.26–5.57	0.00	–	100.0	70
	5.43	5.16–5.59	0.00	–	100.0	67
0.1	5.44	5.12–5.57	0.07	0.05	100.0	63
	5.45	5.37–5.77	0.07	0.05	100.0	60
0.3	5.47	5.22–5.70	0.18	0.06	46.7	47
	5.47	5.26–5.67	0.17	0.05	53.3	48
0.0	7.05	–	0.00	–	100.0	71
	7.05	–	0.00	–	100.0	71
	7.05	–	0.00	–	100.0	73

concentrations. While larval survival after 14 days in most treatments was 90 to 100%, with 0.5 mg Al l^{-1} at pH 4.2 and 4.6, percents surviving were 0 and 61%; with 0.3 mg Al l^{-1} at pH 4.4, 77%. In treatments with over-saturated Al solutions, 0.5 mg Al added l^{-1} at pH 5.2 and 0.3 mg l^{-1} at pH 5.5, no larvae and 50% of the larvae survived the 14 day period, respectively.

3.4. SENSITIVITY OF BROOK TROUT POSTLARVAE (Figure 6, Table V)

Brook trout during the postlarval period were also very tolerant of low pH levels. All experimental units with no Al added at all pH levels tested (4.2 to 5.5) had 95% survival or greater. Addition of 0.1 mg Al l^{-1} at pH 4.9 and 5.5 resulted in no decrease in postlarval survival. Aluminum affected survival only at levels of 0.2 mg l^{-1} and greater. At a given level of Al added, Al toxicity to postlarvae varied less with pH than in experiments with brook trout larvae (Figure 5). With 0.2 mg l^{-1}, survival after 14 days was 86 and 68% at pH levels 4.6 and 5.2, respectively; with 0.3 mg l^{-1}, 58, 39, and 45% at pH levels 4.4, 4.9, and 5.5, respectively; and with 0.5 mg l^{-1}, 6, 18, and 0% at pH levels 4.2, 4.6, and 5.2 respectively. These differences in the responses of brook trout during the larval period vs postlarval period may reflect an additive adverse effect of low pH and elevated Al levels, and a greater sensitivity of larvae to low pH.

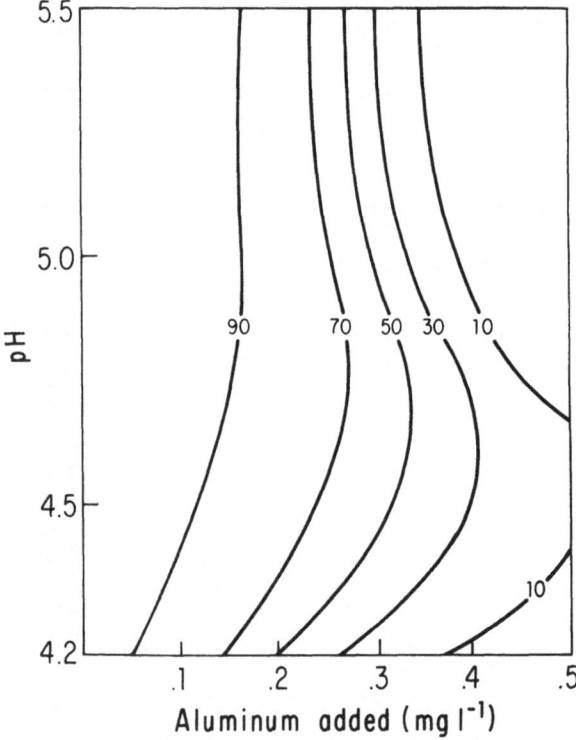

Fig. 6.   Isopleths of % survival of brook trout postlarvae after 14 days in acidic waters with Al.

TABLE V

Survival and growth of brook trout postlarvae exposed to acidic waters with Al, 10 to 24 February, 1979

Al added (mg l⁻¹)	pH		Average measured Al concentrations (mg l⁻¹)		Percent survival	Mean weight (mg)
	Mean	Range	Total	Monomeric		
0.0	4.22	4.16–4.30	0.00	–	100.0	77
	4.22	4.15–4.29	0.00	–	92.0	82
0.5	4.22	4.15–4.26	0.53	0.52	0.0	–
	4.24	4.17–4.31	0.53	0.53	12.0	72
0.0	4.40	4.33–4.48	0.00	–	100.0	86
	4.42	4.34–4.57	0.00	–	92.0	90
0.3	4.43	4.34–4.49	0.31	0.29	52.0	74
	4.43	4.33–4.49	0.31	0.30	64.0	65
0.0	4.62	4.50–4.72	0.00	–	92.0	91
	4.61	4.50–4.75	0.00	–	100.0	84
0.2	4.62	4.58–4.73	0.20	0.19	84.0	72
	4.62	4.57–4.71	0.20	0.20	87.5	68
0.5	4.61	4.53–4.76	0.51	0.49	20.0	73
	4.61	4.53–4.72	0.51	0.48	16.7	81
0.0	4.95	4.81–5.24	0.00	–	100.0	96
	4.93	4.83–5.07	0.00	–	96.0	95
0.1	4.92	4.79–5.08	0.10	0.08	96.0	80
	4.91	4.77–5.24	0.10	0.08	100.0	81
0.3	4.90	4.76–4.95	0.31	0.27	36.0	92
	4.90	4.74–4.97	0.30	0.27	41.7	84
0.0	5.21	5.12–5.40	0.00	–	100.0	106
	5.19	5.09–5.43	0.00	–	100.0	107
0.2	5.16	5.02–5.30	0.18	0.15	60.0	84
	5.17	5.00–5.32	0.18	0.14	76.0	80
0.5	5.16	5.08–5.25	0.40	0.26	0.0	–
	5.15	5.06–5.27	0.40	0.30	0.0	–
0.0	5.47	5.32–5.65	0.00	–	100.0	105
	5.47	5.31–5.69	0.00	–	100.0	102
0.1	5.48	5.22–5.80	0.09	0.05	96.0	89
	5.48	5.24–5.88	0.08	0.04	100.0	88
0.3	5.48	5.37–5.62	0.15	0.06	44.0	73
	5.47	5.37–5.68	0.14	0.09	45.8	67
0.0	6.99	6.99–7.00	0.00	–	100.0	114
	6.99	–	0.00	–	96.0	115
	6.99	–	0.00	–	100.0	107

TABLE VI

Survival and growth of white sucker larvae in acidic waters with Al, 22 June to 5 July, 1979

Al added (mg l⁻¹)	pH		Average measured Al concentrations (mg l⁻¹)		Percent survival	Mean weight (mg)
	Mean	Range	Total	Monomeric		
0.0	4.60	4.59–4.61	–	–	0.0	–
	4.60	4.59–4.61	–	–	0.0	–
0.1	4.59	4.56–4.63	0.08	0.09	0.0	–
	4.59	4.56–4.62	0.08	0.09	0.0	–
0.3	4.63	4.58–4.69	0.30	0.28	0.0	–
	4.63	4.58–4.69	0.29	0.28	0.0	–
0.0	4.83	4.79–4.89	–	–	0.0	–
	4.84	4.79–4.88	–	–	0.0	–
0.2	4.78	4.75–4.84	0.18	0.18	0.0	–
	4.80	4.76–4.85	0.20	0.19	0.0	–
0.5	4.80	4.78–4.83	0.47	0.45	0.0	–
	4.81	4.78–4.84	0.47	0.44	0.0	–
0.0	5.02	4.93–5.17	–	–	40.0	8
	5.02	4.96–5.18	–	–	56.7	9
0.1	5.03	4.99–5.08	0.09	0.08	20.7	4
	5.03	4.99–5.10	0.08	0.09	0.0	–
0.3	4.98	4.94–5.05	0.26	0.25	0.0	–
	4.99	4.94–5.06	0.26	0.22	0.0	–
0.0	5.22	5.13–5.32	–	–	96.7	17
	5.21	5.12–5.29	–	–	100.0	14
0.2	5.17	5.17–5.18	0.17	0.12	0.0	–
	5.18	5.17–5.19	0.17	0.13	0.0	–
0.5	5.11	5.10–5.12	0.37	0.23	0.0	–
	5.15	5.10–5.17	0.38	0.25	0.0	–
0.0	5.38	5.32–5.52	–	–	93.7	14
	5.38	5.28–5.54	–	–	96.8	17
0.1	5.40	5.38–5.42	0.04	0.05	0.0	–
	5.41	5.38–5.44	0.05	0.05	0.0	–
0.3	5.40	5.36–5.48	0.16	0.08	50.0	4
	5.40	5.37–5.50	0.14	0.07	43.3	4
0.0	5.59	5.47–5.74	–	–	100.0	16
	5.59	5.49–5.75	–	–	96.7	19
0.2	5.60	5.54–5.73	0.10	0.05	60.7	6
	5.59	5.54–5.74	0.11	0.03	43.3	4
0.5	5.55	5.46–5.73	0.29	0.07	0.0	–
	5.61	5.53–5.72	0.32	0.04	0.0	–
0.0	7.06	6.89–7.16	–	–	80.0	20
	7.06	–	–	–	96.7	21
	7.06	–	–	–	83.3	21

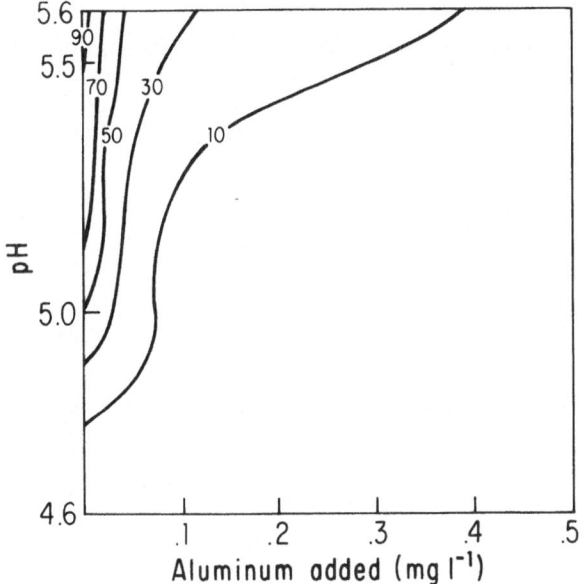

Fig. 7.   Isopleths of % survival of white sucker larvae after 13 days in acidic waters with Al.

3.5. Sensitivity of white sucker larvae (Figure 7, Table VI)

White sucker larvae were very sensitive to low pH levels. A pH of approximately 5.0 appeared critical for survival. At pH levels below 5.0 (with and without Al), all larvae died within 146 hr. At pH levels above 5.0 (with no Al), 80 to 100% of the larvae survived the 13-day experiment. Addition of Al to acidic solutions further decreased survival and

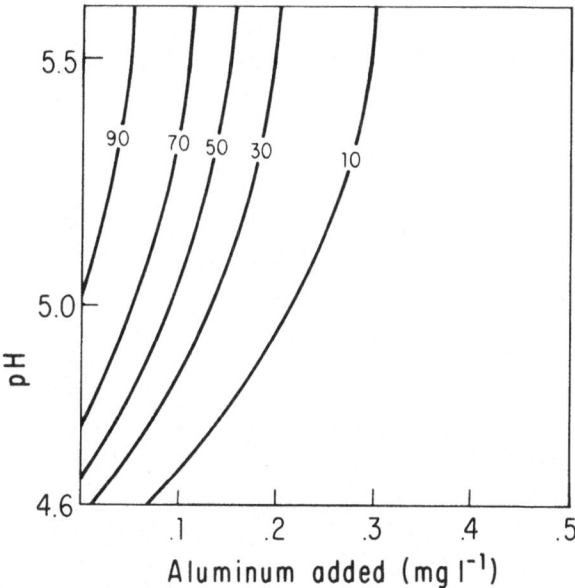

Fig. 8.   Isopleths of % survival of white sucker postlarvae after 13 days in acidic waters with Al.

TABLE VII

Survival and growth of white sucker postlarvae in acidic waters with Al, 10 to 23 July, 1979

Al added (mg l^{-1})	pH		Average measured Al concentrations (mg l^{-1})		Percent survival	Mean weight (mg)
	Mean	Range	Total	Monomeric		
0.0	4.63	4.51–4.85	0.00	–	15.8	14
	4.62	4.51–4.76	0.00	–	68.4	22
0.1	4.63	4.62–4.64	0.10	0.07	0.0	–
	4.63	4.62–4.65	0.09	0.08	0.0	–
0.3	4.62	4.61–4.63	0.28	0.29	0.0	–
	4.61	4.60–4.62	0.28	0.28	0.0	–
0.0	4.79	4.63–5.15	0.01	–	80.0	27
	4.80	4.60–5.12	0.01	–	95.0	33
0.2	4.80	4.77–4.85	0.20	0.15	0.0	–
	4.83	4.79–4.91	0.19	0.16	0.0	–
0.5	4.79	4.76–4.82	0.44	0.42	0.0	–
	4.81	4.79–4.82	0.45	0.41	0.0	–
0.0	4.98	4.75–5.22	0.01	–	86.4	40
	4.97	4.76–5.26	0.01	–	100.0	44
0.1	4.98	4.78–5.14	0.09	0.08	10.5	17
	5.02	4.91–5.18	0.09	0.07	75.0	23
0.3	4.99	4.96–5.01	0.22	0.22	0.0	–
	5.00	5.00–5.01	0.22	0.24	0.0	–
0.0	5.19	4.99–5.38	0.01	–	95.0	40
	5.22	5.08–5.40	0.01	–	100.0	44
0.2	5.23	5.17–5.27	0.12	0.08	35.0	16
	5.21	5.14–5.29	0.12	0.08	42.9	21
0.5	5.17	5.15–5.20	0.33	0.18	0.0	–
	5.17	5.14–5.24	0.37	0.16	0.0	–
0.0	5.36	4.89–5.60	0.00	–	95.0	42
	5.36	5.01–5.57	0.00	–	94.4	48
0.1	5.40	5.33–5.47	0.07	0.06	80.0	34
	5.41	5.33–5.49	0.08	0.05	80.0	31
0.3	5.44	5.43–5.44	0.15	0.04	0.0	–
	5.44	5.43–5.45	0.15	0.04	0.0	–
0.0	5.55	5.22–5.82	0.00	–	94.7	52
	5.54	5.09–5.80	0.00	–	100.0	46
0.2	5.57	5.38–5.70	0.09	0.05	63.2	30
	5.59	5.45–5.71	0.08	0.04	80.0	40
0.5	5.57	5.49–5.70	0.29	0.04	10.0	17
	5.57	5.51–5.70	0.32	0.04	0.0	–
0.0	6.74	–	0.00	–	95.2	54
	6.74	–	0.00	–	90.0	52
	6.74	–	0.00	–	100.0	54

growth. Larvae survived after 13 days in treatments with Al only at pH 5.0 with 0.1 mg Al added l^{-1} (10%), pH 5.4 with 0.3 mg l^{-1} (47%), and pH 5.6 with 0.2 mg l^{-1} (52%). Growth was substantially reduced in these treatments relative to treatments with similar pH levels but without Al. Thus, even very low concentrations of Al (0.1 mg l^{-1}) had a substantial adverse effect on larvae of white suckers.

3.6. SENSITIVITY OF WHITE SUCKER POSTLARVAE (Figure 8, Table VII)

White suckers during the postlarval period were also sensitive to low pH levels. A pH level of approximately 4.6 (with no Al) appeared critical for survival. At pH levels above 4.6, 80% or more of the postlarvae survived the 13-day experiment. At pH 4.6, 16 and 68% of the postlarvae survived in two experimental units. Addition of Al to acidic solutions further decreased survival. At high levels of Al (0.3 to 0.5 mg l^{-1}) at all pH levels, all postlarvae died within 75 hr except for 2 postlarvae at pH 5.6. Addition of 0.1 to 0.2 mg Al l^{-1} resulted in complete mortality within 145 hr at low pH levels (4.6–4.8). At higher pH levels with 0.1 to 0.2 mg Al added l^{-1}, survival increased with increasing pH level, but was also substantially less than in treatments at similar pH levels but with no Al added. Thus, for white sucker postlarvae as with white sucker larvae, even the lowest level of Al tested (0.1 mg l^{-1}) had a significant adverse effect on fish survival and growth.

4. Discussion

Both low pH levels and elevated inorganic Al concentrations can be toxic to fish. Of primary interest is the relative importance of H^+ vs Al as toxic factors at different pH values and to different susceptible fish species and life history stages. For the range of pH levels (4.2 to 5.6), and fish species (white suckers and brook trout) and life history stages (eggs, larvae, and postlarvae) studied, sensitivity to low pH levels decreased with increasing age (Figure 9). The relationship between sensitivity of fish to Al and fish age was, however, more complex, and varied with pH level. At low pH levels (4.2 to about 4.8), the presence of Al was generally beneficial to egg survival through the eyed stage. For both brook trout and white suckers, survival of embryos through the eyed state at pH levels below 5.0 was significantly ($p \le 0.01$) better in treatments with Al than without Al. The effect of elevated H^+ concentrations on fish eggs is also mitigated by the presence of other cations, particularly Ca (Wright and Snekvik, 1978). It is possible that factors controlling the permeability of the chorion or vitelline membrane respond generally to polyvalent cations. In contrast, in experiments with white sucker and brook trout larvae and postlarvae, Al concentrations of 0.1 mg l^{-1} (for white suckers) or 0.2 mg l^{-1} (for brook trout) and greater resulted in measurable reductions in survival and growth at all pH levels. Reasons for the reversal of Al effects at low pH levels in larvae versus eggs are not known. Differences in membrane characteristics of gill surfaces vs egg chorion or perivitelline membranes may be important.

In general, Al was most toxic in over-saturated solutions at pH levels 5.2 to 5.4. Observations suggest the mechanism for toxicity under these conditions involves precipi-

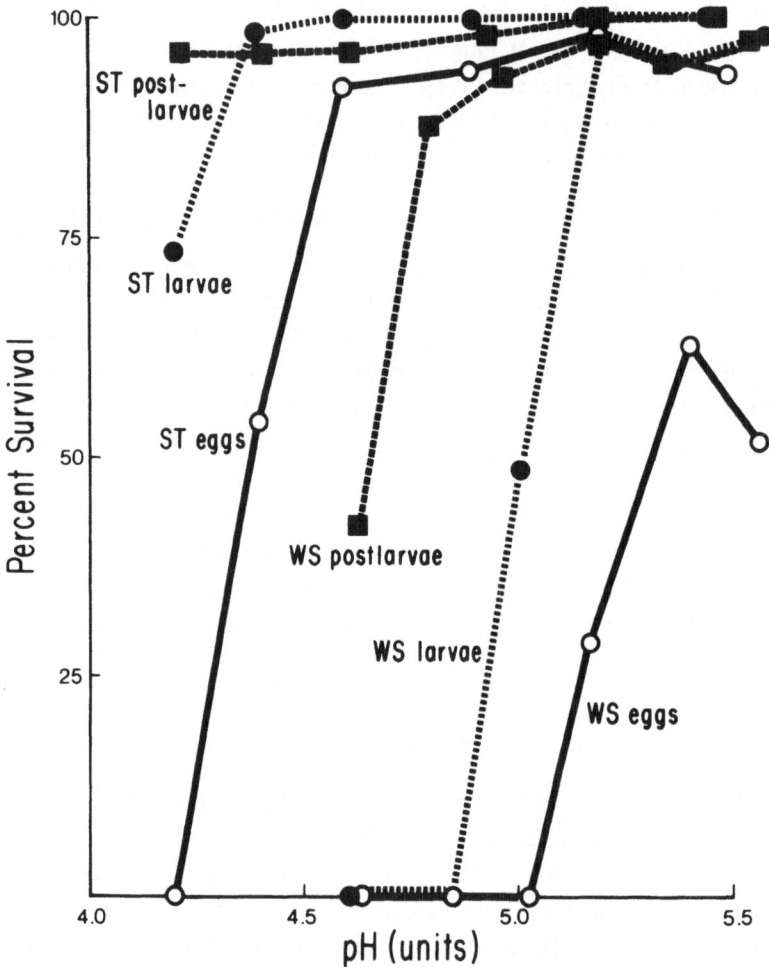

Fig. 9. Percent survival of brook trout (ST) and white sucker (WS) embryos, larvae, and postlarvae in solutions with no Al added.

tation and coagulation of Al hydroxide on gill or other surfaces, or adsorption and nucleation of Al polymers at surface interfaces. Larvae and postlarvae of both brook trout and white suckers and also white sucker eggs were very susceptible to these elevated Al levels at pH 5.2 to 5.4 and the resultant conditions of Al over-saturation. Brook trout embryos prior to hatching were less sensitive. At higher pH levels (5.5 and above), Al toxicity declined, presumably because the initial super-saturation of Al resulted in rapid precipitation of $Al(OH)_{3(s)}$ out of the water column.

Mucous accumulation and gill damage are commonly reported reactions of fish when exposed to elevated metal concentrations in laboratory bioassays (e.g., Freeman and Everhart, 1971; Mattiessen and Brafield, 1973; Mount, 1966; Muniz and Leivestad, 1980). Decreased larval survival and growth and reduced egg hatchability have also been noted as a result of exposure to iron hydroxide suspensions (Smith *et al.*, 1973; Sykora *et al.*, 1972; Smith and Sykora, 1976; Kinne and Rosenthal, 1967). The exact physiological cause of death is, as yet, not clearly defined and is probably a function of many simulaneous physiological disturbances.

Two examples are available suggesting that waters with over-saturated levels of Al may also be toxic to fish under natural conditions. Grahn (1980) reported fish kills of cisco (*Coregonus albua*) related to increasing pH levels in the summers of 1978 and 1979 in two moderately oligotrophic lakes in Sweden. As a result of a long period of dry, sunny weather, pH levels in the epilimnion of the lakes increased from 4.9 and 5.4 to 5.4 and 6.0, respectively. Grahn (1980) hypothesized that this increase in pH level caused precipitation of $Al(OH)_{3(s)}$ in epilimnetic water. Cisco, migrating into surface waters to feed, were exposed to these conditions of increasing pH and potential precipitation of Al trihydroxide. Large numbers of dead cisco were observed with mucous accumulations and gill damage similar to that described by Schofield and Trojnar (1980) for Al toxicity bioassays. Aluminum concentrations on gill surfaces of dead cisco collected after the mortality episode were 40 to 47 mg kg^{-1} as compared to 2 to 7 mg kg^{-1} for fish from reference lakes. Likewise, Dickson (1978) noted that immediately after liming of acidic waters (pH values increased to 5.5 and above), lake waters were still toxic to trout. Despite raised pH levels, concentrations of Al remained high and, presumably, Al would be actively precipitating out of solution.

Elevated levels of Al were also highly toxic to larvae and postlarvae at low pH levels (4.2 to 4.6). Within this pH range, the sensitivity of fry to Al tended to increase slightly with decreasing pH level. Gill damage and mucous accumulation associated with Al toxicity were less severe than in Al solutions at pH levels 5.2 to 5.6. Mortality occurred at times without visible gill damage. Both elevated Al concentrations and low pH levels have been reported to interfere with osmoregulatory functions (Muniz and Leivestad, 1980). Thus, stress on fry from both elevated H$^+$ and Al concentrations may be additive. The more sensitive the organism to low pH levels (white sucker larvae > white sucker postlarvae > brook trout larvae > brook trout postlarvae), the stronger the toxic responce to elevated Al levels with decreasing pH levels.

5. Conclusion

The effects of inorganic Al on survival and growth of brook trout and white suckers during developmental periods as eggs, larvae, and postlarvae, at pH levels 4.2 to 5.6, are complex. The presence of Al was shown to be both detrimental and beneficial to fish survival, depending on life history stage. Sensitivity of fry to Al increased with both increasing and decreasing pH levels, owing to changes in Al chemistry with pH level and to apparent differences in fry sensitivity to different Al forms. However, it is obvious that in assessing the potential effect of acidification of surface waters on survival of fish, the simultaneous increase in Al concentration must be taken into account.

The pH level in lakes and streams may affect fish survival both as a direct toxicant and by controlling the concentration of inorganic Al. Levels of H$^+$ and inorganic Al are highly positively correlated in acidic surface waters (Driscoll et al., 1980; Schofield and Trojnar, 1980; Dickson, 1978; Wright et al., 1980). This relationship between pH and inorganic Al concentration plus the sensitivities of different fish species and life history stages to low pH and elevated Al levels determines minimum pH levels necessary to prevent substantial reductions ($\geq 15\%$) in fish survival in acidic waters (Table VIII).

TABLE VIII

pH level necessary to prevent measurable ($\geq 15\%$) reductions in survival of early life history stages of brook trout and white suckers assuming no Al or Al concentrations typical for an Adirondack surface water at that pH level[a]

Species	Life history stage	Desired pH level	
		With no Al	Accounting for Al levels in natural water
Brook trout	Embryo development through eyed stage	≥ 4.5	≥ 4.1
	Embryo development through hatching	≥ 4.5	≥ 4.1
	Larvae	≥ 4.2	≥ 4.5
	Postlarvae	≥ 3.7	≥ 4.8
White sucker	Embryo development through eyed stage	> 5.6	> 5.6 or 4.6 to 4.8
	Embryo development through hatching	> 5.6	> 5.6
	Lavae	≥ 5.3	> 5.6
	Postlarvae	≥ 4.9	≥ 5.4

[a] Based on Figures 1 to 8 and a regression of inorganic Al concentration as a function of pH level from data for Adirondack lakes and streams in Driscoll (1980).

References

Almer, B., Dickson, W., Ekstrom,C., Hornstrom, E., and Miller, U.: 1974, *Ambio* **3**, 30.

Ashton, W. D.: 1972, *The Logit Transformation With Special Reference to its Use in Bioassay*, Hafner Pub. Co., New York, N.Y., 88 pp.

Baker, J.: 1981, 'Aluminum Toxicity to Fish as Related to Acid Precipitation and Adirondack Surface Water Quality', Ph.D. Thesis, Cornell University, Ithaca, N.Y., 441 pp.

Beamish, R. J. and Harvey, H. H.: 1972, *J. Fish. Res. Board Canada* **29**, 1131.

Cronan, C. and Schofield, C.: 1979, *Science* **204**, 304.

Davis, R., Smith, M., Baily, J., and Norton, S.: 1978, *Verh. Internat. Verein. Limnol.* **20**, 532.

Dickson, W.: 1975, *Rep. Inst. Freshw. Drottningholm* **54**, 8.

Dickson, W.: 1978, *Verh. Internat. Verein. Limnol.* **20**, 851.

Dillon, P., Jefferies, D., Synder, W., Reid, R., Yan, N., Evans, D., Moss, J., and Scheider, W.: 1978, *J. Fish. Res. Board Canada* **35**, 809.

Driscoll, C.: 1980, 'Chemical Characterization of Some Dilute Acidified Lakes and Streams in the Adirondack Region of New York State', Ph.D. Thesis, Cornell University, Ithaca, N.Y., 309 pp.

Driscoll, C. T., Jr., Baker, J. P., Bisogni, J. J., and Schofield, C. L.: 1980, *Nature* **284**, 161.

Freeman, R. A. and Everhart, W. H.: 1971, *Trans. Amer. Fish. Soc.* **100**, 644.

Gjessing, E., Henriksen, A., Johannessen, M., and Wright, R.: 1976, 'Changes in the Chemical Compositions of Lakes', in F. H. Braekke (ed.), *Impact of Acid Precipitation on Forest and Freshwater Ecosystems in Norway*, Research Report FR 6/76, SNSF, Oslo, Norway, pp. 65–85.

Grahn, O.: 1980, 'Fish Kills Due to High Aluminum Concentrations in Lake Water', in *Proceedings of the International Conference on the Ecological Impact of Acid Precipitation, March 1980*, SNSF Project Report, Oslo, Norway, p. 122.

Hutchinson, T. C., Gizyn, W., Havas, M., and Zorbens, V.: 1978, 'Effect of Longterm Ignite Burns on Arctic Ecosystems at the Smoking Hills, N.W.T.', in D. D. Hemphill (ed.), *Trace Substances in Environmental Health*-XII, University of Missouri, Columbia, Mo., pp. 317–332.

Jensen, K. and Snekvik, E.: 1972, *Ambio* **1**, 223.

Johnson, N. M.: 1979, *Science* **204**, 497.

Kinne, O. and Rosenthal, H.: 1967, *Marine Biology* **1**, 65.

Leivestad, H., Hendrey, G., Muniz, I. P., and Snekvik, E.: 1976, 'Effects of Acid Precipitation on Freshwater Organisms', in F. H. Braekke (ed.), *Impact of Acid Precipitation on Forest and Freshwater Ecosystems in Norway*, SNSF Project Report, FR 6, Oslo, Norway, pp. 87–111.

Matthiessen, P. and Brafield, A.E.: 1973, *J. Fish. Biol.* **5**, 607.

Mount, D. I.: 1966, *Air and Wat. Pollut. Int. J.* **10**, 49.

Muniz, I. P. and Leivestad H.: 1980, 'Toxic Effects of Aluminum on Brown Trout (*Salmo trutta*, L.)', in *Proceedings of the International Conference on the Ecological Impact of Acid Precipitation*, March 1980, SNSF Project Report, Oslo, Norway. p. 130.

Oden, S.: 1968, 'The Acidification of Air and Precipitation and Its Consequences on the Natural Environment', *Swedish Nat. Sci. Res. Council. Ecology Committee Bull.* **1**, 68 pp.

Robson, D. S.: 1980, Private Communications.

Schofield, C.: 1976, *Ambio* **5**, 228.

Schofield, C. and Trojnar, J. R.: 1980, 'Aluminum Toxicity to Brook Trout (*Salvelinus fontinalis*) in Acidified Waters', in T. Y. Toribara, M. W. Miller, and P. E. Morrow (eds.), *Polluted Rain*, Plenum Press, N.Y., pp. 341–363.

Smith, E. J., Sykora, J. L., and Shapiro, M. A.: 1973, *J. Fish. Res. Board Canada* **30**, 1147.

Smith, E. J. and Sykora, J. L.: 1976, *Trans. Am. Fish. Soc.* **105**, 308.

Stumm, W. and Morgan, J.: 1970, *Aquatic Chemistry*, Wiley-Interscience, New York, N.Y., 583 pp.

Sykora, J. L., Smith, E. J., and Synak, M.: 1972, *Water Res.* **6**, 935.

Wright, R. and Gjessing, E.: 1976, *Ambio* **5**, 219.

Wright, R. and Snekvik, E.: 1978, *Verh. Internat. Verein. Limnol.* **20**, 765.

Wright, R. F., Conroy, N., Dickson, W. T., Harriman, R., Henriksen, A., and Schofield, C. L.:1980, 'Acidified Lake Districts of the World: A Comparison of Water Chemistry in Southern Norway, Southern Sweden, Southwestern Scotland, the Adirondack Mountains of New York, and Southeastern Ontario', in *Proceedings of the International Conference on the Ecological Impact of Acid Precipitation, March 1980*, SNSF Project Report, Oslo, Norway, p. 89.

LIMING OF ACIDIFIED LAKES:
INDUCED LONG-TERM CHANGES

HANS HULTBERG and INGVAR B. ANDERSSON

Swedish Water and Air Pollution Research Institute,
P.O. Box 5207, S-402 24 Gothenburg, Sweden

(Received July 14, 1981; Revised January 4, 1982)

Abstract. This study presents data concerning long-term trends after neutralization of four acidified lakes in two regions on the Swedish west coast. Neutralization was achieved by a di-Ca-silicate with 52% CaO and about 11.5% MgO. Between 61 and 74% of the spread lime product dissolved during a 5 to 7 yr period.

The liming increased pH, from a range of 4.5 to 5.2 to near neutral and restored alkalinity in the range of 0.2 to 0.3 meq l^{-1} and the Ca-content became 3 to 4 times higher than before liming. In two lakes transparency decreased significantly presumably due to changed phytoplankton composition. These changes successively declined due to dilution and continuous acid loading.

The changes in water chemistry and development of stocked brown trout (*Salmo trutta*) populations initiated biotic changes. Phyto- and zooplankton communities reacted both instantly and later with successions in species composition. Changes of benthic macroinvertebrate species occured over several years, but some pelagic species, e.g. corixids were rapidly reduced due to predation of fish.

Observed changes were predominantly due to expanding populations of species present at very low abundances even during acid state of the lakes. Some organisms found during preacid state of the lakes did not establish new populations and this process may need a prolonged time with favorable conditions. Reacidification towards the end of the study period significantly stressed the brown trout population and also favored expansion of the filamentous alga *Mougeotia sp.* and *Sphagnum sp.* that almost vanished during the first year after liming.

Decreasing concentration of total P was not influenced by neutralization and may be mostly dependent on negative changes in the soils surrounding the lakes. If generally valid, this process may be an important factor for the oligotrophication of lakes in areas where acid deposition is high.

1. Introduction

The consequences of acid deposition to lakes and water-courses in Sweden were described by Odén in 1968. Since then a comprehensive number of investigations and surveys have established the profound changes of aquatic biota being in progress, and the vast regions threatened or already damaged in the Northern Hemisphere (Beamish and Harvey, 1972; Jensen and Snekvik, 1972; Almer *et al.*, 1974; Grahn *et al.*, 1974; Wright and Snekvik, 1978; Overrein *et al.*, 1980).

On the Swedish westcoast, acidification of oligotrophic forest lakes was noticed comparatively early (Schmul, 1969; Hultberg and Stenson, 1970). Watersheds dominated by coniferous forest in this region are vulnerable to acid deposition because of underlaying granitic-gneissic bedrock mostly covered by thin podzolic soils. Impact of the increasing acid deposition (Odén, 1968) could be traced back to the fifties and early sixties (Hultberg, 1980) and this was also confirmed by diatom stratigraphy of lake sediments (Renberg and Hellberg, 1981).

Liming of lakes, which would be one mean to counteract acidification, was already used by sportfishing associations in western Sweden during the early sixties. The S

Water, Air, and Soil Pollution **18** (1982) 311–331. 0049–6979/82/0183–0311$03.15.
Copyright © 1982 by D. Reidel Publishing Co., Dordrecht, Holland, and Boston, U.S.A.

TABLE I

Characteristic climatic and hydrological data for the studied lakes

Lake	Level above sea (m)	Yearly precip. (mm)	Mean temp. °C	Specific run-off ($m^3\,s^{-1}\,km^{-2}$)	Drainage area $m^2\,10^6$	Lake area $m^2\,10^6$	Lake volume $m^3\,10^6$	Mean depth (m)	Max. depth (m)	Retention time (yr)	Acid deposition Excess S ($g\,m^{-2}$)	Acid deposition H^+ ($meqv.\,m^{-2}$)
Gårdsjön (B)	113	≈950	6.6	0.0168	2.11	0.312	1.50	4.9	18.5	≈1.3	≈3.0	≈150
Bredvatten (B)	113	″	″	″	0.888	0.273	0.92	3.8	8.5	≈2.0	″	″
Lysevatten (B)	113	″	″	″	1.215	0.407	1.91	4.7	18.0	≈3.0	″	″
Örvattnet (V)	275	≈880	4.5	0.0166	3.120	0.720	6.02	8.4	36.0	≈3.7	≈1.5	≈65
Stensjön (V)	261	≈960	4.5	0.0192	2.700	0.340	2.29	6.7	22.5	≈1.4	″	≈70
Skitjärn (V)	147	≈800	5.0	0.0136	1.240	0.156	1.87	12.0	28.0	≈3.5	″	≈65

(B) = Province of Bohuslän.
(V) = Province of Värmland.

emissions in Sweden and Northern Europe and the observed adverse effects in limnic systems caused by the S deposition enforced a Swedish plan on S reduction in oil to $\leq 1\%$. The debate on S deposition and surveys showing several thousand acidified lakes in many parts of Sweden triggered an increased liming activity in acidified lakes. The governmental concern resulted in a 5 yr lake-liming program. From initiation, in 1977, some 300 objects (often including several lakes) have been treated with a total of 200 000 tonne of different commercial neutralizing products, predominantly ground limestone (National Board of Fisheries and National Swedish Environment Protection Board, 1981; also cf. Bengtsson et al., 1980).

1.1. BACKGROUND TO THE LONG-TERM PROJECT

In order to study methods and costs, physico-chemical and biological effects and duration of liming as well as effects of lake acidification a long-term project was initiated in 1973 (Grahn et al., 1974; Andersson et al., 1975). The project included six lakes, three in the province of Värmland and three in the province of Bohuslän (cf. Hultberg and Grahn, 1975, Figure 1).

Table I summarizes climatic and hydrological data and gives estimated values for acid deposition (Granat and Söderlund, 1975; Dickson, 1978b). The different treatment of each lake was done as shown in Table II. Intentional duration was 5 yr for the Värmland

TABLE II

Summary of treatments in the different lakes

Lake	Liming amount in metric ton	Rotenone treatment	Restocking with fish		Introduction of benthic organisms and plants	Fertilization with phosphorus
			brown trout	brook trout		
Gårdsjön (B)	–	×	×	×		
Bredvatten (B)	40.0	×	×	×	×	×
Lysevatten (B)	60.0	×	×	×		
Örvattnet (V)						
Stensjön (V)	88.0	×	×		×	
Skitjärn (V)	46.0	×	×			

(B) = Province of Bohuslän
(V) = Province of Värmland

lakes and 10 yr for the Bohuslän lakes. The neutralizing product used was primarily a di-Ca-silicate with 52% CaO and about 11.5% MgO (Grahn and Hultberg, 1975).

2. Results and Discussion

2.1. TRANSPARENCY

Increased transparency of acidified lakes has been observed (Almer et al., 1974, 1978; Dickson, 1975; Schindler, 1980) and suggested casual factors include: decreased color of humic substances at decreased pH (Gjessing, 1976), enhanced precipitation of humic substances caused by Al (Dickson, 1978a), changes in phytoplankton composition or bacterioplankton biomass (Hörnström, 1980), and decreased color of humic substances when complex-bound to Al (Lee, 1981).

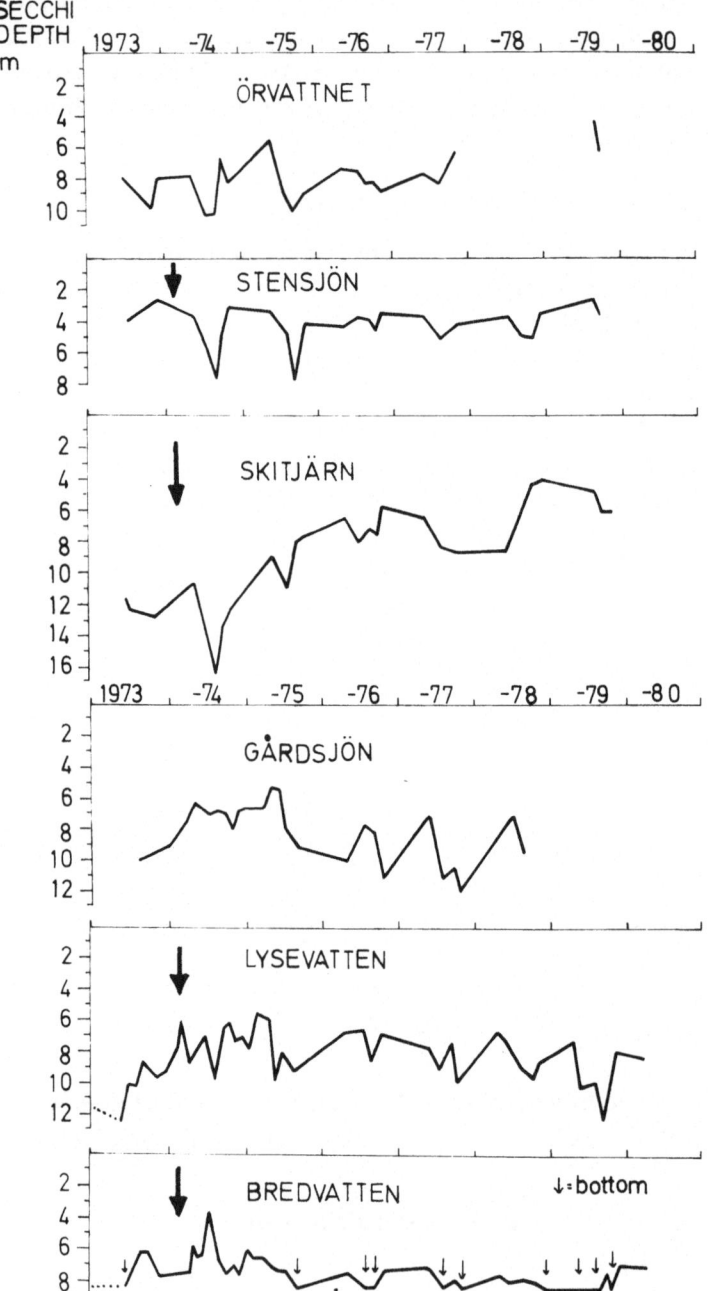

Fig. 1.   Secchi disc transparency in the four limed lakes (Stensjön, Skitjärn, Lysevatten and Bredvatten) and two acid reference lakes (Örvattnet and Gårdsjön). Arrows indicate time for liming.

After liming, transparency decreased in Lake Lysevatten. The secchi-depth declined from preliming values of 8.5 to 12.5 m to a range of about 6.5 to 8.5 m during 4 yr after the liming (Figure 1). In Lake Skitjärn a Secchi-depth of 11.5 to 12.5 m before liming was followed by a significant increase to 17 m immediately after liming, whereafter a continuous decrease occurred. During the period 2 to 6 yr after values varied between 4 to 8 m. A changed phytoplankton composition probably was the primary reason for decreased transparency in Lake Skitjärn and Lake Lysevatten since no change in water color occurred. (See also 2.2). In Lake Bredvatten and Lake Stensjön long-term transparency changes were not apparent. The decreased transparency in Lake Bredvatten during the summer 1974 was caused by resuspension of filamentous algae and detritus from littoral bottoms by storms.

Our results as well as results of others show that changes in transparency have been variable both in acidified and limed lakes. The direction of change is probably determined by the reaction of the factor that is most important for transparency. Anyhow, either increased or decreased transparency influence heat balance and light availability, and this probably is of significance for biotic changes in the lakes.

2.2. pH, ALKALINITY AND CALCIUM

Some older pH values for the lakes are given by Grahn *et al.* (1974), showing preacidification pH of > 6.0 in all the lakes. The pH in Lake Lysevatten varied between 4.4 to 4.7 the years before liming in April 1974. Lime was administered around the lake on shallow (0 to 1 m) rocky and sandy bottoms. The increase in pH (Figure 2) was rapid in epilimnion but due to stratification the whole lake neutralization was achieved first at fall overturn. Regular seasonal fluctuations in both epi- and hypolimnion were consistent, and from 1977 pH began to decrease. Notable was the sharp pH depression in epilimnion at snowmelt 1979, resulting in the first pH value < 5.5. Throughout 1980 no pH > 6.0 was measured. The original alkalinity was lost due to acidification but using the formula of Henriksen (1980) could be estimated to have been about 0.1 meq l^{-1}. However, this must be used with precaution since Dickson (1975) has shown that Ca-concentrations in acid lakes at the westcoast have increased. Alkalinity due to liming reached about 0.2 meq l^{-1} and then successively decreased to almost 0 in 1980. The negative trend for Ca was less rapid since this element was only reduced by dilution, and by 1980 Ca still was about twice the pre-liming concentration. The difference in long-term decrease between HCO_3^- and Ca was caused by the H^+ input to the lake (titration of HCO_3^-). This was used by Wilander and Ahl (1972) to estimate the duration of a lake liming and the acid loading to the lake.

The response in pH, alkalinity and Ca after liming in Lake Skitjärn differed somewhat from Lake Lysevatten. While pH increased as fast as in Lake Lysevatten, alkalinity and Ca reached their highest values early in the summer of 1976, 2 yr after the treatment. Lake Skitjärn was limed on ice in March-April 1974 while Lake Lysevatten was limed from a boat. Another difference between the treatments is that the lime generally was deposited at a greater depth than in Lake Lysevatten due to deeper water (1 to 5 m) near the shoreline in Lake Skitjärn. Thus the lime was not exposed to wave action and dissolved

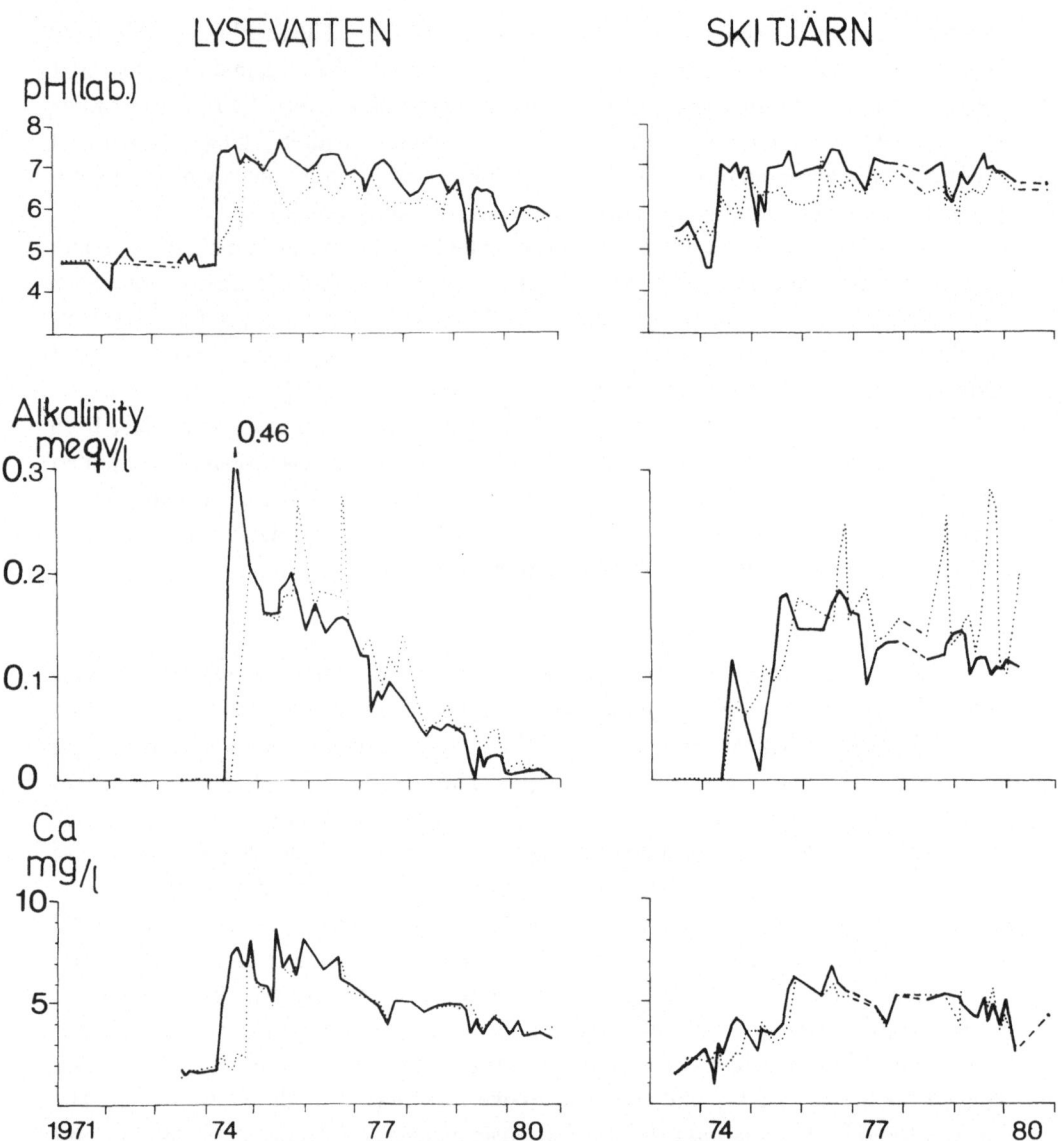

Fig. 2.   pH, alkalinity and Ca-concentration in Lake Lysevatten and Lake Skitjärn. Sampling depths are;
line 1 m and dotted line 1 m above lake bottom.

slowly, resulting in a prolonged increase in Ca and HCO_3^- concentration of the lake. The
combination of differences in liming strategy and lower acid deposition thus has resulted
in a quite different response to the liming in Lake Skitjärn compared to Lake Bredvatten.
(Figure 1).

In both lakes, production of alkalinity was apparent in bottom water during the
stagnant periods. This increased with time after liming in Lake Skitjärn but it decreased
in Lake Lysevatten. Low oxygen levels seemed to enhance this production, which was
independent of water pH, since alkalinity once was found in the bottom water of the acid

Lake Gårdsjön. In these three lakes increased color, increased ammonium and decreased sulphate and nitrate concentrations were observed. The observations indicate that processes, such as ammonification, denitrification and sulphate reduction, which reduce H^+ and produce HCO_3^- at least have influenced restricted parts of hypolimnion during stratification periods. Bioturbation by an increasing profundal population of the crustacean *Pallacea quadrispinosa* in Lake Skitjärn may also have had a significant effect by indirectly having increased decomposition in the sediment. With the possible exception of Lake Skitjärn, these processes generally have been of small significance compared to the great inflow of acid substances to the studied lakes as is the case for most oligotrophic dimictic Swedish lakes where acid deposition is high as shown by Tirén (1981). On the other hand, Schindler *et al.* (1980) found sulphate reduction and precipitation of FeS to be a significant process counteracting their experimental acidification of ELA lake 223. This lake differed, however, by incomplete circulation and thus had a more stagnant hypolimnion.

TABLE III

Balance for Ca dissolved in lake water

Lake		Total amount of Ca spread in each lake (mol 10^3)	Amount of Ca dissolved at fall turnover the first year (% of total)	Lost in out-flow after 7 yr (%)	Remaining in lake water (%)	Total amount dissolved after 7 yr (%)	Unrecognized share (%)
Bredvatten	(B)	372.5	37.5	48.9	11.8	60.7	39.3
Lysevatten	(B)	537.5	47.2	56.2	14.6	70.8	29.2
Stensjön	(V)	817.5	35.9	59.8	5.6	65.6	34.4
Skitjärn	(V)	427.5	22.2	47.6	26.2	73.8	26.2

(B) = Province of Bohuslän
(V) = Province of Värmland

The total dose of lime in each lake is given in Table II. Based on analyses of Ca in lake water throughout the study period and monthly calculations of run-off the amount of dissolved lime have been estimated. As can be seen from the table the amount of lime (Ca) dissolved during the first year was between 22.2 and 47.2%. At the end of the study period (1980) dissolved amounts ranges from 60.9 to 73.8% which is somewhat low compared to the easily soluable fraction (about 80%) of the used lime product.

2.3. LONG-TERM TREND PHOSPHORUS

Phosphorus is the limiting nutrient in many of the acid oligotrophic lakes (Dillon *et al.*, 1979; Almer *et al.*, 1978; Dickson, 1978a; Broberg and Persson, 1981).

In all our lakes, the long-term trend leads to lower total P concentration in epilimnion (Figure 3). In the period from 1973 to 1980 the P concentration has been substantially reduced in the lakes, both limed and acid. In laboratory experiments Dickson (1978a) showed that increased Al content in water caused a precipitation of P and humus. In a recent paper, Broberg and Persson (1981) suggest that the high Al-concentrations (1

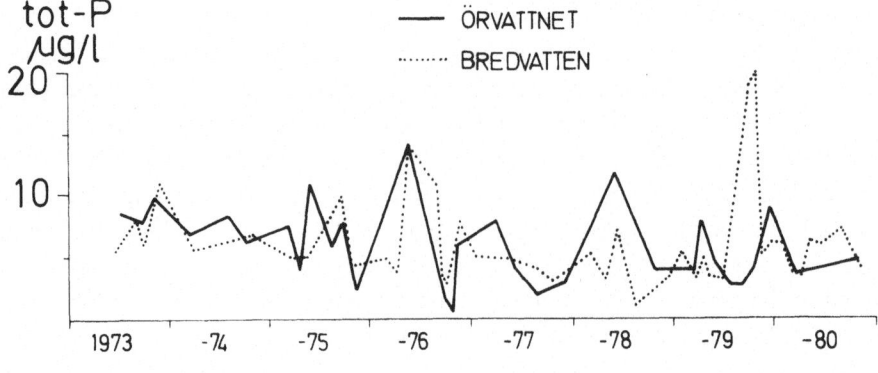

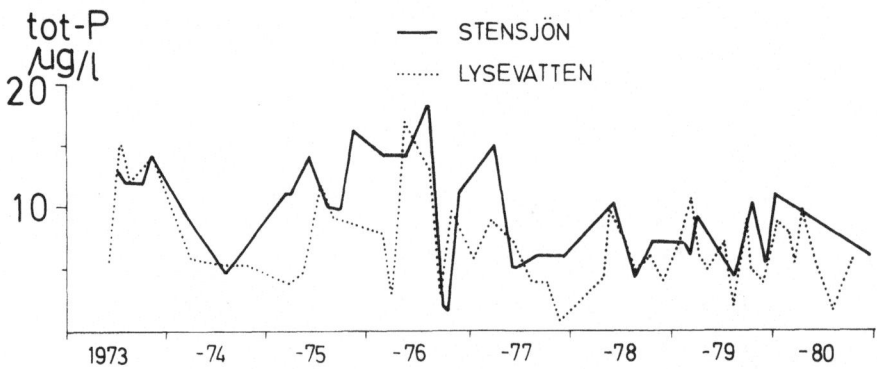

Fig. 3.   Total P-content in three limed lakes. (Bredvatten and Lysevatten in Bohuslän and Stensjön in
Värmland, and one acid reference lake (Örvattnet in Värmland).

to 3 mg l^{-1}) in soil- and groundwater in the watershed around the acid lake Gårdsjön
probably cause a retention of either Al-P complexes or Al-humus-P complexes in the
B-horizon of the soil. These mechanisms might be responsible for the observed long-term
decrease of total P concentrations in the lakes (Figure 3).

Since June 1979 Bredvatten has been fertilized with commercial products. At three
occasions each summer 1979 and 1980 a total of about 40 kg P has been added.
Theoretically this could have increased the P-concentration with 40 μg P l^{-1}. An increase
occurred in autumn 1979 but levels in 1980 were only slightly higher than before. Inflow
from the surrounding watershed to Lake Bredvatten is acid and contains elevated
Al-concentrations which may reach up to 2.6 mg l^{-1} in groundwater (Hultberg and
Johansson, 1981) which together with absorption to the lake sediments (Persson et al.,
1977) may have caused a rapid loss of the added P from the lake water to the sediment.

Changes in P induced by liming were insignificant in our lakes. Apparently none of
the P bound in the lime product became available in the water after treatment. The
dissolved amount of lime would thereotically have increased the P-concentration with
about 10 μg P l^{-1}. National Board of Fisheries and National Environment Protection
Board (1981) reported increased P content at high lime doses in humic lakes and after

liming on land and along small streams. The observed long-term decreasing concentration of total P may therefore be mostly dependent on increased retention in the watershed surrounding the lakes. This process together with precipitation of the P by Al in the lake water may lead to an oligotrophication of the acidified lakes, and, as long as treatment is restricted to the lake proper, the process would continue even in limed lakes.

2.4. MACROPHYTES AND FILAMENTOUS ALGAE

In acidified lakes in western Sweden invasion and expansion of *Sphagnum spp.* is a typical feature as are benthic algal mats (Hultberg and Grahn, 1975; Grahn, 1977; Lazarek, 1980; also cf. Hendrey, 1976; Hendrey and Vertucci, 1980). Vast littoral bottoms in some of our lakes were also covered by a gelatinous felt-like sediment, which reached a thickness of 20 mm and mainly comprised integrated algal filaments, detrital matter, and *Sphagnum*. Observed decrease of submerged vegetation like *Isoëtes spp.* and *Lobelia dortmanna* was probably due to both light competition and direct damage caused by iron encrustments precipitated in the redox gradient that is established when the plants are covered by *Sphagnum* or the felt-like structure. Impact may be substantial since coverage was found to be extensive (Hultberg and Grahn, 1975; Grahn, 1977).

After liming, change was most profound during the first summer. On littoral bottoms where lime primarily was spread the *Sphagnum* died during the 1974 growth season. At depths of 5 to 10 m mossplants still lived 1975, but most were dead by 1976. In Lake Lysevatten *Sphagnum* survived in the depth zone 3 to 10 m throughout the limed period, but growth was inhibited compared to growth under acid conditions. Thus growth rate was reduced from about 8 to 10 cm yr^{-1} to less than 1 cm yr^{-1} (Grahn, 1981). Decreases of pH, alkalinity and Ca were continuous from 1976, and finally growth conditions became favorable for *Sphagnum*, since their growth rate increased in 1979 and 1980 coinciding with increased growth of *Mougeotia sp.* in littoral areas. The felt-like structure was completely disintegrated in Lake Bredvatten during the summer of 1974, indicating also that the benthic blue-green algae (Lazarek, 1980) that formed the 'uniting network' within the felt-like structure, were negatively affected by the liming. These changes were followed by a clearance of littoral bottoms predominantly caused by resuspension and allocation of the finer fractions of detrital matter earlier integrated in the felt-like structure. Cleared littoral bottoms were later successively colonized by both *Isoëtes spp.* and *Lobelia dortmanna*. One other macrophyte, *Potamogeton sp.*, responded positively to the liming and, have expanded continuously in Lake Lysevatten.

2.5. REACTIONS OF PLANKTON COMMUNITIES

Phyto- and zooplankton communities found in acidified lakes are classified as poor in species compared to circumneutral oligotrophic lakes (Hörnström *et al.*, 1973; Almer *et al.*, 1974; Hendrey and Wright, 1976). A successive decrease in number of species correlated to decreasing pH have been noticed in synoptic surveys. Phytoplankton biomasses were often dominated by the peridineans: *Peridinium inconspicuum* and *Gymnodinium uberrimum*, and in humic lakes *Merismopedia tenuissima* (cyanophyta) often was important (Hörnström *et al.*, 1973). Zooplankton species like *Bosmina coregoni* (cladocera), *Eudiaptomus gracilis* and *Cyclops spp.* (copepoda) were, together with a few rotatorian genera, i.e. *Keratella, Kellicottia* and *Polyarthra*, the dominant or sometimes the only species found. The acidified lakes of the long-term project confirmed these general features, although there were minor differences between lakes. Hence, due to lower pH and more prolonged acid state, peridineum algae were more dominant in the Bohuslän lakes. Throughout the study period the acid reference lakes showed the same general picture, in Lake Örvattnet in Värmland, there even was a tendency to an increased dominance of peridineans (Figure 4).

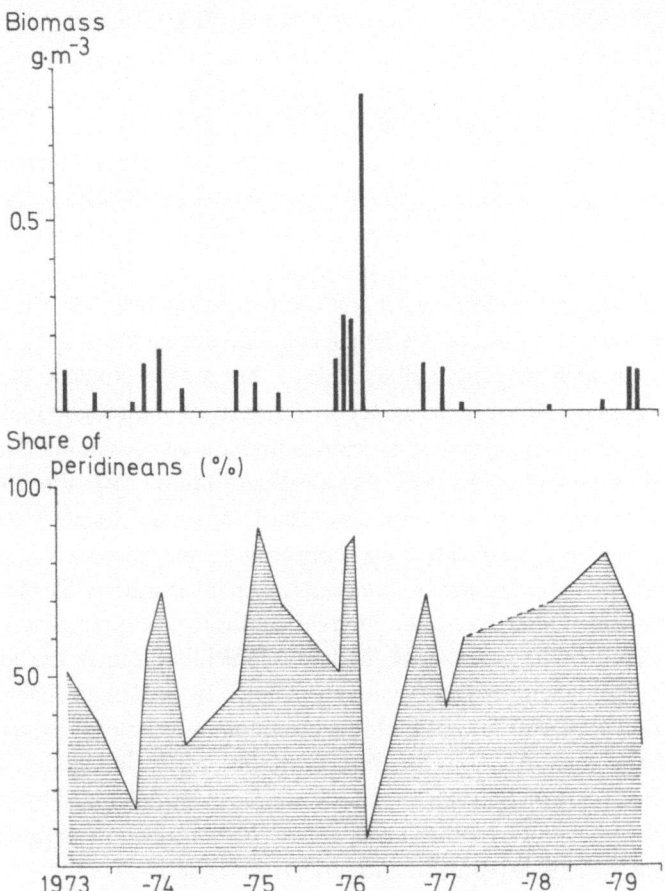

Fig. 4. Total phytoplankton biomass (g × m^{-3}) and percentage share of peridineans in the acid reference Lake Örvattnet in Värmland.

The different time of liming between Bohuslän and Värmland was reflected in phytoplankton reaction. In Lake Lysevatten and Lake Bredvatten spring development had already resulted in increased biomass, which decreased and one month after liming was significantly lower. Changes included compositional shifts and different reactions on species level. Thus in Lake Lysevatten *Gymnodinium sp.* decreased due to liming but again increased to a preliming level in July (1974) and biomass of *Peridinium inconspicuum* was unaltered in Lake Lysevatten but decreased in Lake Bredvatten. During the first summer after liming other changes observed were the appearance of new chlorophycean

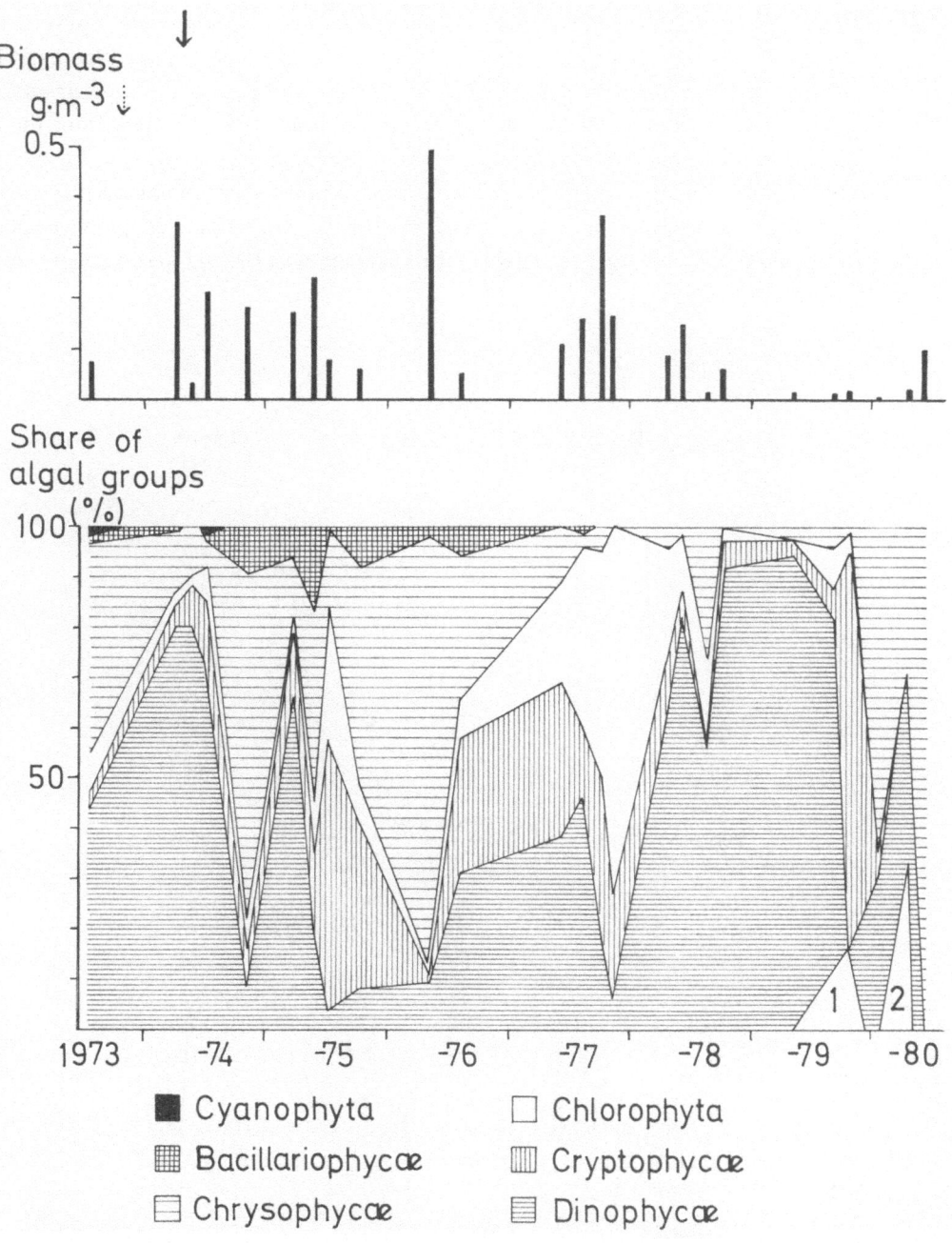

Fig. 5. Total phytoplankton biomass (g × m⁻³) and percentage share of different algal groups in Lake Lysevatten. Dotted arrow indicates time for rotenone treatment and the other arrows time for liming. (1) = *Goniostomum semen*, (2) = *unidentified species*.

species, especially in Lake Bredvatten. In both lakes diatoms increased their share of the total biomass. In the longer perspective compositional changes were more pronounced in Lake Lysevatten (Figure 5) with an increased dominance of small chrysophycean species and also *Dinobryon spp.* as a result. Peridineans dominated biomass in lake Bredvatten until 1979; however, successive shifts of *Peridinium* species were observed, *P. inconspicuum* was replaced by *P. willei* and *P. cinctum*. After the repeated liming in Lake Bredvatten chlorophycean species like *Oocystis sp.*, *Sphaerocystis schroeteri* and small chloromonads dominated the phytoplankton biomass. An influence from the started fertilization seems reasonable (cf. discussion by Dillon *et al.*, 1979).

In Lake Skitjärn the phytoplankton reactions were similar to those observed in the Bohuslän lakes. Successively the dominance of small cryptomonads and chrysophycean species became relatively larger. Parallelling this change was the marked decrease in transparency (cf. above).

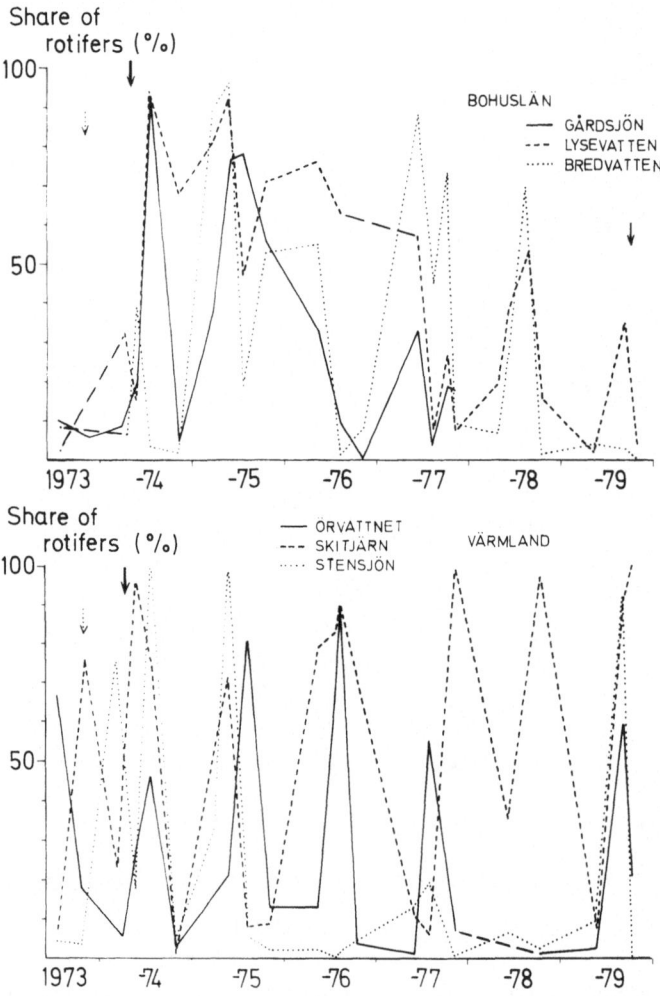

Fig. 6.   Percentage share of rotifers of total zooplankton biomass. Dotted arrow indicates time of rotenone treatment and the other arrows time for liming.

In Lake Stensjön (humic 30 to 40 mg Pt l^{-1}) the liming caused an immediate and almost complete disappearance of *Merismopedia sp.* In this lake peridineans even increased their share of the biomass. During the 2nd and 3rd year after the liming diatoms dominated the biomass. Later on the chrysophycean *Uroglena sp.* dominated the summer biomass.

Immediate zooplankton reactions after the liming are difficult to evaluate because the preceding rotenone treatment virtually exterminated standing crops of cladocera and copepods and the biomasses were dominated by ciliates and rotifers. This effect was observed for 2 yr in Lake Gårdsjörn, but disappeared earlier in the limed lakes. The observed rotifer dominance in Lake Lysevatten and Lake Bredvatten following the treatments (Figure 6) was probably due to food composition and delayed re-establishment of the crustaceans. Dominating rotifer species were partially the same as was observed during acidic state but successively new species like *Asplanchna priodonta* and *Conochilus unicornis* became important. From about the 2nd summer after the liming successions of cladocerans was observed in the lakes. Usually a single species was totally dominant, either occasionally or throughout a whole season. *Daphnia* species, assumed to be sensitive to acidification (Hörnström *et al.*, 1973), were found in all the limed lakes towards the end of the study period. Other species that became important in one or more of the lakes were *Ceriodaphnia quadrangula* and *Holopedium gibberum*. Significant changes within copepods were not observed and throughout the study *Eudiaptomus gracilis* was the most important species.

The general trend in the limed lakes was a decrease in phytoplankton biomasses after liming. Observed species successions after liming cause an unstable phytoplankton community. Also the high biomasses of zooplankton, predominantly rotifers, and cladocerans are indications of an unstable species composition after the liming, as well as the observed successions. The algae *Fragillaria crotonensis* and *Ceratium hirundinella* found before acidification in Lake Lysevatten, however, were never encounted during the study period. Maybe a prolonged period with circumneutral conditions is needed before lost species returns and establish new populations. The decrease in P shown in this study was presumably also important for changes in plankton communities. De Costa and Preston (1980) have shown the importance of P by experiments where they demonstrated that indigenous algae responded to additions of P in both acid conditions and after neutralization.

2.6. CHANGES OF MACRO-INVERTEBRATES

Changes in composition of macro-invertebrates, both benthic forms and actively swimming pelagic species, include decreased species number and biomass of sensitive taxa, such as ephemeroptera, plecoptera, mollusca and crustacae, but also increased biomass of corixids, odonate larvae and certain coleopterans (Grahn *et al.*, 1974; Leivestad *et al.*, 1976; Raddum, 1980). Stenson *et al.* (1978) and Henrikson *et al.* (1980) suggested that observed changes were significantly depending on selective predation, which changed when fish populations were damaged by acidification. Several invertebrates, however, have been shown to be sensitive to decreasing pH, as e.g. crustaceans like *Lepidurus*

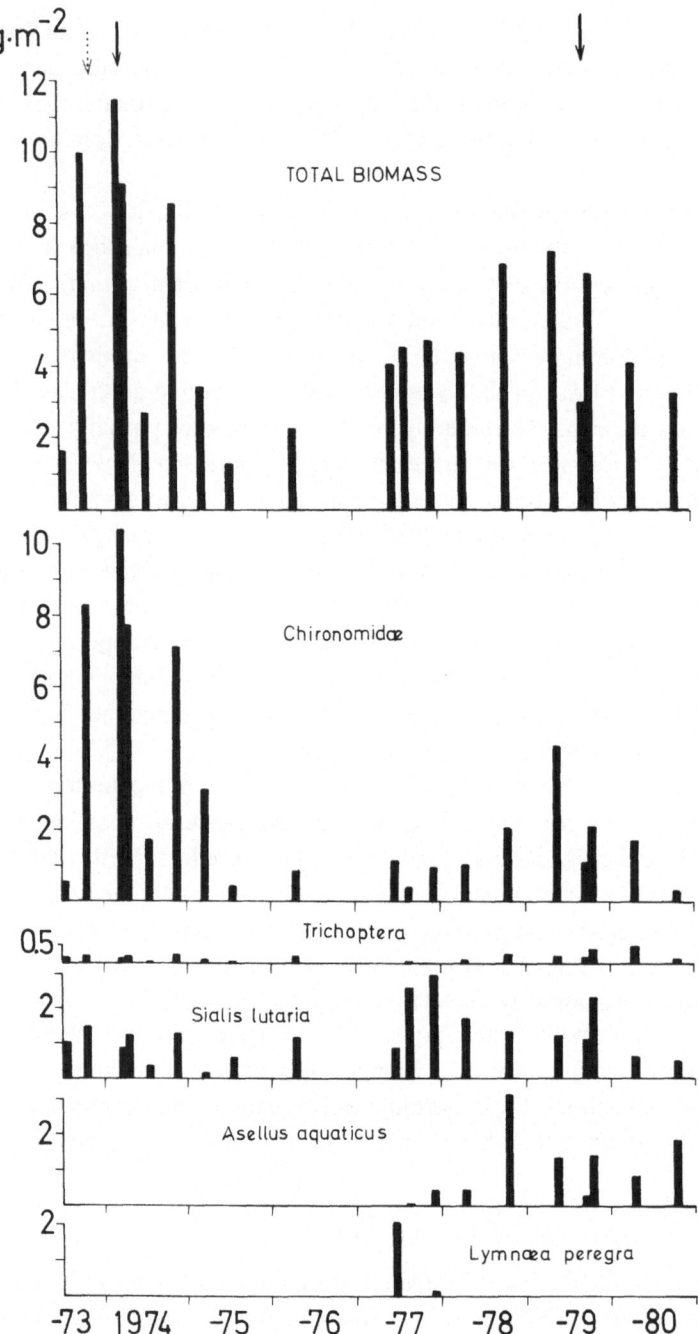

Fig. 7.   Total biomass (g × m^{-2}) and biomass of important benthic organisms on the profundal bottom in Lake Bredvatten. Dotted arrow indicates time for rotenone treatment and the others time for liming.

arcticus and *Gammarus lacustris* (Borgström and Hendrey, 1976; Økland and Økland, 1980) gastropods and ephemerids like *Ephemera vulgata* (Grahn and Hultberg, 1974; Mossberg and Nyberg, 1976; Økland, 1979). The lakes from the two regions had different macro-invertebrate populations, damages were more severe in the Bohuslän lakes, e.g. no *Asellus aquaticus* was found in benthic samples neither on littoral nor on profundal bottoms, and corixids were more abundant.

Populations on littoral bottoms reacted faster after liming. A dense population of *Asellus aquaticus* was found in patches of detritus on shallow sand bottoms in Lake Lysevatten already during 1975, while it was not found on profundal bottoms until 1978. Much the same was observed in Lake Bredvatten. In this lake an introduced gastropod, the snail *Lymnaea peregra*, colonized the littoral bottoms in 1 yr but was observed on profundal bottom only 1977 (Figure 7) when acid melt water had reduced the littoral population to almost zero density.

Biomasses of chironomids decreased substantially (from about 10 g $\times$ m^{-2} to about 1 g $\times$ m^{-2}) in Lake Bredvatten both on profundal (Figure 7) and littoral bottoms. The

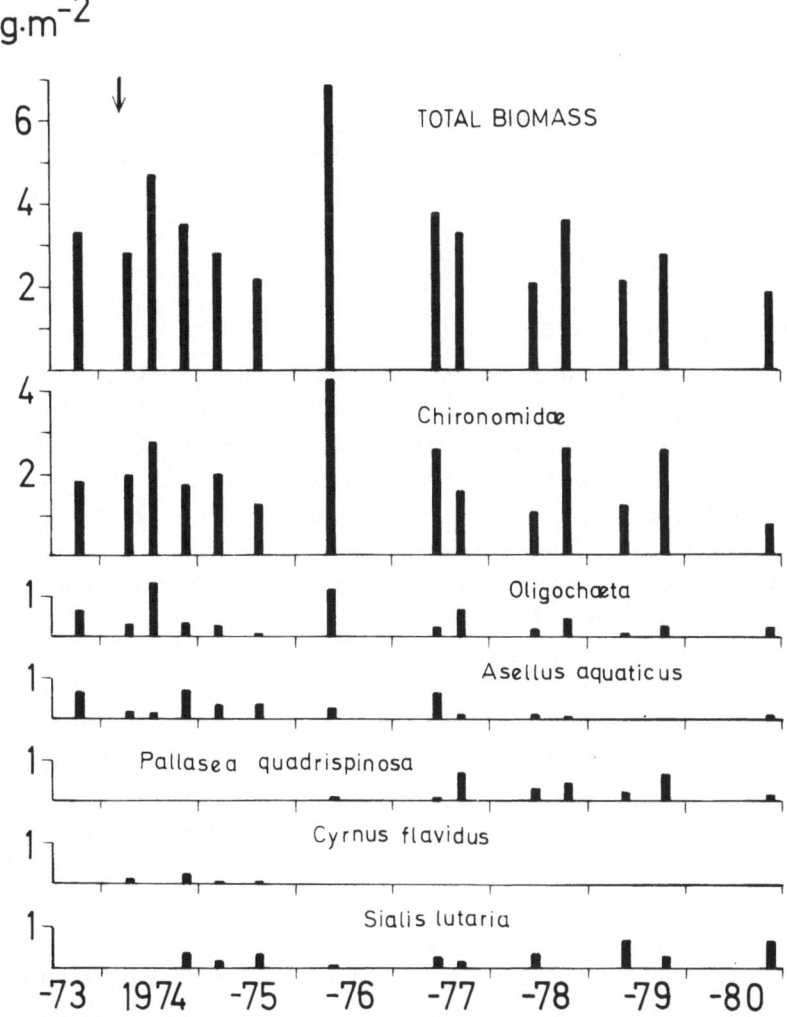

Fig. 8.   Total biomass (g $\times$ m^{-2}) and biomass of important benthic organisms on the profundal bottom in Lake Skitjärn. The arrow indicates time for liming.

opposite was observed in Lake Stensjön where the biomass of *Chironomini* was more than doubled (to about 2 g × m^{-2}) on littoral bottoms the year after the liming. In Lake Skitjärn a fairly stable *Asellus aquaticus* population on profundal bottoms decreased in 1977 (Figure 8) coinciding with an increase of *Pallasea quadrispinosa*, which before liming only was known from fish stomachs. As seen from Figure 7 *Sialis lutaria* and *Asellus aquaticus* increased on profundal bottoms in Lake Bredvatten in 1977, a corresponding increase was also observed in Lake Lysevatten and in this lake also included ephemeropterans and *Chaoborus sp.*

After liming introductions were performed (*Lymnaea peregra*, see above). Cray-fish (*Potamobius astacus*) was successfully introduced to Lake Bredvatten and Lake Stensjön as was *Ephemera vulgata* to Lake Bredvatten. Trials to introduce first *Gammarus pulex* and then *G. lacustris* to lake Bredvatten completely failed for unknown reasons. In general, increase of acid sensitive macro-invertebrate populations proceeded throughout the study period. The best example of this was the expanding population of *Pallasea quadrispinosa*. During the reacidifluation in Lake Bredvatten and Lake Lysevatten this trend did not reverse significantly. The chironomids in Lake Bredvatten, however, seemed to respond to increasing acidity. The population began to increase in 1978 but decreased again after the repeated liming 1979. The lack of fast response to reacidification may be due to an intrinsic inertia of sediment and sediment-living organism but the predation of an unchanged fish population may also be important in this context (cf. Henrikson *et al.*, 1980).

2.7. REACTIONS OF STOCKED FISH

Five lakes were treated with Rotenone (Table II) in order to make possible a study of restocked populations of non-reproducing fish with approximately known density and their effect on macroinvertebrates. A primary restocking with brown trout (*Salmo trutta*) and brook trout (*Salvelinus fontinalis*) (Table II) was done about 1 yr after liming. The amount of fish was based on an estimate of productive littoral bottoms in the lakes. Repeated stockings were later done in each limed lake.

In Lake Gårdsjön one brown trout was caught one year after stocking and thus part of the stocked brown trout had survived but the growth was weak (from 10.5 to 15.6 cm). On the other hand, brook trout grew as well as in the limed lakes. In Lake Gårdsjön and Lake Bredvatten brook trout disappeared during 1976 and 1977, respectively, while the species managed to reproduce in Lake Lysevatten where spawning was successful in an area with seepage from a glaciofluvial aquifer. This upset the planned experiment somewhat, but the aim of comparing fish predation under about equal fish densities in Lakes Bredvatten and Lysevatten then was achieved by yearly stockings of brown trout (fry) in Lake Bredvatten. However, this artificial population increase was hard to match with variable reproductive success of brook trout, and until about 1979 total fish density probably was higher in Lake Lysevatten.

Growth of stocked brown trout strains in Lake Bredvatten and Lake Lysevatten is shown in Figure 9 and 10. At the top of each figure is also a calculated condition factor (f = weight (g) × length^{-3}(mm) × 10^5) for each separate stocking at time of two yearly

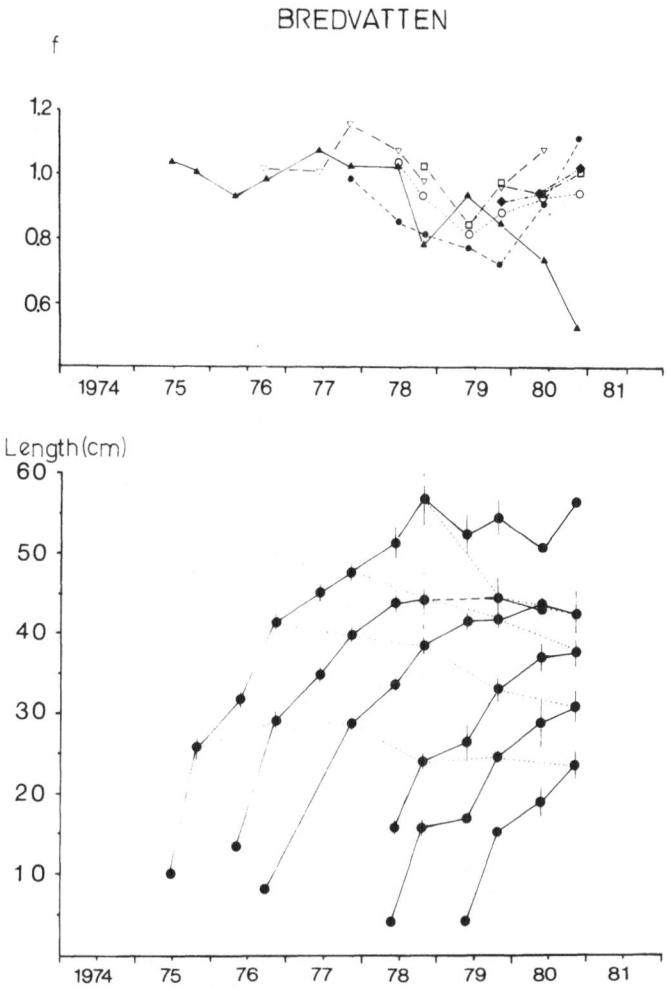

Fig. 9. Growth curves (length in cm) of successive restockings of trout (*Salmo trutta*) in Lake Bredvatten. Dotted lines join fish of the same age. A condition factor was calculated according to the formula f = weight (g) × length^{-3} (mm) × 10^5.

survey fishings. The three first identical stockings showed much the same length increment and condition (f) between the lakes until spring 1978; within lakes there were differences. Growth rate decreased in Lake Bredvatten when comparing the first and third stocking, the Gullspång strain, at the same age (dotted connection lines in Figure 9 and 10). In Lake Lysevatten the difference was contrary with better growth of the third stocking. From autumn 1978 the growth rate and condition of stockings decreased in Lake Bredvatten, but again increased after the repeated liming in 1979 except for the first stocking. In Lake Lysevatten a similar decrease of condition appeared during 1979 and with the exception of the last stocking continued to be low in 1980. Apparently, these changes also were accompanied by decreased swimming activity since the catch by both sportfishers and the gill nets was substantially reduced. Compared to general gill net catches of about 5 to 8 brown trout per net the numbers caught in Lake Bredvatten before

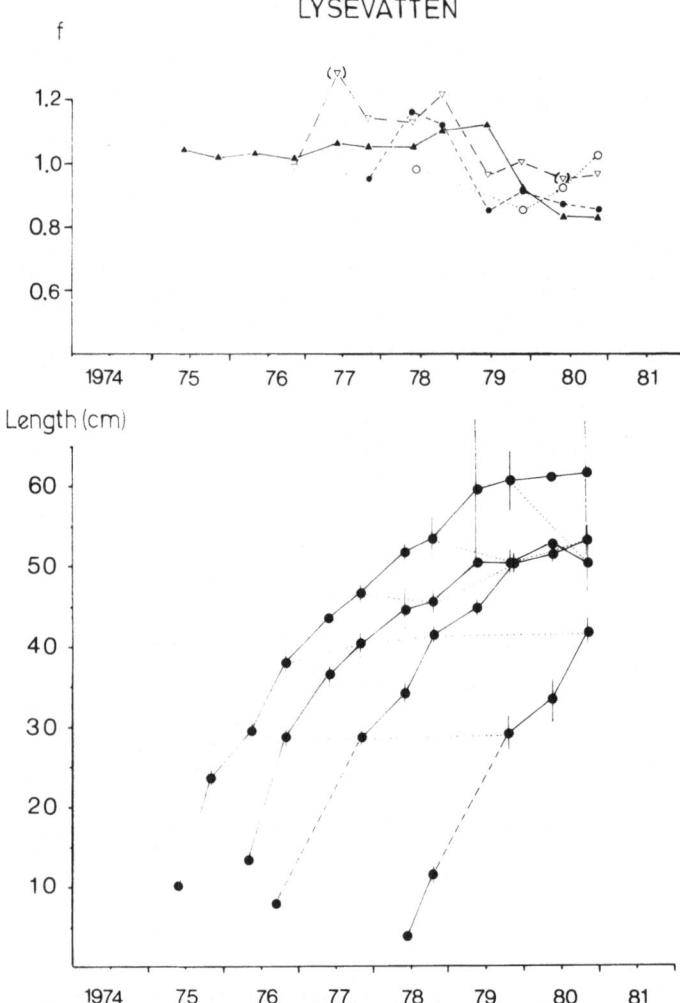

Fig. 10.   Growth curves (length in cm) of successive restockings of trout (*Salmo trutta*) in Lake Lysevatten. Dotted lines join fish of the same age. Also shown is the calculated condition factor *f* (cf. text in Figure 9),

repeated liming were only about 1 fish per net; however, 1 mo after the liming it was about 7 fish per net and continued to be high during 1980. The same phenomenon was observed during the Fall of 1980 in Lake Lysevetten when the catch was only 1 to 2 brown trout per net compared to about 4 to 6 fish per net usually caught.

The observed reduction of growth rate and condition may be dependent on intrinsic differences between discrete brown trout strains, increased competition for food and climatic variations between years. There are, however, indications that beginning reacidification was the most important casual factor. Firstly, brown trout in Lake Bredvatten displayed decreased condition and growth at a lower fish density and when fish food organisms showed no generally decreased populations. Condition of brown trout in Lake Bredvatten increased shortly after repeated liming whereas no other changes were observed. Secondly, the same reaction of brown trout in Lake Lysevatten appeared not until 1980. At this time, competition for food in Lake Lysevatten was substantially reduced since brook trout recruitment almost had stopped. Changes were observed in both lakes when pH was 6.0 to 5.5 and remaining alkalinity was low – (0 to 0.02 meq l^{-1}). At these values inflow from temporary brooks and seepage with pH 3.8 to 4.2 and Al-concentrations of 620 to 730 μg l^{-1} may have influenced the whole lake. The Al-concentration in lake water increased from < 100 μg l^{-1} to about 200 μg l^{-1} during this period. In laboratory tests Al values of about 150 μg l^{-1} have been shown to be acutely toxic (Muniz and Leivestad, 1980) and in other experiments reduced growth of brook trout has been observed (Muniz and Leivestad, 1979). According to this, the increased Al-concentration in the lake water might have caused the reduced growth and condition of brown trout.

If proven valid, the proposed mechanism may be an important factor determining the neutralization effort in areas where acid deposition is high. To avoid the negative effect of Al neutralization should be effective in the sense that a constant pH (> 6) and a certain alkalinity must be present after a treatment and prolonged by repeated treatments before critically low values are reached, altenativly inflow of Al mus be lowered.

3. Summary and Conclusions

The use of lime products is a good and effective means to neutralize acidified lakes if properly administered. In general, effects on aquatic biota are positive. Successions triggered by the neutralization proceed towards a composition of the aquatic ecosystem that probably comes close to the original preacidified state. But, the initiated processes are time dependent and may need substantial time under favorable conditions to reach an equilibrium. Re-establishment of vanished populations of some organisms may be highly improbable due to migratory restrictions. In some cases a return may be possible or timespan shortened by introduction of the organisms, if this is feasible.

There are, however, factors counteracting the positive effects of neutralization. In areas where acid deposition is high or have proceeded for long time changes in watershed soils may be significant. In our project we have notified the trend of decreasing P in lake water and the continuous loading of elevated Al-concentrations from the watershed. The liming of lake water does not exclude these negative factors.

If no measures are taken the acidification will cause a tremendous impoverishment of limnic ecosystems over vast regions. With this in mind, one would be inclined to plea for liming, but in our view liming should only be regarded as a countermeasure to protect sensitive and valuable freshwaters awaiting the necessary and most urgent reduction of acid deposition.

Acknowledgment

This work has been carried out with funds from National Swedish Board for Technical Development (STU), the Central Operating Management (CDL), (a joint organization of major Swedish power producers), Swedish Water and Air Pollution Research Institute (IVL), the National Swedish Environment Protection Board (SNV) and the National Board of Fisheries. We are also grateful to our colleague Olle Grahn who has taken part in much of the work since the start of this long-term study.

References

Almer, B., Dickson, W., Ekström, C., Hörnström, E., and Miller, U.: 1974, *Ambio* **3**, 30.

Almer, B., Dickson, W., Ekström, D., and Hörnström, E.: 1978, 'Sulfur Pollution and the Aquatic Ecosystem', in J. O. Nriagu (ed.), *Sulfur in the Environment*. Part II. Ecological Impacts, J. Wiley and Sons, Inc., New York, p. 271.

Andersson, I., Grahn, O., Hultberg, H., and Landner, L.: 1975, *Comparative Studies of Different Techniques for the Restoration of Acidified Lakes*, Swedish Water and Air Poll. Res. Inst., Stockholm, STU Rep. 73-3651, 48 p.

Beamish, R. J. and Harvey, H. H.: 1972, *J. Fish. Res. Board Can.* **29**, 1131.

Bengtsson, B., Dickson, W., and Nyberg, P.: 1980, *Ambio* **1**, 31.

Borgström, R. and Hendrey, G. R.: 1976, 'pH tolerance of the first larval stages of *Lepidurus arcticus* (Pallas) and adult *Gammarus lacustris*', G. O. Sars (ed.), SNS IR 22/76, 37 p.

Broberg, O. and Persson, G.: 1981, *Närsaltstillgång i försurade sjöar*, Rep. from the Limnological institution, University of Uppsala, 55 pp.

De Costa, J. and Preston, C.: 1980, *Hydrobiologia*, **70**, 39.

Dickson, W.: 1975, *Inst. Freshwater Res. Drottningholm*, Sweden, Rep. No. 54, p. 8–20.

Dickson, W.: 1978a, *Verh. Int. Verein. Limnol.* **20**, 851.

Dickson, W.: 1978b, *Sura vatten – kalkade vatten och kalkningsbehovet för svenska vatten*, National Swedish Environment Protection Board Publication series, PM 1010, 7 p.

Dillon, P. J., Yan, N. D., Scheider, W. A., and Conroy, N.: 1979, *Arch. Hydrobiol. Beih. Ergebn. Limnol.* **13**, 317.

Gjessing, E. T.: 1976, *Physical and Chemical Characteristics of Aquatic Humus*, Ann Arbor Science Publ., Ann Arbor.

Grahn, O.: 1977, *Water, Air and Soil Pollut.* **7**, 295.

Grahn, O.: 1981, pers. comm.

Grahn, O. and Hultberg, H.: 1974, *Effect of Acidification on the Ecosystem of Oliogotrophic Lakes – Integrated Changes in Species Composition and Dynamics*, Swedish Water and Air Poll. Res. Inst., Gothenburg, Medd. No. 2, 12 p.

Grahn, O., Hultberg, H., and Landner, L.: 1974, *Ambio* **3**, 93.

Granat, L. and Söderlund, R.: 1975, *Atmospheric Deposition Due to Long and Short Distance Sources*, Dept. of Meteorology, University of Stockholm. Rep. HC-32, 147 p.

Hendrey, G. R.: 1976, *Effects of pH on the Growth of Periphytic Algae in Artificial Stream Channels*, SNSF, IR 25/76, 50 p.

Hendrey, G. R. and Wright, R. F.: 1976, *J. Great Lakes Res.* **2**, (Suppl. 1), 192.

Hendrey, G. R. and Vertucci, F.: 1980, in Drabløs, D. and Tollan, A. (eds.), *Ecological Impact of Acid Precipitation*, SNSF-project, Oslo-Ås, p. 314.

Henriksen, A.: 1980, in Drabløs, D. and Tollan, A. (eds.), *Ecological Impact of Acid Precipitation*, SNSF-project, Oslo-Ås, p. 68.

Henrikson, L., Oscarsson, H. G., and Stenson, J. A. E.: 1980, in Drabløs, D. and Tollan, A. (eds.), *Ecological Impact of Acid Precipitation*, SNSF-project, Oslo-Ås, p. 330.

Hultberg, H.: 1980, *Metalleffekter i samband med försurning av sjöar*, Svenska Gruvgöreningen, Meddelande No. 153.

Hultberg, H. and Stenson, J.: 1970, *Fauna Flora* **1**, 11.

Hultberg, H. and Grahn, O.: 1975, Effects of Acid Precipitation on Macrophytes in Oligotrophic Swedish Lakes', *Proc. 1st. Spec. Symp. on Atmospheric Contribution to the Chemistry of Lake Waters*. Int. Assoc. Great Lakes Res., Orilla, Ontario, 28 Sept.–1 Oct. 1975, p. 208.

Hultberg, H. and Johansson. S.: 1981, *Nordic Hydrology* **12**, 51.

Hörnström, E., Ekström, C., Miller, U., and Dickson, W.: 1973, *Försurningens inverkan på västkustsjöar.* *Information från Sötvattenslaboratoriet,* Drottningholm, Sweden, No. 4, 81 p.

Hörnström, E.: 1980, *Kalknings- och försurningseffekter på växtplankton i tre västkustsjöar,* National Swedish Environment Protection Board, PM 1220, 55 p.

Jensen, K. W. and Snekvik, E.: 1972, *Ambio* **1**, 223.

Lazarek, S.: 1980, *Naturwissenschaften* **67**, 97.

Lee, Y. H.: 1981, pers. comm.

Leivestad, H., Hendrey, G., Muniz, I. P., and Snekvik, E.: 1976, in Braekke, F. (ed.), *Impact of acid precipitation on forest and freshwater ecosystems in Norway,* SNSF Res. Rep. No. 6/76: p. 87.

Mossberg, P. and Nyberg, P.: 1976, *Försurningseffekter på bottenfauna och fisk o västra Skälsjön. Information från Sötvattenslaboratoriet,* Drottningholm, Sweden, No. 9, 23 p.

Muniz, I. P. and Leivestad, H.: 1979, *Langtidseksponering av fisk til surt vann. Forsøk med bekkerøye Salvelinus fontinalis Mitchill,* SNSF IR No. 44/79, 32 p.

Muniz, I. P. and Leivestad, H.: 1980, in Drabløs, D. and Tollan, A. (eds.), *Ecological Impact of Acid Precipitation,* SNSF-porject, Oslo-Ås, p. 320.

National Board of Fisheries and National Environment Protection Board: 1981, *Kalkning av sjöar och vattendrag 1977–1981,* Inf. Inst. Freshwater Res. Drottningholm, Sweden, No. 4, 201 p.

Odén, S.: 1968, *Nederbördens och luftens försurning – deposition, förlopp och verkan i olika miljöer,* Stockholm, 86 p.

Overrein, L. N., Seip, H. M., and Tollan, A.: 1980, *Acid Precipitation – Effects on Forest and Fish,* SNSF-project, Oslo-Ås, Res. Rep. No. 19/80, 175 p.

Persson, G., Holmgren, S. K., Jansson, M., Lundgren, A., Nyman, B., Solander, D., and Ånell, C.: 1977, 'Phosphorus and Nitrogen and the Regulation of Lake Ecosystems. Experimental Approaches in Subarctic Sweden', in *Proc. from 'Circumpolar conference on Northern Ecology',* Sept. 1975, Ottawa, Can.

Raddum, G.: 1980, in Drabløs, D. and Tollan, A. (eds.), *Ecological Impact Of Acid Precipitation,* SNSF-project, Oslo-Ås, p. 340.

Renberg, I. and Hellberg, T.: 1981, *The pH-history of Lakes in SW Sweden, as Deduced from the Diatom Flora of the Sediments,* Manuscript, Inst. for Ecological Botany, University of Umeå, Sweden.

Schindler, D. W.: 1980, in Drabløs, D. and Tollan, A. (eds.), *Ecological Impact of Acid Precipitation,* SNSF-project, Oslo-Ås, p. 370.

Schindler, D. W., Wagemann, R., Cook, R. B., Ruszczynski, T., and Prokopowich, J.: 1980, *Can. J. Fish Aquat. Sci.* **37**, 342.

Schmul, R.: 1969, *Sjö och fiskeriundersökningar i Härskogsreservatet,* Report for Göteborgsregionens Kommunalförbund.

Stenson, J. A. E., Bohlin, T., Henriksson, L., Nilsson, B. I., Nyman, H. G., Oscarsson, H. G., and Larsson, P.: 1978, *Verh. Int. Verein. Limnol.* **20**, 794.

Tirén, T.: 1981, *Den interna svavel- och kväveomsättningens betydelse för försurning av limniska ekosystem,* National Swedish Environment Protection Board. PM 1376, 68 p.

Wilander, A. and Ahl, T.: 1972, *Vatten* **5**, 431.

Wright, R. F. and Snekvik, E.: 1978, *Verh. Int. Verein. Limnol.* **20**, 765.

Økland, J.: 1979, *Fauna* (Oslo) **32**, 96.

Økland, J. and Økland, K. A.: 1980, in Drabløs, D. and Tollan, A. (eds.), *Ecological Impact of Acid Precipitation,* SNSF-project, Oslo-Ås, p. 326.

STRUCTURE AND PRODUCTIVITY OF EPIPHYTIC ALGAL COMMUNITIES ON *LOBELIA DORTMANNA* L. IN ACIDIFIED AND LIMED LAKES

S. LAZAREK

Institute of Limnology, University of Lund, Sweden

(Received 10 July, 1981; Revised 28 October, 1981)

Abstract. The epiphytic algal community growing on *Lobelia dortmanna* L. was studied in two acidified lakes in southwestern Sweden from May through October 1980; Lake Gårdsjön (pH 4.3 to 4.7) and L. Högsjön (pH 6.3 to 6.7 after liming in 1978/79).

In both lakes a layer of firmly attached diatoms, *Eunotia veneris* and *E. rhomboidea*, covered the *Lobelia* leaves. Scanning electron microscope study revealed a mucoid matrix, and in L. Gårdsjön, heavy cover of detritus. The spring period in L. Gårdsjön was characterized by red alga *Batrachospermum sp.*, which was followed by the appearance of *Binuclearia sp.*, and *Mougeotia spp.* Blue-green algae appeared during the warmer period of the year. The spring period in L. Högsjön was characterized by the development of desmids. Diatoms dominated the community in the summer while green algae appeared in autumn.

The biomass in L. Gårdsjön showed spring and late summer maxima, while in L. Högsjön it increased gradually reaching a maximum in autumn. Chlorophyll *a* concentration was highest in L. Gårdsjön in late summer.

Primary productivity rates calculated per unit substrate surface area at 1.0 m depth were highest in early summer and decreased in late summer.

The results suggest that the liming of L. Högsjön caused significant structural changes in the epiphytic community, favoring diatoms and green algae, especially desmids. Blue-green algae were characteristic for L. Gårdsjön despite that lake's low pH. Productivity rates at 1.0 m depth show similar trends in both lakes.

1. Introduction

An increase in biomass of benthic and epiphytic algae has been observed in many lakes undergoing acidification. Reduced microbial decomposition at low pH (Bick and Drews, 1973) and reduced invertebrate grazing (Hendrey and Wright, 1975) have been proposed as possible mechanisms responsible for this increase. Hendrey (1976) and Müller (1980) both observed low diversity of periphytic communities in their experiments with artificial acidification, but the mechanisms remain unexplained. Acidification apparently changes the structure of algal communities and may, as was observed in phytoplankton (Yan, 1979), reduce their biomass, at least at the beginning of the acidification process. However, this reduction is a transient phenomenon, while the changes in community structure may remain permanent as acidification progresses.

Lime ($CaCO_3$) has been applied to acidified lakes to restore their buffering capacity. The rapid transition from acid to neutral conditions may be harmful to algal communities, however knowledge of their response is poor.

The littoral zone forms a links between terrestrial and aquatic ecosystems. These shallow parts of lakes have a potential to influence the whole lake metabolism by altering the chemical composition of water entering a lake. Organic compounds and metabolities

Water, Air, and Soil Pollution **18** (1982) 333–342. 0049–6979/82/0183–0333$01.50.

released by the macrophyte-epiphyte associations in the littoral zone are added to inflowing water. The role of the littoral zone and its communities was documented by Allen (1971) and reviewed by Wetzel (1975). As was pointed out by Wetzel (1975), a large ratio of the littoral to the pelagial zone in many lakes coincides with a high contribution of attached algae to the total autochthonous carbon production. This should be particularly true of shallow, acidified lakes in southwestern Sweden. In addition, the increase of light penetration and accumulation of organic material on bottoms of acidified lakes may favor the growth of benthic and epiphytic algae.

In situ studies of the ecology of epiphytic algae growing on natural substrate in acidified lakes have not been performed. This study draws attention to the seasonal changes in the structure, biomass and rate of photosynthetic activity of epiphytic algae growing on *Lobelia dortmanna* L. (isoetides) in acidified and limed lakes in southwestern Sweden.

2. Study Area

Studies were performed in southwestern Sweden on the acidified Lake Gårdsjön and on Lake Högsjön which was limed in 1978 and 1979. Some limnological data on these lakes are given in Table I. The lakes are situated 40 km north and 70 km southeast of Gothenburg. Granite rocks form the bedrock of L. Gårdsjön. L. Högsjön region is dominated by gneiss. In general, southwestern Sweden is characterized by soils with weak buffering capacity, and lakes whose watersheds contain mixed deciduous and coniferous forests and a high percent of peat bogs. The littoral zones are dominated by *Sphagnum spp.* and the isoetides *Isoetes lacustris*, and *Lobelia dortmanna*. *Lobelia* often dominates the submerged macrophyte communities in the oligotrophic, ion-poor lakes of northern Europe and America. In the studied lakes, *Lobelia* is found in the littoral zone from 0.0 to 4.0 m, with a maximum development at approximately 0.5 to 1.0 m depth. Vast areas of the bottom of L. Gårdsjön are covered with felt-like cyanophytan mats (Lazarek, 1980). *Littorella uniflora* is common in L. Högsjön.

3. Material and Methods

Quantitative and qualitative sampling of epiphytic algae on *Lobelia* was conducted at approximately monthly intervals during the period May to October 1980. Productivity rates of epiphytic algae were studied between June and October 1980. A structural study of epiphytes was made using a scanning electron microscope (Cambridge Mark 11). Epiphytes on *Lobelia* leaves were fixed in 10% glutaraldehyde fixative with a phosphate buffer. The material was dehydrated with methanol and freon before critical point drying.

Species composition was determined and percent contribution of each species to the total epiphyte biomass was estimated on the basis of cell number. The Sedgewick-Rafter counting cell was used for loosely attached algae. Collodium 'peel-off' technique (Lazarek, in prep.) in combination with a square-marked occular and phase-contrast microscopy was used for counting firmly attached algae. Epiphytes for biomass, chlorophyll *a* and 14-C determinations were removed by agitation and with a soft brush.

TABLE I

Some limnological data on the studied lakes

	Gårdsjön 1980	Högsjön[a]	
		1978	1980
Lake area (km²)	0.3		2.5
Mean depth (m)	7.9		9.0
Transparency (m)	9.5	5.5 –7.5	3.0 –5.8
Colour (mgPt × l⁻¹)	2 –6	10 –40	25 –50
pH	4.3–4.7	0.00	4.2 –5.3	6.3 –6.7
Alkalinity (meq × l⁻¹)	6.2–8.1	0.00	0.13–0.19
Conductivity (mS × m⁻¹)	1.86	5.4 –6.9	6.1 –7.2
Ca (mg × l⁻¹)	1.21	2.2	5.6
Mg (mg × l⁻¹)	0.26	2.0	4.8
Al (mg × l⁻¹)	10.94	0.08–0.11	0.01–0.09
SO₄ (mg × l⁻¹)	0.25	9.6	9.6
SiO₂ (mg × l⁻¹)	3.7	0.17–0.32	0.20–0.30
PO₄–P (μg × l⁻¹)	6.1	3.0	—
Total–P (μg × l⁻¹)	380	3 –13	6 –11
Total–N (μg × l⁻¹)		360 –500	380 –870

[a] Unpublished data provided by Länsstyrelsen, Halmstad. L. Högsjön was limed in 1978 and 1979.

The total biomass of epiphytes was determined as ash-free dry weight (110 and 550 °C) per unit substrate surface area. Chlorophyll a per square cm substrate surface area, corrected for pheophytin a, was determined spectrophotometrically according to Lorenzen (1967) using reagent grade 90% methanol. Calculations were made in accordance to Marker et al. (1980), with the absorption coefficient of chlorophyll a in methanol equal to 77.0. The extraction period was 30 min. in darkness at room temperature, with a final brief boiling. All readings were corrected for turbidity at 750 nm.

Primary productivity rates were based on in situ 14-C measurements. Lobelia has a relatively simple morphology and unlike other submerged phanerogams satisfies its CO_2 requirements by absorption from sediment (Steeman Nielsen, 1960; Wium Andersen, 1971). Therefore $^{14}CO_2$ injected into the water was assumed assimilated by the epiphytes and not by the host-plant. Incubation chambers were made from 3 mm plexiglas. Dark chambers were made from black plexiglas and were covered with aluminum foil. Chambers were placed over Lobelia plants and $NaH^{14}CO_3$ was injected via serum bottle stoppers using 2 ml syringes. It was assumed that pumping the syringe several times provided a homogenous distribution of 14-C in the chamber. After a 4 h incubation, the plant material was stored over formaldehyde in a desiccator for transport to the laboratory. Removed epiphytes were filtered onto a prewetted 0.45 μm Sartorius cellulose acetate membrane filter. Lobelia leaves were rinsed and dried on blotting paper. Leaf surface area was determined and the leaves were placed on prewetted membrane filters. All samples were fumed approx. 10 min above conc. HCl and placed in liquid scintillation vials after 10 min exposure to the air. Root cuts were sampled to determine if 14-C had diffused below the sediment and entered the *Lobelia* leaves via the root system. The material was digested in scintillation vials approx. 5 h at 45 °C with 1.0 ml Soluene-350 (Packard Inst. Co. Inc.). Brown coloring was removed by the addition of 200 μl isopropanol and 200 μl 30% H_2O_2 and subsequent warming at 45 °C for 20 min. Dimilume-30 (Packard Inst. Co. Inc.) was added (9 ml), and counting done in an Intertechnique SL-31 liquid scintillation counter after a 48 h delay to allow for chemo-luminescence decay. Dark uptake of 14-C was subtracted from the light values. Quench correction was calculated by external standard channels ratio using a quenched series prepared with blank membrane filters. The 14-C was added as hexadecane standard (Radiochemical Centre, Amersham, England) and acetone was used as the quenching agent. The efficiency of counting ranged from 75 to 95% in epiphyte samples.

Dissolved inorganic carbon (DIC) in the water was determined in an IR Beckman Carbon Analyzer. Water samples were injected into 15 ml Vacutainers (Becton-Dickinson Co., New Jersey) and kept at −5 °C in darkness until analyses were performed.

Photosynthetic rates in this study were calculated for one depth (1.0 m) as follows: milligrams of C fixed = (sample activity/added activity) × (milligrams of DIC avail-able) × (1.06 correction for isotope discrimination).

Data on hourly and daily solar radiation measured at Landvetter airport in Gothen-burg were utilized in the calculation of light conditions at both lakes. Underwater measurements of light intensity during the incubation period were made with a quantum meter (LI-185, Lambda Inst. Co.), combined with an underwater silicon photodiode quantum sensor. It was assumed that photosynthetically available radiation (PAR = 350 to 700 nm) equals 46% of the total solar radiation. A constant 10% loss of PAR due to surface reflection was assumed during the incubation.

4. Results and Discussion

4.1. COMMUNITY STRUCTURE

As *Lobelia* is a perennial hydrophyte, its leaves are available to epiphyte colonization throughout the year. Studies by Moeller (1978) showed that *Lobelia* is a slow-growing plant which results in a more even distribution of epiphytes on its leaves. In both lakes *Lobelia* leaves were covered by a layer of firmly attached acidophilic, morphologically similar diatoms; *Eunotia veneris* (Kütz.) O. Müller and *E. rhomboidea* Hust. (Figure 1a). A mucoid matrix with embedded epiphytes was observed on older leaves of *Lobelia* in both lakes (Figure 1b), and was similar to that proposed by Allen (1971) and described later by Allanson (1973), but due to the acid conditions in the lakes it was not impregnated by calcium carbonate. Nevertheless, it gives mechanical support to holdfasts of fila-mentous green algae. These in turn serve as substrate for other sessile organisms, particularly diatoms (Figure 1c) and epiphytic bacteria (Figure 1f). The epiphytic com-munity in L. Gårdsjön contained a great amount of detrital material as reported earlier (Lazarek, 1979). The liming of L. Högsjön gradually brought new species to the firmly attached community (e.g. ubiquitous *Achnanthes minutissima* Kütz. and alkaliphilic

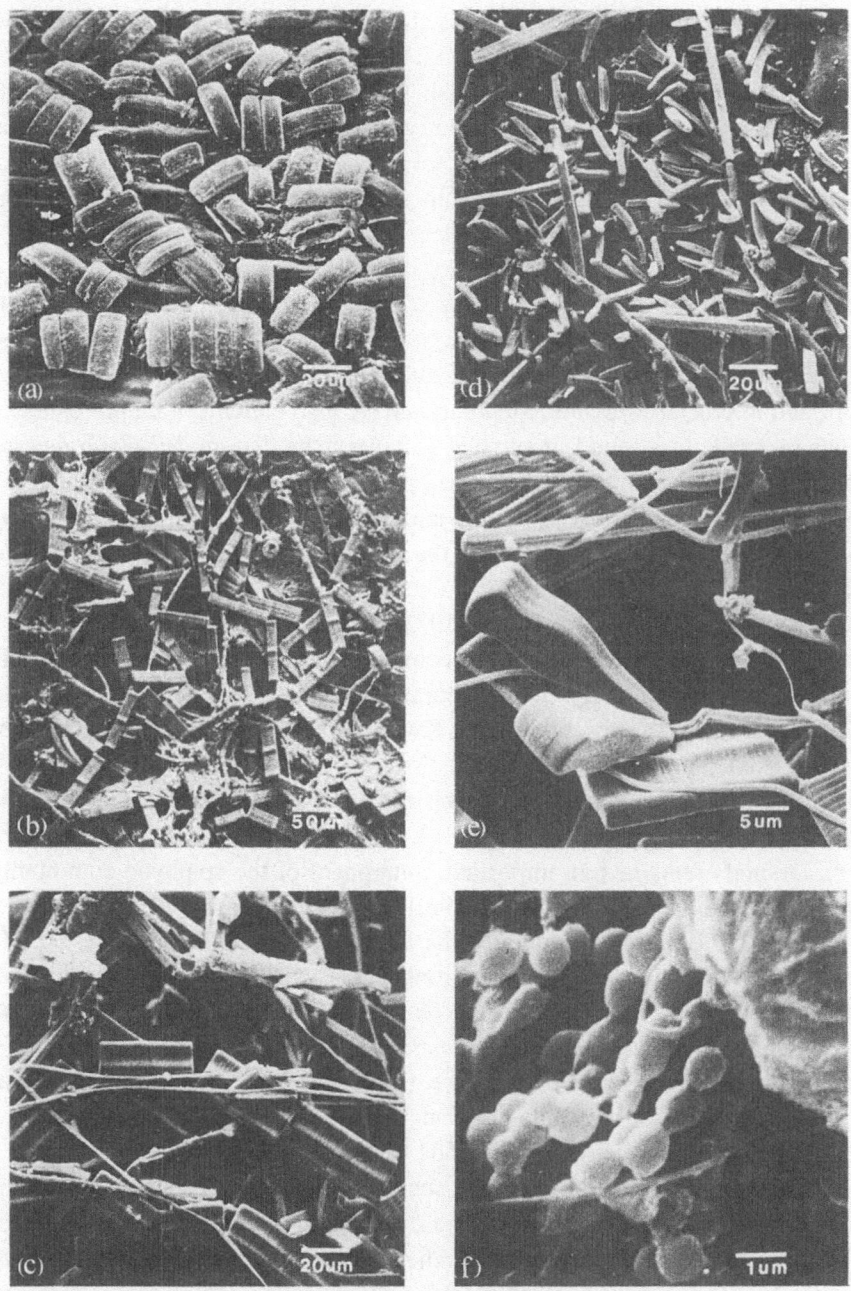

Fig. 1.   Some characteristic epiphytes on *Lobelia* in the studied lakes. (a) a layer of firmly attached diatoms *Eunotia veneris* (Kütz.) O. Müller and *E. rhomboidea* Hust. in L. Gårdsjön; (b) cells of *Tabellaria fenestrata* (Lyngb.) Kütz. embedded in a mucoid matrix in L. Gårdsjön; (c) loosely attached cells of *T. flocculosa* (Roth) Kütz. in L. Gårdsjön; (d) a layer of firmly attached cells of *Achnanthes minutissima* Kütz. in L. Högsjön; (e) *Gomphonema constrictum* Ehrenb., *T. flocculosa* and *Synedra spp.* in L. Högsjön; (f) epiphytic bacteria in L. Högsjön. Scanning electron microscopy (SEM), 1980.

Gomphonema constrictum Ehrenb.) (Figure 1d, e). In contrast, studies in Sweden (Bengtsson *et al.*, 1980) show that two years after liming, planktonic communities were still dominated by the same species.

There was a marked seasonal pattern within the epiphytic communities in the studied lakes. In L. Gårdsjön, an initial dominance of the diatoms *Eunotia veneris* and *E. rhomboidea* in May–June (28% of the total cell number) was observed. Colonies of *Batrachospermum vagum* (Roth) Ag. and filaments of long pectin-rich cells (spring form) of *Binuclearia tectorum* (Kütz.) Berger in Wichm. were abundant (altogether 21% of the total cell number). Single filaments of blue-green alga *Homeothrix sp.* were characteristic. In summer, loosely attached *Mougeotia spp.* filaments, together with firmly attached colonies of *Bulbochaete spp.*, dominated the community (58% of the total cell number). As water temperature increased to 19 °C in July, colonial blue-green algae like *Aphanothece spp.* and *Microcystis spp.* became abundant. A high content of empty planktonic *Peridinium sp.* shells was found in summer samples. The diatom *Tabellaria fenestrata* (Lyngb.) Kütz. and filaments of *Binuclearia* (autumn form) became important components of the autumn community. The only important epiphytic desmid in L. Gårdsjön was *Penium sp.* forming from 2 to 7% of the total cell number throughout the study period.

In contrast to the statement of Brock (1973) that blue-green algae disappear below pH 5, they were abundant in the epiphytic community in L. Gårdsjön, composing 46% of the total cell number in July. The appearance of blue-green algae in such low pH conditions agrees with the observations of Kwiatkowski and Roff (1976), and Conroy *et al.* (1976) from Canadian lakes.

Spring time in L. Högsjön was characterized by a great number of desmids with dominating acidophilic *Hyalotheca dissiliens* J. E. Smith ex Bréb. (30% of the total cell number). Desmids remained an important component of the epiphytic community in L. Högsjön, throughout the study period. The diatom *Achnanthes minutissima,* which was rarely found in 1979 samples, dominated the community in summer, 1980 (42% of the total cell number). Stockner and Armstrong (1971) also reported the midsummer maxima of this diatom in ELA lakes of Ontario. Other diatoms like *Gomphonema constrictum, Synedra rumpens, Frustulia rhomboides* and *Tabellaria flocculosa* were common in the summer. Filamentous green algae (*Oedogonium spp.*) and desmids prevailed in the autumn. The planktonic blue-green alga *Anabaena lemmermannii* P. Richt. was found in samples during the warmer periods of the year.

The influence of environmental factors, such as the pH-CO_2-bicarbonate system, P and N, and toxicity of Al on the growth and distribution of epiphytes in acidified lakes is practically unknown. Experimental studies similar to that undertaken by Moss (reviewed in 1973) are necessary for the understanding of algal responses to acidification and liming.

4.2. BIOMASS

Both lakes showed a similar range for ash-free dry weight values (Figure 2). Spring and late summer maximum were noticed in L. Gårdsjön (1.0 mg cm^{-2}). The biomass of

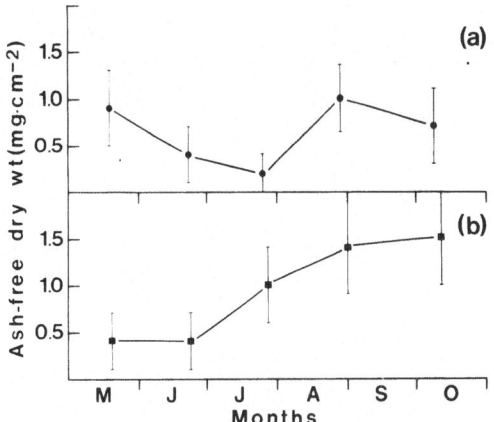

Fig. 2. Mean seasonal values of the dry weight biomass of epiphytes on *Lobelia* in L. Gårdsjön (a), and L. Högsjön (b), expressed as milligrams ash-free dry weight per square cm of the leaf surface area. Vertical bars equal two standard errors of the mean ($n = 5$).

epiphytes in L. Högsjön increased gradually to its maximum (1.5 mg cm^{-2}) in the autumn. The values presented are probably too high because of the accumulation of detrital material and dead algal cells within the epiphytic community. Obtained epiphyte biomass values are higher than reported from artificially acidified streams in Norway (Hendrey, 1976), and the periphyton biomass values reported for epiphytes in Lake George in the Adirondack Mountains, N.Y. (Sheldon and Boylen, 1976). Hargraves and Wood (1967) reported that in an oligotrophic pond (pH 4.0 to 4.9) attached algae peaked twice, in June due to the growth of green algae, and in September due to accumulation of blue-green algae. No clear relationship was found between changes in biomass and the occurrence of any of the main algal groups in these two lakes.

The highest chlorophyll *a* content was found in L. Gårdsjön in autumn (3.4 µg cm^{-2}), Figure 3. Chlorophyll *a* content followed the dry weight changes in this lake in summer

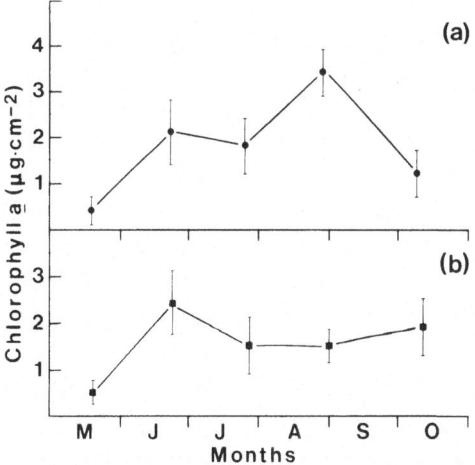

Fig. 3. Mean seasonal values of chlorophyll *a* content of epiphytes on *Lobelia* in L. Gårdsjön (a) and L. Högsjön (b), expressed as micrograms per square cm of the leaf surface area. Vertical bars equal two standard errors of the mean ($n = 5$).

and autumn. Low values observed in May suggest a high contribution of detrital material
to the epiphyte biomass. In L. Högsjön chlorophyll *a* values are relatively low in summer
and autumn, despite high dry weight values and abundance of green algae during these
periods.

4.3. PRIMARY PRODUCTIVITY RATES

Average epiphytic productivity rates, presented in Figure 4, indicate early summer and
autumn maxima for both lakes. Comparative values for algae growing on glass slides,

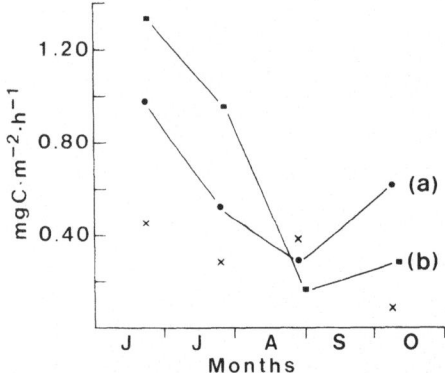

Fig. 4.   Seasonal changes in the productivity rates of epiphytes on *Lobelia* at 1.0 m depth. (a) In L. Gårdsjön,
(b) in L. Högsjön. (x) represents comparative values for the productivity rates of algae growing on glass slides
suspended for one month in L. Gårdsjön. Values expressed in miligrams carbon-14 assimilated per square
meter of leaf surface area, per hour. Duplicate estimations with a correction for dark uptake.

suspended for 1 mo in L. Gårdsjön show relatively low productivity rates. Uptake of
14-C in the dark was negligible, ranging from 1 to 3% of light fixation, thus indicating
low heterotrophic metabolism within the epiphytic communities. No activity was
detected in root-cuts.

Productivity was measured under light saturation conditions. Approximately 62% of
(PAR) reaches 1.0 m depth in the littoral zone of L. Gårdsjön, and 33% in L. Högsjön.
Water temperature during the incubation time varied between 18 °C in June and 12 °C
in October. The observed drop in productivity in both lakes in late August coincided with
reduced radiation during the incubation period and may indicate an ability of the
epiphytes in these lakes to utilize the high intensity light. Other factors, like nutrient
depletion (particularly P), may have been responsible for this drop in productivity but
no chemical analysis of the littoral water during the incubation period was made.

Average DIC concentration at the sampling site in L. Gårdsjön was 1.2 mg C l^{-1}, and
in L. Högsjön 4.9 mg C l^{-1}. These C concentrations were not considered a limiting
factor in the photosynthesis of epiphytes in these lakes. Schindler *et al.* (1980) reported
a concentration range from 1.4 to 0.8 mg C l^{-1} in the ELA lakes of Ontario, and found
that phytoplankton photosynthesis was not C limited. Extremely low concentrations of
DIC (less than 0.6 mg C l^{-1}) in the Canadian Shield lake were not C limiting for
photosynthesis, when N and P were in sufficient supply (Schindler *et al.*, 1973).

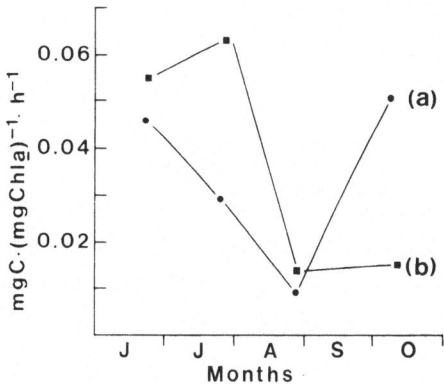

Fig. 5. Seasonal variation in specific productivity of epiphytes on *Lobelia* expressed as the ratio of assimilated carbon-14 to chlorophyll *a* content, per hour. (a) In L. Gårdsjön, (b) in L. Högsjön.

Specific productivity, calculated as a ratio of assimilated 14-C to chlorophyll *a* (Figure 5), follows the changes in productivity rates. There was no evidence of a close relationship between productivity and chlorophyll *a* content in the epiphytes at 1.0 m depth in L. Gårdsjön. Productivity in L. Högsjön was correlated with chlorophyll *a* values ($r = 0.58$).

Productivity rates of epiphytes in the limed lake, L. Högsjön, were not significantly higher (t-test on means at 95% confidence level) than in L. Gårdsjön. Presented productivity rates cannot, in any case, be representative for the studied lakes, due to the heterogeneity of the littoral zones and the limited number of replicas. The obtained results do show that the epiphytic productivity in the studied lakes at 1.0 m depth, is similar to that reported in oligotrophic lakes (Hooper and Robinson, 1976).

Arce and Boyd (1975) reported that the productivity of phytoplankton in limed ponds was generally higher than it was in untreated ponds. Bicarbonate could be a substantial source of CO_2. Also Sheldon and Boylen (1975) reported that in C limited conditions of Lake George, an addition of bicarbonate stimulated the growth of epiphytes by 30% in a 2 h period.

Significant productivity changes, both in acidified and limed lakes, may not be apparent for a long time due to the complexity of factors involved. However, the effect of acidification and/or liming on the structure of epiphytic communities cannot be ignored. As most algae have their origin in attached habitats (Wetzel, 1975), changes in the epiphytic community structure must also cause changes in phytoplankton. This study shows that epiphytes in the limed lake, L. Högsjön, are part of a more dynamic system of significant structural changes during the post-liming period.

There have been few attempts to measure the *in situ* rates of photosynthesis in epiphytic algae on their natural substrate. Technical problems and the uneven distribution of algae on host-plants make such attempts often difficult. This study demonstrates that some of these problems can be overcome. The choice of the host-plant is an important factor considering the existence of the intimate metabolic interactions between the plant and its associated epiphytes (Allen, 1971). The extent to which epiphytes utilize inorganic compounds (O_2, PO_4, CO_2) and organic micronutrients released by macrophytes, is unknown. Further study of epiphytic communities in acidified and limed lakes requires an experimental design which includes the influence of P and Al on the growth of epiphytes.

5. Summary

The main purpose of this study was to provide some characteristics of the seasonal changes in the structure and productivity of epiphytes in acidified and limed lakes. The study was also methodological endeavoring to develop a more convenient method for in situ epiphyte studies and to encourage further investigation of the Lobelia-epiphyte complex.

Acknowledgments

This study was carried out within the integrated 'Lake Gårdsjön project' funded by the Swedish National Environmental Protection Board (SNV), the Fishery Board of Sweden and the Swedish Council for Planning and Coordination of Research (FRN).

I would like to express appreciation to Professor S. Björk, Drs G. Andersson, W. Granéli, and A. Södergren for their encouragement during this study. Dr M. F. Coveney offered invaluable help in improving my methods of productivity measurements and critically discussing results. Finally, thanks to my wife T. M. Press,. for her patient assistance during the field work.

References

Allanson, B. R.: 1973, Freshwater Biol. 3, 535.
Allen, H. L.: 1971, Ecol. Monogr. 41, 97.
Arce, R. G. and Boyd, G. E.: 1975, Trans. Amer. Fish. Soc. 2, 308.
Bengtsson, B., Dickson, W., and Nyberg, P.: 1980, Ambio 9, 1.
Bick, H. and Drews, E. W.: 1973, Hydrobiologia 42, 393.
Brock, T.: 1973, Science 179, 480.
Conroy, N., Hawley, K., Keller, W., and LaFrance, C.: 1976, J. Great Lakes Res. 2 (suppl. 1), 146.
Hargraves, P. E. and Wood, R. D.: 1967, Int. J. Oceanol. Limnol. 1, 55.
Hendrey, G. R. and Wright, R. F.: 1975, J. Great Lakes Res. 2, 192.
Hendrey, G. R.: 1976, Research report, SNSF project, Ås-Norway.
Hooper, N. M. and Robinson, G. G.: 1976, Can. J. Bot. 54, 2810.
Kwiatkowski, R. E. and Roff, J. C.: 1976, Can. J. Bot. 54, 2546.
Lazarek, S.: 1979, Research report, ISSN 0348-0798.
Lazarek, S.: 1980, Naturwissenschaften 67, 97.
Lorenzen, C. J.: Limnol. Oceanogr. 12, 343.
Marker, A. F. H., Nush, E. A., Rai, H., and Riemann, B.: 1980, Arch. Hydrobiol. Beih. 14, 91.
Moeller, R. E.: 1978, Can. J. Bot. 56, 1425.
Moss, B.: 1973, J. Ecol. 61, 157.
Müller, P.: 1980, Can. J. Fish. Aquat. Sci. 37, 342.
Schindler, D. W., Brunskill, G. J., Emerson, S., Broecker, W. S., and Peng, T. H.: 1973, Science 177, 1192.
Schindler, D. W., Wagemann, R., Cook, R. B., Ruszczynski, T., and Prokopowich, J.: 1980, Can. J. Fish. Aquat. Sci. 37, 342.
Sheldon, R. B. and Boylen, C. W.: 1975, Appl. Microbiol. 30, 657.
Steemann Nielsen, E.: 1960, 'Uptake of CO_2 by the Plant', in Handbuch der Pflanzenphysiologie 5, 70.
Stockner, J. G. and Armstrong, F. A. J.: 1971, J. Fish. Res. Bd. Canada 28, 215.
Wetzel, R. G.: 1975, Limnology, W. B. Saunders Co., Philadelphia, 743 pp.
Wium Andersen, S.: 1971, Physiol. Plant. 25, 245.
Yan, N. D.: 1979, Water. Air. and Soil Pollut. 11, 43.

THE EFFECT OF pH AND CALCIUM ON FISH AND FISHERIES

D. J. A. BROWN

CERL Freshwater Biology Unit, Midlands Region Scientific Services Department,
Ratcliffe-on-Soar, Nottingham, NG11 OEE, England

(Received 11 August, 1981; Revised 28 October, 1981)

Abstract. The combined effect of Ca and pH on fish and fisheries is considered for both laboratory and field studies. It can be seen that at concentrations less than 100 $\mu eq\,l^{-1}$, Ca can exert a significant influence on survival times of fish, and similarly in the field, the number of fishless lakes and the number of fish species found in lakes are less dependent on H^+ concentration at low concentrations of Ca than at high Ca levels. The limited historical field data available suggest that alongside any increase there may have been in surface water acidity, Ca concentrations have also increased, and the latter may have offset to some extent the deleterious biological effects of this increased acidity. Nevertheless, details of seasonal and spatial variations in these important water quality factors will need to be considered before a full understanding of the response to acidity of a fishery can be reached.

1. Introduction

There are many factors which have been shown to affect the survival of fish in acid waters; these include biotic factors such as species (e.g. Grande *et al.*, 1978), size and/or age (e.g. Kwain, 1975) and genetic strain (e.g. Gjedrem, 1976); physical factors such as temperature (e.g. Robinson *et al.*, 1976) and season (e.g. Falk and Dunson, 1977); and chemical factors such as trace metals (e.g. Schofield and Trojnar, 1980) and major ions (e.g. Brown, 1981). Of these chemical factors, the influence of the most important trace metal, Al, has been reviewed recently (Hall *et al.*, 1982; Baker and Schofield, 1982) and the purpose of this paper is to consider the combined effects of pH and major ions, in particular Ca, on fish and fisheries.

2. Laboratory and Hatchery Studies

The importance of the overall ionic concentrations to survival of fish at low pH was first demonstrated by Lloyd and Jordan (1964) who found that the short term (up to four days) toxicity of acidified dilutions of hard water with constant low free CO_2 to rainbow trout (*Salmo gairdnerii*) was related to the extent of the dilution. More recently the mortality of brook trout (*Salvelinus fontinalis*) and white sucker (*Catostomus commersoni*) fry at low pH has been found to be lower in high conductivity water (Trojnar, 1977a, b) and higher mortalities of eggs of brown and sea trout (*Salmo trutta*) and salmon (*Salmo salar*) have been reported in half diluted stream water compared with undiluted medium when both were acidified to pH 4.0 (Carrick, 1979). Also addition of salts to acid waters has been shown to lead to improved survival of various salmonids in a Norwegian hatchery (Bua and Snekvik, 1972).

Water, Air, and Soil Pollution **18** (1982) 343–351. 0049–6979/82/0183–0343$01.20.
Copyright © 1982 by D. Reidel Publishing Co., Dordrecht, Holland, and Boston, U.S.A.

In a series of experiments designed to determine the effects of individual ions on the survival of yearling brown trout at pHs between 3.5 and 4.0, Brown (1981) showed that of the major cations tested (Na, K, Ca, and Mg) only Ca was universally effective at prolonging survival time compared with that in deionized water. The physiological responses of fish to low pH have also been shown to be strongly influenced by the external concentration of Ca (McWilliams and Potts, 1978; McDonald et al., 1980).

In studies where the combined effects of pH and Ca have been investigated over a range of concentrations, it is possible to construct isopleths to show where different combinations of the two ions produce the same effect. The survival time data of Lloyd and Jordan (1964) (Figure 1) suggest that throughout the ranges of concentrations used

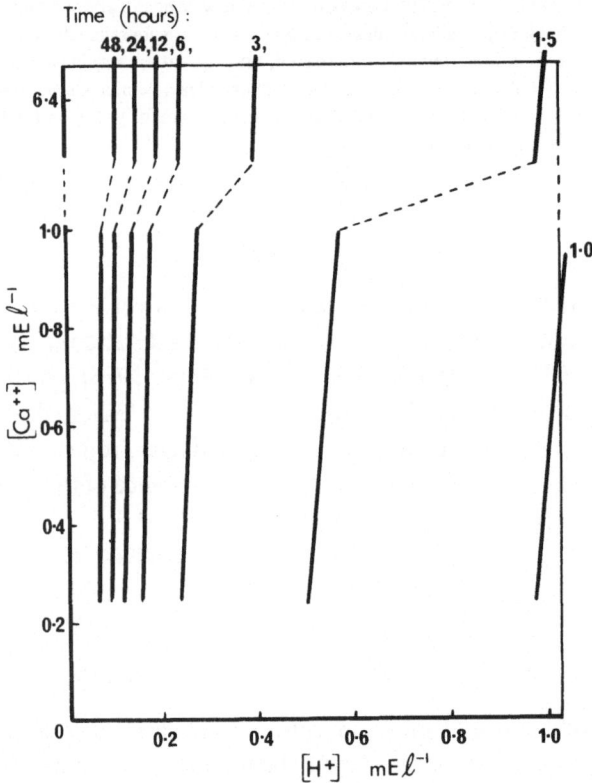

Fig. 1. The median of survival (hr) of rainbow trout (*Salmo gairdnerii*) in solutions with a range of Ca and H[+] concentrations (data of Lloyd and Jordan, 1964).

in their experiments (60 to 1000 μeq H l^{-1} and 240 to 6400 μeq Ca l^{-1}) the effect of Ca is relatively unimportant compared with that exerted by the H^{+} concentration. The data of Brown (1981) (Figure 2) covering a lower range (100 to 300 μeq H l^{-1} and 0 to 400 μeq Ca l^{-1}), however, show that at concentrations less than 100 μeq l^{-1} Ca can exert a significant influence on survival times. In Figure 3 are shown some more recent results on survival times of brown trout in concentrations of both Ca and H^{+} of less than 100 μeq l^{-1}. These were obtained using methods similar to those described by Brown

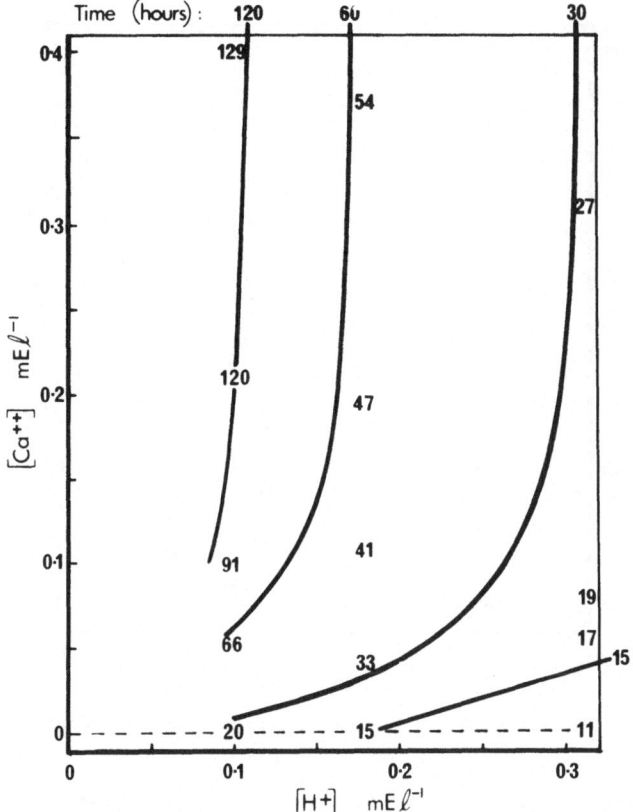

Fig. 2. The median period of survival (hr) of brown trout (*Salmo trutta*) in solutions with a range of Ca and H⁺ concentrations (data of Brown, 1981).

(1981) but with the following basic differences. A temperature of 10 °C, a constant concentration of 40 μeq l⁻¹ of Na and chloride in all treatments and the experimental animals were fry from a hard water hatchery. These differences explain why the values obtained do not correspond exactly with the extrapolations from Figure 2, but the shape of the relationship is undoubtedly similar and confirms the influence of Ca at low concentrations.

3. Field Studies

Although in this paper only the relationships that exist between pH, major ions and fish in the field situation will be considered, it should be remembered that there are many factors in addition to chemical ones that affect fisheries. For example, lake morphometry is known to be an important factor in determining fish productivity in lakes (Rawson, 1955). Indeed, in the acid La Cloche lakes, mean depth has been shown to be the best single predictor of growth of rock bass (*Ambloplites rupestris*) (Ryan and Harvey, 1977).

The fact that the conductivity of the water affects the response of a fishery to acidity has been recognized by, amongst others, Muniz and Leivestad (1980). In a sample of

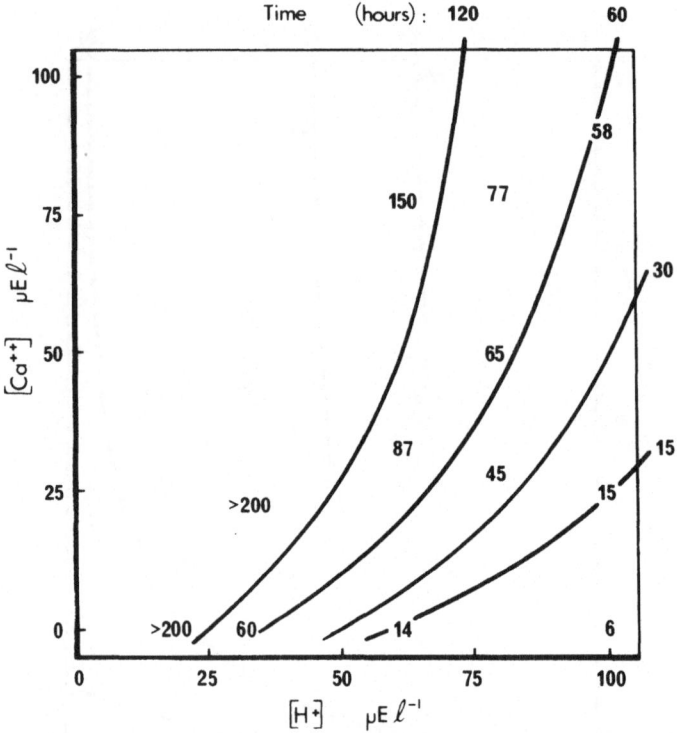

Fig. 3.   The median period of survival (hr) of brown trout (*Salmo trutta*) in solutions with a range of Ca and H⁺ concentrations.

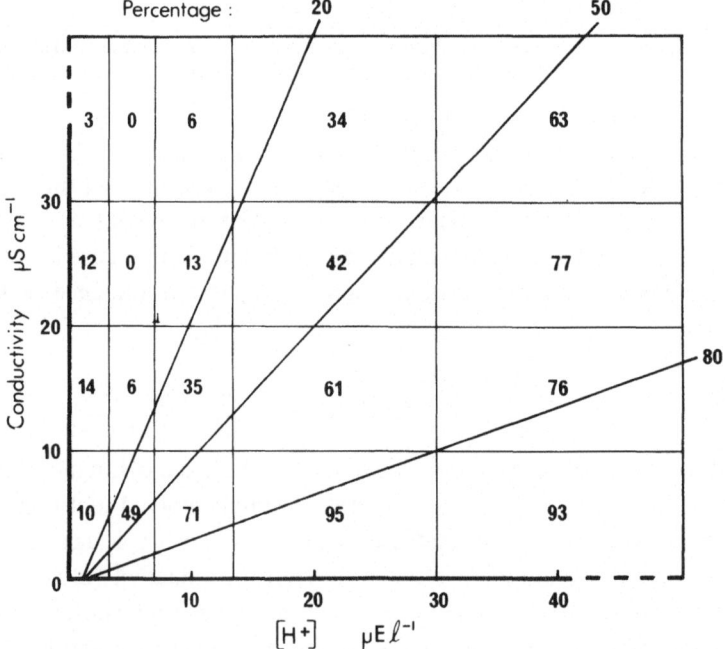

Fig. 4.   The percentages of lakes without fish populations in various categories of conductivity and H⁺ concentration from a total of 1034 lakes in southernmost Norway (data of Muniz and Leivestad, 1980).

1000 lakes in southermost Norway, they showed that the proportion of lakes at a particular pH which have no fish is lower in lakes with a high salt content. The data are represented in Figure 4 where isopleths of similar effects (i.e. percentage of fishless lakes) are suggested.

From a very detailed study of chemistry and fishery status of 700 lakes in the same region, Wright and Snekvik (1978) were able to demonstrate that, as in the laboratory studies on fish, the specific ion responsible for the increased acid tolerance of fish populations at higher conductivity levels was Ca. Their data for the percentage of lakes which have no fish populations over the ranges of Ca and H^+ concentrations are represented in Figure 5. In common with the results for the survival of fish in laboratory

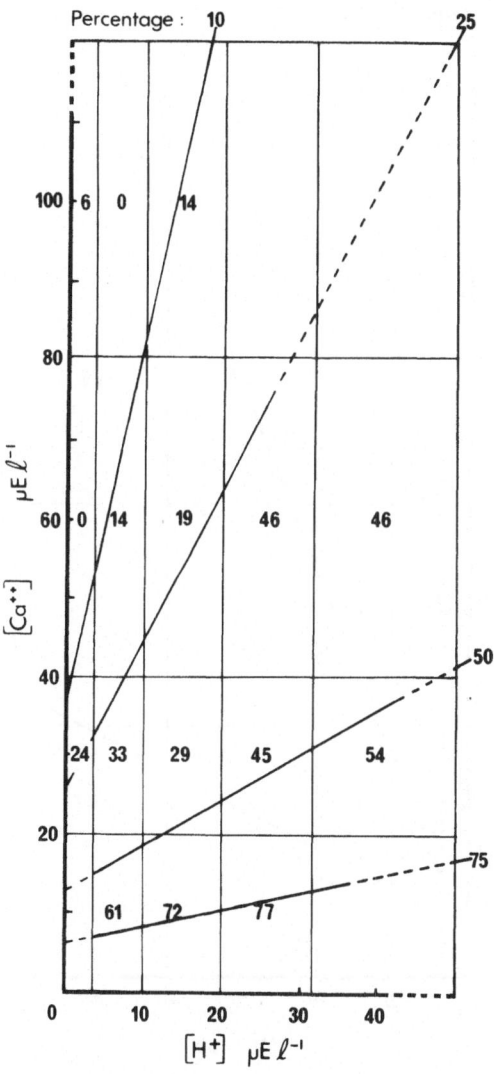

Fig. 5.   The percentages of lakes without fish populations in various categories of Ca and H^+ concentrations from a total of 700 lakes in southernmost Norway (data of Wright and Snekvik, 1978).

experiments, the percentage of fishless lakes is less dependent on H^+ concentration at low concentrations of Ca than at high Ca levels.

Whereas the fish populations in the Norwegian lakes described above are predominantly of a single species (brown trout), a variety of species is often found in North American lakes. Harvey (1975) investigated the relationship between fish diversity and various chemical and morphometric features of 66 La Cloche lakes. His data for the number of fish species, H^+ concentration and Ca hardness are shown in Figure 6. Although there are missing data blocks, the evidence suggests that the pattern of effects produced by the two ions is similar to that previously described.

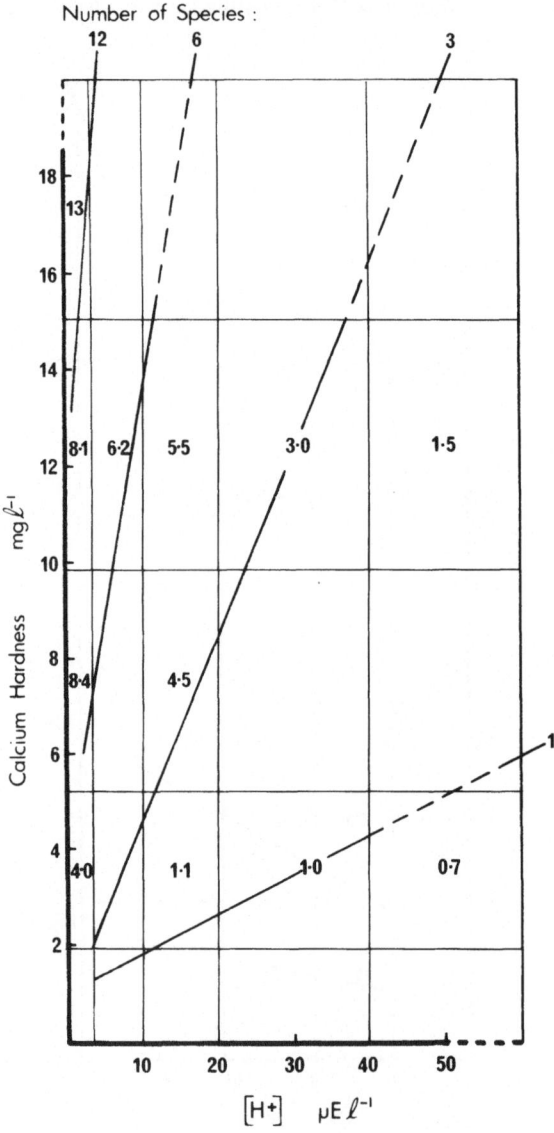

Fig. 6.   The mean number of fish species in various categories of Ca hardness and H^+ concentration from a total of 66 lakes in Ontario (data of Harvey, 1975).

4. Discussion

The biological effects of combinations of Ca and H^+ described above are undoubtedly relevant to discussions on acid precipitation because the oligotrophic lakes which are thought to be susceptible have concentrations of Ca in the range that has been shown to be important for fish and fisheries. For example, over 50% of the 700 lakes sampled by Wright and Snekvik (1978) (Figure 5) had Ca concentrations less than 40 μeq l^{-1}. Further, the limited historical field data available suggest that alongside any change there may have been in surface water acidity, Ca concentrations have also increased. For example, Dickson (1980) gives information on a subset of six acid lakes in the Swedish west coast region which in 1947–52 had H^+ concentrations of 2 to 13 μeq l^{-1} and non-marine Ca plus Mg concentrations of 20 to 100 μeq l^{-1}. In 1974–79 the same lakes had concentrations of H^+ and non-marine Ca plus Mg of 13 to 80 μeq l^{-1} and 85 to 230 μeq l^{-1}, respectively. Also the data presented by Gjessing et al. (1976) for the changes in pH and total hardness in seven rivers in southernmost Norway during the period 1965–75 show the average increase in H^+ concentration during this period was 3 μeq l^{-1} whereas the average hardness increased by 21 μeq Ca l^{-1}. Similarly, Watt et al. (1979), comparing data on the chemistry of 12 lakes in Novia Scotia sampled in 1955 with more recent (1977) analyses, found average increases in H^+ and Ca concentrations of 16 and 23 μeq l^{-1}, respectively.

In addition to this historical data, there is circumstantial evidence available from regional surveys of lakes where higher Ca concentrations have been found in acid lakes relative to unaffected lakes on similar geologies (Almer et al., 1978; Dillon et al., 1979). Also in an analysis of the data of Wright and Snekvik (1978), Brown et al. (1980) found significant trends towards both higher H^+ and Ca concentrations with higher sulphate concentrations in lakes in southernmost Norway. The effects of the higher Ca on the fish populations may counteract the effect of the higher H^+ concentrations because the fishery status of these Norwegian lakes, on an area of fairly uniform geology, is independent of sulphate concentration (Brown and Sadler, 1981).

Thus, the hypothesis that concentrations of Ca in surface waters may have increased at the same time as any change in their acidity, and thereby offset to some extent the deleterious biological effects of this increased acidity is worthy of further investigation. However, it must be pointed out that water chemistry is known to vary seasonally and with flow, and yet many of the relationships described between fisheries and water chemistry are based on a single water sample. There are occasions when Ca and H^+ concentrations in lake water may not change in the same direction. Johannessen et al. (1980) showed that during snowmelt, Ca concentrations in the run-off water were frequently reduced at the same time that H^+ concentrations were elevated. The effect of such run-off on lake water chemistry will depend on such factors as the ratio of the snowmelt volume to the lake volume and stratification. With a rapid turnover, snowmelt water could induce conditions deleterious to fish, but only a few instances of direct fish kills have been reported (Leivestad and Muniz, 1976). Indeed, examples are known where fish have gathered in niches with better water quality during acid incidents (Muniz and Leivestad, 1980). This could be the reason why lake morphometric features have been shown to be important to fisheries, with larger lakes offering a greater diversity of conditions.

It can be seen therefore that details of seasonal and spatial variations in the important water quality factors will need to be considered before a full understanding of the reponse to acidity of a fishery can be reached, and the results presented in this paper justify the conclusion that Ca concentration is one of these important water quality factors.

Acknowledgments

The assistance given by Mrs S. Lynam with the laboratory study is gratefully acknowledged, as are the helpful discussions with Dr G. D. Howells and Dr K. Sadler.

This paper is published by permission of the Central Electricity Generating Board.

References

Almer, B., Dickson, W., Ekstrom, C., and Hornstrom, E.: 1978, 'Sulphur Pollution and the Aquatic Ecosystem', in Nriagu, J. O. (ed.), *Sulphur in the Environment. Part II Ecological Impacts*, J. Wiley and Sons, New York, pp. 271–311.

Baker, J. and Schofield, C.: 1982, *Water, Air, and Soil Pollut.* **18**, 289 (this issue).

Brown, D. J. A.: 1981, *J. Fish Biol.* **128**, 31.

Brown, D. J. A., Sadler, K., Howells, G. D., and Kallend, A. S.: 1980, 'Fish and Freshwater Chemistry', in Drablos, D. and Tollan, A. (eds.), *Proc. Int. Conf. Ecol. Impact Acid Precip.*, Norway, SNSF project, pp. 280–281.

Brown, D. J. A. and Sadler, K.: 1981, *J. Appl. Ecol.* **18**, 433.

Bua, B. and Snekvik, E.: *Vann.* **7**, 86.

Carrick, T. R.: 1979, *J. Fish Biol.* **14**, 165.

Dickson, W.: 1980, 'Properties of Acidified Water', in Drablos, D. and Tollan, A. (eds.), *Proc. Int. Conf. Ecol. Impact Acid Precip.*, Norway, SNSF project, pp. 75–83.

Dillon, P. J., Yan, N. D., Scheider, W. A., and Conroy, N.: 1979, *Archs. Hydrobiol. Ergeb. Limnol.* **13**, 317.

Falk, D. L. and Dunson, W. A.: 1977, *Water Res.* **11**, 13.

Gjedrem, T.: 1976, *Genetic Variation in Tolerance of Brown Trout to Acid Water*, SNSF project FR 5/76, 1432 As–NLH, Norway.

Gjessing, E. T., Henriksen, A., Johannessen, M., and Wright, R. F.: 1976, 'Effects of Acid Precipitation on Freshwater Chemistry', in *Impact of Acid Precipitation on Forest and Freshwater Ecosystems in Norway*, SNSF project FR 6/76, pp. 64–85, 1432 As–NLH, Norway.

Grande, M., Muniz, I. P., and Anderson, S.: 1978, *Verh. Internat. Verein Limnol.* **20**, 2076.

Hall, R., Pratt, J. M., and Likens, G. E.: 1982, *Water, Air, and Soil Pollut.* **18**, 273 (this issue).

Harvey, H. H.: 1975, *Verh. Internat. Verein Limnol.* **19**, 2406.

Johannessen, M., Skartveit, A., and Wright, R. F.: 1980, 'Streamwater Chemistry Before, During and After Snowmelt', in Drablos, D. and Tollan, A. (eds.), *Proc. Int. Conf. Ecol. Impact Acid Precip.*, Norway, SNSF project, pp. 224–225.

Kwain, W. H.: 1975, *J. Fish. Res. Board Can.* **32**, 493.

Leivestad, H. and Muniz, I. P.: 1976, *Nature* **259**, 391.

Lloyd, R. and Jordan, D. H. M.: 1964, *Int. J. Air Wat. Poll.* **8**, 393.

McDonald, D. G., Hobe, H., and Wood, C. M.: 1980, *J. Exp. Biol.* **88**, 109.

McWilliams, P. G. and Potts, W. T. W.: 1978, *J. Comp. Physiol.* **126**, 277.

Muniz, I. P. and Leivestad, H.: 1980, 'Acidification – Effects on Freshwater Fish', in Drablos, D. and Tollan, A. (eds.), *Proc. Int. Conf. Ecol. Impact Acid Precip.*, Norway, SNSF project, pp. 84–92.

Rawson, D. S.: 1955, *Verh. Internat. Verein Limnol.* **12**, 164.

Robinson, G. D., Dunson, W. A., Wright, J. E., and Mamolito, G. E.: 1976, *J. Fish Biol.* **8**, 5.

Ryan, P. M. and Harvey, H. H.: 1977, *J. Fish. Res. Board Can.* **34**, 2079.

Schofield, C. L. and Trojnar, J. R.: 1980, 'Aluminium Toxicity to Brook Trout (*Salvelinus fontinalis*) in Acidified Waters', in *Proc. Conf. Polluted Rain*, Plenum, New York.

Trojnar, J. R.: 1977a, *J. Fish. Res. Board Can.* **34**, 262.

Trojnar, J. R.: 1977b, *J. Fish. Res. Board Can.* **34**, 574.

Watt, W. D., Scott, D., and Ray, S.: 1979, *Limnol. Oceanogr.* **24**, 1154.

Wright, R. F. and Snekvik, E.: 1978, *Verh. Internat. Verein. Limnol.* **20**, 765.

ACIDIFICATION OF NOVA SCOTIA LAKES I:
RESPONSE OF DIATOM ASSEMBLAGES IN THE HALIFAX
AREA

HAGUE H. VAUGHAN, JOHN K. UNDERWOOD

Nova Scotia Department of the Environment, P.O. Box 2107, Halifax, Nova Scotia B3J 3B7

and

J. GORDON OGDEN, III

Department of Biology, Dalhousie University, Halifax, Nova Scotia

(Received 7 July, 1981; Revised 8 October, 1981)

Abstract. A comparison is made of diatom remains in surficial lake sediments collected in 1971 and 1980. Changes of similar magnitude towards more acid tolerant assemblages are observed in four lakes studied in detail despite widely differing aquatic environments. No attempt is made to reconstruct past chemistry, the significant point being that at the most fundamental level, local lacustrine biological systems are being stressed to the point of alteration. The pattern of that alteration with emphasis on 'simplification' is discussed with the suggestion that rates of acidification may be as important as net changes.

1. Introduction

Studies of the modern chemistry of lakes in the Halifax area have demonstrated changes attributable to induced acidification through comparison with similar studies performed in the recent past (e.g., Watt *et al.*, 1979). This research attempts a similar demonstration through the use of diatom assemblages with the additional object of developing a technique for the manipulation of diatom data which may be used in sediment cores to document in detail the biological effects of acidic inputs. While such a demonstration has not previously been attempted locally, the techniques involved have been successfully used in Europe.

Based on the grouping of diatoms according to their pH preference, Nygaard (1956) introduced a quantative treatment which emphasized the ecological significance of the extreme groups. He derived three indices which he used to investigate the development of the Danish lake Store Gribsö.

According to this method, diatoms are divided into five groups:

alkalibiontic (alkb): occurring at pH values above 7.

alkaliphilous (alkf): occurring at pH values about 7 and with widest distribution at pH above 7.

circumneutral (circ): occurring around pH 7.

acidophilous (acf): occurring at about pH 7 and with widest distribution at pH values below 7.

acidobiontic (acb): occurring at pH values below 7; optimum distribution at pH 5.5 or less.

Nygaard's resulting indices are:

$$\text{Index } \alpha: \quad = \frac{\text{acid units}}{\text{alkaline units}}$$

$$\text{Index } \omega: \quad = \frac{\text{acid units}}{\text{number of acid taxa}}$$

$$\text{Index } \varepsilon: \quad = \frac{\text{alkaline units}}{\text{number of alkaline taxa}}$$

where 'acid units' for example are computed by adding the relative frequency of the acidophilous taxa to the relative frequency of the acidobiontic taxa multiplied by five. The arbitrary coefficient five emphasizes the ecological significance of the acidobiontics.

Meriläinen (1967) showed that Nygaard's indices could be clearly related to pH in 14 lakes and ponds in Finland. Recently van Dam and Kooyman-van Blokland (1978) have used the analysis of diatom assemblages by these pH groups to demonstrate rain induced acidification in some Dutch moorland pools through comparison of modern samples with samples collected between 1916 and 1929.

More recently, Davis and Berge (1980) and Dickman *et al.* (1981) have used similar techniques to infer the pH history of two Norwegian and two Quebec lakes respectively.

The Nova Scotia Department of the Environment has attempted to use this approach to clarify, rather than strictly quantify, recent trends. In doing so, it was first necessary to show that the pH preferences and the technique generally, being based on European assemblages were valid locally before any investigation of whether induced acidification has caused shifts in the spectra so produced. Results from investigations in the Halifax area are presented in this report. Similar studies in areas of Nova Scotia more distant from local emissions are underway.

2. Methods, Applicability and Investigated Lakes

A part of a 1971 study of Halifax area lakes, lists of the diatom remains observed in the surface sediments were prepared by G. H. Yezdani but were neither published nor utilized in subsequent interpretation of local water quality (Ogden, 1972). Beyond Yezdani's notations of 'common' and 'very common', no information was given as to the distribution of individuals amongst the species listed.

Based on the pH preferences listed by Meriläinen (1967), van Dam and Kooyman-van Blockland (1978), Berge (1975), and van Dam (1979), an average 91.7% of the species listed could be assigned to pH groups for 26 Metropolitan area lakes. Species described as 'common' and 'very common' (average 3.1 per lake) were weighted by the arbitrary factors 2.5 and 3.5 respectively.

Represented as weighted percent of species the spectra so produced were found to be most closely related to alkalinity rather than mid-summer pH (Figure 1). This probably reflects the common situation where diatom blooms occur in the spring and fall when

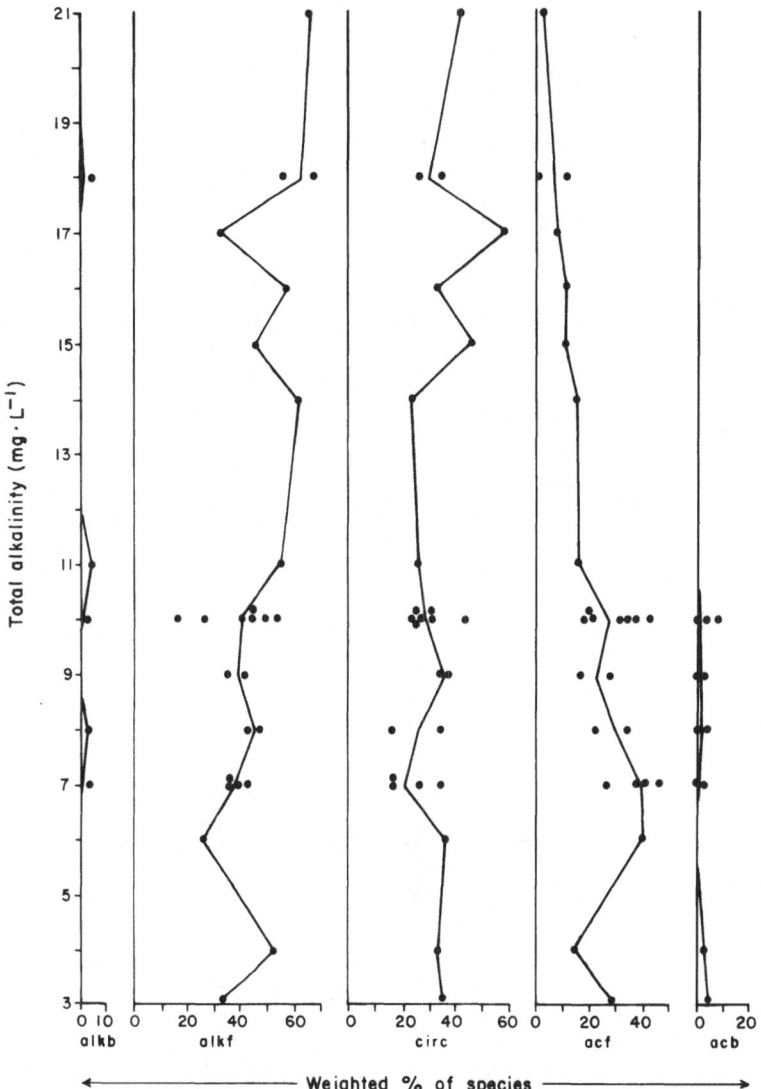

Fig. 1.   Diatom pH groups vs total alkalinity for 26 Halifax metropolitan are lakes (1971).

the stress induced by anthropogenic acid inputs is greatest. Alkalinity, as a measure of the resistance to such stress, appears to parallel between-lake variation in the extent to which pH might be depressed at those times in these dilute waters. This suggest that in waters of low alkalinity, mid-summer pH will not adequately describe the acidic conditions affecting seasonally varying populations represented in sediment samples.

It was observed that those spectra which were not consistent with the general trends were those subject to excessive human disturbance, or were brown-water lakes with low pH and alkalinity.

Because Figure 1 represents such a diversity of lake conditions, the lakes were subdivided into four broad categories to lower the effects of this variability. Results are

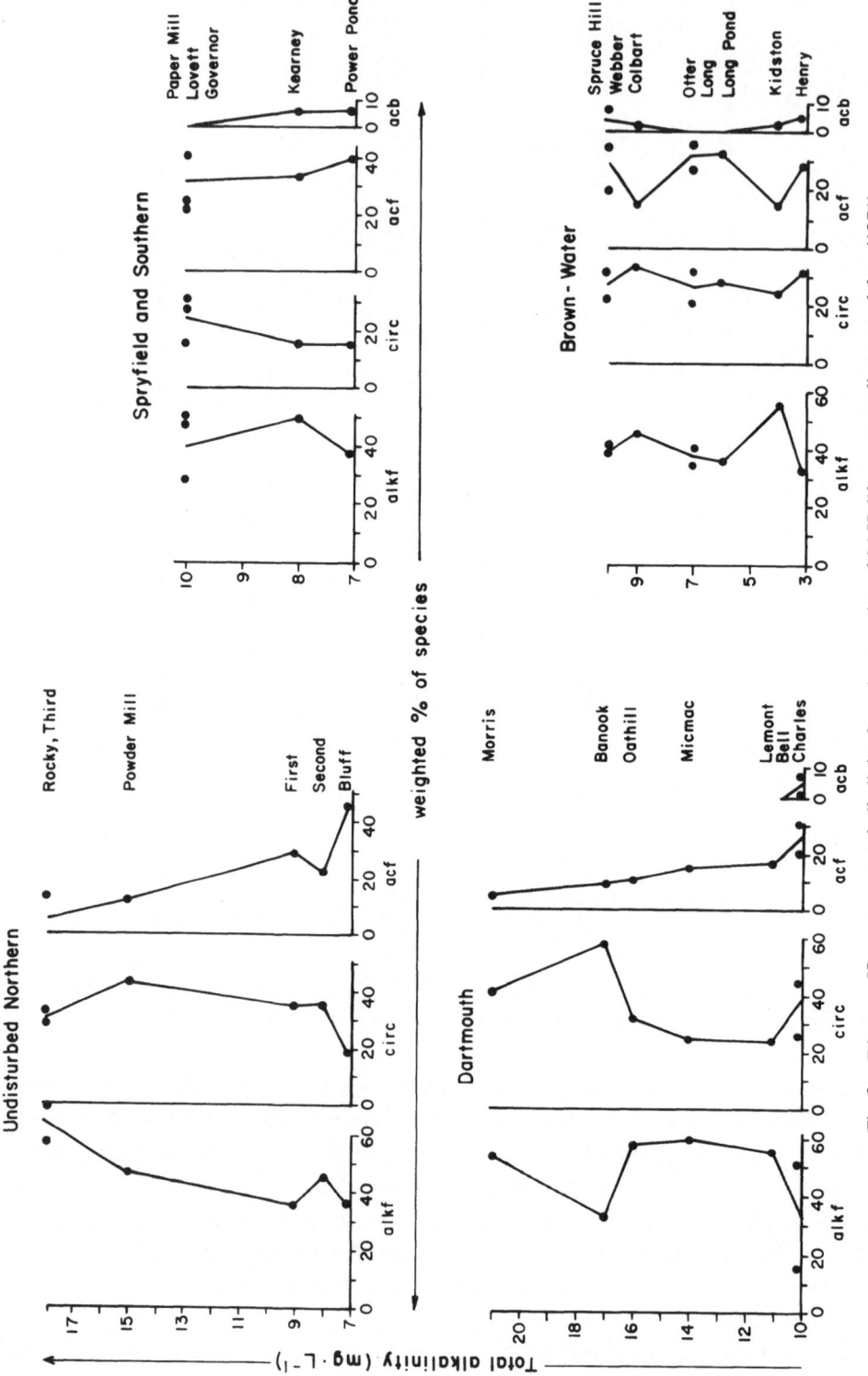

Fig. 2. Diatom pH groups vs total alkalinity for subcategories of 26 Halifax metropolitan area lakes (1971).

shown in Figure 2 for the relatively undisturbed oligotrophic systems to the north of the city (e.g., Third Lake), moderately disturbed (road salt influence) systems south and west of Halifax (e.g., Kearney Lake), moderately disturbed (nutrient influence) systems in the Dartmouth area adjacent to Halifax (e.g., MicMac Lake) and relatively undisturbed brown-water/dystrophic systems (e.g., Otter Lake).

With the possible exception of the latter, this limited classification better reveals the extent to which the diatom spectra, particularly the increases in *acf* and *acb* groups, may be related to total alkalinity.

The applicability of the technique having been demonstrated, surficial sediments were collected with an Ekman dredge in late summer 1980 from the deep areas of Third, MicMac, Otter and Kearny Lakes. Each lake represents one of the four categories shown on Figure 2 and was chosen because alkalinity was in the mid to upper range and there was minimal alteration in the pattern of local human disturbance since 1971. Subsamples were taken in the field of the flocculent surface material and the surface of the consolidated sediment. The latter is assumed to contain sediment from approximately 1975 to 1977 but is hereafter referred to an 197X.

Samples were boiled in H_2O_2 for 20 to 30 min. After adding H_2SO_4 (conc.) and leaving for 20 to 30 min, the material was again boiled for 5 to 10 min. When necessary, this oxidation was repeated. In one case, successive decantations were used to remove clays.

Mounted samples were examined with a minimum of 25 species or 150 individuals being noted for each in order to be compatible with the 1971 data.

Results were treated as the 1971 data being weighted for commonness and classified according to the pH groups.

3. Results and Discussion

The results of these analyses, as seen in Figure 3, reveal that all four lakes have undergone similar and pronounced shifts in diatom assemblages: the percentages of alkaliphilous species have fallen since 1971 while acidophilous species have become more common. These systematic shifts have occurred despite the fact that the four lakes are measureably different in almost every respect. Supported by chemical results of lacustrine and precipitation samples in the Halifax area, these shifts may be ascribed to the influx of air-borne pollutants.

In order to be compatible with the 1971 data, Figure 3 is based only on weighted presence/absence data so Nygaard's indices are not applicable. An approximate measure of the shift may, however, be seen in the authors' index where $I_A = \%\ acf + 2\ (\%\ acb)$.

While these results and studies of changes in local lake chemistry are mutually supportive, the important point is that local rates of acidification are stressing the basic biological components of lacustrine systems to the point where such changes are taking place. However, speculative at this time, some insight into the pattern of such changes is desirable.

Data on the distribution of individuals in the upper samples (1980 and 197X) were analyzed for Shannon-Weiner diversity (H'), redundancy (R), percentage dominance of

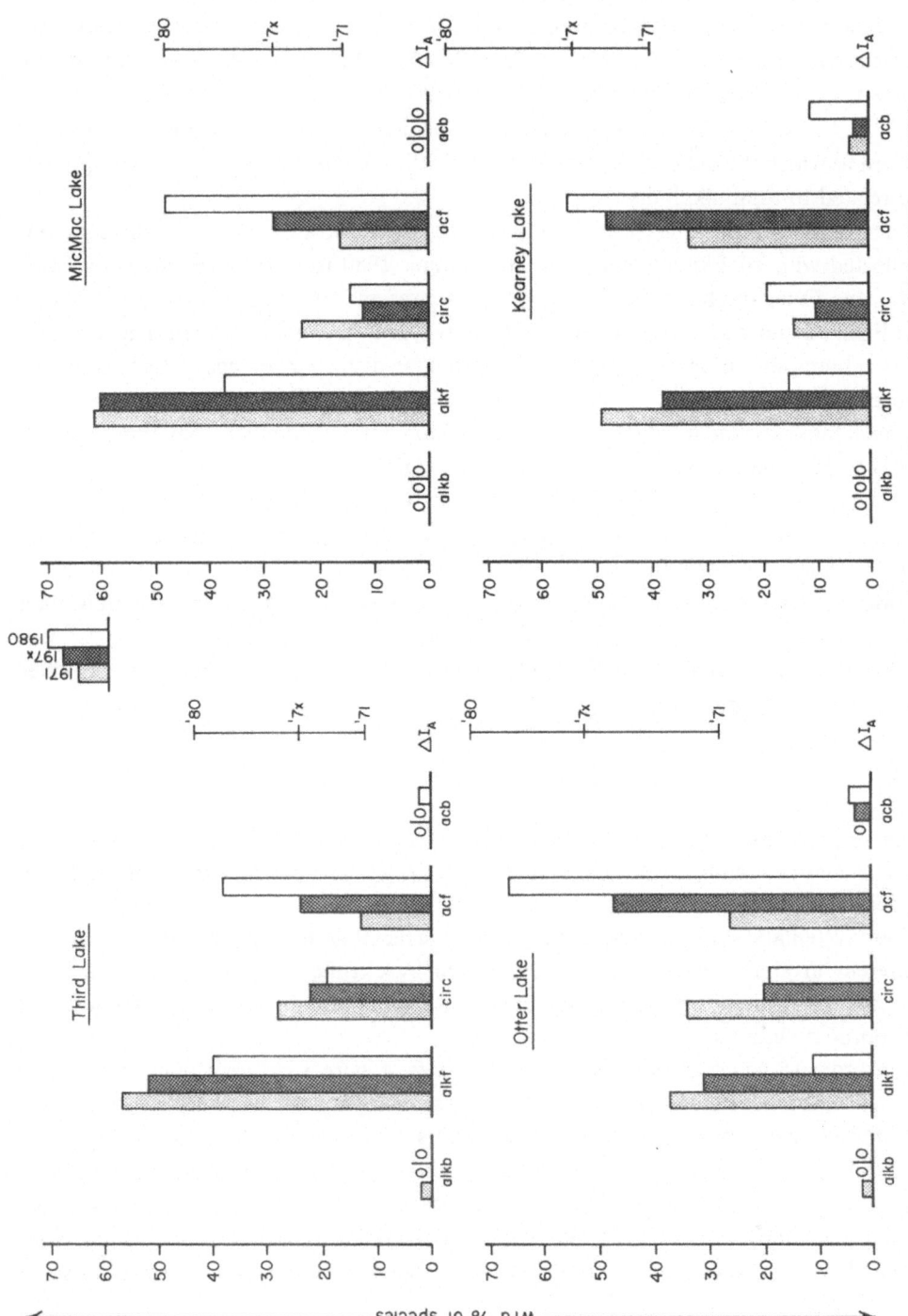

Fig. 3.   Diatom pH spectra and authors' acidic index (ΔI_A) for four Halifax metropolitan area lakes: 1971, 197X, and 1980.

Tabellaria fenestrata and Nygaard's indices (α, ε, ω). Some observed changes were similar in all four lakes between 197X and 1980:

(A) Increases in both I_A and α.

(B) Decreases in ε, a measure of the weighted mean relative frequency of the alkaline preferring species.

(C) Increases in the dominance of *T. fenestrata* consistent with a pattern of 'simplification' of communities under induced acidification.

Changes in the other parameters are less consistent. It should be noted that H' and R represent only the diversity and redundancy observed for each sample and are at best indicative of conditions within living assemblages.

In an attempt to find some pattern in these changes, the various parameters were plotted against Nygaard's α index (Figure 4). Changes in this index have been found to indicate changes in pH though not in a linear manner, perhaps partly because of the effect of low alkalinities hypothesized previously. Though each of Nygaard's indices may be related to pH (Meriläinen, 1967), the use of ω or ε as an ordinate here or as an indicator of past pH in lacustrine cores (Berge, 1975; Dickman *et al.*, 1981) tends to exclude the possibility that a diverse diatom community could exist at low pH's.

In Figure 4, effective acidity increases to the right and the changes in the parameters for each lake between 197X and 1980 are shown. Since all four lakes are subject to similar inputs of anthropogenic acids, there is some justification to treating these results as a continuum by empirically fitting curves. Doing so suggests a general pattern where the maximum diversity occurs around neutrality, the other parameters being consistent with this. It also suggests that at slightly alkaline conditions acid preferring species are present at high mean frequencies which decline as neutrality is approached, i.e. the acid preferring individuals are more evenly distributed amoungst an increased number of species. While this is not surprising, it appears that the alkaline preferring species do not react similarly under slightly acidic concidition: ε does not increase. While this may be a common pattern, it is possible that it reflects the local rate of acidification rather than solely the level of acidity as approximated by α. The observation of a higher than expected diversity in brown-water Otter Lake in 197X, though perhaps an artifact, would tend to support such a hypothesis since it would suggest that a higher diversity is possible at lower pH's when communities or species are subject to more constant conditions.

Preliminary results from similar analyses for lakes in south-central Nova Scotia are consistent with this hypothesis in that diverse diatom communities are observed at lower pH values than would be expected from the Halifax data. These brown-water lakes are removed from the influence of Halifax though subject to long-range transport of acidic pollutants (pH of rain < 4.5).

Though the proportion of acidic deposition self-generated in the Halifax area is unresolved, these results suggest that the process of 'simplification' depends not only on pH but also on the rate of change and that diatom communities have not responded in a linear fashion to the local increase in deposition. This implies that if deposition of acids in Nova Scotia increases in the future, it may result in relatively excessive effects on lacustrine biological communities.

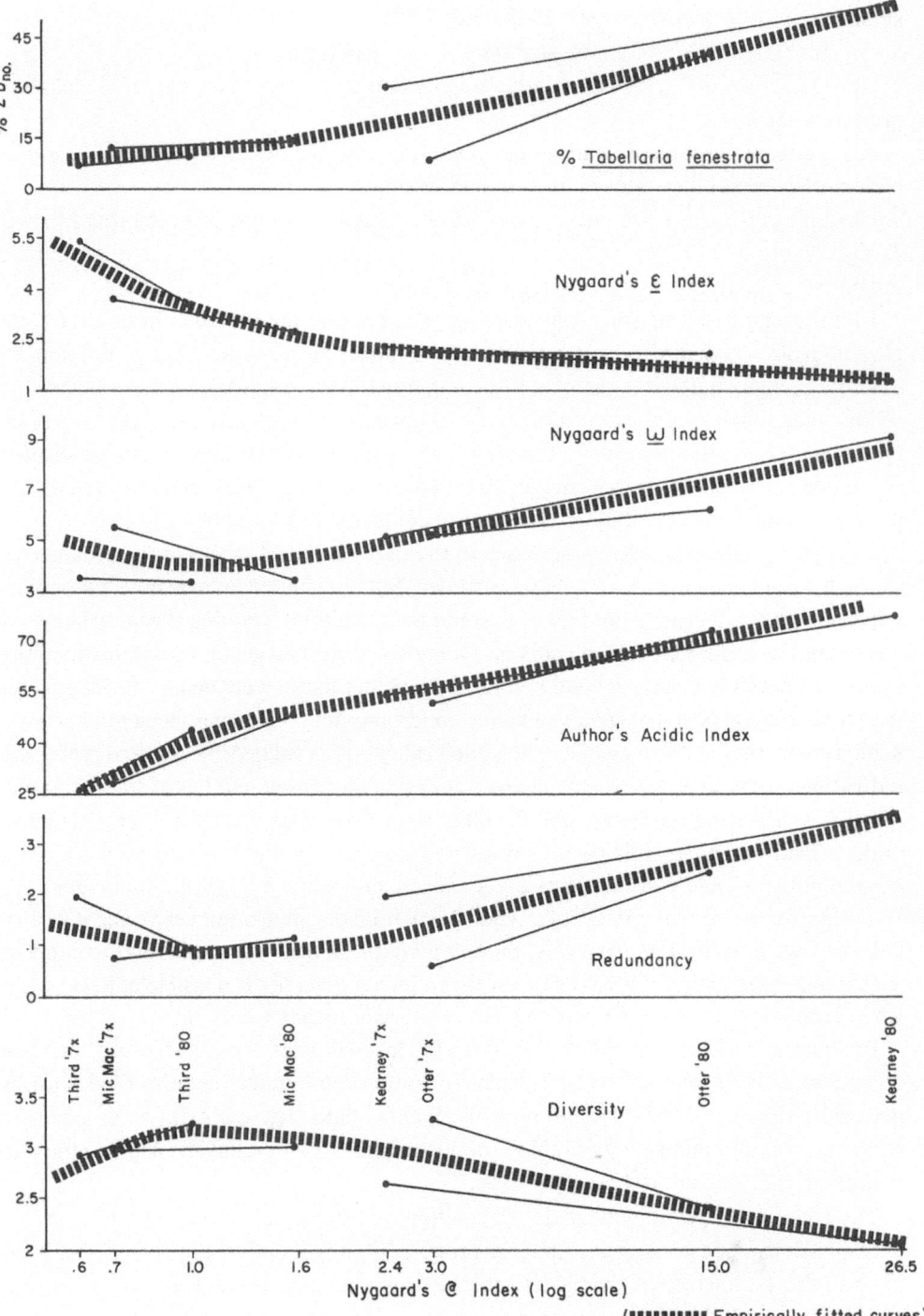

Fig. 4. Various diatoms indices vs Nygaard's α for four Halifax metropolitan area lakes: 197X and 1980.

Acknowledgments

We gratefully acknowledge the assistance of S. Hollett and D. Hirtle in the preparation of this manuscript.

References

Berge, F.: 1975, 'pH Changes and Sedimentation of Diatoms in Lake Langtjern, Norway', SNSF Internal Report, IR 11/75, Oslo-Aas, Norway.

Davis, R. B. and Berge, F.: 1980, 'Atmospheric Deposition in Norway During the Last 300 Years as Recorded in SNSF Lake Sediments II. Diatom Stratigraphy and Inferred pH', in Drablos, D. and Tollan, A. (eds.), *Ecological Impact of Acid Precipitation,* SNSF Project, Oslo-Aas, Norway, 383 pp.

Dickman, M., Cox, W., and Ouellet, M.: 1981, 'The Use of Fossil Diatoms as Indicators of the Evolution of Lake Acidity', paper presented at 3rd meeting, Canadian chapter, International Society for Theoretical and Applied Limnology, Montreal, Jan. 7.

Meriläinen, J.: 1967, *Ann. Bot. Fenn.* **4**, 51.

Nygaard, G.: 1956, *Folia. Limnol. Scand.* **8**, 32.

Ogden, J. G.: 1972, 'Water Quality Survey of Selected Metropolitan Area Lakes', Report to Metropolitan Area Planning Committee, Halifax, N.S.

van Dam, H.: 1979, *Hydrobiol. Bull.* **13**, 13.

van Dam, H. and Kooyman-van Blockland, H.: 1978, *Int. Rev. Ges. Hydrobiol.* **63**, 587.

Watt, W. D., Scott, D., and Ray, S.: 1979, *Limnol. Oceanogr.* **24**, 1154.

OXIDANT EFFECTS ON FOREST TREE SEEDLING GROWTH IN THE APPALACHIAN MOUNTAINS

STEPHEN F. DUCHELLE, J. M. SKELLY, and BORIS I. CHEVONE

Laboratory for Air Pollution Impact to Agriculture and Forestry, Department of Plant Pathology and Physiology, Virginia Polytechnic Institute and State University, Blacksburg, VA 24061, U.S.A.

(Received 12 June, 1981; Revised 28 September, 1981)

Abstract. Long range transport of episodic concentrations of O_3 into the Appalachian Mountains of Virginia was recorded in the summer season of 1979 and 1980. Continuous monitoring of O_3 indicated monthly averages of $\simeq 0.05$ ppm O_3 and several periods averaged $\simeq 0.08$ ppm O_3. Open-top chambers were used to test the effect of ambient doses of the pollutant on the growth of 8 planted forest tree species native to the area. Height growth was suppressed for all species at the end of the second growing season when grown in open plots (no chamber) and ambient chambers compared to those grown in charcoal-filtered air supplied chambers. Height growth trends of open < ambient chamber < filtered air chamber were consistent. Virginia pine and green ash were significantly taller ($p = 0.10$) when grown within filtered air chambers. Tulip poplar and green ash manifested purple stippling on the adaxial leaf surface and sweetgum developed purple coloration under open or ambient chamber conditions; other species exhibited no visible injury.

1. Introduction

The deleterious effect of air pollutants on numerous plant species has been well documented over the past several decades. The majority of this research has involved exposing certain species of plants (mainly field crops) to known concentrations of pollutants in the laboratory and evaluating the effects based on the observation of macroscopic symptoms.

Ozone was first identified as injuring plant tissue when Richards *et al.* (1958) verified it as the incitant of grape stipple in the San Bernardino Valley of California. One year later, O_3 was identified as being phytotoxic to tobacco in the eastern United States (Heggested and Middleton, 1959). Since these early studies, much information has accumulated regarding the phytotoxic effects of O_3 on various plant species.

Rich (1964) listed 57 different plant species exhibiting sensitivity to O_3. In this listing, *Pinus* was the only forest tree genus mentioned. Since the early 1900's, a disease of eastern white pine (*Pinus strobus* L.), often called 'needle blight', has been reported throughout the range of the species. The incitant of the disease was eventually demonstrated to be O_3 (Berry, 1961; Linzon, 1960).

Ponderosa pine needle damage was first noticed in California in 1953 (Asher, 1956). Later, Parmeter *et al.* (1962) described the 'chlorotic decline' of ponderosa pine as characterized by a decrease in height and diameter growth, loss of all but the current seasons needles, yellow mottling of needles, reduced size of needles, deterioration of feeder roots, and eventual death of the tree. Ozone was eventually determined to be the cause of this disorder (Miller *et al.*, 1963).

Following these initial studies, numerous reports of foliar effects on forest tree species

Water, Air, and Soil Pollution **18** (1982) 363–373. 0049–6979/82/0183–0363$01.65.

fumigated under artificial conditions with various doses of O_3 became available (Davis and Wood, 1972; Berry, 1971; Barnes, 1972; Wood and Coppolino, 1972; Davis and Coppolino, 1976; Jensen and Masters, 1975; Harward and Treshow, 1975; Kress and Skelly, 1981).

However, very little is known about the effects of ambient doses of pollutants on the growth of forest trees and other native plant species occurring in their native habitat. This is especially true in the eastern United States where vegetation grows under moderately humid conditions.

Treshow and Stewart (1973) conducted 150 field fumigation studies in grassland, oak, aspen, and conifer communities in Utah. They determined injury thresholds to O_3 for many native species to be below 0.15 ppm O_3 for a single 2 h exposure. The authors postulated that low O_3 concentrations at prolonged or repeated exposures may impair growth and affect community vigor and stability.

Community changes in a deciduous forest in the Ohio River Valley exposed to airborne chloride, SO_2, and fluoride were reported to include reductions in species, richness, evenness, and Shannon diversity index (McClenahen, 1978).

Houston and Dochinger (1977) studied the effects of ambient pollution concentrations (mainly SO_2) on cone, seed, and pollen characteristics of eastern white and red pines (*Pinus resinosa* Ait.) and reported significantly lower values for seeds per cone, 100 seed weight, and percent pollen germination when eastern white pines were growing on sites with high pollution doses compared to sites with lower pollution doses.

The long distance transport of photochemical oxidants and/or their precursors from urban areas to rural areas has been recorded in various parts of the United States. Miller *et al.* (1972) showed evidence of photochemical oxidant transport from an urban area in the San Joaquin Valley to a higher elevation valley in the Sierra Nevada foothills. Cleveland and Kleiner (1975) reported that rural areas downwind of the Camden-Philadelphia urban complex frequently had higher O_3 concentration than the urban area itself. Cleveland and Kleiner (1976) also found O_3 concentrations in Connecticut and Massachusetts were higher when winds came from the New York City Metropolitan area. Samson and Ragland (1977) and Chung (1977) reported that high surface O_3 concentrations were typically found on the western side of high pressure systems in the midwest and southern Ontario. Air mass trajectory analysis indicated that stagnant air over the eastern midwest accumulated O_3 and O_3 precursors which were subsequently transported northward, on the western edge of anticyclones, into Wisconsin, Minnesota and southern Ontario. Wolf *et al.* (1977) have reported concentrations above the current NAAQS in forested areas of the mid-western United States. They reiterated that long distance transport of O_3 or its precursors was responsible for high O_3 concentrations in areas remote from urban sources.

Hayes and Skelly (1977) associated high O_3 concentrations at three rural sites in the western mountains of Virginia with winds from the north and northeast and stationary high pressure systems; concentrations equalled or exceeded 0.08 ppm for 104 h at one of the rural sites. Decreases in O_3 concentrations were correlated with the passage of frontal systems. Skelly and Johnston (1978) reported that O_3 concentrations in western

Virginia during July of 1977 were above 0.08 ppm 30% of the time. During July 14–24, an oxidant air pollution episode occurred during which they recorded a peak hourly average of 0.166 ppm O_3. Synoptic meteorological conditions indicated that the air mass which resulted in this pollution episode probably originated from the midwestern region of the United States.

The Shenandoah National Park in Virginia is an important resource that provides numerous opportunities for recreational activities and aesthetic enjoyment for visitors to the Blue Ridge Mountains. With the realization of long range transport of pollutants into the Blue Ridge Mountains, adverse effects on native eastern white pines have become evident (Skelly and Johnston, 1978). The specific objectives of this current study were to monitor the concentrations of air pollutants (O_3, SO_2) found naturally occurring in the Shenandoah National Park, and to determine their impact on the growth and symptom expression of 8 forest trees native to the park.

2. Materials and Methods

2.1. POLLUTANT MONITORING

Ozone concentrations at the Big Meadows site were monitored continuously with a Bendix Model 8002 chemiluminescent O_3 analyzer (Bendix Corporation, Lewisburg, WV). The O_3 monitor was calibrated at 2- to 4-week intervals using a Bendix Model 8852 Dynamic Calibration Unit as a transfer standard. The primary calibration standard was a Dasibi UV spectrophotometric O_3 analyzer located at the Quality Assurance Branch of EPA at Research Triangle Park, NC. Sulfur dioxide concentrations were monitored using a TECO Model 43 pulsed fluorescent SO_2 analyzer (Thermo Electron Corporation, Hopkinton, MA). An NBS traceable source of bottled SO_2, obtained from Airco Industrial Gases (Murray Hill, NJ) was used in conjunction with the Bendix Calibration Unit as a primary calibration standard for the SO_2 analyzer. Ozone and SO_2 concentrations were recorded continuously with a strip chart recorder and expressed as hourly concentrations.

The efficiency of the charcoal filtration units in removing ambient O_3 concentrations from air supplied to the filtered open-top chambers was determined periodically throughout the growing seasons in 1979 and 1980. Efficiency was maintained at 85 to 90% for the duration of the study.

2.2. FOREST TREE SPECIES

Within a 2 ha meadow community located in the Big Meadows area of the Shenandoah National Park, twelve 3 m diameter plots were established 3 m apart on May 9, 1979. Treatments were randomized over the area and four plots received charcoal-filtered air and four received non-filtered air via blowers attached to open-top chambers (Heagle *et al.*, 1973); four additional plots were left undisturbed except for placement of a 5 cm high steel hoop around the plot border. Chambers operated continuously in 1979 from May 9 to October 9 and in 1980 from April 24 to September 15.

In April of 1979, the following year-old, wild-type seedlings were obtained from the Virginia Division of Forestry:

Tulip poplar	*Liriodendron tulipifera* L.
Sweetgum	*Liquidambar styraciflua* L.
Green ash	*Fraxinus pennsylvanica* Marsh.
Black locust	*Robinia pseudoacacia* L.
Table mountain pine	*Pinus pungens* Lamb
Eastern white pine	*Pinus strobus* L.
Virginia pine	*Pinus virgininana* Mill.
Eastern hemlock	*Tsuga canadensis* (L.) Carr.

Seedlings were transplanted one per pot to a 1 : 1 : 1 : 1 (v/v) peat : vermiculite : soil potting mix in 15 cm diameter peat pots. Seedlings were maintained outside at Blacksburg, Virginia prior to planting in the field. Seedlings were arranged within species into 2 groups – tall and short and a random numbers table was used to pick one representative species from each group for planting in each of the 12 plots at the Big Meadows Area. All species were planted in their peat pots during May of 1979 after removing the bottom of the pots.

The planted trees were observed for foliar symptoms at monthly intervals during respective growing seasons. Hardwood species were examined for stipples and flecking on the supper leaf surface of all leaves. Sweetgums were evaluated for premature reddening of the leaves as noted previously by Stone (1973). Conifers were evaluated for chlorosis, banding, mottling, and necrosis. Symptom evaluation consisted of determining, by plot, the proportion of trees within a species exhibiting a characteristic symptoms and the percent leaf area affected for each symptomatic plant. Height measurements to terminal bud were taken at the time of planting and on September 15, 1980.

Height growth of each species, by treatment, was regressed on several soil parameters including pH, Ca, Mg, K, Al, and percent base saturation. A simple linear regression model was used (Barr et al., 1979). Soil pH was determined to the be the only parameter. that had a significant effect on height growth. Because the pH variation within the soil that existed between plots was determined by regression analysis to affect height growth, it became necessary to eliminate this variation before a statistical comparison between treatments could be made. An analysis of covariance was done on the height growth data of each species using pH as the continuous variable (the covariate) (Barr et al., 1979). If the covariate (pH) was significant (alpha = 0.05), a general linear model (GLM) procedure with a P option was used to find the residual value of each observation. The residual value was the observation adjusted for the covariate (pH). Three different TTESTS were executed on each species using either the original observations or the residual values (when the pH was significantly affecting height growth).

(1) A TTEST to test the hypothesis that the height growth of the charcoal-filtered chambers and non-filtered chambers was equal.

(2) A TTEST to test the hypothesis that the height growth means of the non-filtered chambers and open plots were equal.

(3) Initially, the height growth means of the open and non-filtered chambers were compared to determine the chamber effect. Residuals were used in this comparison if the covariate (pH) was significant. The difference between the two means (the average chamber effect) was added or subtracted from each height observation in the open plot in order to obtain values adjusted for the chamber effect. A TTEST was then executed to test the hypothesis that the height growth means of the charcoal-filtered chambers and open plots (adjusted values) were equal (Duchelle, 1981).

3. Results

3.1. POLLUTANT MONITORING

Ambient SO_2 concentrations at Big Meadows ranged from undetectable to 0.03 ppm and were considered insignificant relative to plant productivity loss. The monthly average and peak hourly average ambient O_3 concentrations recorded at Big Meadows during the daylight hours from 1000 to 1800 h during 1979 and 1980 are presented in Table I. The highest monthly average and the peak hourly average for 1979 and 1980 respectively were 0.055, 0.100 ppm and 0.059, 0.102 ppm.

TABLE I

Monthly average and peak hourly O_3 concentrations (ppm) monitored at Big Meadows, Shenandoah National Park, for the daylight hours of 1000 and 1800 during 1979 and 1980

Month	1979		1980	
	Monthly average	Peak hourly average	Monthly average	Peak hourly average
January	0.041	0.061	0.020	0.035
February	—	—	0.026	0.038
March	0.060	0.097	0.034	0.045
April	0.055	0.100	0.042	0.068
May	0.051	0.082	0.048	0.088
June	0.055	0.082	0.058	0.088
July	0.046	0.094	0.059	0.087
August	0.053	0.074	0.052	0.102
September	0.047	0.095	0.045	0.087
October	0.043	0.082	0.034	0.076
November	0.039	0.071	0.035	0.056
December	0.028	0.045	0.041	0.057
Yearly avg.	0.047		0.041	

Ozone episodes occurred frequently during both 1979 and 1980 (Table II). During these episodes, peak hourly O_3 concentrations often exceeded 0.09 ppm and hourly

TABLE II

Ozone episodes recorded at Big Meadows, Shenandoah National Park, during the
spring and summer of 1979 and 1980

Date	Duration (h)	Mean hourly concentration (ppm)	Peak hourly concentration (ppm)
1979, April, 11–12	31	0.083	0.100
1979, June, 5	10	0.079	0.082
1979, July, 19	7	0.081	0.094
1979, Sept., 12	20	0.079	0.091
1980, May, 29–30	18	0.084	0.100
1980, June, 13–15	38	0.081	0.093
1980, July, 14–15	22	0.080	0.089
1980, July 31	17	0.074	0.084
1980, August, 29–30	27	0.077	0.102

concentrations throughout the duration of the episodes averaged about 0.08 ppm. The
regional aspect of these ozone episodes was demonstrated by comparing ambient O_3
concentrations at three rural monitoring sites in the mountainous areas of western
Viriginia (Table III). The Rocky Knob monitoring station was located approximately
260 km south-southwest of Big Meadows and the Salt Pond Mountain monitoring
station approximately 230 km southwest of Big Meadows. Only the O_3 episode occurring
on July 31, 1980 at the Big Meadows site did not conform to a regional pattern.

TABLE III

Peak hourly O_3 concentrations (ppm) recorded at three monitoring stations in western Virginia
during O_3 episodes in the spring and summer of 1979 and 1980

Date	Monitoring site		
	Big meadows	Rocky Knob	Salt pond mt.
1979, April, 11–12	0.100	0.091	0.086
1979, June, 5–6	0.082	0.110	–[a]
1979, Sept., 12	0.096	0.065	–
1980, May, 29–30	0.100	0.072	0.069
1980, June, 13–15	0.093	0.076	–
1980, July, 14–15	0.089	0.072	–
1980, July, 31	0.084	0.049	0.043
1980, August, 29–30	0.102	0.113	0.129

[a] Data not available.

3.2. FOREST TREE GROWTH

During the 1979 growing season, symptoms of O_3 injury were difficult to evaluate due
to the stress that resultated from transplanting shock. The leaves of many hardwood

species became unthrifty in appearance and developed some necrosis. Conifers exhibited mild chlorosis. However, no characteristic O_3 symptoms developed.

During the 1980 growing season, symptoms characteristic of O_3 induced injury developed on tulip polar, green ash, and sweetgum when growing in non-filtered air chambers and open plots. Symptoms were not observed on green ash and sweetgum grown in charcoal-filtered chambers. No characteristic O_3 symptoms developed on black locust or on any of the conifers.

Symptoms were first observed on tulip poplar in the beginning of August 1980. Symptoms of purple stippling on the adaxial leaf surface were noted on several trees in the non-filtered chambers and open plots (Table IV). Symptoms were more frequent on older leaves.

TABLE IV

Symptom expression of tulip poplar, green ash, and sweetgum on September 9, 1980 for the charcoal filtered chambers (CFC), non-filtered chambers (NFC) and open plots (OP), at Big Meadows, SNP

Species	Treatment	Proportion of plants with symptoms[a]	Range of leaf area exhibiting symptoms[b]
Tulip poplar	CFC	1 of 8	20–40
	NFC	4 of 8	23–84
	OP	5 of 8	19–69
Green ash	CFC	0 of 8	0
	NFC	4 of 8	13–75
	OP	4 of 7	13–45
Sweetgum	CFC	0 of 8	0
	NFC	5 of 8	10–84
	OP	6 of 6	9–78

[a] Tulip poplar and green ash leaves were evaluated for presence of purple stipples while sweetgum leaves were evaluated for presence of red pigmentation.
[b] Ranges for all leaves from each symptomatic plant.

Symptoms were first observed on green ash during the middle of August 1980. Purple stipples on the upper leaf surface were observed on several trees in the open plots and non-filtered chambers.

Premature red pigmentation of sweetgum leaves was also first observed in the middle of August 1980 on trees receiving ambient air. Symptoms were more severe on older leaves.

In all 3 tree species that exhibited characteristic O_3 symptoms, the incidence and severity of symptoms were much greater when plants were grown in non-filtered chambers and open plots compared to charcoal-filtered chambers.

The average height growth increase of seven forest trees has been presented in Table V. All seven species obtained their largest average height growth increase when grown in charcoal filtered air. Sweetgum height growth data were not considered accurate because

TABLE V

Average height growth (cm) increase of 7 forest trees from May 9, 1979 to October, 1980 established at the Big Meadows Research Site in charcoal filtered and non-filtered chambers, and open plots

Species	Significance of covariate[a] (pH)	Treatments		
		Charcoal filtered chambers (cm)	Non-filtered chambers (cm)	Open plots (cm)
Tulip poplar	0.46	18.1(6)[b]	10.7(6)	5.9(3)
Green ash	0.51	21.8(6)	7.3(8)	7.6(2)
Black locust	0.04	114.8(7)	94.5(8)[c]	35.2(5)
Virginia pine	0.0001[a]	39.8(8)	31.2(8)	27.3(8)
Eastern white pine	0.01	15.7(8)	12.5(8)	11.6(8)
Table mountain pine	0.04	20.1(8)	17.6(8)	12.2(8)
Eastern hemlock	0.43	19.2(7)	14.7(8)	7.6(7)

[a] If the covariate was significant at $p = 0.05$, height growth data was adjusted for pH.
[b] Denotes number of trees used in computing averages; remainder of the trees either died or the tops broke off during the winter of 1979.
[c] Two trees in one chamber grew $220+$ cm.

there was extensive terminal bud die-back during the winter of 1979. Green ash and Virginia pine were the only 2 species that exhibited a significant treatment effect ($p = 0.10$) when height growth in charcoal-filtered chambers was compared to non-filtered chambers. When height growth in charcoal-filtered chambers was compared to open plots (adjusted for the covariate) only green ash demonstrated a significant treatment effect ($p = 0.10$). Tulip poplar, black locust, and table mountain pine height growth was significantly increased when grown in open-top chambers ($p = 0.10$).

4. Discussion

The fact that the 1980 monthly average O_3 concentrations increased from January to June, stabilized during the summer months, and then decreased again around October demonstrates a typical O_3 season. A similar trend was also noted when O_3 concentrations were monitored in the Blue Ridge Mountains during 1977 and 1978 (Skelly *et al.*, 1979). This type of seasonal variation of O_3 concentrations has also been reported in the San Bernardino Mountains in southern California (USEPA, 1977). The 1979 monthly average O_3 concentrations should be considered atypical on a seasonal basis since the two highest monthly averages occurred in March and April. The fact that monthly average O_3 concentrations were slightly higher when only the hours of 10 am to 6 pm were considered indicates there was a diurnal variation in O_3 concentrations with the daylight hours having the higher concentrations. Ozone concentrations monitored at various times of the day have been reported to show consistent patterns (USEPA, 1977). Normally, O_3 concentrations are low during the night, begin to rise in the morning, peak in early afternoon, and then drop again at sunset. The monthly average O_3 concentrations which considered the hours of 10 am to 6 pm may be the most important in terms of pollutant effects because these are the hours of peak photosynthesis and pollutant uptake is more likely to occur.

Comparison of the early O_3 measurements of Berry (1964) and Costonis (1971) with the more recent measurements of Hayes and Skelly (1977), Skelly and Johnston (1978), Skelly *et al.* (1979), and those reported here indicates a substantial increase in O_3 concentrations at remote forested areas in the Blue Ridge and Southern Appalachian Mountains.

Suppressed growth of seven forest tree species occurred within the Big Meadows area of the Shenandoah National Park when the trees were grown in open plots or ambient air open-top chambers compared with charcoal-filtered air. Growth reductions most often occurred without the expression of O_3 induced foliar symptoms. Characteristic O_3 symptoms developed on only two out of the seven tree species showing growth suppression. Tulip poplar and green ash both manifested purple stippling on the adaxial leaf surface when grown in non-filtered chambers and open plots. This type of injury would be expected to reduce the photosynthetic capacity of the plants since such stippling results from tissue death.

The height growth data indicate that plants can be adversely affected by air pollutants without evidence of foliar injury and support results obtained by many other researchers who have worked with trees and found adverse effects that could not be explained on the basis of symptom expression (Treshow *et al.*, 1967; Kress, 1978; Stone and Skelly, 1974; Houston and Dochinger, 1977; Keller, 1977). Ozone exposures have been demonstrated to cause reductions in apparent photosynthesis of many pine species (Barnes, 1972; Botkin *et al.*, 1972; Miller *et al.*, 1969; Yang, 1980). According to Barnes (1972), a depression of photosynthesis would be expected to result in a reduction of the carbohydrates available for growth. Decreased photosynthesis and increased respiration could very well be altered physiological processes occurring in the tree species at Big Meadows during exposure to long term low concentrations of O_3. All tree species obtained their greatest average height growth increase after two growing seasons when grown in charcoal-filtered chambers. Therefore, it appears that the ambient concentrations of O_3 which occurred during the 1979 and 1980 growing season were sufficient to induce growth suppressions in forest tree seedlings. Despite verification statistically ($P = .010$) with only green ash and Virginia pine, the trends of growth suppression in ambient air were quite evident. Tulip poplar and green ash exhibited growth reductions of 44 and 77%, respectively, when grown in non-filtered chambers compared to charcoal-filtered chambers. The other five species exhibited growth reductions ranging from 13% for table mountain pine to 23% for eastern hemlock. If height growth was used as the criteria for evaluating O_3 sensitivity among species tulip poplar and green ash should be regarded as very sensitive while black locust, Virginia pine, eastern white pine, table mountain pine, and eastern hemlock should be regarded as sensitive.

The results of these studies have shown that ambient concentrations of O_3 within the Shenandoah National Park are capable of inducing growth reductions in eastern forest tree species grown under natural field conditions. Ozone concentrations monitored at Big Meadows during the two growing seasons never exceeded the current NAAQS of

0.12 ppm O_3. This fact alone demonstrates the importance of determining the effects of low doses of O_3 on plant growth over an entire growing season. Forest trees are perennial and long-lived in nature and so are exposed repeatedly during the year and over several years to atmospheric pollutants. Low concentrations of pollutants over extended time periods may be more important in terms of overall tree growth and productivity than high concentrations of pollutants over a short time period. The percent of total hours monitored above 0.06 and 0.08 ppm O_3 may be the most important indicators for determining growth effects. These two concentrations were both exceeded over extended periods of time during the 1979 and 1980 growing season.

Acknowledgment

The authors wish to thank the U.S. Department of the Interior, National Park Service, Division of Air Quality for supporting this research. Cooperative Agreement No. CX001-9-0011. Manuscript is part of Master of Science Thesis of senior author.

References

Asher, J. E.: 1956, 'Observation and Theory on 'X' Disease or Needle Dieback', File Report. Arrowhead Dist., San Bernardino Nat. Forest, California.

Barnes, R. L.: 1972, *Environ. Pollut.* **3**, 133.

Barr, A. J., Goodnight, J. H., and Sall, J. P.: 1979, 'SAS User's Guide', 1979 edition. SAS Institute Inc., Raleigh, NC, 494 p.

Berry, C. R.: 1961, 'White Pine Emergence Tipburn, a Physiogenic Disturbance', USDA For. Serv., Southeastern Forest Exp. Station Paper No. 130, 8 p.

Berry, C. R.: 1964, *J. of the Air Poll. Control Assoc.* **14**(6), 238.

Berry, C. R.: 1971, *Phytopathology* **61**, 231.

Botkin, D. B., Smith, W. H., Carlson, R. W., and Smith, T. L.: 1972, *Environ Pollut.* **3**, 273.

Chung, Y. S.: 1977, *J. Appl. Metero.* **16**, 1127.

Cleveland, W. S. and Kleiner, B.: 1975, *Envir. Sci. and Tech.* **9**, 869.

Cleveland, W. S. and Kleiner, B.: 1976, *Science* **191**, 179.

Costonis, A. C.: 1971, *Phytopathology* **61**, 717.

Davis, D. D. and Coppolino, J. B.: 1976, *Plant Dis. Rep.* **60**, 876.

Davis, D. D. and Wood, F. A.: 1972, *Phytopathology* **62**, 14.

Duchelle, S. F.: 1981, 'The Response of Indigenous Vegetation in the Blue Ridge Mountains of Virginia to Photochemical Oxidant Air Pollution', M.S. Thesis, Virginia Polytechnic Institute and State University, Blacksburg, Virginia, 122 p.

Harward, M. and Treshow, M.: 1975, *Environ. Conserv.* **2**, 17.

Hayes, E. M. and Skelly, J. M.: 1977, *Plant Dis. Rep.* **61**, 778.

Heagle, A. S., Body, D. E., and Heck, W. W.: 1973, *J. Environ. Qual.* **2**, 365.

Heggestad, H. E. and Middleton, J. T.: 1959, *Science* **129**, 208.

Houston, D. B. and Dochinger, L. S.: 1977, *Environ. Pollut.* **12**, 1.

Jensen, K. F. and Masters, R. G.: 1975, *Plant Dis. Rep.* **59**, 760.

Keller, T.: 1977, 'SO_2 Concentration upon Photosynthesis of Conifers', Proc. IV Internat. Clean Air Congr. Tokyo, p. 81–83.

Kress, L. W.: 1978, 'Growth Impact of O_3, SO_2, and NO_2 Singly and in Combination on Loblolly Pine (*Pinus taeda* L.) and American Sycamore (*Platanus occidentalis* L.)', Ph.D. Dissertation. Virginia Polytechnic Institute and State University, Blacksburg, VA, 202 p.

Kress, L. W. and Skelly, J. M.: 1981, 'Effect of Low Dose Exposure to O_3 and O_3 + NO_2 on Ten Forest Tree Species' (in press, *Plant Dis.*).

Linzon, S. N.: 1960, *Can J. Bot.* **38**, 153.

McClenahen, J. R.: 1978, *Can. J. For. Res.* **8**, 432.

Miller, P. L., McCulchan, M. H., and Milligan, H. P.: 1972, *Atmos. Envir.* **6**, 623.

Miller, P. R., Parmeter, J. R., Jr., Flick, B. H., and Martinez, C. W.: 1969, *J. Air Poll. Contr. Assoc.* **19**, 435.

Miller, P. R., Parmeter, J. R., Jr., Taylor, O. C., and Cardiff, E. A.: 1963, *Phytopathology* **53**, 1072.

Parmeter, J. R., Jr., Bega, R. V., and Neff, T.: 1962, *Plant Dis. Rep.* **46**, 269.

Rich, S.: 1964, *Annu. Rev. Phytopathology* **2**, 253.

Richards, R. L., Middleton, J. T., and Hewitt, W. B.: 1958, *Agron. J.* **50**, 559.

Samson, P. J. and Ragland, K. W.: 1977, *J. Appl. Metero.* **16**, 1101.

Skelly, J. M., Duchelle, S. F., and Kress, L. W.: 1979, 'Impact of Photochemical Oxidant Air Pollution on Eastern White Pine in the Shenandoah, Blue Ridge Parkway, and Great Smoky Mountains National Parks', Proc. II Conf. on Sci. Res. in National Parks, San Francisco, CA.

Skelly, J. M. and Johnston, J. W.: 1978, 'Oxidant Levels and Their Associated Impact to *Pinus strobus* in the Blue Ridge Mountains of Virginia', in *Proceedings of the 3rd International Congress on Plant Pathology,* Munich, West Germany, 1978, 341 p.

Stone, L. L.: 1973, 'The Effects of Oxidants on the Growth of Forest Trees and Tobacco Cultivars', M.S. Thesis, Virginia Polytechnic Institute and State University, Blacksburg, VA, 84 p.

Stone, L. L. and Skelly, J. M.: 1974, *Phytopathology* **64**, 773.

Treshow, J., Anderson, F. K., and Harner, F.: 1967, *For. Sci.* **13**, 114.

Treshow, M. and Stewart, D.: 1973, *Biol. Conserv.* **5**, 209.

USEPA: 1977, 'Photochemical Oxidant Air Pollution Effects on a Mixed Conifer Ecosystem – A Progress Report', 339 p. USEPA Ecol. Res. Series EPA-600/3-77-104, Corvallis, OR.

Wolff, T. T., Lioy, P. J., Wight, G. D., Meyers, R. E., and Cederwall, R. J.: 1977, *Atmos. Env.* **11**, 797.

Wood, F. A. and Coppolino, J. B.: 1972, *Wien* **97**, 233.

Yang, Y. S.: 1980, (Abstr.). *Phytopathology* **70**, 694.

EFFECTS OF ACIDIC PRECIPITATION AND ACIDITY ON SOIL MICROBIAL PROCESSES

A. J. FRANCIS

*Department of Energy and Environment, Brookhaven National Laboratory,
Upton, NY 11973, U.S.A.*

(Received 10 July, 1981; Revised 14 October, 1981)

Abstract. Effects of soil acidity on microbial decomposition of organic matter and transformation of N in an acid forest soil were investigated. In the oak-leaf-amended pH-adjusted acid soils, CO_2 production in 14- and 150-day preincubated samples decreased by about 6 and 37%, respectively. In the control (unamended) acidified soils, reductions in CO_2 production of 14% in 14-day preincubated samples and of 52% in 150-day samples were observed.

Ammonia formation in the pH-adjusted acid soil was about 50% less than in the naturally acid soil. Increased rates of ammonification and nitrification were observed in the pH-adjusted neutral soil. Little autotrophic and heterotrophic nitrifying activity was detected in naturally acid and acidified forest soils.

The rate of denitrification was rather slow in acid soils, and at greater acidities N_2O was the predominant end product.

The abundance of N-fixing free-living bacteria was very low in acidic and acidified forest soils, and N gains by asymbiotic bacterial fixation in an acid forest ecosystem may be insignificant. These results suggest that further acidification of acid forest soils by addition of H_2SO_4 or by acid precipitation may lead to significant reductions in the leaf litter decomposition, ammonification, nitrification, and denitrification and thus reduce nutrient recycling in the forest ecosystem.

1. Introduction

The rain and snow falling on much of northern Europe and the eastern United States is becoming more acidic because of the presence in the atmosphere of H_2SO_4, HNO_3, and HCl (Cogbill and Likens, 1974; Galloway *et al.*, 1976). These strong acids are formed from gaseous man-made pollutants, such as SO_2 and NOx resulting from the combustion of fossil fuels.

Soil microorganisms play an important role in nature and are critical to ecosystem function and the well-being of plants, animals, and humans. They are responsible for transformations of various elements and occupy a significant place in C, N, P, and S cycles. If acidic precipitation has a significant impact on soil microbial processes, then there is a potential for reduced soil fertility and economic loss, primarily in unmanaged range and forest soils. Recently, Alexander (1980a, b) has discussed the effects of acidity and acid precipitation on microorganisms and biochemical processes in soil.

The objective of this study is to obtain comprehensive baseline information on the microbial activity in a forest soil in order to begin an evaluation of the impact of pollutants resulting from energy related activities. Since most of the soils on Long Island, New York, are poorly buffered and may respond to acid precipitation more quickly than other soils, it was decided to study the effect of acidity on various soil microbial processes in relation to nutrient cycling and forest productivity.

Water, Air, and Soil Pollution **18** (1982) 375–394.   0049–6979/82/0183–0375$03.00.

1.1. ORGANIC MATTER DECOMPOSITION

A diverse group of microorganisms participates in the decomposition of natural organic materials in soil. Although an increase in soil acidity due to acidic precipitation may enable certain groups of organisms to proliferate, most microorganisms are sensitive to acidity. Depending on the nature of organic materials (type of vegetation), soil type, pH, temperature, moisture, etc., differences are to be expected in the rate and extent of organic matter decomposition (Abrahamsen *et al.*, 1980; Roberts *et al.*, 1980). Acidic rain application significantly reduced organic matter decomposition in forest soils where increases in soil acidity have been observed (Abrahamsen *et al.*, 1980; Baath *et al.*, 1980).

1.2. NITROGEN TRANSFORMATION

Nitrogen is the major nutrient limiting plant growth in nature. Higher plants are known to assimilate N in the form of nitrate and ammonia. Ammonification, nitrification, N fixation, and denitrification are affected to varying degrees by soil acidity.

1.2.1. *Nitrification*

The process of nitrification is the sequential oxidation of ammonia to nitrite and then to nitrate by autotrophic and heterotrophic microbial communities. Nitrification occurs optimally at neutral to slightly alkaline pH (Focht and Chang, 1975; Focht and Verstraete, 1977). Some soils nitrify at pH 4.5, others do not; the difference may possibly be attributed to acid-adapted strains or to chemical differences in the two habitats. Accumulations of nitrate in four acid soils of pH 3.9 to 4.7 were reported by Weber and Gainey (1962). Ishaque *et al.* (1971) found nitrate accumulation in an acid laterite soil of pH 4.2 but were unable to detect the presence of autotrophic ammonium- and nitrite-oxidizers. However, Walker and Wickramasinghe (1979) isolated pure cultures of ammonium-oxidizing but not nitrite-oxidizing autotrophic, nitrifying bacteria from acid soils at pH 4.0 to 4.5. Dancer *et al.* (1973) found that soil pH 4.7 to 6.6 did not affect the rates of ammonification appreciably, but that it had a significant effect on nitrification rates, with the acid soil exhibiting substantial reduction in activity.

In a humus sample of pH 4.4 ammended with ammonium sulfate, Hovland and Ishac (1975) found negligible amounts of nitrate even though nitrifying bacteria were present. They also observed that the simulated acid rain of pH 4 stimulated the production of nitrate, but that further increase in the acidity seemed to reduce the nitrification rate. Abrahamsen *et al.* (1976) reported that nitrification did not occur in unlimed soils of pH 4.4 to 4.1 but did occur with liming.

In a beech forest soil with pH about 3, very small numbers of nitrite- and nitrate-forming organisms were found (Niese, 1971); however, in that soil substantial amounts of nitrate were detected (Runge, 1971). The possible formation of nitrate by heterotrophic nitrification in acid soils has been suggested by Abrahamsen *et al.* (1976) and Ishaque and Cornfield (1976). Acidification of soil by addition of either powdered sulfur or sulfuric acid decreased CO_2 evolution (decomposition) by soils, whereas it increased the amount of mineral-N (ammonification) in the sample, but lowered the amount of nitrate (Tamm *et al.*, 1976).

1.2.2. *Denitrification*

Soil bacteria are known to reduce nitrates to nitrogen gas under anoxic conditions in the presence of available C, and the process is called denitrification. Of particular interest to soil and atmospheric scientists is the biogenic evolution of N_2O and its subsequent effect in the depletion of the atmospheric O_3. Soil pH is known to affect the rate and the composition of the gaseous end products of denitrification. It has been shown that denitrification is favored by relatively high pH values, and that at pH values below 6 the reduction of N_2O is often strongly inhibited (Wijler and Delwiche, 1954; Nommik, 1956; Bremner and Shaw, 1958; Bollag *et al.*, 1973).

1.2.3. *Nitrogen Fixation*

Nitrogen-fixing microorganisms differ in tolerance to acidity. Among bacteria *Azotobacter, Beijerinckia,* and *Clostridium* have been extensively studied. Although the increase in N in soil by nonsymbiotic fixation is low, its ecological significance cannot be overlooked.

Soil acidification affects symbiotic nitrogen fixation in legumes. Although several physical and chemical factors contribute to efficient nitrogen fixation in legume-*Rhizobium* symbioses, soil acidity affects (1) plant growth, (2) survival of *Rhizobia,* and (3) the symbiotic relationship. Reductions in nodulation and plant growth due to reduced N fixation have been reported (Shriner, 1977; Evans *et al.*, 1980). Toxicity resulting from Fe or Al in acid soils also has a profound effect upon N fixation.

Reductions in total microbial populations and the activities of the enzymes urease and dehydrogenase were found in acidified forest soil (Francis *et al.*, 1980, 1981). Also investigated were the effects of the pesticides 2,4-D, Cacodylic acid, Dylox, Methoxychlor, Sevin, and Paraquat on soil microbial activity and the fate of ^{14}C-labeled 2,4-D, Sevin, and Paraquat in naturally acid and pH-adjusted acid and neutral soils (Francis and Olson, manuscript in preparation). The data suggest that the addition of these compounds to soils did not have any effect on soil respiration. However, variations in the rate and degradation of 2,4-D, Sevin, and Paraquat were observed. Paraquat was not degraded in acid soils.

2. Materials and Methods

2.1. SAMPLE COLLECTION AND PREPARATION

A Riverhead sandy loam soil sample up to 15 cm deep was collected, after removal of the surface leaf debris, from an oak-pine forest at Brookhaven National Laboratory. The soil was air dried, thoroughly mixed, and passed through a 2.4-mm sieve. To study the effect of soil acidity due to acid rain, 1500 g of air-dried forest soil was adjusted to pH 3.0 by the addition of 1 N H_2SO_4 or to pH 7.0 by the addition of $Ca(OH)_2$. The natural soil, which received only distilled water, served as a control. The pH-adjusted and the natural soil samples were incubated in the dark at room temperature. The moisture content of tne soil samples was maintained at 28% by periodic addition of distilled water. After thorough mixing of the incubated soils, samples were taken for use in the various experiments described below.

2.2. Organic matter decomposition

The rates of decomposition of organic material by natural acid and pH-adjusted acid and neutral soils which had been preincubated for 14 and 150 days were determined by monitoring CO_2 evolution in a constant aeration apparatus described by Parr and Smith (1969). Dried oak leaves were ground in a Waring blender and passed through a number 30 sieve. One gram of ground leaf material was added to 49 g of soil in a 250-ml Erlenmeyer flask. Control samples (50 g soil) received no amendment. The moisture contents of the 14- and 150-day preincubated soils were 19 and 15%, respectively. Samples were run in either duplicate or triplicate for each soil treatment. Carbon dioxide was removed from purified air by passing the air through a series of gas-washing bottles, two containing $2 N$ NaOH and two containing distilled water. The CO_2-free air was passed through the incubation vessel, and the effluent gas was bubbled through $0.5 N$ NaOH to trap the CO_2 produced by the samples. The gas flow rate was maintained at 20 ml min^{-1} to give an effective flow rate to volume ratio of 5.0 h^{-1}. Carbon dioxide production was determined by precipitating the carbonate with 15% $BaCl_2$ and titrating the excess base with standard HCl, with phenolphthalein as an indicator.

2.3. Ammonification and nitrification

Ammonification and nitrification in natural acid and pH-adjusted acid and neutral soils were determined by monitoring ammonia and nitrate formation in incubated soils on a periodic basis.

2.3.1. *Autotrophic and Heterotrophic Nitrification: Soil Perfusion Study*

Nitrification by autotrophic and heterotrophic organisms in the neutral and pH-adjusted acidic and neutral forest soil samples was determined by the soil perfusion technique (Longden and Clairidge, 1976). Soil samples that had been preincubated for 60 days were air dried, 50-g samples were placed into percolation units, and 100 ml of percolate solution was added to each. The percolate solutions consisted of (a) distilled water, (b) distilled water plus urea to give a final concentration of 100 ppm of urea-N (w/w soil), and (c) distilled water, 100 ppm of urea-N (w/w soil), and 25 ppm (w/w soil) of N − Serve [2-chloro-6-(trichloromethyl)pyridine]. N-Serve, supplied by Dow Chemical Co., is an inhibitor of autotrophic nitrification (Shattuck and Alexander, 1963). All treatments were performed in duplicate. Percolate solutions were collected at intervals, tested for pH, and analyzed for nitrite and nitrate according to Montgomery and Dymock (1961, 1962). At the conclusion of the experiment, percolate samples were analyzed for total Kjeldahl-N and ammonia. In addition, the population distributions of *Nitrosomonas* sp. and *Nitrobacter* sp. were determined by the most probable number (MPN) technique (Alexander and Clark, 1965).

2.4. DENITRIFICATION

Samples (25 g) of natural and pH-adjusted soil were transferred to 60-ml glass bottles. The forest soil sample received 7 ml of distilled water with KNO_3 to give a final concentration of 11 ppm of NO_3^-–N. Control samples received only distilled water. Reaction bottles were fitted with serum stoppers, evacuated, flushed with helium several times, and finally filled with helium to atmospheric pressure. Triplicate samples were prepared and incubated at 26 °C. A 0.2-ml headspace gas sample was withdrawn with a gastight syringe at time 0 and at regular intervals and analyzed for N_2O and CO_2 by gas chromatography. A Perkin-Elmer Model 3920 gas chromatograph, equipped with a thermal conductivity detector, was used to detect N_2O and CO_2. A 12' × 1/8″ stainless steel column packed with Porapak QS (80/100 mesh) was used to separate the various components. The detector was operated at 260 °C with a filament current of 175 mA. The injector temperature was 80 °C. The column temperature was held initially at 70 °C for 1 min and increased to 120 °C at the rate of 16° min^{-1}. Helium was used as carrier gas at a flow rate of 35 ml min^{-1}.

The nitrate-reducing and denitrifying bacterial populations in natural and pH-adjusted soil samples were determined by the MPN technique (Focht and Joseph, 1973).

2.5. NITROGEN FIXATION

Samples (15 g) of air-dried freshly collected forest soil and of soil incubated for about 4 weeks were transferred to 20-ml serum bottles. Distilled water containing glucose (4 ml) was added to give a final concentration of 1% glucose. Unamended samples received 4 ml of distilled water. Samples were preincubated either aerobically or anaerobically for 96 h. Aerobic sample bottles were fitted with foam plugs. Anaerobic sample bottles were fitted with serum caps, evacuated, flushed several times with N_2, and finally filled with N_2 to atmospheric pressure and incubated at 26 °C in the dark.

After 96 h preincubation, the samples were assayed for nitrogen fixation by the acetylene reduction technique.

Nitrogen fixation by free-living aerobic bacteria was assayed after the foam plug of the aerobically preincubated sample was replaced with a serum cap. The atmosphere was replaced with He, 0.22 atm of O_2, and 0.1 atm of acetylene.

Nitrogen fixation by anaerobic bacteria was determined by replacing the N_2 atmosphere with He and 0.1 atm of acetylene. Control samples were included to detect the presence of ethylene-producing organisms in the soil.

Samples were incubated at 26 °C in the dark for 2 h. At the end of the incubation period, a 0.2-ml gas sample was withdrawn and analyzed for ethylene by means of a gas chromatograph (Perkin-Elmer Model 3920) equipped with a flame ionization detector and fitted with a 9-ft-long, 1/8-in.-diam stainless steel column packed with Porapak R (100/120 mesh). The column, detector, and injector temperatures were 50°, 60°, and 75 °C, respectively.

2.6. CHEMICAL ANALYSES OF SOILS

The natural and pH-adjusted acid and neutral soil samples which had been incubated under moist conditions at room temperature were periodically removed, air dried, and analyzed for organic matter content, pH, exchangeable H^+, total N, NH_3-N, NO_3^--N, P, K, Mg, Ca, Mn, Fe, and Al. The soil chemical analyses were performed according to Greweling and Peech (1965).

3. Results and Discussion

The chemical properties of the natural soil and the soils preincubated for 30 days are given in Table I. Increased concentrations of extractable Fe and Al have been observed in the acidified soil throughout the incubation period (Francis et al., 1981). The elevated levels of Fe and Al in soil solution may inhibit soil microbial activities or reach toxic levels for plant growth. Furthermore, these elements will be leached from soil and consequently appear in increased concentrations in ground and surface waters.

TABLE I

Chemical analysis of natural, pH-adjusted acid and neutral soils[a]

Constituent[b]	Natural soil	Acidified soil[c]	Neutral soil[d]
Organic matter (%)	3.8	2.6	3.4
pH	4.9	3.2	7.1
Ex. H (meq. 100 g^{-1})	10	26	6
P	1.0	2.0	1.0
K	32.5	3.25	25.0
Mg	12.5	22.5	12.5
Ca	67.5	57.5	1275
Mn	13.5	20.5	7.0
Fe	25.5	368	5.5
Al	117.5	1150	35.0
NO_3^--N	2.5	22.5	10.0
NH_3-N	25.5	14.0	29.5
Total N (%)	0.09	0.09	0.09

[a] Chemical analyses were performed on soils preincubated under moist conditions for 30 days in the laboratory.
[b] Concentration in ppm w/w soil except as noted.
[c] Soil pH was altered by addition of H_2SO_4.
[d] Soil pH was altered by addition of $Ca(OH)_2$.

3.1. ORGANIC MATTER DECOMPOSITION

The rates of CO_2 production by 14- and 150-day preincubated soils with and without leaf amendments are shown in Figures 1 and 2. In all cases, regardless of the soil preincubation history, an increase in CO_2 production was observed with an increase in soil pH.

In the control soil samples (preincubated for 14 days), the total CO_2 production by the pH-adjusted acid soil was 14% less than that of the natural acid soil. An increase

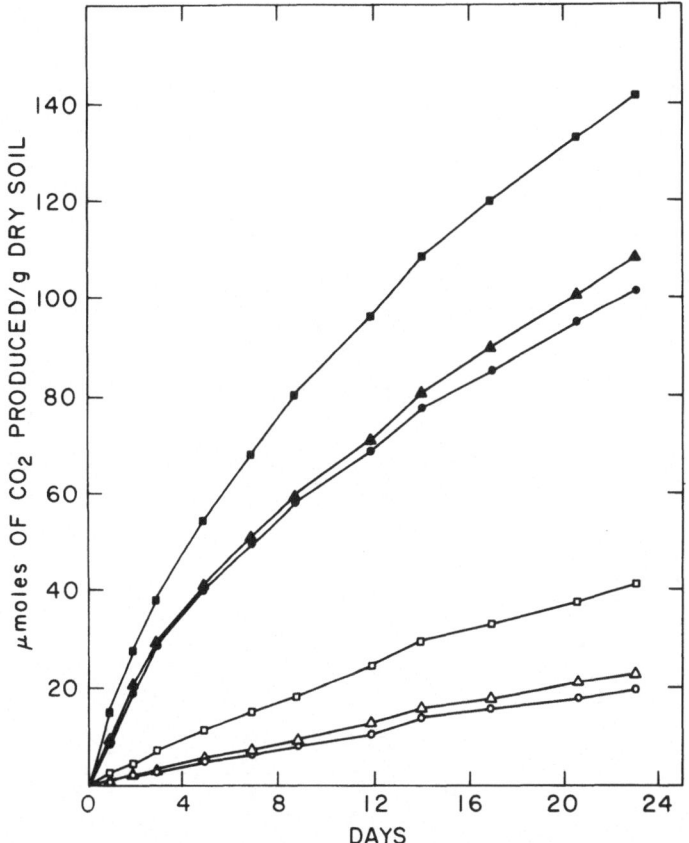

Fig. 1. Effect of soil pH on organic matter decomposition by 14-day preincubated forest soil. Open symbols – control soils; closed symbols – amended soils. $\bigcirc$, $\bullet$ – pH 3.5; $\triangle$, $\blacktriangle$ – pH 4.6 (natural soil); $\square$, $\blacksquare$ – pH 7.0.

in CO_2 production, about 83%, was observed in the pH-adjusted neutral soil. Highly significant differences in the rates of CO_2 production were observed between natural and pH-adjusted soils amended with organic material. A reduction of 6% in total CO_2 production by the acid soil and an increase of 30% in that by the pH-adjusted neutral soil (relative to production by the natural acid soil) were observed (Figure 1).

In the 150-day preincubated control soils, there was a 52% decrease in total CO_2 production by the pH-adjusted acid soil and a 1% increase in that by the pH-adjusted neutral soil relative to production by the natural control soil. Highly significant differences ($p < 0.01$) in the rates of CO_2 production were seen between the soils amended with organic material. A 37% reduction in the total CO_2 evolution by the pH-adjusted acid soil and an increase of 11% in that by the pH-adjusted neutral soil relative to production by the natural acid soil were noted (Figure 2).

Differences between the total amounts of CO_2 produced by the 14- and 150-day preincubated soils were observed in the decomposition study. In general, the 150-day control soils showed a 72% decrease in CO_2 evolution compared with the control soils

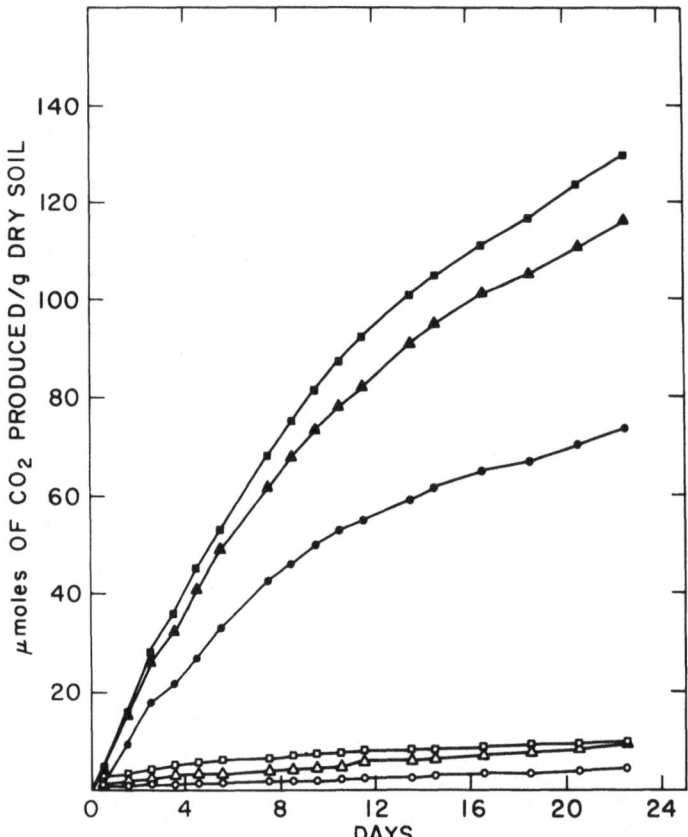

Fig. 2.   Effect of soil pH on organic matter decomposition by 150-day preincubated forest soil. Open symbols
– control soils; closed symbols – amended soils. $\bigcirc$, $\bullet$ – pH 3.5; $\triangle$, $\blacktriangle$ – pH 5.0 (natural soil); $\square$, $\blacksquare$ –
pH 6.4.

preincubated for 14 days. Comparison of the 14- and 150-day preincubated leaf-amended samples showed dramatic changes in the effects of soil pH on CO_2 evolution. At 14 days, the pH-adjusted acid soil, compared with the natural acid soil, showed a 6% decrease in total CO_2 production, whereas the 150-day sample showed a 52% decrease. The 14- and 150-day-old pH-adjusted acid soils differed only slightly in pH, and the differences between their CO_2 productions are probably due to the nature and the decomposition status of the natural organic material and to the makeup of the soil microbial community (including the pH associated toxic effects on microorganisms).

At the conclusion of the decomposition experiment, the initial and final pH and the exchangeable H^+ content of the soil samples were measured. In general, the final pH values of the amended soils were 0.3 to 0.7 units lower than those of the control soils. This was probably due to the acidity contributed directly by the added oak leaves and also to that from their decomposition products. The amount of exchangeable H^+ present in the pH-adjusted acid soil is substantially greater at 150 days' preincubation than at 14 days' preincubation. Among the soils tested (control and amended), there was a highly significant ($p < 0.01$) correlation ($r = -0.8160$) between the relative amount of CO_2 produced and the exchangeable H^+ content of the soil (Figure 3).

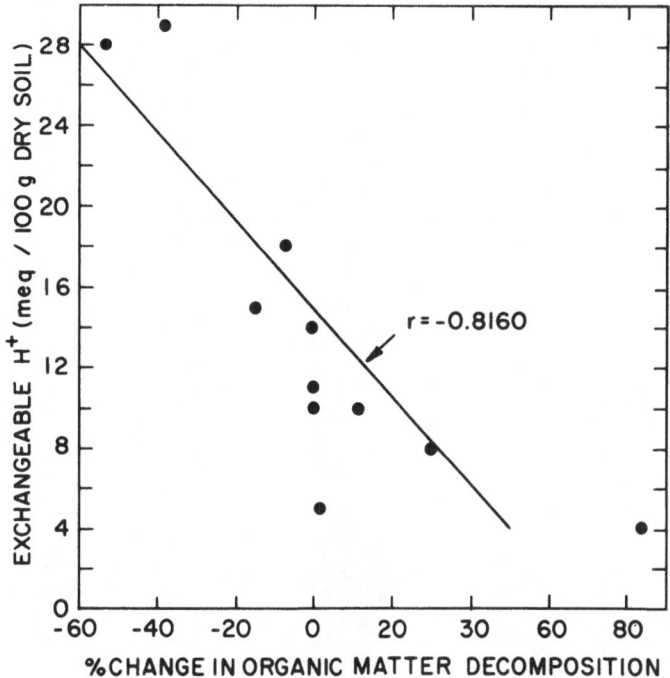

Fig. 3.   Effect of exchangeable H^+ concentration on organic matter decomposition in a forest soil.

The ammonium and nitrate contents of the 150-day preincubated control and amended soil samples were analyzed at the termination of the decomposition study. The results are presented in Figure 4. The initial ammonium and nitrate contents (Figure 4A) of each soil were compared with those of aerobically incubated control (Figure 4B) and amended (Figure 4C) soil samples. Loss of substantial quantities of ammonium was noted in the pH-adjusted acid amended soil and the natural amended soil (Figure 4C), and of nitrate in the pH-adjusted neutral amended soil (Figure 4C). The data indicate that ammonification is affected by soil pH as evidenced by a decrease in ammonium formation as the soil acidity increased. The data also indicate that nitrification is very sensitive to acidic conditions, since little nitrate accumulated in the acidified and natural acid control soils compared with the pH-adjusted neutral soil.

These results suggest that further acidification of acid forest soils by acid precipitation may lead to a significant reduction in the rate of organic matter decomposition and N mineralization.

3.2. AMMONIFICATION AND NITRIFICATION IN INCUBATED SOILS

3.2.1. *Natural soil*

Natural forest soil (pH 4.6) incubated under moist conditions (28% moisture) at room temperature was used to determine the changes in ammonium, nitrite, and nitrate

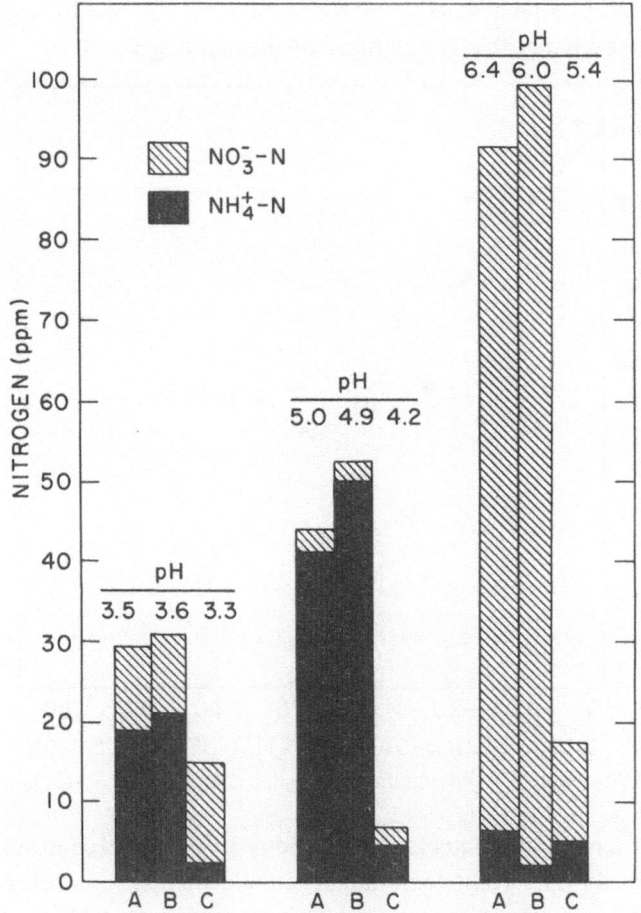

Fig. 4.   Ammonium and nitrate content of control and organic-matter-amended 150-day preincubated soil.
(A) initial (day 0); (B) final control soil (day 24); (C) final amended soil (day 24).

concentrations over a period of time. At intervals, soil samples removed from the
incubation vessel after thorough mixing were air dried and analyzed for total Kjeldahl
N, ammonium, nitrite, and nitrate. The ammonium and nitrate concentrations are shown
in Figure 5. No significant changes in the total N and nitrate-N were observed in the
natural incubated soil over a period of 180 days, but the ammonium concentrations
increased throughout the incubation period. Although ammonification occurs in the
natural acid soil, this soil shows no indication of formation and accumulation of nitrate.
Normally, about 3 to 8 ppm of nitrate is detected. Possibly nitrate was formed in this
soil in low amounts but was reutilized and/or denitrified (Bremner and Blackmer, 1978).

3.2.2. *Acidic Soil*

Ammonium and nitrate formation in pH-adjusted acidic soil is shown in Figure 6.
Ammonium formation in this soil was rather slow and was about 50% less than in the

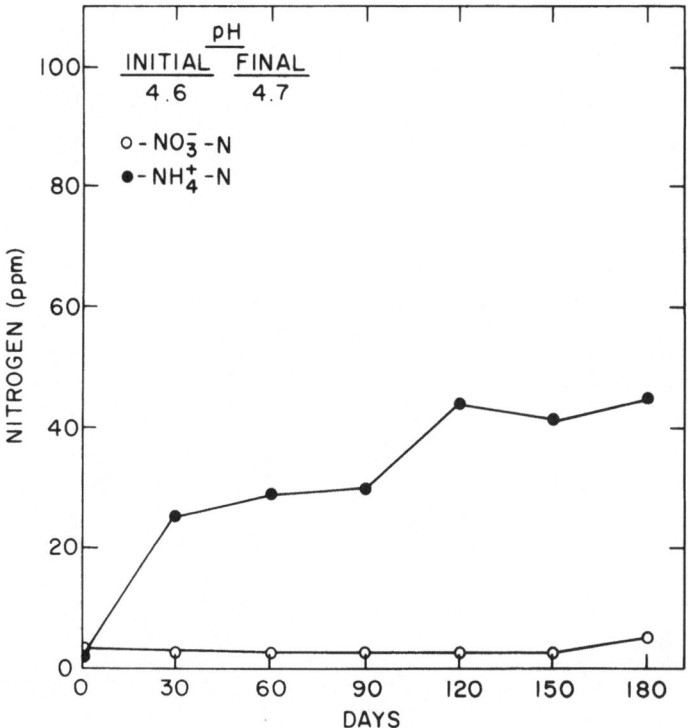

Fig. 5. Ammonification and nitrification in incubated natural forest soil.

natural acid soil. The nitrate concentration, however, increased rapidly from day 0 to day 30, and leveled off from 60 days onwards to about 10 ppm. The nitrate in this very acidic soil was two to three times as high as in the natural acid soil, which may have been due to heterotrophic nitrification, although the soil perfusion data with N-Serve do not lend strong support to this process. The accumulation of nitrate in the most acidic soil may have been due to a decrease in microbial assimilation as well as to slow and incomplete denitrification.

TABLE II

Populations of autotrophic nitrifiers in incubated soils

	Initial soil pH	MPN/g dry soil			
		40	200	240	300
				(days)	
Nitrosomonas sp.	3.2	0	0	0	0
Nitrobacter sp.		0	9.6×10^1	3.9×10^1	2.2×10^0
Nitrosomonas sp.	4.8	0	0	0	0
Nitrobacter sp.	(natural soil)	0	3.4×10^3	9.9×10^1	2.2×10^0
Nitrosomonas sp.	7.1	0	0	1.8×10^1	1.8×10^1
Nitrobacter sp.		6.8×10^1	1.0×10^4	4.4×10^2	2.4×10^2

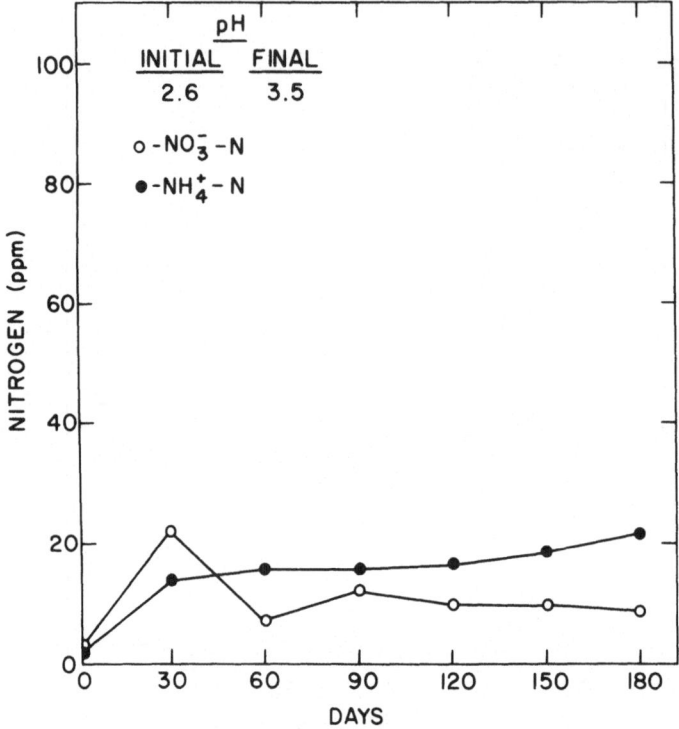

Fig. 6.   Ammonification and nitrification in incubated acidified forest soil.

3.2.3. *Neutral Soil*

Ammonification and nitrification in the pH-adjusted neutral soil (pH 7.2) are shown in Figure 7. As ammonium was released from the nitrogenous organic materials, it was rapidly oxidized to nitrite and then to nitrate. Nitrite was not detected in this soil. The oxidation of ammonium is due primarily to the autotrophic nitrifiers *Nitrosomonas* sp. and *Nitrobacter* sp. whose activities increase because the soil pH is conducive to their

TABLE II

Populations of denitrifiers in incubated soils

	Initial soil pH	MPN/g dry soil			
		40	200	240 (days)	300
Nitrate reducers	3.2	1.8×10^3	1.1×10^3	6.4×10^2	2.6×10^3
Denitrifiers	3.2	5.3×10^1	2.8×10^1	1.7×10^1	1.2×10^1
Nitrate reducers	4.8	9.8×10^3	4.5×10^3	5.4×10^3	3.1×10^4
Denitrifiers	4.8	5.3×10^2	3.2×10^2	3.2×10^2	1.4×10^3
Nitrate reducers	7.1	1.1×10^4	1.1×10^3	1.3×10^5	1.7×10^5
Denitrifiers	7.1	2.3×10^3	2.7×10^3	1.0×10^4	1.4×10^4

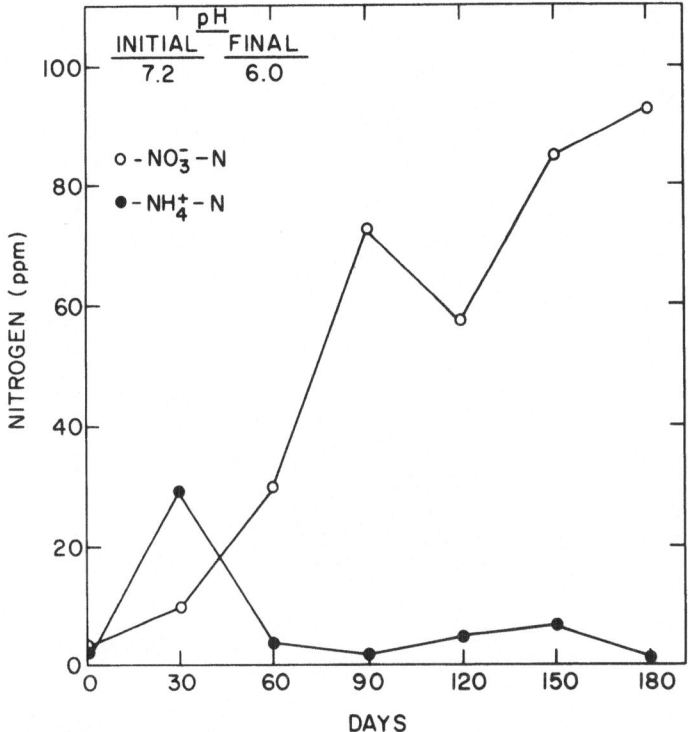

Fig. 7.   Ammonification and nitrification in incubated pH-adjusted neutral forest soil.

growth. Furthermore, low numbers of *Nitrosomonas* sp. and much higher numbers of *Nitrobacter* sp. were present in the neutral soil (Table II). In natural acid and acidified soils, only *Nitrobacter* sp. were detected. This may be due to the availability of nitrite generated by heterotrophic nitrate-reducing bacteria. The populations of denitrifiers and nitrate reducers in incubated soils are presented in Table III. The concentration of ammonium in the neutral soil was about 30 ppm on day 30, then started to drop until it reached a level of 4 ppm, where it remained constant for up to 180 days of incubation. The nitrate levels started to increase at a steady state and reached about 93 ppm on day 180. The results of this study demonstrate that ammonification (to a lesser degree) and nitrification (to a greater degree) are affected by increases in soil acidity.

3.3. AUTOTROPHIC AND HETEROTROPHIC NITRIFICATION: SOIL PERFUSION STUDY

Nitrification in natural and pH-adjusted forest soils that had been preincubated for 60 days was determined by the soil perfusion technique. Nitrate formation in control soils that received only distilled water is shown in Figure 8. Little nitrate was detected in natural and acidified soils during 35 days of incubation. However, in the neutral soil about 38 ppm of nitrate-N was present initially, which increased to 60 ppm by day 14, remained at that level until day 38, and then began to decrease. The increase in nitrate levels in pH-adjusted neutral soil indicates the activity of autotrophic nitrifiers.

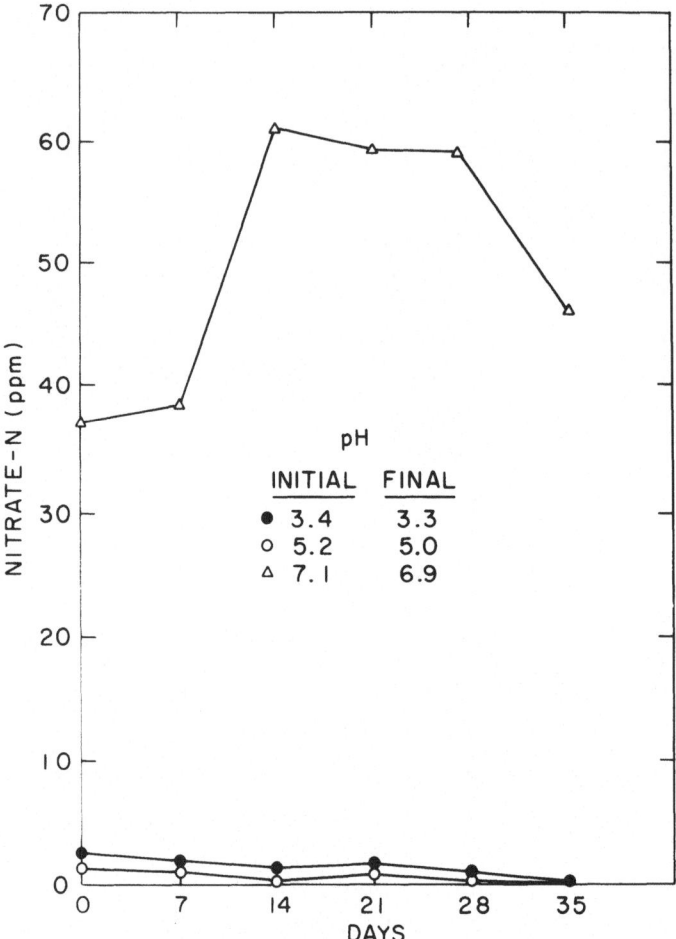

Fig. 8. Nitrification in natural acid and pH-adjusted forest soils.

Nitrification in soils amended with 100 ppm of urea-N is shown in Figure 9. Appreciable amounts of nitrate were formed in pH-adjusted neutral soil after day 7, which indicated the release of ammonium from urea and subsequent oxidation of ammonium to nitrite and then to nitrate. Maximum nitrate concentrations (150 ppm) were detected on day 21. Little nitrate was detected in the 100-ppm urea-amended natural and acidified soils. Since the ammonium formation in these amended and unamended soils was not monitored on a periodic basis, the rates of ammonification are not known, but the data indicate that at least 78% of the urea added to the pH-adjusted acid soil was ammonified. In nature, a wide variety of microorganisms is known to participate in the ammonification process, and the process does not appear to be drastically affected by soil acidity. Autotrophic ammonia-oxidizing bacteria are very sensitive to acidic pH as evidenced by little nitrate formation in natural and acidified soils. Nitrite was not detected in any of these soils. The decrease in nitrate concentration, especially in the neutral soils, may be due to cellular metabolism and denitrification.

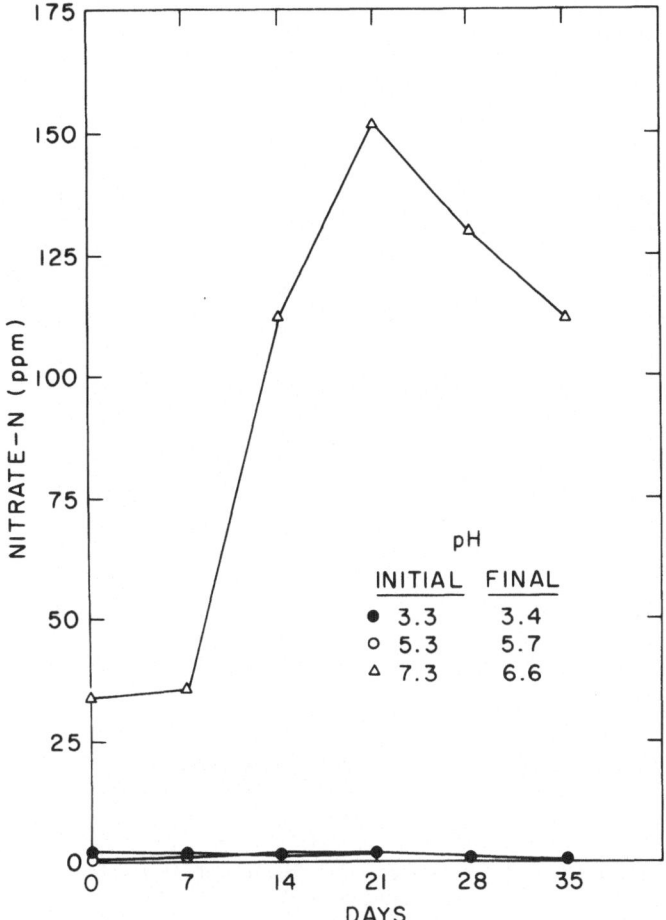

Fig. 9.   Nitrification in natural acid and pH adjusted forest soils amended with 100 ppm urea-N.

To distinguish between the autotrophic and heterotrophic nitrification processes in preincubated natural and pH-adjusted acid and neutral soils, 25 ppm of N-Serve, an inhibitor of autotrophic nitrifiers, was added to these soils along with 100 ppm of urea-N. These soils were incubated in the perfusion apparatus, and liquid samples were withdrawn periodically for nitrite and nitrate analysis. The results are presented in Figure 10. At soil pH 7.2, the initial nitrate concentration was about 20 ppm, at day 7 all the initial nitrate-N was utilized, and no significant nitrate was formed thereafter up to 35 days of incubation. Little nitrate (< 2 ppm) was detected in natural and pH-adjusted acid soils, and nitrate remained more or less at the same level throughout the course of the experiment. The absence of nitrate formation in neutral soil containing 100 ppm of urea-N and 25 ppm of N-Serve strongly indicates that the nitrification process in the neutral soil without N-Serve (Figures 8 and 9) is due to autotrophic nitrifying organisms, *Nitrosomonas* sp. and *Nitrobacter* sp. The natural and the acidic soils did not show an increase in nitrate formation and were similar to control and amended samples (Figures 8

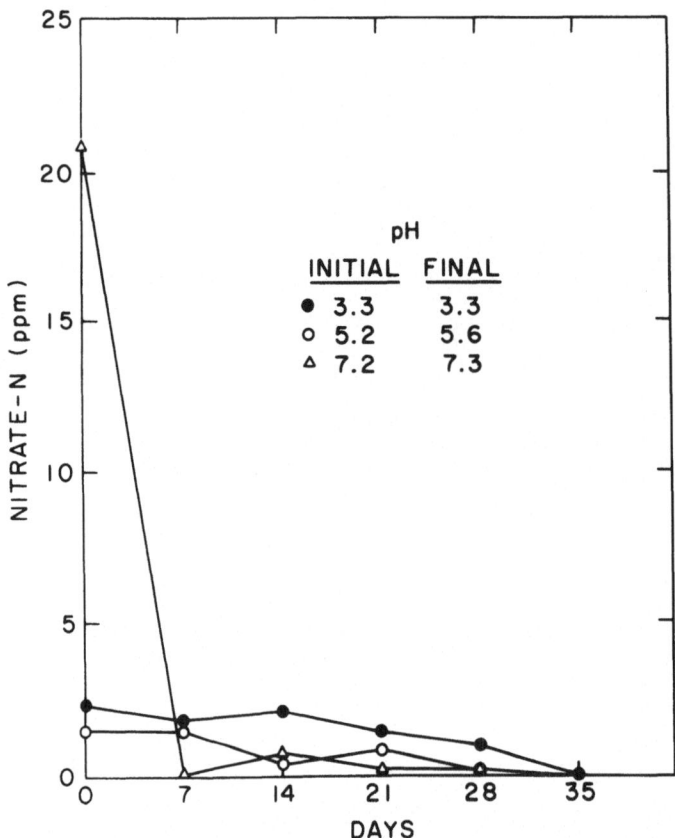

Fig. 10. Heterotrophic nitrification in natural acid and pH-adjusted forest soils amended with 100 ppm
urea-N and 25 ppm N-Serve.

and 9). The N content and the autotrophic nitrifier populations in the perfusate were
determined at the conclusion of the experiments, and the results indicate that the
autotrophic organisms in the acidic forest soil are probably responsible for nitrate
formation. Although the autotrophic nitrifiers are very sensitive to acidic conditions, they
are probably present in low numbers in the natural acid soil, as evidenced by the increased
nitrification rate in pH-adjusted neutral soil, and they may be active in microsites of the
soil where the pH is higher than the measured soil pH. Furthermore, the experimental
results indicate that the contributions of nitrate by heterotrophic nitrifiers in these soils
are insignificant or negligible. In natural ecosystems, however, the significance of nitrate
formation by the heterotrophic nitrification process still remains small.

3.4. DENITRIFICATION

Nitrous oxide production by 2- and 5-week preincubated soils amended with nitrate is
shown in Figure 11. The rate of N_2O production and disappearance at pH 6.6 and 5.8
was very rapid in soils preincubated for 2 weeks. The natural soil, pH 4.6, and
pH-adjusted soil, pH 4.3, followed the same pattern. However, at pH 3.2, N_2O produc-

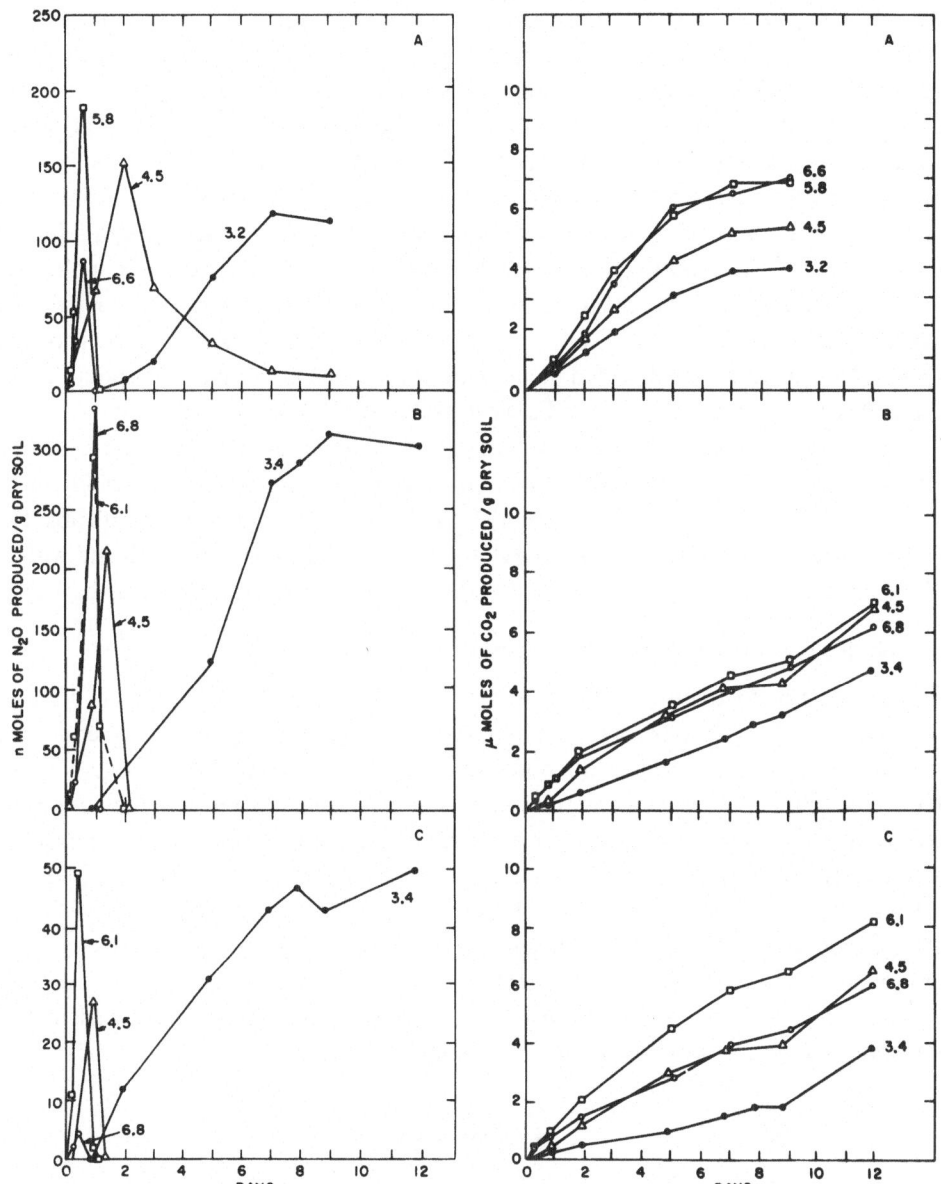

Fig. 11. Effect of pH on N_2O and CO_2 production in a preincubated forest soil. (A) Soil preincubated for 2 weeks and then amended with 11 ppm of $NO_3^- - N$. (B) Soil preincubated for 5 weeks and then amended with 11 ppm of $NO_3^- - N$. (C) Soil preincubated for 5 weeks receiving no nitrate and serving as control. N_2O was not detected from 2-week preincubated control soil that received no nitrate.

tion was rather slow and did not disappear as rapidly as in higher pH soils (Figure 11A).

The pH of the soils preincubated for 5 weeks increased gradually by about 0.2 units. Nitrous oxide production at pH 6.8 and 6.1 followed the same pattern as in 2-week incubated soil samples (Figure 11B). The rate of production and disappearance of N_2O at pH 4.5 was more rapid, and after 2 days no N_2O was detected. The most acidic soil, pH 3.4, produced N_2O at a slow rate, which accumulated in the incubation vessel. In control samples that received no nitrate, N_2O was detected only in 5-week preincubated soils (Figure 11C). Nitrous oxide disappeared within 2 days in soils of pH 4.5, 6.1, and 6.8; in a pH 3.4 sample N_2O was still produced but not further reduced to N_2 gas.

Carbon dioxide evolution by the pH-adjusted soils preincubated for 2 and 5 weeks followed more or less the same pattern. The acidic soil of pH 3.2 produced about 4 µmoles of CO_2 g^{-1} dry soil, whereas pH 6.8 and 4.5 soils produced 7 and 5 µmoles of CO_2 g^{-1} dry soil, respectively, at day 9. Some minor differences were seen in the rates of CO_2 production by different pH-adjusted and preincubated soils. The control soils produced as much CO_2 as the soils amended with 11 ppm of NO_3^-–N, as shown in Figure 11. Little CO_2 was detected in autoclaved control soils with and without nitrate. There was no apparent correlation between N_2O and CO_2 production.

The formation and subsequent accumulation of N_2O in acid soils indicate that denitrifying bacteria are present in low numbers and are active most probably in the microsites of the soils where the pH is conducive to bacterial activity (Table III). Furthermore, the absence of N_2O generation in autoclaved soils, both with and without added nitrate, supports the opinion that N_2O evolution in these acid soils is biological.

3.5. Nitrogen fixation by free-living bacteria

Nitrogen fixation by free-living bacteria in freshly collected soil samples and in naturally acid and pH-adjusted soils (pH 3.6 and 6.4) that had been incubated for several weeks was studied by the acetylene reduction technique. At pH 3.6 and 4.7, no activity was detectable in freshly collected samples or in samples preincubated with glucose under aerobic and anaerobic conditions. At pH 5.8, only samples amended with glucose and preincubated under aerobic or anaerobic conditions exhibited some activity (0.25 and 0.04 µmoles of ethylene/g dry soil/h). Anaerobic soil samples amended with glucose at pH 6.4 produced 37.5 µmoles of ethylene/g dry soil/h, whereas other samples at pH 6.4 showed no measurable activity. These results suggest that the distributions of free-living N-fixing bacteria in acidic forest soil samples are very low, and N input by asymbiotic bacterial fixation in forest soils may be insignificant.

4. Conclusion

To date, studies with simulated acidic rain indicate overall reductions in several soil microbial processes. In some cases stimulatory effects on microbial activity have been observed; but these have been assumed to be temporary. The physical and chemical characteristics of the soil and its response to environmental pollutants often determine the type, abundance, and activities of soil microorganisms. No generalizations can be made because of the diversity and complexity of these systems. Therefore, from the existing data, it is not clear at what rate and to what extent the current acidic precipitation is affecting soil microbial processes. Further acidification of acid forest soils by acidic rain may be a very slow process, and many years may be required for acidic rain to change the soil pH. Rapid adaptability of microbial populations to changing physical and chemical environments and substantial differences in the measured soil pH versus the actual pH in the microsite environments make accurate monitoring of short-term changes that might be caused by acidic precipitation very difficult. Slow acidification may affect soil microbial communities in a way that may gradually result in the selection of acid-resistant or tolerant organisms or in total elimination of certain species. On a long-term basis, acidic rain may affect certain key processes catalyzed by soil micro-organisms such as organic matter decomposition, nitrogen transformation, and ulti-mately the nutrient cycling in the forest ecosystem.

Acknowledgments

I thank D. Olson and R. Bernatsky for their technical assistance.

This work was performed under the auspices of the U.S. Department of Energy under Contract No. DE-ACO2-76CH00016.

References

Abrahamsen, G., Horntvedt, R., and Tveite, B.: 1976, 'Impacts of Acid Precipitation on Coniferous Forest Ecosystems', in Dochinger, L. S. and Selinga, T. A. (eds.), *Symp. on Acid Precipitation and the Forest Ecosystem*, USDA Forest Service General Tech. Rep. NE-23, p. 991.

Abrahamsen, G., Hovland, J., and Hagvar, S.: 1980, 'Effects of Artificial Acid Rain and Liming on Soil Organisms and the Decomposition of Organic Matter', in Hutchinson, T. C. and Havas, M. (eds.), *Effects of Acid Precipitation on Terrestrial Ecosystems*, Plenum, New York, p. 341.

12345Alexander, M.: 1980a, 'Effects of Acidity on Microorganisms and Microbial Processes in Soil', in Hutchinson, T. C. and Havas, M. (eds.), *Effects of Acid Precipitation on Terrestrial Ecosystems*, Plenum, New York, p. 363.

Alexander, M.: 1980b, 'Effects of Acid Precipitation on Biochemical Activities in Soil', in Drabløs, D. and Tollan, A. (eds.), *Proc. Int. Conf. Ecol. Impact Acid Precip.*, SNSF Project, Norway, p. 47.

Alexander, M. and Clark, F. E.: 1965, 'Nitrifying Bacteria', in Black, C. A. (ed.), *Methods of Soil Analysis, Part 2*, Amer. Soc. of Agronomy, Madison, WI.

Baath, E., Berg, B., Lohm, U., Lundgren, B., Lundkvist, H., Rosswall, T., Soderstrom, B., and Wiren, A.: 1980, 'Soil Organisms and Litter Decomposition in Scots Pine Forest – Effects of Experimental Acidifi-cation', in Hutchinson, T. C. and Havas, M. (eds.), *Effects of Acid Precipitation on Terrestrial Ecosystems*, Plenum, New York, p. 375.

Bollag, J. M., Drzymala, S., and Kardos, L. T.: 1973, *Soil Science* **116**, 44.

Bremner, J. M. and Blackmer, A. M.: 1978, *Science* **199**, 295.

Bremner, J. M. and Shaw, K.: 1958, *J. Agric. Sci.* **51**, 40.

Cogbill, C. V. and Likens, G. E.: 1974, *Water Resour. Res.* **10**, 1133.

Dancer, W. S., Peterson, L. A., and Chesters, G.: 1973, *Soil Sci. Soc. Amer. Proc.* **37**, 67.

Evans, L. S., Lewin, K. F., and Vella, F. A.: 1980, *Plant and Soil* **56**, 71.

Focht, D. D. and Joseph, H.: 1973, *Soil Sci. Soc. Amer. Proc.* **37**, 698.

Focht, D. D. and Chang, A. C.: 1975, *Adv. Appl. Microbiol.* **19**, 153.

Focht, D. D. and Verstraete, W.: 1977, *Adv. Microbial Ecol.* **1**, 135.

Francis, A. J., Olson, D., and Bernatsky, R.: 1980, 'Effect of Acidity on Microbial Processes in a Forest Soil', in Drabløs, D. and Tollan, A. (eds.), *Proc. Int. Conf. Ecol. Impact Acid Precip.*, SNSF Project, Norway, p. 166.

Francis, A. J., D. Olson, and Bernatsky, R.: 1981, 'Microbial Activity in Acid and Acidified Forest Soils', BNL 51379, Brookhaven National Laboratory, Upton, NY.

Galloway, J. N., Likens, G. E., and Edgerton, E. S.: 1976, *Science* **194**, 722.

Greweling, T. and Peech, M.: 1965, 'Chemical Soil Tests', Cornell U. Agric. Exp. Sta., NY State College of Agriculture, Ithaca, Bull. No. 960.

Hovland, J. and Ishac, Y. Z.: 1975, 'Effects of Simulated Acid Precipitation and Liming on Nitrification in Forest Soil', SNSF-Project 1R/14, 15 pp.

Ishaque, M. and Cornfield, A. H.: 1976, *Tropic. Agric.* (Trinidad) **53**, 157.

Ishaque, M., Cornfield, A. H., and Cawse, P. A.: 1971, *Plant and Soil* **34**, 201.

Longden, A. R. and Clairidge, C. A.: 1976, *Appl. Environ. Microbiol.* **32**, 188.

Montgomery, H. A. C. and Dymock, J. F.: 1961, *Analyst* (London) **86**, 414.

Montgomery, H. A. C. and Dymock, J. F.: 1962, *Analyst* (London) **87**, 74.

Niese, G.: 1971, *Ecol. Stud.* **2**, 119.

Nommik, H.: 1956, *Acta Agr. Scand.* **6**, 195.

Parr, J. F. and Smith, S.: 1969, *Soil Science* **107**, 271.

Roberts, T. M., Clarke, T. A., Ineson, P., and Gray, T. R.: 1980, 'Effects of Sulfur Deposition on Litter Decomposition and Nutrient Leaching in Coniferous Forest Soils', in Hutchinson, T. C. and Havas, M. (eds.), *Effects of Acid Precipitation on Terrestrial Ecosystems,* Plenum, New York, p. 381.

Runge, M.: 1971, *Ecol. Stud.* **2**, 191.

Shattuck, G. E. and Alexander, M.: 1963, *Soil Sci. Soc. Amer. Proc.* **27**, 600.

Shriner, D. S.: 1977, *Water, Air, and Soil Pollut.* **8**, 9.

Tamm, C. O., Wiklander, G. W., and Popovic, B.: 1976, 'Effects of Application of Sulfuric Acid to Poor Pine Forest', in Dochinger, L. S. and Selinga, T. A. (eds.), *Proc. First Int. Symp. on Acid Precipitation and the Forest Ecosystem,* USDA Forest Service General Tech. Rep. NE-23, p. 1011.

Walker, N. and Wickramasinghe, K. N.: 1979, *Soil Biol. Biochem.* **11**, 231.

Weber, D. F. and Gainey, P. L.: 1962, *Soil Science* **94**, 138.

Wijler, J. and Delwiche, C. C.: 1954, *Plant and Soil* **5**, 155.

EFFECTS OF ACIDITY IN PRECIPITATION ON TERRESTRIAL VEGETATION

LANCE S. EVANS

Laboratory of Plant Morphogenesis, Manhattan College, The Bronx, NY 10471 U.S.A.

and

LAND AND FRESHWATER ENVIRONMENTAL SCIENCES GROUP

Department of Energy and Environment, Brookhaven National Laboratory, Upton, New York 11973 U.S.A.

(Received 10 July, 1981; Revised 8 October, 1981)

Abstract. Over the last several decades rain in the Northeastern United States has become more acidic presumably as a result of anthropogenic inputs of SO_x and NO_x to the atmosphere and their conversion to H_2SO_4 and HNO_3. Present experimental results suggest that acidic precipitation would initially affect organisms on leaf surfaces and epidermal cells of leaves of higher plants. More internal cell layers would be affected with increasing duration or frequency of exposure. Differences in responses of plant foliage among plant species to acidic precipitation appear to be due to the degree of leaf wetting and differences in responses of leaf cells to low pH rain. Moreover, within the same plant, particular structures or cell types may be more sensitive than others. If the United States is to utilize coal reserves for electric power generation that might increase rainfall acidity in the future, an assessment of the impact that acidic rain might have on terrestrial vegetation is necessary. In one experiment, field-grown soybeans were exposed to short duration rainfalls of either pH 4.0, 3.1, 2.7, or 2.3 to provide inputs of 50, 397, 998, or 2506 µeq·of H^+, respectively, above ambient levels throughout the growing season. Control plots received only ambient rainfalls. These additional H^+ decreased seed yield, 2.6, 6.5, 11.4, and 9.5%, respectively. A treatment response function determined between H^+ treatments and seed yield was $y = 21.06 - 1.01 \log x$ had a correlation coefficient of -0.90. Researchers must design additional experiments with adequate experimental controls to assess the impact that acidic rain, at the present pH levels of 3.0 to 4.0 or at anticipated worst-case levels, that could occur if the acidity of rain should increase. Only a holistic view of the impacts that acid precipitation may have on vegetation will enable optimal energy and environmental policy decisions to be made.

1. Brief Literature Review – Effects of Acid Precipitation on Terrestrial Vegetation

The full impact that acidic precipitation is having on terrestrial vegetation has yet to be determined. Research to date has shown that plant responses can vary greatly depending on species, environment, and method of exposure. Due to this variability, it has been necessary to study impacts of acidic precipitation on a plant-by-plant basis. From this approach an overall picture of which plant groups are most sensitive and the types of injury which occur is beginning to emerge.

1.1. ACID RAIN INDUCED LESIONS ON VEGETATION OCCUR MOSTLY ON LEAVES AND REPRODUCTIVE STRUCTURES

Simulated rain at pH 3.4 is about the highest pH at which visible lesions on leaves have been observed. Lesions are located preferentially at bases of trichomes, in guard cells, and in epidermal cells above veins (vascular tissues). A large percentage of the leaf area may exhibit lesions. [1–8]

Water, Air, and Soil Pollution **18** (1982) 395–403.   0049–6979/82/0183–0395$01.35.
Copyright © 1982 by D. Reidel Publishing Co., Dordrecht, Holland, and Boston, U.S.A.

1.2. PHOTOSYNTHESIS AND RESPIRATION IN FOLIAGE MAY BE INFLUENCED BY EXPOSURE TO ACIDIC PRECIPITATION

Leaves exposed to low pH simulated rain (pH 2.3) had lower sugar and stretch contents compared with plants exposed to smulated rain of pH 5.7. [4] Foliage exposed to acidic rain may be stressed but short-term measurements of photosynthesis and respiration may not be sensitive enough to detect the degree of stress. The relationship between the changes in carbohydrate status, on long-term bases, with loss of plant and seed biomass must be more firmly established.

1.3. ACIDIC PRECIPITATION INFLUENCES NUTRIENT LEACHING FROM PLANT SURFACES

Acidic rain can cause an increased rate of foliar leaching. Wood and Bormann [9] demonstrated that K^+, Ca^{2+}, and Mg^{2+} were leached from leaves of pinto beans more rapidly at pH levels of 3.0 and 3.3 than at pH levels of 4.0 and 5.0. Ca^{3+} leached faster from foliage of sugar maple than K^+ or Mg^{2+} at pH 3.0 than at pH 3.3. Leaching rates of K^+ and Mg^{2+} were similar at all pH levels between 3.0 and 3.3. However, the leaching rate at pH 3.0 was significantly higher than at pH 4.0 for K^+ and Mg^{2+}. In tobacco leaves, Ca^{2+} leached faster from foliage exposed to simulated rain of pH 3.0 than foliage exposed at pH 6.7. [10] In pinto beans exposed to simulated rain for 5 days, more Ca, nitrate, and sulfate were leached from foliage at pH levels of 2.7 and 2.9 than at a control pH of 5.7. [11] In contrast, the amount of K leached was greater from leaves exposed to pH 5.7 than leaves exposed to pH 2.7 or 2.9. The amounts of ammonium, Mg, and Zn leached were the same at all pH levels tested.

In addition to removal of plant nutrients from foliage, nutrients may also be removed from tree bark and crows via stem flow. [12] Measurements of electrolytic conductivity of bark extracts of forest trees were more sensitive than measurements of pH or sulfate to inputs of sulfate and/or acidic rain. [13] Bark surfaces may be better indicators of nutrient accumulations than wood samples taken from trees. [14] Although seasonal variations in buffering capacity were present, bark extracts taken from Cracow City were more acidic than bark extracts taken from locations more remote from industrial areas. Tree barks that are naturally less acidic have a rough texture, such as linden and ash, are most suitable to detect pollutant accumulations. [14]

Since the relationship(s) between the exchange of cations and anions and the physiological state of cellular metabolism remains unclear, only continued efforts in this area will demonstrate the impact(s) of acidic precipitation on nutrient leaching and plant growth.

1.4. ACIDIC PRECIPITATION MAY AFFECT GAS EXCHANGE IN PLANTS

Individual epidermal cells are injured upon initial exposure to acidic rain. [6,7] Moreover, acid rain may cause alterations in the cuticle and/or functions of guard cells. [15] Foliage exposed to acidic rain may be more subject to wilting or water stress as well as becoming more sensitive to gaseous air pollutants. Knowledge of the effects of acidic rain in combination with gaseous pollutants such as O_3 and SO_2 is needed in order to understand the total impacts acidic rain might have on vegetation in nature.

1.5. ACIDIC PRECIPITATION CAN AFFECT TOTAL PLANT PRODUCTIVITY

Significant reductions (19% and 11%) in dry weight of trifoliate leaves and dry weights of pods and seeds, respectively, of pinto beans occurred after exposure to acidic mists with no visible leaf injury. [5] Wood and Bormann [3] showed that simulated acid mists of pH 3.0 did not reduce plant growth rates of yellow birch even though all leaves exhibited leaf pitting and curling. However, with many crop plants, yield may be determined by development and survival of reproductive organs as well as by cumulative injury to foliage.

Simulated acidic rain of pH 3.1 and below decreased the dry mass of seeds, leaves, and stems of pinto beans grown under greenhouse conditions. [15] On a percentage-mass basis the decrease in seed yield was comparable with reductions in biomass of leaves and stems. The decrease in yield of pinto beans by simulated acidic rain was attributed to both (1) a decrease in number of pods per plant and (2) a decrease in number of seeds per pod. In soybeans, simulated acidic rain decreased dry mass of both stems and leaves. Seed yield also decreased after treatment with rain of pH 2.5. However, an increase in seed yield occurred when plants were exposed to rain of pH 3.1. A larger dry mass per seed was responsible for the larger dry mass of seed per plant. Lee *et al.* [17] showed that marketable yield was reduced for 5 crops (radish, beet, carrot, mustard greens, and broccoli) and stimulated for 6 crops (tomato, green pepper, strawberry, alfalfa, orchard grass, and timothy) when they were treated with acidified rain in field chamber tests. No consistent effects were observed for 16 other groups.

1.6. ACIDIC PRECIPITATION AFFECTS UNICELLULAR ORGANISMS

Most multicellular organisms that have evolved from aqueous environments onto dry land have evolved specialized surfaces to retard desiccation. Organisms with protective coverings may be less sensitive to acidic rain than organisms that lack these coverings. Bacteria and other single-celled organisms are very sensitive to pH changes. Procaryotic organisms seem to survive within a smaller range of pH levels than most multicellular organisms. Motility of most procaryotic organisms is greatest between pH 6.8 and 9.0. [18–20] Heoninger [21] showed that a pH decrease results in an increase in abnormal wave motions in flagella or *Proteus*. Below pH 5.0 almost all motility ceased.

Fertilization and spermatozoid motility in moss and fern gametophytes is very sensitive to low pH and additions of sulfate, nitrate, and chloride. [22–24] Acidity inhibited one or more critical stages during reproduction. Fern fertilization is inhibited by pH levels below 5.7 and additions of sulfate and nitrate (43µm). These results suggest that the acidity of ambient precipitation in the north-eastern United States inhibits reproduction of ferns. [24]

1.7. ACIDIC PRECIPITATION CAN AFFECT MICROORGANISMS THAT INHABIT SUR-
FACES OF HIGHER PLANTS

Simulated acidic rain produced an 86% inhibition of the number of telia of *Cronartium
fusiforme* on willow oak *(Quercus phellos)* and a 29% decrease in the percentage leaf area
affected by bean rust *Uromyces phaseoli)* on *Phaseolus vulgaris*. [25] Halo blight caused
by *Pseudomonas phaseolicola* infections of leaves of *Phaseolus vulgaris* was stimulated and
inhibited by simulated acidic rain depending upon the timing of application. Simulated
acidic rain inhibited initial infection but stimulated pathogen development if applied after
infection began.

1.8. EFFECTS ON FOREST PRODUCTIVITY

There have been conflicting reports as to whether acidic precipitation can influence forest
productivity. Tamm [26] concluded that except in areas where forest trees exhibit visible
pollution symptoms, ambient acidic precipitation or other types of atmospheric acidity
have no effect on tree growth. Other researchers found a statistically significant difference
in tree growth between areas exposed to acidity and areas more remote from transported
atmospheric acidity. [2, 27] They concluded that there was no reason to attribute growth
reduction to anything other than acidic rain. Researchers in Norway [28] and the United
States [29] have not detected any consistent decrease in tree growth which could be
attributed to acid deposition.

2. Soybean Field Experiment at Brookhaven National Laboratory

An important area of interest is to determine the effects of acidic precipitation on yields
of agronomic crops under field conditions since foliage of broad-leaved angiosperms
appear most sensitive to simulated acidic rainfalls. [8] Experiments of this nature must
have large sample sizes, adequate randomization of treatment plots to avoid local soil
problems, and appropriate statistical analyses. Experiments described herein were
performed with field-grown soybeans at Brookhaven National Laboratory during the
summer of 1979. A preliminary experiment was performed the preceding year at the same
site to estimate the most appropriate plot design and statistical analyses.

Soybeans were seeded to provide six Latin Squares. Five treatments (ambient rainfalls
only, and ambient rainfalls with simulated rainfalls [6] of pH levels of 4.0, 3.1, 2.7, and
2.3) replicated five times in each Latin Square were used to produce a total of 30 plots
per treatment (Figure 1). Individual plots were 2.43 m in length and contained three rows
of plants. Rows were 45.7 cm apart. Only the center row of plants of each plot was
exposed to simulated rain. Only 15 plants in the central portion of the center row were
harvested. Table I shows the most important aspects of the experimental design.

Results (Figure 2) show that seed mass per plant decreased with an increase in amount
of H^+ applied. In addition to H^+ in ambient rainfall, additions of 50 (pH 4.0), 397
(pH 3.1), 998 (pH 2.7) and 2506 (pH 2.3) μeq H^+ per plant decreased seed yields 2.6,
6.5, 11.4, and 9.5%, respectively.

This decrease in seed mass per plant with an increase in H^+ added resulted from a

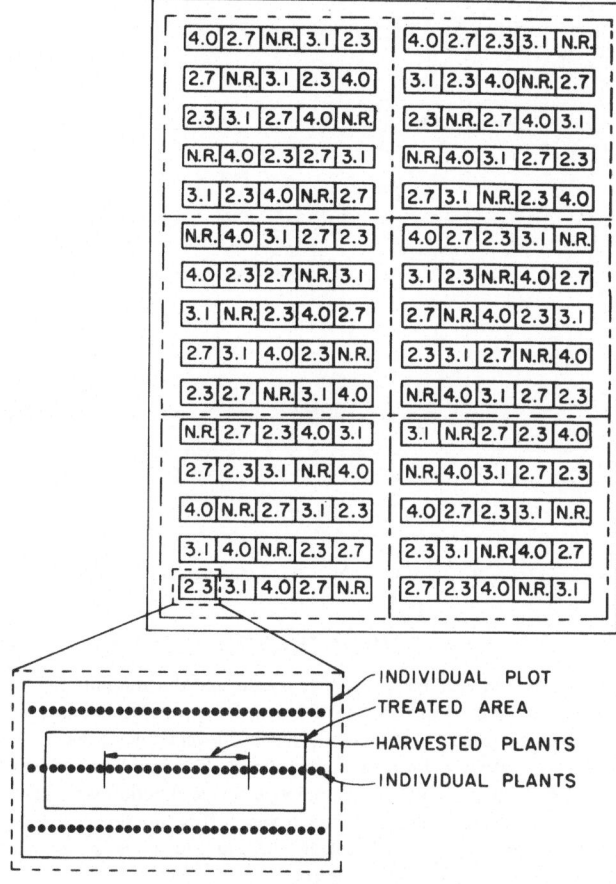

Fig. 1. Diagram of the field plot design with six Latin Squares and detail of the location of harvested plants in each plot are shown.

TABLE I

Experimental design of the soybean field experiment at Brookhaven national laboratory

Duration of experimental treatments:	14 weeks 6 June–7 September.
Duration of the experiment:	22 weeks 22 May–18 October.
Duration of rainfall events:	spray to wet foliage only.
Number of rainfalls:	41
Frequency of rainfalls:	3 week^{-1}
Number of plots per treatment:	30
Number of harvested plants per plot:	15

Treatments	Sulfate concen. (mg l^{-1})
– No additional rainfall	0
– Simulated rain of pH 4.0	1.4
– Simulated rain of pH 3.1	28.3
– Simulated rain of pH 2.7	83.0
– Simulated rain of pH 2.3	265.0

Nitrate concentration of the simulated rainfalls = 3.9 mg l^{-1}.

Soybean (*Glycine max*, cv Amsoy 71) seeds were inoculated with *Rhizobium japonicum* immediately prior to planting.

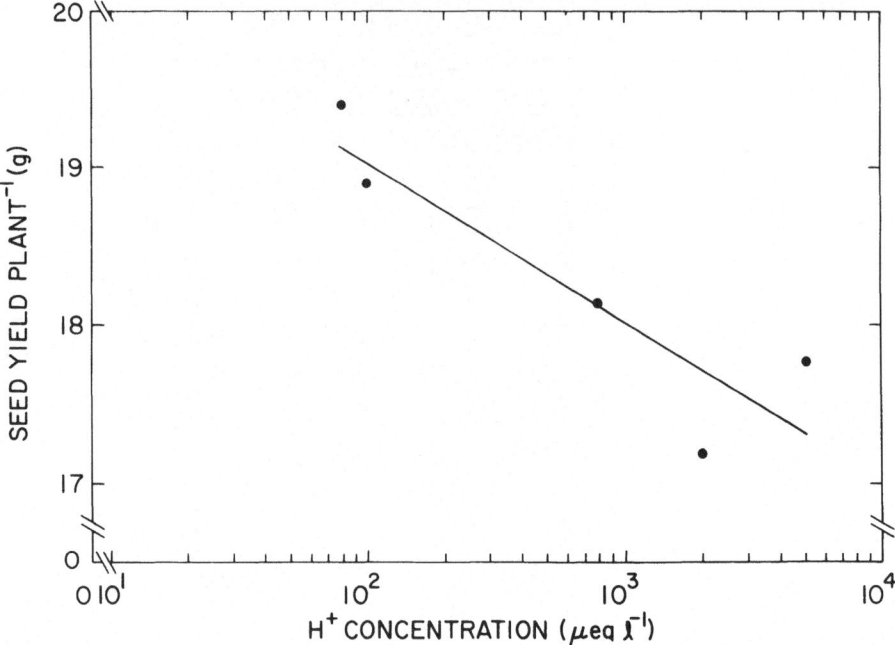

Fig. 2.   Relationship between seed yield per plant of soybeans and the amount of H$^+$ applied per plant to field-grown soybean plants. An analysis of variance of the means gave a probability value of less than 0.10. A Dunnetts' test [30] showed a significant difference at the 0.05 level between the 998 μeq and 0 μeq of H$^+$ added treatments. Probabilities derived from two-tailed T-tests [30] are: 0 vs 998 μeq H$^+$: p = 0.01; 0 vs 2506 μeq: p = 0.07; 50 vs 998 μeq: p = 0.05. A treatment-response function was determined between the hydrogen ion concentration of the treatments and seed yield. This function, expressed by the equation, $y = 21.06 - 1.01 \log x$, has a correlation coefficient of -0.90 and its slope is significantly different from zero.

TABLE II

Seed yield characteristics of field-grown soybeans exposed to simulated acidic rain

Treatment	Pod number plant^{-1}	Seed number plant^{-1}	Seed number pod^{-1}	Mass seed^{-1}
Rain pH 2.3	32 ± 1[a]	87 ± 2[b]	2.7 ± 0.05[c]	0.21 ± 0.00[d]
Rain pH 2.7	32	82	2.5	0.21
Rain pH 3.1	34	89	2.6	0.21
Rain pH 4.0	36	89	2.6	0.21
Ambient rainfall only (AR)	36	91	2.5	0.22

[a] Mean and standard error of all means, respectively. An analysis of variance of the means give a probability of 0.05. Probabilities derived from two-tailed T-tests are: AR vs pH 2.7: p = 0.01; AR vs pH 2.3: p = 0.01

[b] Means and standard error of all means, respectively. An analysis of variance of the means showed no significant differences among the treatment means.

[c] Mean and standard error of all means, respectively. An analysis of variance of the means gave a probability value of 0.09. Probabilities derived from two-tailed T-tests are: AR vs pH 2.3: p = 0.01; pH 2.7 vs pH 2.3: p = 0.3.

[d] Mean and standard error of all means, respectively. An analysis of variance of the means showed no significant difference among the treatment means.

decrease in the number of pods per plant. The decrease in number of seeds per plant is a direct reflection of the number of pods per plant because the number of seeds per pod (2.6) did not vary among the treatments (Table II). Moreover, the mass of individual seeds per plant (0.21 g) did not vary among the experimental treatments. As a result, the decrease in soybean plant yield was attributed to a decrease in the number of mature pods retained at maturity. This decrease may result from a decrease in flower pollination or pod retention. Similar results were obtained when greenhouse-grown soybean plants were exposed to simulated acidic rain. [16]

These results show that additions of small amounts of simulated acidic rain to soybeans decreased the number of pods per plant. This decrease in the number of pods per plant produced a small but significant decrease in seed mass. Yield decreases (2.6, 6.5, 11.4, and 9.5% at simulated rain pH levels of 4.0, 3.1, 2.7, and 2.3, respectively) were obtained from soybeans exposed to low pH ambient rainfalls at Brookhaven National Laboratory.

We assume that the presence of the small amount of water applied had a negligible effect. The amount of water added was only 6.2% of the total volume of ambient rainfalls. Plants were wetted by 29 ambient rainfall events that encompassed 135 h during the treatment period 16 June through 7 September, 1979). Furthermore, from dewpoint calculations it was determined that 100% relative humidity was present for more than 2 h for an additional 65 times (a total of 755 h) during the treatment period. We assume that dew formed on foliage if the above conditions were met. Therefore, the number of simulated rainfall events (41) comprised 30% of the total number of times (135) that the foliage was wet. While there is a possibility that this difference may have some slight and unknown effect on the crop yields, our results do not give any indication of this happening, nor do we have any more reason to believe that such a difference would decrease yields than that it would actually increase them.

These documented changes in seed yield and nutrient contents occurred in an acidic rain impacted area. The average pH of ambient rainfalls determined for the time period in which the simulated rainfalls occurred was 4.1. This result suggests that rainfall with a mean pH of 4.0 has an effect on soybean yields.

The experimental results indicate that acidic rainfalls which occur over the course of the growing season may decrease soybean seed yields of soybean seeds under field conditions. The results also indicate that experiments to accurately determine the impacts of acidic rain on growth and yield of field crops must be performed under standard agronomic practices with a large number of plots per treatment, adequate randomization of treatment plots to counteract local variations, and appropriate statistical analyses of the data. Experiments must be performed with other crop plants in order to assess the degree of damage various crops will incur in nature after exposure to acidic precipitation.

3. Summary

At the present time there is little unequivocal evidence that acidic precipitation, at ambient levels, is having deliterious effects upon terrestrial vegetation within the United

States. The above statement is made more from a lack of concrete information than from an evaluation of large amounts of experimental data. Experiments that have been conducted in laboratory or greenhouse environs have yielded some information that may point to mechanisms of injury. However, most field experiments have not yielded definitive results. Although decreases in yield have been demonstrated in controlled-environment experiments at pH levels below 3.0, only well-designed field experiments will document changes in plant productivity or survival that may be expected from actual acidic rain exposure.

Acknowledgment

This research was performed under the auspices of the United States Department of Energy under Contract No. DE-AC02-76CH00016 to Brookhaven National Laboratory, and in part under Associated Universities Incorporated Contract No. 533972-S to Manhattan College. By acceptance of this article, the publisher and/or recipient acknowledges the U.S. Government's right to retain a nonexclusive, royalty free licence in and to any copyright coverying this paper.

References

[1] Gordon, C. C.: 1972, 'Short-long Conifer Needle Syndrome', Interim Report to the Environmental Protection Agency.

[2] Jonsonn, B. and Sundberg, R.: 1972, 'Has the Acidification by Atmospheric Pollution Caused a Growth Reduction in Swedish Forests? A Comparison of Growth Between Regions with Different Soil Properties', Rapport No. 20, Department of Forest Yield Research, Royal College of Forestry, Stockholm.

[3] Wood, T. and Bormann, F. H.: 1974, *Environmental Pollut.* **7**, 259.

[4] Ferenbaugh, R. W.: 1976, *Amer. J. Bot.* **63**, 283.

[5] Hindawi, I. J., Rea, J. A., and Griffis, W. L.: 1977, 'Response of Bush Bean Exposed to Acid Mist', *70th Annual Meeting of the J. Air Pollut. Control Assoc.,* Abstract 77–30.4.

[6] Evans, L. S., Gmur, N. F., and Da Costa, F.: 1977, *Amer. J. Bot.* **64**, 903.

[7] Evans, L. S., Gmur, N. F., and Da Costa, F.: 1978, *Phytopathology* **68**, 847.

[8] Evans, L. S. and Curry, T. M.: 1979, *Amer. J. Bot.* **66**, 953.

[9] Wood, T. and Bormann, F. H.: 1975, *Ambio* **4**, 169.

[10] Fairfax, F. A. and Lepp, N. W.: 1975, *Nature* **225**, 324.

[11] Evans, L. S., Curry, T. M., and Lewin, K. F.: 1981, *The New Phytol.* **88**, 403.

[12] Tamm, C. O.: 1951, *Physiol. Plant.* **4**, 184.

[13] Härtel, V. O. and Grill, D.: 1972, *Eur. J. For. Path.* **2**, 205.

[14] Grodzinska, K.: 1977, *Water, Air, and Soil Pollut.* **8**, 3.

[15] Evans, L. S. and Hendrey: 1980, 'Effects of Acid Precipitation on Vegetation, Soils, and Terrestrial Ecosystems', A Report of the International Workshop held at Brookhaven National Laboratory, June 11–14, 1979. BNL Report 51195.

[16] Evans, L. S. and Lewin, K. F.: 1981, *Environ. Exper. Botany* **21**, 103.

[17] Lee, J. J., Neely, G. E., Perrigan, S. C., and Grothans, L. C.: 1981, *Environ. Exper. Botany* **21**, 171.

[18] Ogiuti, K.: 1936, *Jap. J. Exp. Med.* **14**, 19.

[19] Shoesmith, J. G.: 1960, *J. Gen. Microbiol.* **22**, 528.

[20] Adler, J. and Templeton, B.: 1967, *J. Gen. Microbiol.* **46**, 175.

[21] Hoeninger, J. F.: 1965, *J. Bacteriol.* **90**, 275.

[22] Mahlberg, P. G. and Yarus, S.: 1977, *J. Exp. Bot.* **28**, 1137.

[23] Evans, L. S.: 1979, *J. Air Pollut. Control Assoc.* **29**, 1145.
[24] Evans, L. S. and Conway, C. A.: 1980, *Amer. J. Bot.* **67**, 866.
[25] Shriner, D. S.: 1978, *Phytopathology* **68**, 213.
[26] Tamm, C. O.: *Ambio* **5**, 235.
[27] Jonsonn, B.: 1977, *Water, Air, and Soil Pollut.* **7**, 497.
[28] Abrahamsen, G., Horntvedt, R., and Tveite, B.: 1977, *Water, Air, and Soil Pollut.* **8**, 57.
[29] Cogbill, C. V.: 1977, *Water, Air, and Soil Pollut.* **8**, 90.
[30] Steel, R. and Torrie, J.: 1960, *Principles and Procedures of Statistics,* McGraw-Hill, New York.

POTENTIAL EFFECTS OF ACID PRECIPITATION ON SOIL NITROGEN AND PRODUCTIVITY OF FOREST ECOSYSTEMS

J. D. ABER

Department of Forestry. University of Wisconsin, Madison, WI 53706, U.S.A.

G. R. HENDREY

Department of Energy and Environment. Brookhaven National Laboratory, Upton, N.Y. 11973, U.S.A.

D. B. BOTKIN

Environmental Studies Program. University of California-Santa Barbara, Santa Barbara, CA 93016, U.S.A.

A. J. FRANCIS

Department of Energy and Environment. Brookhaven National Laboratory, Upton, N.Y. 11973, U.S.A.

and

J. M. MELILLO

Ecosystems Center, Marine Biological Laboratory, Woods Hole, MA 02543, U.S.A.

(Received July 10, 1981; Revised November 30, 1981)

Abstract Numerous field and laboratory studies have shown measureable effects of soil acidification on soil processes and yet there is no indication to date that forest production is being affected even in heavily impacted areas. A discussion of possible reasons for this apparent contradiction is presented along with results of two computer simulations of possible responses to acid rain induced changes in (a) N availability and (b) soil organic matter decomposition rate. The first simulation shows a direct relationship between N availability and forest production. The second indicates the possibility for a more complex response with changes in total soil organic matter more than compensating for changes in decomposition rate and producing an inverse relationship between decomposition rate and N availability.

1. Introduction

A number of effects of increased acidity of precipitation on productivity of forest ecosystems have been proposed. These include direct damage to living plants (Tamm and Cowling, 1977) and indirect effects through altered soil chemistry and biology (Norton, 1980; McFee *et al.*, 1977; Francis *et al.*, 1980). The impact of direct damage effects has been evaluated by field (Abrahamsen *et al.*, 1980) and computer simulation (Botkin and Aber, 1979) methods using current and forseeable levels of precipitation acidification. Results indicate no reduction in forest productivity. Effects on soils can take many forms (Table I). Certain of these processes have shown significant sensitivity to changes in pH, including CO_2 evolution from humus (Tamm *et al.*, 1977; Abrahamsen *et al.*, 1980; Francis *et al.*, 1980) and leaching of Al from soils (Cronan and Schofield, 1979; Meyer and Ulrich, 1977; Norton, 1980; Johnson, 1979). Base saturation and soil pH appear less sensitive to acid precipitation effects (Abrahamsen *et al.*, 1980; Singh *et al.*, 1980; Sposito *et al.*, 1980).

TABLE I

Some potential effects of acid precipitation on soil chemistry and nutrient availability

---

A. Direct effects of increased H^+ concentration

   1. Decreased base saturation
   2. Reduced availability of cations
   3. Increased solubility of Al, Fe, heavy metals, precipitation of PO_4

B. Indirect effects of increased H^+ concentration

   1. Reduced decomposition rate, and alterations in soil microbial populations and soil chemistry
   2. Reduced root activity

C. Effects of increased NO_x input

   1. Increased availability of N
   2. Increased decomposition rate

We have concentrated on N availability because it is the only nutrient which has been shown, by fertilizer trials, to be growth-limiting in the northern hardwood forests of the eastern United States which are the subject of this report. We feel that alterations in productivity, if they occur, will result from changes in N availability which in turn can be affected by any modification in the composition of the soil solution (e.g. concentrations of H, Al, Ca, heavy metals, etc.).

The purpose of this paper is to evaluate the potential effects of acid precipitation on N availability in forest ecosystems and to test the sensitivity of northern hardwood forests to these potential changes through a series of computer simulations.

2. Acid Precipitation, Nitrogen Availability and Tree Growth

To date, no chages in forest production have been reported even under severe, although short term, adicifications in both lab (Wood and Bormann, 1977) and Field (Tamm *et al.*, 1977; Abrahamsen, 1980) trials. Actual increases are occasionally reported (Wood and Bormann, 1977; Tviete and Abrahamsen, 1980). Attempts to measure changes in tree growth through time under natural conditions have also shown little (Jonnson, 1977) or no (Cogbill, 1977) effect. However, it has been shown for northern hardwoods that any change in the total amount of N becoming available in a year will cause a nearly proportional change in productivity (Aber *et al.*, 1980). Matching the latter result with laboratory incubation studies on the effect of pH on organic matter decomposition (Tamm *et al.*, 1977; Francis *et al.*, 1980), by which N is made available, indicates that changes in soil pH should reduce production.

Five factors could delay or counteract the appearance of a growth response. (1) Acid precipitation may not cause acidification of soils. Peterson (1980), Frink and Voigt (1977) and McFee *et al.* (1977) all suggest that H^+ inputs in precipitation may be minimal compared with the generation of organic acids and protons from soil processes, particularly litter decay, nitrification and plant uptake. However, Mayer and Ulrich (1977) emphasize that pH changes in the top few centimeters would require smaller H^+ inputs before reducing process rates. Tamm (1977) also states that effects at colloid surfaces or in microsites might be important well before changes in bulk properties occur. Other workers stress that acid precipitation will essentially increase the rate of podzolization, a long term process of soil horizon development and surface acidification (Peterson, 1980; Norton, 1977). Thus, long term measurement may be required to document these changes. In the first mention of at least middle term results, Norton (1980) reports measureable differences in pH and base saturation over 8 and 19 yr periods. They also report a spatial sequence representing a time-intensity sequence showing foliar changes in Ca : Al, Mn : Al, Pb, and Zn. They conclude that effects can now be seen in forest litter and shallow inorganic soils. Finally, extreme, short term field trials have shown increased leaching of cations and decreased soil pH (Abrahamsen, 1980), with the former more extreme than the latter. The question of long term acidification remains crucial and unanswered.

(2) pH changes in soils may not cause changes in total N mineralization. This would be counter to effects observed in lab incubations and generally accepted notions of the effects of pH on microbial activity. However, in situ incubations under undisturbed conditions over a wide range of natural soil pH conditions will be required to answer this question completely.

(3) Decreases in N mineralization may be offset by increases in N inputs in acid precipitation. Such inputs can total over 20 kg ha^{-1} yr^{-1} in a heavily impacted area (Heinrich and Mayer, 1977; Likens et al., 1978). This may be half of the total annual requirement for N in conifer stands such as those studied by Abrahamsen (1980) and Tamm et al. (1977). An imporant failing of many acid precipitation forest studies has been the exclusion of the N component. Controlling pH alone with H_2SO_4 is misleading.

(4) Short term changes in N availability may be buffered by drawing on an internal pool of this nutrient in the plant. Nitrogen content of leaf/needle litter fall is roughly half that of green tissues on the tree. This retranslocated N is stored until the next growing season. Forty percent of total above ground N demand can be met in this way in hardwood stands, more in conifer stands. This would act as a buffer, delaying the appearance of deficiency symptoms.

(5) The vegetation could be tolerant of low N availability. This is the case for the conifer stands studied by Abrahamsen (1980) and Tamm et al. (1977). No experimental acidification of soils in hardwood stands has been reported. This vegetation type is more sensitive to changes in N availability and should therefore be more susceptible to acid precipitation induced changes in its availability (Aber et al., 1980). Hardwood forests also occupy most of the heavily impacted areas in the United States.

There is a substantial difference between these five processes in terms of long term impact on forest production. The second and third would suggest that no such effects will occur as no long term changes in N availability will occur. The first, fourth and fifth would represent delays in the expression of symptoms which would eventually appear. Thus it is crucial to sort out the relative importance of each.

The following is our working hypothesis of the interaction and relative importance of these and other factors in determining acid precipitation effects on N availability in hardwood forests in the northeastern U.S.

Changes in soil pH will generally depress rates of soil organic C and N mineralization (rejecting number 2 above). All laboratory studies associated with acid precipitation research have reported decreased CO_2 evolution at lower pH (Abrahamsen et al., 1980; Tamm et al., 1977; Francis et al., 1980). Apparent increases in N availability (measured as leachate from soil incubations, e.g. Tamm et al., 1977) must be a short-term phenomonon resulting from the death and lysing of microbial cells and displacement of ammonium ions from exchange sites. Long-term increases in N mineralization and decreases in C mineralization could occur only through a drastic increase in the C : N ratio in soils. This is very unlikely, but if it did occur, increased C : N ratio would almost certainly result in a further decrease in mineralization rate. Likewise, number 5 has been shown by field and simulation studies to be inapplicable to these hardwood forests (Auchmoody and Filip, 1973; Mitchell and Chandler, 1939; Aber et al., 1980). Nitrogen is an important limiting nutrient in most northeastern hardwood forests (see analysis by Mitchell and Chandler, 1939). Internal storage through retranslocation (number 4) should provide only short term buffering. Fertilizer studies show a five year carry over effect on growth, at least partially due to this mechanism. This could help to explain a lack of growth response in short term, extreme acidification studies such as those carried out in Scandinavia, but would not provide significant long term protection.

We are left, then, with the interaction of increased N inputs and potentially decreased mineralization rates as important acid precipitation effects. The effect of the former is fairly easy to predict. Most of these inputs are as mineral N directly available for plant uptake.

The rate and effect of soil acidification remains the most important unknown. This is affected not only by precipitation pH but also by initial soil properties (e.g. CEC, base saturation, organic matter content, texture) and rates of plant uptake and recycling of N and other, non-limiting nutrients (Ca, Mg, K, P, Fe) and elements (e.g. Al). All of these affect the content of the soil solution and hence microbial dynamics. Interactions of the soil solution with primary and secondary minerals, and chemical precipitates (e.g. oxides and phophates of Fe and Al) will also affect this solution. The nature of the soil solution determines the environment of microbial activity and hence organic matter catabolism and N mineralization. We need to know much more about the interactions of these processes with precipitation inputs of different pH before soil acidification estimates can be accurately made.

Within the limitations of this discussion, we have carried out a series of computer simulations on the sensitivity of northern hardwood forest ecosystems to potential acid precipitation induced changes in N availability.

3. Methods

The methods used in this study have been reported in detail previously (Aber et al., 1981) and will only be summarized briefly here. We used two different models representing two levels of complexity in the simulation of forest ecosystems. The first is a modified version of the JABOWA model (Botkin *et al.*, 1972; Aber *et al.*, 1979) which predicts species dynamics and forest productivity using species-specific responses to a set of fixed environmental parameters, one of which is N availability. In this experiment, N availability ranged from 40 to 100 kg ha^{-1} yr^{-1} with 80 serving as the control valve. The second model is FORTNITE (Aber *et al.*, 1982; Aber and Melillo, 1981). This contains species-specific responses similar to those in JABOWA but also allows the vegetation to alter the availability of N through changes in the quantity and quality of litter produced. For this experiment, decomposition of litter and soil organic matter varied from 0.5 to 1.2 times control values. Thus the first model tests the direct or immediate effects of possible changes in N availability on production. The second deals with the potential for species or soil responses to offset or enhance the initial responses. Both models have been well validated for northern hardwood conditions. All simulations begin with a complete clear cutting.

4. Results

Again, the results of this study have been reported in detail elsewhere (Aber *et al.*, 1981) and will only be summarized here.

The two different sets of simulations yield somewhat different results. The modified JABOWA simulations show a direct correspondence between altered N availability and rate of biomass accumulation (Figure 1). Thus we would predict that if acid rain alters

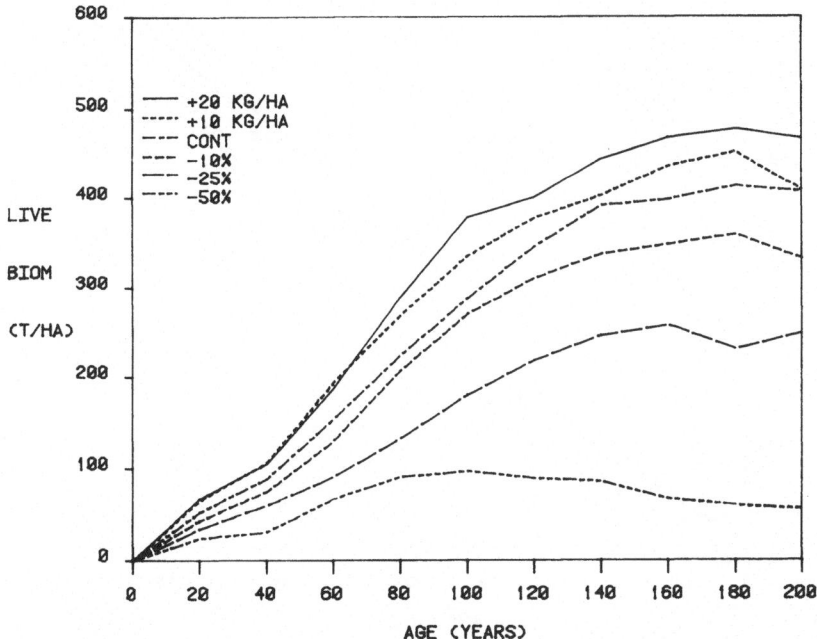

Fig. 1. Live biomass accumulation for northern hardwood forests following clearcutting with different values of nitrogen availability. Control (cont) is 80 kg ha^{-1} yr^{-1}. Results from modified JABOWA simulations.

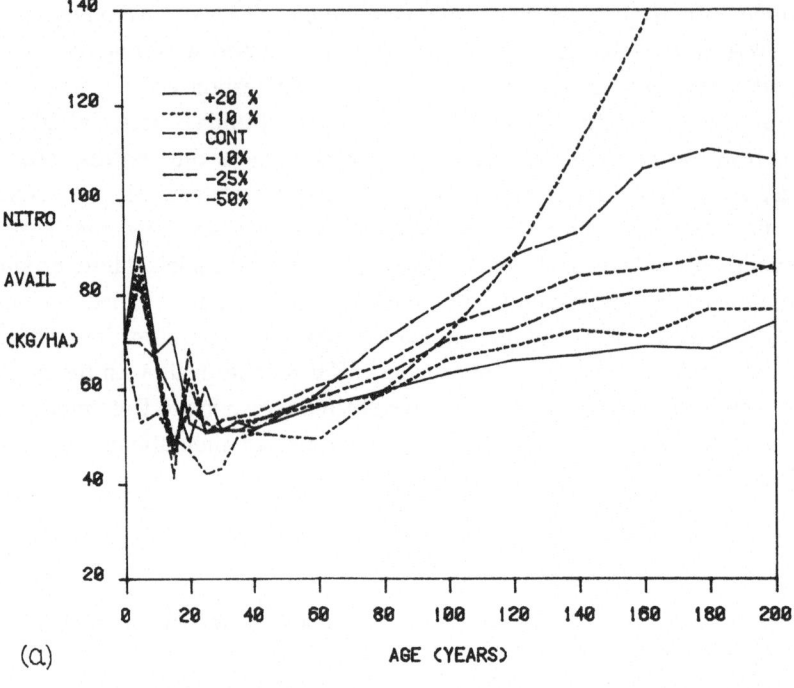

(a)

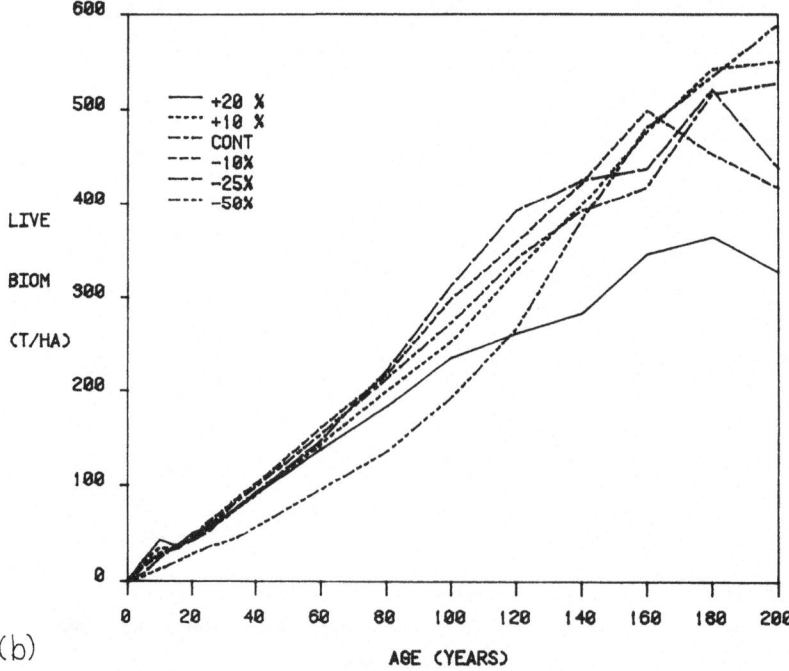

(b)

Fig. 2. Nitrogen availability (a) and live biomass accumulation (b) following clearcutting in northern hardwood forests with different rates of organic matter decomposition. Results from FORTNITE simulations.

N availability, there will be an immediate response in forest production. However, the FORTNITE simulations indicate that, if N availability is altered by acid rain induced changes in decomposition rate, then the system can respond with an increased pool of total soil organic matter. In fact, FORTNITE predicts that, after 120 yr, N availability will be inversely related to decomposition rate (Figure 2a). Predicted rates of live biomass accumulation are shown in Figure 2b. While this possibility cannot be ruled out, we feel it is more likely that the FORTNITE model does not represent the decomposition process with sufficient accuracy to be applicable to such a wide range of inherent decomposition rates (see Aber *et al.*, 1981 for more complete discussion).

References

Aber, J. D., Botkin, D. B., and Melillo, J. M.: 1979, *Can. J. For. Res.* **9**, 10.

Aber, J. D., Hendrey, G. R., Botkin, D. B., Francis, A. J., and Melillo, J. M.: 1980, *Simulation of Acid Precipitation Effects on Soil Nitrogen and Productivity in Forest Ecosystems*, Brookhaven Nat. Lab. Inf. Rept. BNL-28658.

Aber, J. D., Hendrey, G. R., Botkin, D. B., Francis, A. J. and Melillo, J. M.: 1981, 'Potential Effects of Acid Precipitation on Soil Nitrogen and Productivity in Forest Ecosystems', in D'Itri (ed.), *Effects of Acid Precipitation on Ecological Systems in the Graet Lakes Region, Ann Arbor Science* (in press).

Aber, J. D. and Melillo, J. M.: 1982, FORTNITE: *A Computer Model of Organic Matter and Nitrogen Dynamics in Forest Ecosystems*, Univ. of Wisconsin Res. Bull. (in press).

Abrahamsen, G.: 1980, 'Impact of Atmospheric Sulphur Deposition on Forest Ecosystems', in Shriner, Richmond, and Lindberg (eds.), *Atmospheric Sulphur Deposition: Environmental Impacts and Health Effects*, Ann Arbor Sci., Ann Arbor, MI, pp. 397–415.

Abrahamsen, G., Hovland, J., and Hagvar, S.: 1980, 'Effects of Artificial Acid Rain and Liming on Soil Organisms and the Decomposition of Organic Matter', in Hutchinson and Havas (eds.), *Effects of Acid Precipitation on Terrestrial Ecosystems*, Plenum Press.

Auchmoody, L. R. and Filip, S. M.: 1973, 'Forest Fertilization in the Eastern United States: Hardwoods', in *Forest Fertilization*, USFS USDA Gen. Tech. Rept. NE-3.

Botkin, D. B. and Aber, J. D.: 1979, *Some Potential Impacts of Acid Rain on Forest Ecosystems: Implications of a Computer Simulation*, Brookhaven Natl. Lab. Rept. BNL-50889.

Botkin, D. B., Janak, J. F., and Wallis, J. R.: 1972, *J. Ecol.* **60**, 849.

Cogbill, C.: 1977, *Water, Air, and Soil Pollut.* **8**, 89.

Cronan, C. S. and Schofield, C. L.: 1979, *Science* **204**, 304.

Francis, A. J., Olson, D., and Bernatsky, R.: 1980, 'Effects of Acidity on Microbial Processes', in *Proc. Int. Conf. on the Ecol. Impact of Acid Precip.*, Sandefjord, Norway.

Frink, C. R. and Voigt, G. K.: 1977, *Water, Air, and Soil Pollut.* **7**, 371.

Heinrichs, H. and Mayer, R.: 1977, *J. Environ. Qual.* **6**, 402.

Johnson, N. M.: 1979, *Science* **204**, 497.

Jonsson, B.: 1977, *Water, Air, and Soil Pollut.* **7**, 497.

Likens, G. E., Bormann, F. H., Pierce, R. S., Eaton, J. S., and Johnson, N. M. 1978, *Biogeochemistry of a forest ecosystem*. Springer-Verlag. 145 pp.

Mayer, M. and Ulrich, B.: 1977, *Water, Air, and Soil Pollut.* **7**, 409.

McFee, W. W., Kelly, J. M., and Beck, R. H.: 1977, *Water, Air, and Soil Pollut.* **7**, 401.

Mitchell, H. L. and Chander, R. F.: 1939, *The Nitrogen Nutrition and Growth of Certain Deciduous Trees of Northeastern United States*, Black Rock For. Bull. 11.

Norton, S. A.: 1977, *Water, Air, and Soil Pollut.* **7**, 389.

Norton, S. A.: 1980, 'The Impact of Acidic Precipitation and Heavy Metals on Soils in Relation to Forest Ecosystems', in *Int. Symp. on Effects of Air Poll. on Mediterranean and Temperate Forest Ecosystems*.

Petersen, L.: 1980, 'Podzolization: Mechanisms and Possible Effects of Acid Precipitation', in Hutchinson and Havas (eds.), *Effects of Acid Precipitation on Terrestrial Ecosystems*, NATO Conf.

Singh, B. R., Abrahamsen, G., and Stuanes, A.: 1980, *Soil Sci. Soc. Amer. J.* **44**, 75.

Sposito, G., Page, A. L., and Frink, M. E.: 1980, *Effects of Acid Precipitation on Soil Leachate Quality: Computer Calculations*, EPA report.

Tamm, C. O.: 1977, *Water, Air, and Soil Pollut.* **7**, 367.

Tamm, C.O. and Cowling, E. B.: 1977, *Water, Air, and Soil Pollut.* **7**, 503.

Tamm, C. O., Wiklannder, L., and Popovic, B.: 1977, *Water, Air, and soil Pollut.* **8**, 75.

Tveite, B. and Abrahamsen, G.: 1980, 'Effects of Artificial Acid Rain on the Growth and Nutrient Status of Trees', in Hutchinson and Havas (eds.), *Effects of acid precipitation on terrestrial ecosystems*. NATO Conf.

Wood, T. and Bormann, F. H.: 1977, *Water, Air, and Soil Pollut.* **7**, 479.

AN ACID DEPOSITION PERSPECTIVE FOR NORTHEASTERN ALBERTA AND NORTHERN SASKATCHEWAN

S. R. SHEWCHUK

Saskatchewan Research Council, 30 Campus Drive, Saskatoon, Saskatchewan, Canada, S7N OX1

(Received 10 July, 1981; Revised 5 October, 1981)

Abstract. The deposition of S and N compounds within the study area is low at the present time, but may increase, mainly because of emissions growth.

In the bedrock geology there is a zone of calcareous material running through the area. Systems lying to the north of this zone exist on the Precambrian Shield and are highly sensitive to acid deposition. Systems lying to the south are influenced by a very complex network of buffering agents and are relatively insensitive.

One-half of the study area consists of freshwater lake systems that have water chemistry parameters which indicate they are highly sensitive to acid deposition. Present day levels of acidity measurements taken from the sensitive lakes indicate that the lake support healthy fish populations.

Fortunately, time appears available to determine the magnitude of any future effects and to permit steps to be taken in time to prevent such occurrence. Recommendations regarding the research needed to guide future action include air and water monitoring programs.

1. Introduction

It is well documented (Smith and Jeffrey, 1975) that airborne emissions of gases and particulates can be transported many hundreds or thousands of kilometres. They may be deposited and then interact with the ecosystems at any point along their trajectory. A study area is selected in northeastern Alberta and northern Saskatchewan. Some potential impacts of acid deposition due to an increase in industrial emissions within the area is to be addressed. The location of the area (650 000 km^2) within Canada is shown in Figure 1.

2. Atmospheric Processes

The atmosphere is considered as a dynamic pathway for natural and man-induced emissions of SO_x and NO_x. These substances are passed to natural systems by a complex serious of deposition mechanisms. The background acid deposition amounts is increased when regional loading of industrial emissions takes place. Whenever elevated levels of acid deposition are reported 60 to 70% of the increase is due to H_2SO_4 primarily from SO_2 emissions, and the remainder is due to HNO_3 (Rosencranz and Wetstone, 1980). By considering the deposition rates of S and N compounds, the degree of impact of anthropogenic acidity may be addressed.

Since S is not accumulating in the atmosphere (Cullis and Hirschler, 1980), there is an increase deposition to the earth's surface with increased emissions, primarily as acid precursors.

The wet deposition of S as sulphate and N as nitrate within the study area and within

Water, Air, and Soil Pollution **18** (1982) 413–419.   0049–6979/82/0183–0413$01.05.
Copyright © 1982 by D. Reidel Publishing Co., Dordrecht, Holland, and Boston, U.S.A.

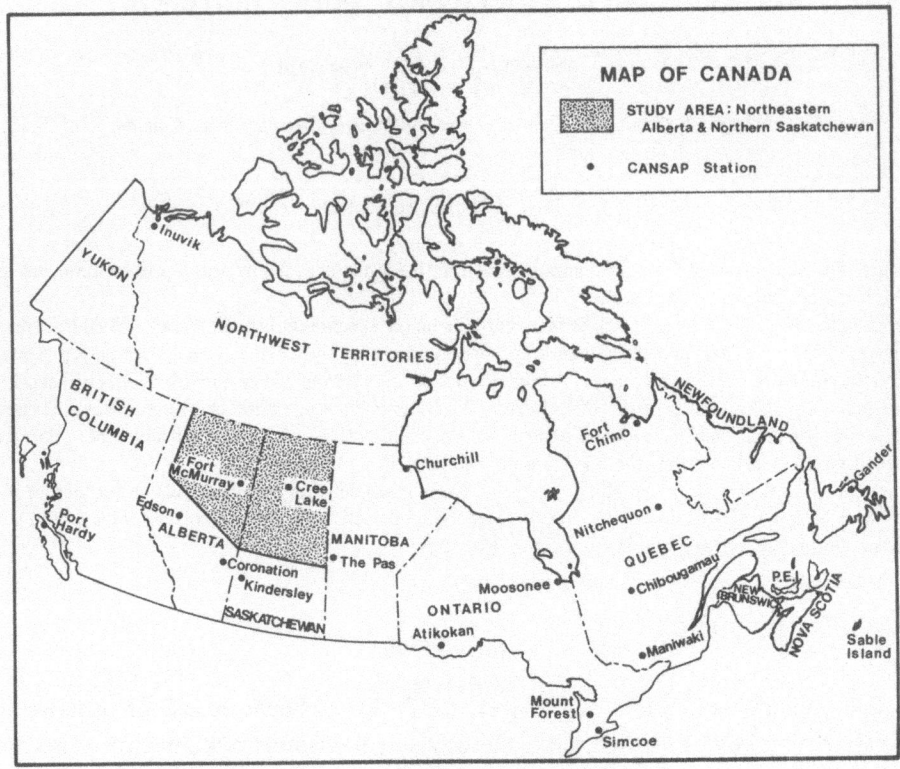

Fig. 1.   Study area within Canada.

various regions of Canada is shown in Table I. All primary data are supplied by the
CANSAP (Canadian Network for Sampling of Precipitation) network of Environment
Canada. The location of a few select CANSAP stations is shown in Figure 1. There are
significant regional differences in wet deposition across Canada.

Locations in northern Canada such as Inuvik, Fort Chimo, and Cree Lake represent
regions where the wet deposition of S and N is lowest. Global background levels of wet
deposition of S as sulphate are not well known, however a value of 2 kg ha^{-1} yr^{-1} is

TABLE I

Wet deposition of S (as sulphate) and N (as nitrate) at selected CANSAP
stations from across Canada (1978)

Location	kg (S–SO$_4$)	kg (N–NO$_3$)
Inuvik	0.9	0.2
Fort McMurray	5.2	1.4
Cree Lake	1.8	0.5
The Pas	3.3	1.3
Mount Forest	23.6	10.9
Maniwaki	9.9	5.1
Fort Chimo	2.6	0.7

taken as representative of unpolluted areas (Frantisak *et al.*, 1980; Almer *et al.*, 1978). Global background levels of wet deposition of N as nitrate are not known. Locations in southern Canada such as Mount Forest and Maniwaki represent regions where S as sulphate and N as nitrate wet deposition is highest. The average value of wet deposition in eastern United States (Galloway and Whelpdale, 1980) is estimated to be 11.9 ± 4.7 kg ha^{-1} yr^{-1}. This value is an average of twenty-five regionally representative sites. In the United Kingdom (Martin, 1980) precipitation deposited S is typically 8 to 12 kg ha^{-1} yr^{-1} at regionally representative sites. The N deposition as nitrate in the rain is given as a matter of record in Table I as well. It seems likely that values of N (as nitrate) in the range of 0.2 to 1.0 kg ha^{-1} yr^{-1} are at background values.

To address dry deposition within the study area concentrations of gaseous and particulate components are required. The values on an ambient regional level are not well documented. However background measurements of SO_2 have been reported as < 0.2 µg m^{-3} day^{-1} within the study area (Fanaki *et al.*, 1979). Particulate ambient concentrations have been reported on a special projects basis (Shewchuk, 1979) as being less than 0.5 µg m^{-3} day^{-1} within the study area. If it is assumed that the annual average concentration for both gaseous and particulate S is approximately equal to these backgrounded values, then dry deposition within the study area in regions remote from existing industry will be less than 1 kg ha^{-1} yr^{-1}. For Europe (OECD, 1977) the annual mean SO_2 concentration ranges from approximately 20 µg m^{-3} day^{-1} in rural areas close to major source regions to 2 µg m^{-3} or lower in remote areas. Annual average particulate S concentrations for Europe are in the order of 5 µg m^{-3}.

The regionally representative value for wet deposition for the study area is estimated to be 2.5 kg ha^{-1} yr^{-1}. This value is derived by assuming Cree Lake and The Pas are regionally representative sites. Total deposition for the study area is 3.5 kg ha^{-1} yr^{-1}. By way of comparison, the lowest total deposition value of Central Ireland and northeast Scotland is 15 kg ha^{-1} yr^{-1} (Martin, 1980). Most of Europe receives a total deposition in access of 25 kg ha^{-1} yr^{-1}.

Sulphur emissions increases due to rapid growth of the energy resource sector of Alberta have been outlined (Webber and Warne, 1979) to the year 2006 A.D. Emissions are to increase from approximately 100 000 tonnes of SO_2 yr^{-1} to 400 000 tonnes of SO_2 yr^{-1}. Most new emissions will be located in the Athabasca Oil Sands Area. This area is located within a few tens of kilometre radius of Fort McMurray, Alberta in the western portion of the study area. Less than 0.3% of the S (Barrie, 1980) emitted by an isolated oil sands extraction plant (located in the Athabasca Oil Sands Area) is deposited within a 25 km radius of the stack. The emissions will be regionally significant to the study area.

The modelling of pollutants from the area has not been studied. However, three day trajectory classifications of air parcels moving into and out of the Athabasca Oil Sands Area has shown that air parcels primarily entered from the northeast to southwest and flowed out from the northeast to southeast (Denison, 1977).

The vital missing factor at present is the linkage between a given emission increase and its effect on regional deposition rates. Conceivably, the deposition increase could range from near zero to being proportional to any emission increase. However if all the increased emission deposit uniformly with the study area, an increase of 2.3 kg ha^{-1} yr^{-1} of S is to be expected.

3. A Description of the Study Area With Respect to Acid Deposition

The bedrock geology of the study area is shown in Figure 2. It can be divided into three principal groups; Cretaceous rocks, Silurian and Devonian rocks, and Precambrian rocks. The Cretaceous rocks which occupy the southwest half of the study area consist of interbedded fine to coarse sand, silt, and clay with local cemented zones, carbonaceous zones, and thin coal beds. The Silurian and Devonian rocks are carbonate-rich rocks, mainly limestones and dolomites. These rocks occur in the northwest and southeast parts of the area and form a zone between the Cretaceous rocks to the southwest and the Precambrian rocks to the northeast. The sensitivity to acid deposition is based primarily on the presence or absence of calcareous material but it also involves the relative amounts

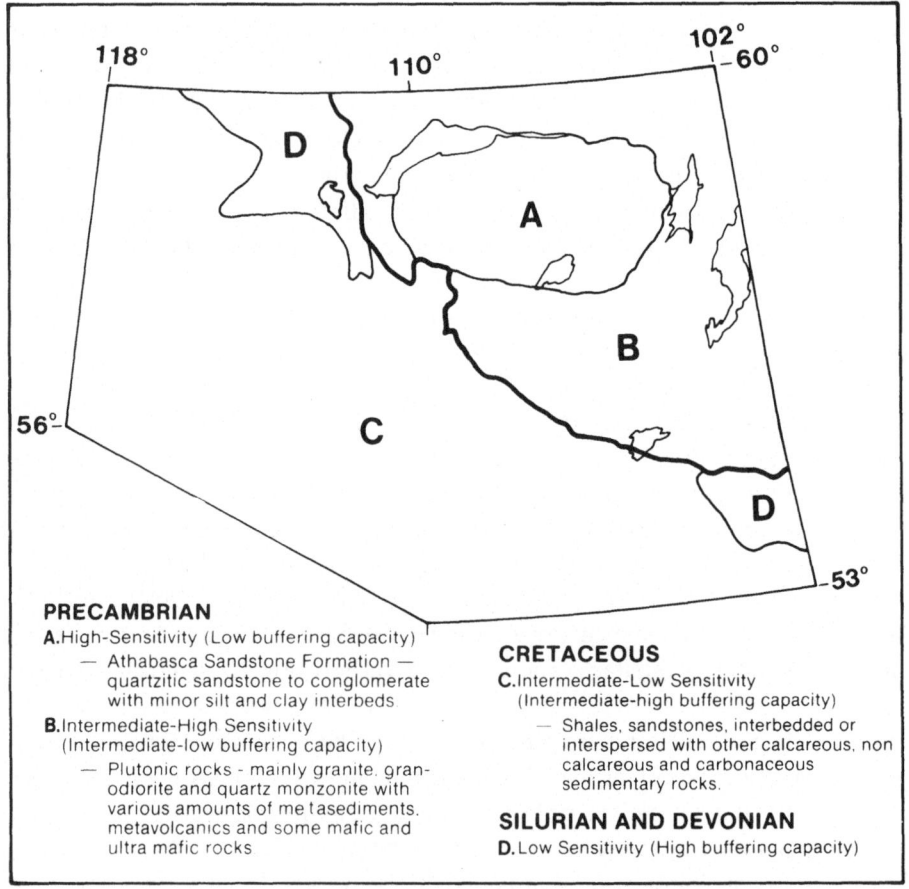

Fig. 2.   Sensitivity of geology with respect to acid deposition.

of mafic minerals, clay minerals and other components. Regions classified as having low sensitivity are underlain by carbonate-rich rocks. Regions classified as high sensitivity are restricted to the Precambrian Shield and, in particular to the Athabasca Sandstone basin, which is composed of quartzitic sandstone, primarily.

A detailed description of the sensitivity of the soils in the study area to acid deposition is given by Holowaychuk *et. al.* (1981). The bulk of the soils which were classified as highly sensitive to acid deposition occurs mainly in the Precambrian Shield. A detail description of the sensitivities of the terrestrial systems for the study area is addressed by Abouguendia and Godwin (1981). However, no satisfactory sensitivity rating system is available at the present time to accomodate the complexities of the natural vegetation variability.

The response of freshwater systems to acid deposition is varied and complex. During the 1960's populations of Walleye and Lake Trout disappeared rapidly from the La Cloche Mountains of Ontario (Harvey, 1979). The lake acidification was primarily attributed to acid deposition from a metal smelter about 100 km away.

The volume of water in Saskatchewan's Precambrian Shield lakes is estimated (Murray, 1980) to be 500×10^9 m^3. Most of the lakes are 'very small', averaging 0.3 km^2 in surface area. Commercially these lakes contribute quite significantly to the provinces economy byway of its natural fish resource.

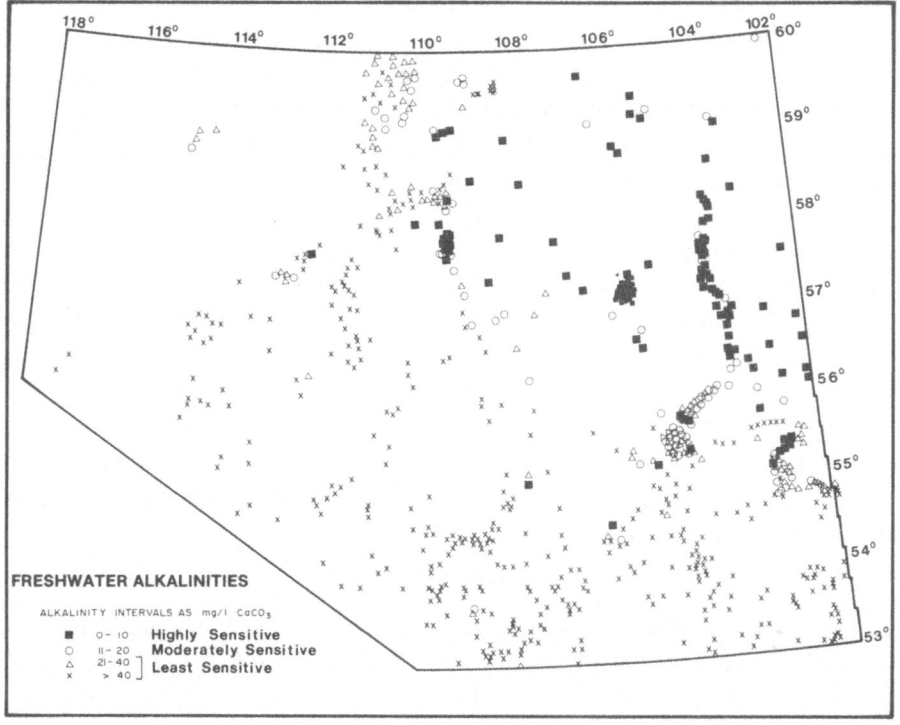

Fig. 3.   Freshwater alkalinity distribution within the study region.

Freshwater lakes within the study area were classified according to alkalinity intervals as shown in Figure 3. Most measurements were taken with inflection-point techniques by potentiometric titration. The classifications were established based on guidelines of the United States–Canada Research Consultation Group (Altshuller and McBean, 1980). Most of the freshwater lakes which fall in the Precambrian Shield portion of the study area have been classified as being highly sensitive to acid deposition. An empirical method has been developed by Henriksen (1979) to evaluate the status of a lake with regard to acidification. On applying this method to data collected from Saskatchewan's Precambrian Shield lakes (Liaw and Atton, 1981), results indicate that none of the lakes chemistry parameters are being influenced by the present levels of acid deposition.

4. Conclusions

At present the deposition rates of S and N compounds are near background levels. Due to a likely four old increase of S emissions within the study area the deposition of acid compounds may be increased. Air parcel trajectories originating from the source region show that a significant proportion of the increase in emissions will be transported over the highly sensitive freshwater lakes of the Precambrian Shield. Under the circumstances the only viable course of action is to ensure that emissions rates do not lead to depositions in excess of the ability of the regional environment to cope with them. It is essential, therefore, to undertake research to help understand this potential acid deposition problem so as to ensure that emission guidelines can be soundly based on scientific evidence and implemented sufficiently early in the development process. Fortunately, there appears to be adequate time to plan, coordinate and undertake a major research program to do this for the region, provided the work is started now.

Accordingly it is recommended that all new developments planned for the region be analyzed for their probable sulphate and nitrate emission levels, and that interim emission controls be devised to guide such development from the outset. It is recommended that an immediate start be made on air, rain and snow sampling via a monitoring network of air pollution stations, precipitation event collectors and snow surveys throughout the region, to establish background levels, identify episode events, provide data on acidification anomalies and to aid in determining the relative roles of dry and wet deposition.

It is recommended that a few small lake be selected as monitoring sites now, that a major small lake inventory survey be undertaken to identify present ranges and sensitive features, and that this survey be used as the basis for a subsequent full scale lake monitoring program.

Acknowledgments

Because of the multidisciplinary approach which was taken in this paper, the author is indebted to several persons for making direct and very significant contributions to this work: B. T. Schreiner, a geologist with the Saskatchewan Research Council; G. A. Padbury, a soil scientist with the Saskatchewan Institute of Pedology; N. Holowaychuk, a soil scientist with the Alberta Research Council; R. Hopkinson and J. Dublin, meteorologists with the Canada government's Atmosperic Environment Service; Z. Abou-guendia and B. Godwin, plant biologists with the Saskatchewan Research Council; F. M. Atton and W. K. Liaw, fish biologists with the Saskatchewan government Department of Tourism and Renewable Resources; and finally, M. Paetz, a fish biologist with the Alberta government Department of Natural Resources. All of these individuals contributed to the baseline data of this paper.

Graphics design was undertaken by E. Perkins and manuscript typing and organization was provided by E. Weegar and her staff, all of the Saskatchewan Research Council.

Major funding for this work was provided by way of a contract from the governments of Alberta, Saskatchewan and Canada.

References

Abouguendia, Z. M. and Godwin, R. C.: 1981, 'Potential Effects of Acidic Deposition on Plants With Special Reference to N. Saskatchewan, paper presented at the Canadian Meteorological and Oceanographic Society Congress, Saskatoon, Saskatchewan, May 27–29.

Almer, B., Dickson, W., Ekstrom, C., and Hornstrom, E.: 1978, 'Sulphur Pollution and the Aquatic Ecosystem', in Nriagu (ed.), *Sulphur in the Environment,* Part II, Wiley-Interscience, pp. 273–311.

Altshuller, A. P. and McBean, G. A.: 'The LRTAP Problem in North America', a preliminary overview prepared by the United States–Canada Research Consultation Group, 1980.

Barrie, L. A.: 1980, *Annals of the New York Academy of Sciences* **338**, 434.

Cullis, C. F. and Hirschler, M. M.: 1980, *Atmospheric Environment* **14**, 1263.

Denison, P. J.: 1977, 'A Climatology of Low-level Air Ttrajectories in the Alberta Oil Sands Area', Alberta Oil Sands Environmental Research Report ME 3.4, 118 pp. Alberta Environment, Edmonton.

Fanaki, F., Hoff, R., Barrie, L. A., Micle, R., Lusis, M., Anlauf, K., Gallant, A., Kovalick, J., Froude, F., Markes, J., Arnold, J., Melnichuk, S., Brymer, D., Gaudenzi, A. G., and Moser, A.: 1979, 'Air System Summer Field Study in the AOSERP Study Area', June 1977. Alberta Oil Sands Environmental Research Program Research Report number 68. Alberta Environment, Edmonton, Alberta.

Frantisak, F., Pelletier, G. H., Bedard, M., and Castonguay, F.: 1980, 'Precipitation Quality in Northwest Quebec-Canada', Air Pollution Control Association, 73rd Annual Meeting, Montreal, Quebec, June 22–27, 1980.

Galloway, J. N. and Whelpdale, D. M.: 1980, *Atmospheric Environment* **14**, 409.

Harvey, H. H.: 1979, 'The Acid Deposition Problem and Emerging Research Need in the Toxicology of Fishes', Proc. Fifth Ann. Aquatic Toxicity Workshop. Fish. Serv. Tech. Rept. 862, pp. 115–128.

Henriksen, A.: 1979, *Nature* **278**, 542.

Holowaychuk, N., Padbury, G., and Schreiner, B.: 1981, 'Sensitivity of Soils to Acidic Deposition in a Selected Region of Western Canada', paper presented at the Canadian Meteorological and Oceanographic Society Congress, Saskatoon, Saskatchewan, May 27–29.

Liaw, W. K. and Atton, F. M.: 1981, 'Sensitivity of Aquatic Ecosystems in Northern Saskatchewan to Acidification', Saskatchewan Fisheries Laboratory, Saskatoon, Saskatchewan (unpublished).

Martin, A.: 1980, *Environmental Pollution* (Series B) **1**, 177.

Murray, A. R.: 1980, 'The Potential Economic Impact of Acid Rain on Northern Saskatchewan Fisheries' (in press).

OECD: 1977, 'The OECD Program on Long Range Transport of Air Pollutants', Summary report, Environment Directorate, Paris.

Rosencranz, A. and Wetstone, G.: 1980, *Environment,* Wash., D.C. **22**, 6.

Shewchuk, S. R.: 1979, 'A Background Study of Potential Sulphur Stresses in West Central Saskatchewan', SRC Report No., pp. 79–6.

Smith, F. B. and Jeffrey, G. H.: 1975, *Atmospheric Environment* **9**, 643.

Webber, H. J. and Warne, G. A.: 1979, 'Sulphur Emissions – Alberta Energy Industries', PNWIS–APCA, Annual meeting 7–9, November 1979, Edmonton, Alberta.

THE EFFECTS OF SIMULATED ACID PRECIPITATION ON PHOTOSYNTHESIS IN THE CARIBOU LICHEN

Cladina stellaris (Opiz) Brodo

MARTIN J. LECHOWICZ

Department of Biology, McGill University, 1205 Avenue Docteur Penfield, Montreal, Quebec, Canada H3A 1B1

(Received 24 July, 1981; Revised 19 October, 1981)

Abstract. The photosynthetic capacity of the caribou lichen, *Cladina stellaris*, was impaired when the lichen was wetted by artificial acid precipitation. When wetted by a solution having a pH = 4.0 and a sulfate concentration = 10.00 mg l^{-1}, the maximal photosynthetic capacity was lowered 27% from normal levels. Moreover, the lichen which is dormant when dry, took 14% longer after wetting to attain even this reduced rate of photosynthesis. The possibility that lichen growth rates may be reduced by acid precipitation with serious implications for many northern ecosystems is discussed.

1. Introduction

The extreme sensitivity of many lichens to atmospheric pollutants such as SO_2 and NO_2 is well known and has been reviewed by Nash (1976) and by Skye (1979). After 3 weeks exposure under field conditions to concentrations of gaseous SO_2 no higher than 250 ppb, photosynthesis in the lichen *Cladina stellaris* was reduced 40% (Moser *et al.*, 1980). Respiration rates of *Usnea hirta,* on epiphytic lichen on ponderosa pine, were reduced about 80% by exposure to 94 ppb SO_2 in the field (Eversman, 1978). The disruption of these critical physiological processes will lead to a decrease in the annual growth rate of lichens as ambient SO_2 levels increase. Taking advantage of the resultant inability of some species to survive at even moderate levels of ambient SO_2, biologists have used lichens as bioindicators for mapping pollution zones around urban and industrial sites (Hawskworth, 1973). Most such studies have reported 'normal' lichen communities at relatively short distances (dozens of km) from a point source of atmospheric pollutants.

Only recently has any threat to lichens from gaseous pollutants in areas more remote from urban and industrial sites been perceived (Puckett, 1979; Gilbert, 1980). The further oxidation of S and N combustion products during long range atmospheric transport results in both dry and wet deposition of strong acids (Dovland and Semb, 1980; Fowler, 1980). The deleterious effects of this acid deposition, even in wilderness areas far from any human activity, is increasingly apparent (Drabløs and Tollan, 1980; Overrein *et al.*, 1980). Considering their known sensitivity to SO_2 and NO_2, the effects of acid deposition on lichens have surprisingly not been studied. The irony of this lack of research is heightened because the metabolic nature of lichens leaves them especially vulnerable to acid precipitation.

Water, Air, and Soil Pollution **18** (1982) 421–430. 0049–6979/82/0183–0421$01.50.
Copyright © 1982 by D. Reidel Publishing Co., Dordrecht, Holland, and Boston, U.S.A.

Unlike most higher plants, lichens are able to survive extended periods of drought at tissue water contents only a few percent of their dry weight (Blum, 1973). In this dry condition lichen metabolism is dormant; respiration occurs at only minuscule rates and photosynthesis not at all (Cowan *et al.*, 1979). Upon wetting of the lichen, its metabolism is rapidly reactivated (Bewley and Krochko, 1981) and the physiological processes controlling growth proceed at rates dependent on ambient conditions of light, temperature, and moisture (Lechowicz, 1981a). Lichens, however, have no specialized structures for the uptake or conservation of water. Thus these periods of metabolic activity are episodic and strongly dependent on the local precipitation and evaporative regimes. A dormant lichen may, for example, be reactivated by a morning rain shower, and dry to dormancy level water content by midafternoon of the same day. During such an episode of metabolic activity, the respiratory losses associated with recovery from dormancy must, on average, be offset by photosynthetic gains if the lichen is to grow and survive in that habitat (Lechowicz, 1981a, b). Any decline in the photosynthetic capacity after wetting by acid rather than normal precipitation will thus have serious implications for lichen growth and survival. Neither the time course of photosynthetic recovery nor its sensitivity to wetting of the lichen by acid rather than normal precipitation, however, has been studied. The purpose of this paper is to describe these critical physiological responses for the caribou lichen *Cladina stellaris* – an abundant and ecologically important species of circumboreal distribution.

2. Methods

Air-dry *Cladina stellaris* was collected in August 1980 in an open spruce-lichen woodland near Schefferville in northern Quebec, Canada (see Figure 1; also Rencz and Auclair, 1978). The collections were flown to Montreal and stored in dry condition at $-20\,°C$ until used in an experiment; Larson (1978) has shown that such frozen storage for up to 3.5 yr had no significant effect on the gas exchange metabolism of the arctic lichen *Alectoria ochroleuca*. Schefferville is the site of a large iron ore mine but processing of the ore occurs elsewhere. Except for the localized disruption of the mining activity itself, the region is representative of the pristine lichen woodlands which occupy much of the Canadian subarctic. Both Lewis and Hrebenyk (1979) and Drake (1980) report recent data on precipitation chemistry for Schefferville suggesting only minor impact of acid precipitation in the area; precipitation pH was typically only slightly below 5. This is in accord with data from the few CANSAP network stations in northern Quebec and Labrador (U.S./Canada Work Group, 1981). The *C. stellaris* used in these experiments has not been selected for resistance to atmospheric pollutants and can provide insight into the potential impact of acid precipitation on lichens in remote regions.

A single podetium of *Cladina stellaris* was used in each experiment. The podetium represents a natural individual unit in the intricately branched and interwoven mat of *Cladina stellaris* (see Figure 1). Yarronton (1975) describes the demographic characteristics of *C. stellaris* podetia. All podetia used in these experiments were in the mature growth phase (Andreev, 1954). Only the living, actively growing tissues from the upper

Fig. 1.   Aspect of the lichen woodland in the vicinity of Schefferville in northern Quebec, Canada. The ground is covered by extensive mats of the caribou lichen *Cladina stellaris*. The inset shows a typical podetium of *C. stellaris* like those used in the experiments reported here; the scale on the inset has mm divisions.

part of the podetium were used. Senescent tissues, recognized by their duller, more gray color, were removed. Each experiment consisted of three essential steps: (1) wetting of a dry, dormant podetium, (2) an interval of incubation of the wetted sample, and (3) radiometric assay using $^{14}CO_2$ of the lichens' photosynthetic capacity after the incubation interval.

In preliminary experiments lichens were wetted by 15 min immersion in distilled water; this is sufficient time to fully hydrate *C. stellaris* (Blum, 1973). In subsequent experiments an artificial rain solution (Shriner, 1978) was used to wet the lichens instead of distilled water. This artificial rain contained trace amounts of Ca^{++}, Mg^{++}, Na^+, K^+, NH_4^+, $SO_4^=$, NO_3^- and Cl^- representative of unpolluted rain. To examine the effects of pH of the wetting solution on lichen metabolism, this artificial rain which had pH 5.6 was acidified by addition of HCl to achieve solutions of pH 2.5 and 4.0. Comparable solutions were also prepared with sulfate concentrations of 10 mg l^{-1} to independently assess possible effects of sulfate ion, an important component of acid precipitation especially in remote regions (U.S./Canada Work Group, 1981). After wetting, the *C. stellaris* was incubated in the dark at room temperature until assayed for photosynthetic activity.

After a fixed interval of up to 400 min from initial wetting, a podetium of *C. stellaris* was removed from the incubation chamber and reimmersed in the wetting solution for 1 min. The sample was centrifuged for 45 s at a standard speed to remove loosely adherent water that can impede gas exchange (Lange, 1980; Snelgar *et al.*, 1981). The lichen was sealed into a cylindrical, plexiglas chamber and exposed to a light intensity of 2000 µEinsteins · m^{-2} · s^{-1}. This irradiance is sufficient to achieve maximal photosynthesis in *C. stellaris* (Lechowicz, 1978). Water was circulated through the double wall of the chamber to maintain the lichen tissue temperature at 20 °C. After exactly 1 min in the light, radioactive air with a CO_2 concentration of 358 µl l^{-1} and a $^{14}CO_2$ specific activity of 1.38 mCi mmol^{-1} was released to the chamber at a flow rate of 0.34 l m^{-1} for exactly one additional minute. This flow rate provided a turnover time of 40 s for the air in the exposure chamber. A squirrel cage fan built in to the chamber assured uniform exposure of the entire lichen to the radioactive air. Immediately after the 1 min exposure to radioactive air, the chamber was flushed with normal air at a flow rate of 3.0 l m^{-1}. After 1 min the lichen was removed and immersed in liquid nitrogen to stop metabolic activity and then killed by rapid drying in a 100 °C oven. This protocol, adapted from the methods of Tieszen *et al.* (1974), provides an estimate of photosynthesis unconfounded by respiratory gas exchange.

The dried lichen was ground to a fine powder in a mortar and pestle to obtain a homogeneous sample representative of the whole podetium. A weighed subsample of the ground, labelled lichen was catalytically combusted in an Intertechnique Oxymat to oxidize all organic material to CO_2 and H_2O. The released CO_2 was trapped directly in a scintillation cocktail containing phenylethylamine and counted on a Beckmann Model LS-200 Scintillation Counter. The samples were quench corrected by the internal standard channels-ratio method (Wang *et al.*, 1975). Calibration standards of known activity were oxidized and counted with each set of samples to maintain a check on machine function. The photosynthetic rate was calculated as mg CO_2 g^{-1} (oven dry weight) h^{-1} (see Tieszen *et al.,* for calculation details).

Between 4 and 7 replicate photosynthetic rates were determined by this method for every combination of the following conditions: (1) pH of 5.6, 4.0, and 2.5, (2) sulfate concentrations of 0.53 and 10.00 mg l^{-1}, and (3) 15, 45, 75, 105, 135, 165, 196, 255, 315, and 375 min since initial wetting. These data were analyzed by least-squares multiple regression (Draper and Smith, 1966) using the Statistical Analysis System computer routines (Barr *et al.,* 1976).

3. Results

The general nature of the recovery of photosynthetic capacity in *Cladina stellaris* after wetting of the dormant lichen by distilled water appears in Figure 2. Photosynthetic rates on the order of 40% of photosynthetic capacity are attained in as little as 30 min after the dormant lichen is wetted. Maximal photosynthetic capacity is attained about 3 to 4 h after wetting. After the maximum photosynthetic rate is attained there may be a slow decline in photosynthesis but this possible trend is confounded by the high variance between replicate podetia.

The photosynthetic time course for *C. stellaris* wetted by an artificial rain solution at pH 5.6, however, also shows a slow photosynthetic decline after maximal photosynthetic capacity is attained (Fig. 3A). By 30 min after wetting 45% of the maximal photosynthetic capacity is attained. At pH 5.6, representative of normal rain, the maximal photosynthetic rate of 0.44 mg CO_2 g^{-1} h^{-1} is attained at just over 4 h after wetting.

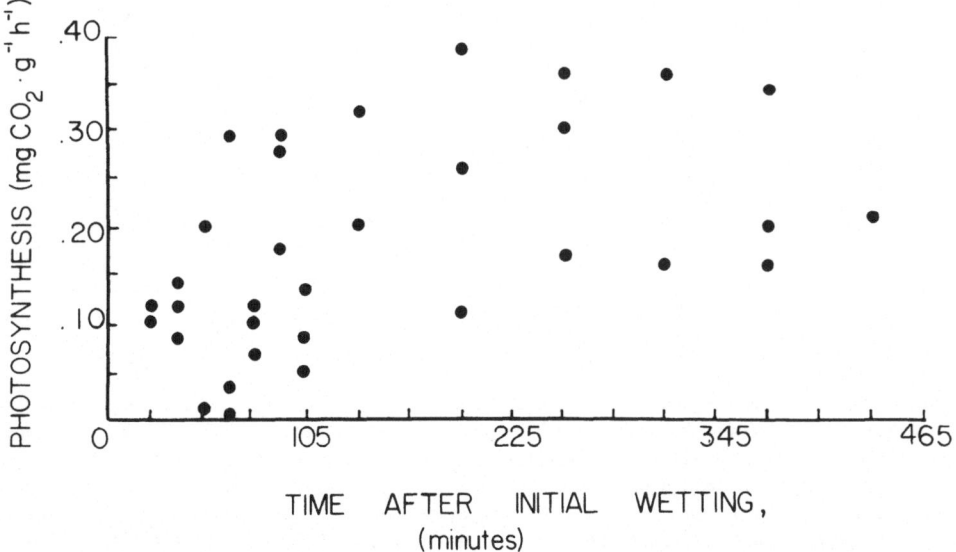

Fig. 2. The time course of photosynthesis of the caribou lichen *Cladina stellaris* after wetting by immersion in distilled water. Each point represents the photosynthetic rate determined radiometrically for a single podetion. Tissue temperature was about 20 °C and tissue water content was near saturation.

These results and the contour graphs in Figure 3 are derived from the fitted regression equation:

$P = 0.000\,404\,92\ TH - 0.000\,031\,62\ TS - 0.000\,000\,14\ T^2H^2 + 0.134\,097\,64$ where P is the photosynthetic rate in mg CO_2 g^{-1} h^{-1}, T is time since wetting in min, H is the

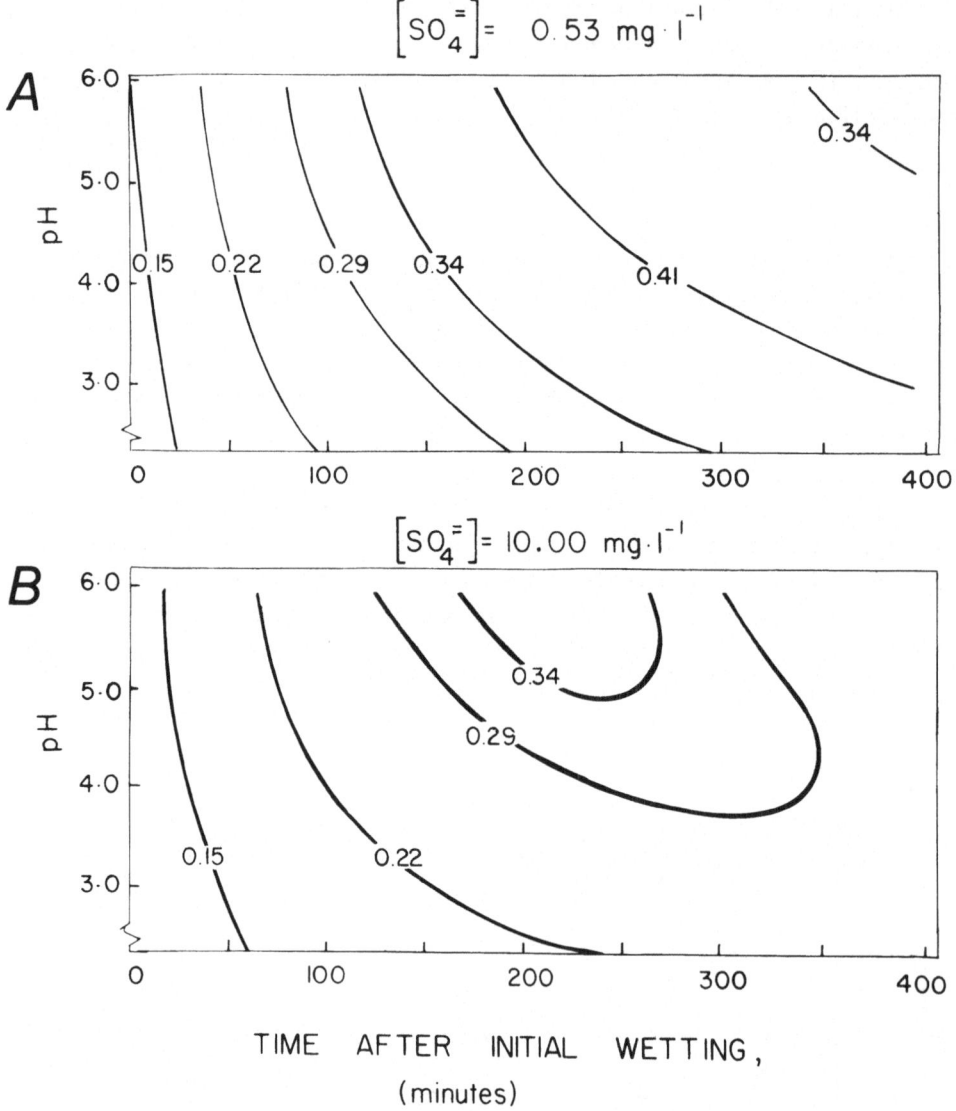

Fig. 3. The time course of photosynthesis of the caribou lichen *Cladina stellaris* after wetting by immersion in normal and simulated acid rain solutions. The contours of photosynthetic rate, mg CO_2 g^{-1} h^{-1}, appear as a function of time since wetting and the pH of the wetting solution. In interpreting this graph it may be helpful to recognize the analogy between these photosynthetic contours and the altitudinal contours on a topographic map. The closer together are two photosynthetic contours, the more rapidly photosynthetic rate is changing with changes in pH and time. The pH of artificial rain solutions was adjusted with HCl at two sulfate concentrations: 0.53 mg l^{-1} representative of normal rain (Figure 3A) and 10.00 mg l^{-1} representative of polluted rain (Figure 3B). Tissue temperature was about 20 °C and tissue water content was near saturation.

pH of the wetting solution, and S is the sulfate concentration in mg l^{-1}. The regression is highly significant as are all its individual terms ($p < 0.0001$). The high replicate variance apparent in Figure 2 is reflected here in an $r = 0.57$ but the residuals show no trend. The fitted model is a sound predictor of the general photosynthetic time course after wetting.

When wetted by artificial acid rain low in sulfate (Figure 3A) *C. stellaris* recovers photosynthetic capacity more slowly than when wetted by normal rain. For example, at pH 4.0 30 min after wetting, photosynthesis is 43% of maximum and the maximal rate is attained only after 6.0 h. When wetted by artificial acid rain high in sulfate (Figure 3B) the maximal rate of photosynthesis attained is depressed. At pH 4.0 a maximum rate of only 0.32 mg CO_2 g^{-1} h^{-1} is attained after 4.9 h. This is 27% less than the maximal rate achieved after wetting by normal rain and it takes 14% longer after wetting to attain. This depression of photosynthetic capacity occurs in the presence of high sulfate concentration even at the normal pH of 5.6. Acid rain high in sulfate will lengthen the time necessary to recover full photosynthetic capacity and depress the maximal rate of photosynthesis achieved.

4. Discussion

The data presented here for the caribou lichen *Cladina stellaris* show that acid precipitation has deleterious effects on lichen photosynthesis. *Cladina stellaris* wetted by artificial acid rain recovers from dormancy more slowly than when wetted by normal rain. When wetted by artificial acid rain high in sulfate concentration not only is photosynthetic recovery slower but the maximal rate of photosynthesis attained is lower. These trends are qualitatively clear but any precise quantitative prediction of the photosynthetic time course after wetting is confounded by the high variance evident between replicates.

This high replicate variance most likely arises from two sources: (1) differences in the relative amounts of vital and senescent tissue included in a sample, and (2) differences in the rates of recovery due to differences in the respiratory substrate available to fuel the recovery process. Both possible explanations arise from poorly understood aspects of lichen biology.

Our knowledge of the temporal dynamics of growth in the caribou lichens is only little extended from Andreev's (1954) early summary. The mature podetia used in the experiments here are growing at the tip and dying at the base; each year a single new whorl of branches is produced. Growth of the annual whorl continues for up to 8 yr after its initiation (Andreev, 1954; Kärenlampi, 1970). Thus about the upper 6 cm of the podetium of *Cladina stellaris* includes the most vital tissues (Figure 1). This region of active growth, distinguished by color from older, senescent tissues toward the podetial base, was used in these experiments. If some senescent tissue is inadvertently included in the sample our weight-based estimate of photosynthesis will be spuriously low. Furthermore Nash *et al.*'s (1980) data suggest that there is an increasing gradient in photosynthetic capacity upward within even the living tissues. Thus inadvertent inclusion of unusually low amounts of older living tissue could raise observed photosynthetic rates. It is necessary to define the tissue age-dependence of photosynthesis within a podetium of *Cladina stellaris.* Then a uniform standard including a fixed number of annual branch whorls could be used to delimit the podetial samples. This improvement in the experimental protocol should remove part of the high replicate variance.

The remainder of the variance will likely be reduced by inclusion of another term in the statistical model: the total polyol content of the sampled podetium. The substantial polyol or sugar alcohol pool in the lichen tissue is utilized as respiratory substrate during recovery from dormancy (Farrar 1976a, b). The rate of recovery after wetting may be proportionate to the available polyol concentration. The high variance in the time course of photosynthetic recovery would then be partly due to combining individuals differing in polyol concentration, and therefore in rates of recovery, as replicates for a particular interval after wetting. Consideration of polyol concentration as well as time since wetting should result in more precise prediction of photosynthetic recovery.

Despite the relatively high variance in the present data, wetting by acid precipitation lengthens the time required for photosynthetic recovery and can depress the rate of photosynthesis attained. Photosynthesis is the only source of the carbon and energy essential for lichen growth. Both phenomena will thus reduce the growth and annual productivity of lichens unless offset by correlated reductions in respiratory rates. Growth rates in the caribou lichens, under natural conditions, range from only 3 to 6 mm yr^{-1} (Andreev, 1954; Scotter, 1963; Pegau, 1968; Ouzilleau and Payette, 1975). Any reduction of these already very low growth rates by acid precipitation will decrease the survival of lichens in northern ecosystems.

The lichens play a substantial role in many northern ecosystems. The caribou lichens, especially *C. stellaris,* constitute a major winter food resource for caribou and reindeer (Bergerud, 1977). Hanson *et al.* (1975), for example, determined that a free-roaming female caribou in northern Alaska ingested 5 kg dry weight of lichens per day during January through April. Rouse (1976) has shown that the extensive mats of *C. stellaris* in subarctic regions control the regional water balance through their mulching effect on the soil surface. Lichen mats, by ameliorating soil surface temperatures and reducing evaporative stress, influence establishment of vascular plant seedlings (Kershaw, 1977, 1978). Through release of organic acids, lichens can inhibit the growth of vascular plant competitors (Rundel, 1978). Finally, nitrogen-fixing lichens can improve the nutritional status of northern ecosystems (Crittenden and Kershaw, 1978). Considering the depression of photosynthetic capacity in lichens wetted by acid precipitation, it is therefore essential that the threat of long term reductions in lichen biomass in northern ecosystems because of acid precipitation receive further study.

Acknowledgments

The comments of Marcia Waterway on the manuscript were greatly appreciated. Thanks to Keith Puckett for advice on the experimental protocol, and to Georgette Verebely for technical assistance. This research was supported by funds from the Canadian Wildlife Service and the Natural Sciences and Engineering Research Council of Canada.

References

Andreev, V. N.: 1954, 'Growth of Forage Lichens and Methods of Regulating It', *Proc. Bot. Inst. USSR Acad. Sci., Series III, Geobotany* **9**, 11–74; translated from Russian as Canadian Wildlife Service Translation CWS-TR-RUS-213.

Barr, A. J., Goodnight, J. H., Sall, J. P., and Helwig, J. T.: 1976, *A User's Guide to SAS-76*, Sparks Press, Raleigh, North Carolina.

Bergerud, A. T.: 1977, 'Diets for Caribou', in Rechcigl, M. (ed.), *Handbook Series in Nutrition and Food Selection, G, Diets, Culture Media, and Food Supplements*, Volume I, CRC Press, Cleveland, Ohio, pp. 243.

Bewley, J. D. and Krochko, J. E.: 1981, *Encyclopedia of Plant Physiology*, in press.

Blum, O. B.: 1973, *Water relations*, in Ahmadjian, V. and Hale, M E. (eds.), *The Lichens*, Academic Press, N.Y., pp. 381.

Cowan, D. A., Green, T. G. A., and Wilson, A. T.: 1979, *New Phytol.* **82**, 489.

Crittenden, P. D. and Kershaw, K. A.: 1978, *Bryologist* **81**, 258.

Dovland, H. and Semb, A.: 1980, 'Atmospheric Transport of Pollutant', in Drabløs, D. and Tollan, A. (eds.), *Ecological Impact of Acid Precipitation*, Proc. Int. Conf., .SNSF project, Oslo, pp. 14–21.

Drabløs, D. and Tollan, A. (eds.): 1980, *Ecological Impact of Acid Precipitation*, SNSF, Oslo.

Drake, J. J.: 1980, *Can. Geographer* **24**, 291.

Draper, N. R. and Smith, H.: 1966, *Applied Regression Analysis*, Wiley, New York.

Eversman, S.: 1978, *Bryologist* **81**, 368.

Farrar, J. F.: 1976a, *New Phytol.* **77**, 105.

Farrar, J. F.: 1976b, *New Phytol.* **77**, 115.

Fowler, D.: 1980, 'Removal of Sulphur and Nitrogen Compounds from the Atmosphere in Rain and Dry Deposition', in Drabløs, D. and Tollan, A. (eds.), *Ecological Impact of Acid Precipitation*, Proc. Int. Conf., SNSF project, Oslo, pp. 22–32.

Gilbert, O. L.: 1980, *Bull. Br. Lichen Soc.* **46**, 1.

Hanson, W. C., Whicker, F. W., and Lipscomb, J. F.: 1975, in Luick, J. R., Lent, P. C., Klein, D. R., and White, R. G. (eds.), *Proceedings of the First International Reindeer and Caribou Symposium*, Biological Papers of the University of Alaska, Special Report 1, Fairbanks.

Hawksworth, D. L.: 1973, 'Mapping Studies', in Ferry, B. W., Baddeley, M. S., and Hawksworth, D. L. (eds.), *Air Pollution and Lichens*, University of Toronto Press, Ontario, Canada, pp. 38.

Kärenlampi, L.: 1970, *Rep. Kevo Subarctic Res. Stat.* **7**, 33.

Kershaw, K. A.: 1977, *Can. J. Bot.* **55**, 393.

Kershaw, K. A.: 1978, *Bryologist* **81**, 294.

Lange, O. L.: 1980, *Oecologia* **45**, 82.

Larson, D. W.: 1978, *Can. J. Bot.* **56**, 2119.

Lechowicz, M. J.: 1978, *Oecologia* **32**, 225.

Lechowicz, M. J.: 1981a, 'Adaptation and the Fundamental Niche: Evidence from Lichens', in Carroll, G. and Wicklow, D. T. (eds.), *The Fungal Community: Its Organization and Role in the Ecosystem*, Marcel Dekker, N.Y., p. 89.

Lechowicz, M. J.: 1981b, *Oecologia* **50**, 210.

Lewis, J. E. and Hrebenyk, B.: 1979, *Climatological Bull.* **26**, 1.

Moser, T. J., Nash III, T. H., and Clark, W. D.: 1980, *Can. J. Bot.* **58**, 2235.

Nash III, T. H.: 1976, *Die Naturwissenschaften* **63**, 364.

Ouzilleau, J. and Payette, S.: 1975, *Naturaliste Can.* **102**, 597.

Overrein, L. N., Seip, H. M., and Tollan, A.: 1980, *Acid Precipitation – Effects on Forest and Fish*, Final report of the SNSF project, 1972–1980, SNSF, Oslo.

Pegau, R. E.: 1968, *Arctic* **21**, 255.

Puckett, K. J.: 1979, 'The Effects of Acid Precipitation on Lichens', Presentation at workshop on *The Effects of Acid Rain on Wildlife*, November 13–14, 1979, Canadian Wildlife Service, Ottawa.

Rencz, A. N. and Auclair, A. N. D.: 1978, *Can. J. For. Res.* **8**, 168.

Rundel, P. W.: 1978, *Biochem. Syst. and Ecol.* **6**, 157.

Rouse, W. R.: 1976, *Arct. Alp. Res.* **8**, 357.

Scotter, G. W.: 1963, *Can. J. Bot.* **41**, 1199.

Shriner, D. S.: 1978, in Heck, W. W., Krupa, S. V., and Linzon, S. N. (eds.), *Methodology for the Assessment of Air Pollution Effects on Vegetation*, p. 1.

Skye, E.: 1979, *Ann. Rev. Phytopathol.* **17**, 325.

Snelgar, N. P., Green, T. G. A., and Wilkins, A. L.: 1981, *New Phytol.* **88** (in press).

Tieszen, L. L., Johnson, D. A., and Caldwell, M. M.: 1974, *Photosynthetica* **8**, 151.

U.S./Canada Work Group: 1981, 1, Impact Assessment, *Memorandum of Intent on Transboundary Air Pollution*, Interim report, February 1981, U.S. Environmental Protection Agency and Environment Canada.

Wang, C. H., Willis, D. L., and Loveland, W. D.: 1975, *Radiotracer Methodology in the Biological, Environmental, and Physical Sciences*, Prentice Hall, Englewood Cliffs, New Jersey.

Yarronton, G. A.: 1975, *New Phytol.* **75**, 99.

ANNOUNCEMENT

International Working Group on Environmental Systems Analysis

During the UNESCO workshop on 'Systems Analysis in Environmental Engineering and Management', held in Udine, Italy, in December 1978, it was decided to create an International Working Group on Environmental Systems Analysis to cater for the need for coordination of teaching and research activities in this field, in particular in Europe and in developing countries.

The aim of the Group is to link together National Institutions and International Societies which are responsible for the organization of research in the area of Environmental Systems Analysis but are not yet coordinated in their efforts. This aim will be achieved by promoting International Research Programmes, by organizing meeting and conferences of various levels, and by holding short international courses. Particular importance will be placed on problem oriented workshops.

The members of the working Group are:

Dr. M. Benarie, *Ist. Nat. Rech. Chim. Appl., Vert-le-Petit, France*
Prof. C. Brebbia, *University of Southampton, England*
Dr. L. Jeftic, *R. Boskovic Institute, Zagreb, Yugoslavia*
Prof. S. Jorgensen, *University of Copenhagen, Denmark*
Prof. J. E. Nash, *University College, Galway, Ireland*
Prof. S. Rinaldi, *Politecnico of Milano, Italy*
Prof. J. Sündermann, *University of Hamburg, Germany*
Prof. C. Vansteenkiste, *University of Gent, Belgium*

Other people from countries not yet represented will joint the Working Group in the near future.

EVENTS PROPOSED BY THE WORKING GROUP in 1982

Application of Mathematical Models in Management of Aquatic Ecosystems

Dubrovnik, Yugoslavia, September–October 1982

An international course sponsored by the Centre for Marine Research, Zagreb, Yugoslavia.

Modelling and Control of River Quality

Milano, Italy, November 23-25, 1982

A postgraudate course sponsored by the Politecnico of Milano.

Further information about the above activities can be obtained from:

International Working Group on Environmental Systems Analysis
c/o Prof. S. RINALDI
Centro Teoria dei Sistemi,
C.N.R., Politecnico di Milano
Via Ponzio 34/5
20100 Milano, Italy

AUTHOR INDEX

Volume 18, Nos. 1, 2, and 3

SUBJECT INDEX

Volume 18, Nos. 1, 2, and 3

SUBJECT INDEX